COMPUTATIONAL STUDIES OF NEW MATERIALS

COMPUTATIONAL STUDIES OF NEW MATERIALS

Editors

Daniel A Jelski
State University of New York, Fredonia

Thomas F George
University of Wisconsin-Stevens Point

World Scientific
Singapore • New Jersey • London • Hong Kong

Published by

World Scientific Publishing Co. Pte. Ltd.

P O Box 128, Farrer Road, Singapore 912805

USA office: Suite 1B, 1060 Main Street, River Edge, NJ 07661

UK office: 57 Shelton Street, Covent Garden, London WC2H 9HE

British Library Cataloguing-in-Publication Data
A catalogue record for this book is available from the British Library.

COMPUTATIONAL STUDIES OF NEW MATERIALS

ISBN 981-02-3325-6

Printed in Singapore.

CONTENTS

Preface   vii

Introduction   1
Daniel A. Jelski and Thomas F. George

Ab Initio Studies of Compound Semiconductor Surfaces   6
Tapio T. Rantala

Molecular-Dynamics Studies of Defects and Impurities in Bulk
Semiconductors   27
Stefan K. Estreicher and Peter A. Fedders

Tight-Binding Molecular Dynamics Study of Structures and Dynamics
of Carbon Fullerenes   74
C. Z. Wang, B. L. Zhang and K. M. Ho

Computations of Higher Fullerenes   112
Zdeněk Slanina, Xiang Zhao and Eiji Osawa

Relaxations of Charge Transfer and Photoexcitation in C_{60} and Polymers   152
Xin Sun, Guoping Zhang, Thomas F. George and R. T. Fu

Ionic Charge Transport in Molecular Materials: Polymer Electrolytes   174
Mark A. Ratner

Computational Approaches in Optics of Fractal Clusters   210
Vadim A. Markel and Vladimir M. Shalaev

Local Fields' Localization and Chaos and Nonlinear-Optical
Enhancement in Composites   244
Mark I. Stockman

Atomic Valences in Aperiodic Crystals Studies by the Bond Valence Method   273
Sander van Smaalen

Surface Light-Induced Drift   295
Michael A. Vaksman

Theoretical Treatment of Surface Adsorbates   334
László Nánai, Csaba Beleznai and Thomas F. George

Phase Conjugation Through Four-Wave Mixing   351
Henk F. Arnoldus and Thomas F. George

Electromagnetic Propagators in Micro- and Mesoscopic Optics   375
Ole Keller

Nanoscale Materials: Conceptual and Computational Challenges   440
Mushti V. Ramakrishna

Index   449

PREFACE

When World Scientific Publishers first approached us three years ago about editing a book on the overall topic of computational materials science, we were flattered but rather awed by the task. Since the project seemed interesting and timely, we agreed to do it. It soon became clear that we would have to narrow down the topic a bit. One possibility, for example, was a book titled *Computational Methods in Materials Science*. However, this would look different from the current volume: it would bristle with computer code, discuss the appropriateness of various basis sets, and be much more of a how-to manual for the computer set. We decided not to edit such a book, partly because we thought it would reach a rather narrow audience.

The title we have chosen, *Computational Studies of New Materials*, puts more emphasis on the materials and somewhat less on the computation. And that is indeed the way this book has evolved. The focus is on the materials, be they fullerenes, fractals or polymers. Of course, computational methods are not far below the surface, and in particular, the volume contains an excellent description of tight-binding methods. But it is our hope that this volume will adorn the shelves of scientists who are not explicitly computationalists.

We have found our contributors to be very good writers, interesting people, and above all, imaginative and talented scientists. Working with them has been a rewarding scientific adventure, and we thank every one of them for the hard work and effort they have devoted to this volume. We also thank Dr. Chi-Wai (Rick) Lee, Scientific Editor at World Scientific, for his guidance and encouragement throughout the course of this project. Finally, we gratefully acknowledge the staff and students in the Office of the Chancellor at UW-Stevens Point for assistance in assembling the index.

The Editors
January 1999

INTRODUCTION

DANIEL A. JELSKI
Department of Chemistry
State University of New York, College at Fredonia
Fredonia, NY 14063
E-mail: jelski@fredonia.edu

THOMAS F. GEORGE
Chancellor and Professor of Chemistry and Physics
Office of the Chancellor
University of Wisconsin - Stevens Point
Stevens Point, WI 54481
E-mail: tgeorge@uwsp.edu

Once upon a time, not too terribly long ago, materials science and solid-state physics were roughly synonymous. In those days a volume such as this would have bristled with terms such as "Brillioun zone" and "Wigner-Seitz cells." Then crystals were periodic, solids stretched infinitely in all directions, polymers consisted of monomer units, and the polarizability was simply proportional to the volume. The world was simple and wonderful.

Nowadays crystals come with prefixes, such as quasi-, and solids are amorphous or have a fractal dimension, or are ridden with defects or "mesoscopic vacancies," or are bounded by reconstructed surfaces, or have shrunk into "nanostructures." Today's monomers have evolved into "functional groups," and the polarizability of classical physics has turned into the chemistry of "charge transfer." The Brillioun zone still makes a cameo appearance in this book, though Wigner-Seitz cells have been vanquished. Materials science has become the study of microscopic interactions, as much a part of chemistry as physics. The world is still wonderful, but it is no longer simple; it is an incredibly complicated and rich place where our fundamental understanding of nature can now be applied to very specific and realistic problems. This last sentence summarizes the goals of modern computational materials science, and constitutes the theme for this book.

The editors thought long and hard about how to best categorize the material we solicited for this book. There are many ways that we could have sliced and diced the material given to us. We went through several different versions of the Table of Contents, considering the pros and cons of Parts, Sections or various other subdivisions. We could, for example, easily have had a section on fullerenes. We could also easily have had a section on tight-binding methods. However, some articles would have had to appear in both sections, and it was problems such as this that led us to discard a structure beyond the chapter level for this book. Instead, the book simply presents our contributors and lets them speak for themselves.

But the chapter order is not random. The rough order is from the basic tools of computational materials science to more specific applications. We begin with Rantala's excellent introduction to local density approximation (LDA) methods. The method is compared to two other techniques – the pseudo-potential plane-wave and LCAO methods – and illustrates dramatically the transition from the old world of materials science to that of today. The chapter is well written, lists the most important references, and is an excellent introduction for the novice to this field. The application presented by Rantala is the reconstruction of solid surfaces, specifically compound semiconductors.

Estreicher and Fedders' article on molecular dynamics of defects and impurities in bulk semiconductors really does deliver what the title promises. But along way, the authors have given us a wonderful review of various methods used in molecular dynamics. These range from classical molecular dynamics, through tight-binding methods and on to local density approximation methods. Using a plane-wave basis set is particularly efficient because one can then easily incorporate Car-Parrinello methods. Problems in molecular dynamics are also considered, e.g., how one can ensure that the relevant region of phase space is being sampled. In our opinion, the most interesting discussion is about amorphous silicon hydride, and how one can best model it. The authors also consider a novel class of nanostructures, namely mesoscopic defects in solids.

Fullerenes are a new class of materials that generated three contributions from our authors. Wang, Zhang and Ho, from Iowa State University, specifically use a tight-binding model applied to fullerenes. Like LDA methods, tight-binding methods are also a traditional tool of solid-state physics, but placed in the hands of our Iowa State colleagues, these are put to quite non-traditional uses. Such a model must be transferable, i.e., must be useful for different phases of the material, in order to be useful, and the Iowa authors demonstrate that their model satisfies this criterion. Indeed, they do a superb job of describing tight-binding methods generally. They put the model through its paces: they show that it reproduces the LDA-calculated lowest energy isomers of C_{84}, that the HOMO-LUMO band gaps are correctly reproduced, and that other phases of carbon, along with carbon chains and rings, are properly accounted for. The larger fullerenes present a problem that recurs frequently throughout computational materials science, namely the necessity to deal with innumerable combinations. The Iowa authors describe the novel face-dual method that is successfully used to generate structures for the larger fullerenes. Among the more interesting applications, the authors consider fragmentation and collision processes of fullerenes, both of which are applications of the molecular dynamics methods mentioned above.

The problem of combinatoric enumeration is central to the article by Slanina, Zhao and Osawa, which deals with the structure of larger fullerenes. These authors provide an excellent review of the fullerene "combinatoric" literature, i.e., papers that try to enumerate the combinations using simple geometry and assuming the isolated pentagon rule. Slanina *et al.* then use semiempirical quantum methods, such as MNDO, AM1, PM3 and SAM1, to probe these structures. They are especially interested in the thermodynamic properties of fullerenes, for which such methods are well suited, and which include the Boltzmann distribution of isomers. The latter can be experimentally checked by NMR experiments.

In the final fullerene paper, Sun, Zhang, George and Fu consider charge transfer in fullerenes, specifically buckminsterfullerene and polymers including buckminsterfullerene. They also use a tight-binding model, although in this case one that includes spin. As with other similar models, it includes a classical repulsive term, but the basis set is severely limited: one orbital per atom. For buckminsterfullerene, however, this is sufficient since the π-band is sufficiently separated from the σ-band. The authors demonstrate that their method reproduces experiment. The primary interest is in calculating the Jahn-Teller effect when buckminsterfullerene is charged, either as an anion or cation. Similar effects are measured for photoexcitation processes. Relaxation pathways are studied for both phenomena.

Molecular dynamics also figures into the paper by Ratner. His paper concerns ion transfer through an amorphous polymer solvent. His first order of business is to point out the widely-varying time scales involved in modeling this phenomenon, a problem that bedevils much of materials science. This particular article is most concerned with what happens at the shortest times, and Ratner describes LDA calculations that are used to discern how the ion is attached to the polymer strand. It is found that the ion is chelated rather than simply bound to the strand, and that it moves along more like an insect than a flea, i.e., one leg at a time rather than by jumps. The venerable hopping model may not apply to these sorts of problems. Ratner's article is precisely the sort of work mentioned above, i.e., it is computational science applied to very specific chemical problems.

The next chapter by Markel and Shalaev is a pleasant contrast to Ratner's piece. It is about the structure of fractal clusters. Instead of trying to describe the behavior of a specific system, as Ratner does, Markel and Shalaev attack a more general problem. A quote from their chapter illustrates this: "The most simple and extensively used model for fractal clusters is a collection of identical spherically symmetrical particles (monomers) that form a self-supporting geometrical structure." This, in a nutshell, is the complete description of the material they discuss. The paper is a discussion of the optical properties of such clusters. It is very well written and is an excellent introduction to that field.

Similarly, Stockman's chapter also concerns the nonlinear optical properties of fractal clusters. The only necessary definition of the material is a representative length (the distance between monomer units) and the polarizability of each monomer. In such a fractal system, nonlinear phenomena are greatly enhanced, and the enhancement is all the greater the higher the nonlinearity. An example presented in the article is a discussion of surface-enhanced Raman spectroscopy.

Van Smaalen's chapter concerns the structure of aperiodic crystals, specifically incommensurate structures. In this case the bond valence method is used, where the valence around an atom is the sum of the contributions of the individual bond valences. The bond valences, in turn, are a function of interatomic distance. This yields a description of the crystal that can be used to predict metal oxidation states. Most interesting is the discussion of charge density wave compounds. These are compounds where the electron density fluctuates periodically. The valence bond model predicts that this is not caused by a variation in the oxidation state of the metal. Indeed, it is found that the metal oxidation state remains relatively constant independent of coordination number.

The chapter by Vaksman is about surface light-induced drift. This is a method by which gas-solid interactions can be studied. Consider a laser tuned to a frequency slightly different from a resonance frequency of the gas. The gas is contained in a cell constructed from the surface of interest. Then, because of the Doppler effect, gas molecules moving in one direction will be preferentially excited. But excited molecules interact with the surface differently than their ground state counterparts, and hence a pressure gradient will develop in the container. This can be measured, and it proves to be a very sensitive measure of gas-surface interactions.

Nánai, Beleznai and George consider methods for modeling surface adsorbates. A number of different techniques are discussed, but the diligent student of previous chapters (notably those by Rantala and by Estreicher and Fedders) will have an easier time following this essay. For example, here it is mentioned that density-functional techniques are useful for describing the relaxation of adsorbed particles. The authors discuss cluster methods at some length.

Two articles in the book are much more specifically about the optical properties of materials. Arnoldus and George, for example, describe the material simply as "a nonlinear crystal," and the phase conjugate effects follow from there. A phase-conjugated mirror involves the reversal of an optical wave front. Keller's article is about the optical properties of microscopic and mesoscopic systems as associated with nanostructures.

Finally, the last chapter in the book, by Ramakrishna, is more an editorial than a review article, but it is none the less interesting on that account. The article is about how nanostructures, specifically of silicon, differ from their bulk counterparts. In a sense, nanostructures form a different phase of matter, and the impact of this on the design and manufacture of silicon chips may become important within the next decade.

In a word, our authors have written wonderful articles about stuff – the stuff of this world. Some of it is detailed and specific, whereas for other articles the "stuff" is briefly or generally described. And speaking of stuff, there is a lot of good stuff between the covers of this book. Enjoy!

AB INITIO STUDIES OF COMPOUND SEMICONDUCTOR SURFACES

Tapio T. RANTALA

Department of Physical Sciences, University of Oulu,
P.O.Box 333, FIN-90571 Oulu, Finland.
E-mail: `Tapio.Rantala@oulu.fi`

A brief review of compound semiconductor surface structures and properties studied with *ab initio* methods is given. First, the methodology based on the density-functional formalism and local-density approximation is described, followed by a description of two alternative computational techniques. Then, two chosen semiconductor surfaces are considered as examples, the $(10\bar{1}0)$ face of a tetrahedrally bonded CdS and the (110) face of a rutile SnO_2. Also, some aspects of the surface chemical activity are discussed.

1 Introduction

Compound semiconductors offer a multitude of possibilities for designing new materials. The number of possible binary compounds formed from the elements of main groups from II to VII is very large, already, let alone the combinations of more than two components. Furthermore, doping and use of layered structures allows one to tune the electrical and optical properties of semiconductor materials at will. A play with crystalline, layered, amorphous and porous materials allows one to select the mechanical and chemical properties, too. In all, both the bulk properties and surface chemical activity of compound semiconductors can be engineered, if the underlying principles and origin of the properties are sufficiently well known. It is this expertise where the theoretical and computational research can contribute most.

It is the surface of solid materials, where the contact and interaction with the environment of a piece of matter takes place. It makes the surface properties of materials of essential importance. Until lately, the main focus of attention in surface science has mostly been in research of simple metal and other elemental surfaces. Though, it has given us much insight to surface phenomena, adsorption, surface diffusion, catalyzed reactions, etc., there still remain details to be uncovered. This is true with complex compound materials, in particular. On the other hand, as pointed out above the compound materials offer good possibilities for materials engineering. This motivates the computational (and experimental) studies of compound semiconductor surfaces, in general[1−7] and in such new applications as gas sensors,[1,2] in particular.

In this text we first present the density-functional formalism, which is the basis of the applied *ab initio* methods in our studies. Next, the two applied computational techniques, one based on the linear-combination-of-atomic-orbitals (LCAO) and the other employing plane waves and pseudopotentials (PWPP) are described. We aim at giving a relatively complete account of these matters with as simple concepts as possible. Therefore, we start from basics but include only the most essential concepts and try to keep the text easy to read. Then, the use of these computational techniques is demonstrated in the subsequent two case studies. Relaxation of the chosen cleavage surfaces of tetrahedrally bonded CdS and rutile structure metal oxide SnO_2 are considered. Some aspects of the surface chemical activity are discussed, too.

2 First Principles Approach

2.1 Many-electron Problem

The vast majority of the properties of matter depend on its electronic structure, the quantum state of the electrons involved. The electrons bind the atoms to molecules or solids and they are responsible for most of the interactions between pieces of matter. Furthermore, it is also the electrons that respond to many external perturbations of matter, e.g. irradiation. The main job left for the atomic nuclei is to provide the charge balancing environment for the electrons to move, but the conformation and dynamics of nuclei, on the other hand, follow the force field given by the electronic structure. It is this interplay between electrons and nuclei that is responsible for the surface chemistry, thermal properties and many other bulk properties of solids. Therefore, we are interested in computationally searching for the properties and dynamics of a system of electrons and nuclei.

The stationary quantum state of the many-particle system (here, electrons and nuclei) is a solution to the Schrödinger equation

$$H\Psi = E\Psi, \tag{1}$$

where the hamiltonian H includes all of the Coulomb pair interactions and the kinetic energies of all particles in the system. The total energy of the system E is obtained as an eigenvalue associated to the wavefunction Ψ. This solution, in principle, provides us with all the information we may wish.

In what follows we do not treat nuclear dynamics quantum mechanically and we limit us to the non-relativistic treatment. Furthermore, the spin of the electrons is not treated explicitly. Thus, the N-electron hamiltonian in the surroundings of M atomic nuclei reads (in atomic units) as

$$H = \sum_i^N \left(-\frac{1}{2}\nabla_i^2 + v_{\mathrm{ne}}(\mathbf{r}_i) \right) + \sum_{i<j}^{N,N} \frac{1}{r_{ij}} + V_{\mathrm{nn}}, \tag{2}$$

where

$$v_{\mathrm{ne}}(\mathbf{r}_i) = \sum_\mu^M \left(-\frac{Z_\mu}{r_{i\mu}} \right), \quad V_{\mathrm{nn}} = \sum_{\mu<\nu}^{M,M} \frac{Z_\mu Z_\nu}{R_{\mu\nu}}, \tag{3}$$

$r_{i\mu} = |\mathbf{r}_i - \mathbf{R}_\mu|$, $r_{ij} = |\mathbf{r}_i - \mathbf{r}_j|$, $R_{\mu\nu} = |\mathbf{R}_\mu - \mathbf{R}_\nu|$ and Z_μ are the nuclear charges. The two terms, $v_{\mathrm{ne}}(\mathbf{r})$ and V_{nn}, Eqs. 3, are the Coulomb potential energy of electrons in the field of nuclei and the mutual Coulomb repulsion energy of nuclei, respectively. They depend on the set of electronic coordinates $\{\mathbf{r}_i\}$ and the set of nuclear coordinates $\{\mathbf{R}_\mu\}$.

It should be noted that it is the nuclear conformation $\{\mathbf{R}_\mu\}$, nuclear charges $\{Z_\mu\}$ and the number of electrons N that suffice to specify the whole quantum state of the electronic system. The calculation procedures which start with this least possible information about the system and use only principles of quantum theory are called *ab initio* or "first principles" methods. Basically both of these terms mean the same concept, though some authors have assigned them to some specific calculation procedures, too.

The solution to the Eq. 1 should provide us with data to compare with the related experiments, and concepts for obtaining physical insight to the system. For these reasons it is helpful to decompose the complex many-electron system (or state) to one-electron states, and correspondingly, describe the many-electron quantities, like total energy, with contributions of single electron states. In fact, the standard solution procedures start with this s.c. one-electron picture: mutually interacting electrons in their own separable eigenstates. These eigenstates ψ_i are solutions to their one-electron Schrödinger equations

$$h_i(\mathbf{r}_i)\,\psi_i(\mathbf{r}_i) = \varepsilon_i\,\psi_i(\mathbf{r}_i). \tag{4}$$

There are two conventional *ab initio* approaches to solve Eq. 1 for Ψ, the wavefunction formalism and the density-functional formalism. The wavefunction formalism starts with the one-electron spin-orbitals ψ_i of the electrons in the system. An antisymmetrized product of spin-orbitals Ψ^{HF}, e.g. a Slater determinant,[8] and the variational total energy minimization leads to a set of one-electron equations of the form of Eq. 4. These are s.c. Hartree–Fock equations, where h_i is called the Fock operator.[8,9] The antisymmetric Hartree–Fock wavefunction Ψ^{HF} includes the exchange interaction of electrons but does not take into account all details of the mutual Coulomb correlations of electrons resulting from the interaction term $\sum_{i<j} 1/r_{ij}$. This deficiency is a limitation of the one-electron picture and it is usually corrected with s.c. "configuration interaction" or "multiconfiguration" scheme, where the correlated wavefunction Ψ^{CI} is written as a linear combination of the ground state and a large number of different excited state Hartree–Fock wavefunctions Ψ^{HF}.

The configuration interaction method offers, in principle, a systematic procedure to the solution of arbitrary high accuracy. However, as the system size or the number of electrons N increases the computational labor, increasing as N^3, soon becomes intolerable. This is due to the increasing number of electron–electron interaction integrals. Furthermore, with infinite systems like solids more fundamental problems arise. The long range of Coulomb interaction of electrons leads to unphysical singularities, unless an additional screening is included into the Hartree–Fock formalism.[10]

In the next section we introduce the other formalism, which circumvents many of these problems with a trade off of losing some of the accuracy. Nevertheless, it is better suited for the treatment of large and infinite system like crystals and surfaces. For finite systems, for example, the computational labor increases only linearly with respect to the number of electrons N.

2.2 Density-functional Formalism

The density-functional theory (DFT) is the other conventional *ab initio* approach to search for a solution to the many-electron Schrödinger equation, Eq. 1. A similar one-electron picture is invoked but the exchange and correlation of electrons is treated differently. Within DFT, the description is based on the electron density of the many-electron system. This is convenient, because the electron density is a well-defined measurable physical quantity and becomes even more relevant as the system size becomes larger. However, it should be kept in mind that the DFT is based on the very same physical preassumptions and first principles as the wavefunction formalism.

The starting point of DFT is the first Hohenberg–Kohn theorem,[11] which says that the external potential is determined, within a constant, by the electron density. This implies that there is one-to-one correspondence between the external potential of electrons $v_{\mathrm{ne}}(\mathbf{r})$, Eq. 3, and the ground state one-electron density $\rho(\mathbf{r})$ for a fixed number of electrons N. As the external potential $v_{\mathrm{ne}}(\mathbf{r})$ in the hamiltonian, Eq. 2, determines the many-electron wavefunction $\Psi(\{\mathbf{r}_i\})$ of Eq. 1, it follows that the one-to-one correspondence extends to the normalized $\Psi(\{\mathbf{r}_i\})$, too. Thus, all the ground state properties of a many-electron system are specified by its ground state one-electron density $\rho(\mathbf{r})$. Note that this is not restricted to Coulomb potentials, only.

Now, for a certain external potential $v_{\mathrm{ne}}(\mathbf{r})$, we can write the total energy of a many-electron system as a functional of its electron density and decompose it formally as

$$E[\rho] = T[\rho] + V_{\mathrm{ne}}[\rho] + V_{\mathrm{ee}}[\rho]. \tag{5}$$

The first term $T[\rho]$ is the kinetic energy of electrons, the second term

$$V_{\mathrm{ne}}[\rho] = \int \rho(\mathbf{r})\, v_{\mathrm{ne}}(\mathbf{r})\, \mathrm{d}\mathbf{r}, \tag{6}$$

is the Coulomb energy of charge density $\rho(\mathbf{r})$ in the external potential $v_{\mathrm{ne}}(\mathbf{r})$ and the third term $V_{\mathrm{ee}}[\rho]$ covers all the electron–electron interactions.

Application of the variational principle to the total energy functional $E[\rho]$ with respect to the function ρ proves the second Hohenberg–Kohn theorem: [11] the total energy is stationary at the ground state density with the energy E_0, or

$$E_0 \le E[\rho]. \tag{7}$$

Thus, it has been demonstrated that solving the N electron problem for E_0 is equivalent with minimizing the total energy functional with respect to the electron density or with searching for the ground state density $\rho(\mathbf{r})$.

Again, a convenient practical approach [12] to search for $\rho(\mathbf{r})$ starts with writing it in terms of non-interacting one-electron states $\psi_i(\mathbf{r})$, for which

$$\rho(\mathbf{r}) = \sum_i^N \langle \psi_i | \psi_i \rangle = \sum_i^N |\psi_i(\mathbf{r})|^2 \tag{8}$$

and $\int \rho(\mathbf{r})\mathrm{d}\mathbf{r} = N$. Hence, we can write

$$T[\rho] = \sum_i^N \langle \psi_i | (-\frac{1}{2}\nabla_i^2) | \psi_i \rangle, \tag{9}$$

$$V_{\mathrm{ne}}[\rho] = \sum_i^N \langle \psi_i | v_{\mathrm{ne}} | \psi_i \rangle \tag{10}$$

and

$$V_{\mathrm{ee}}[\rho] = \frac{1}{2} \sum_i^N \langle \psi_i | v_{\mathrm{H}} | \psi_i \rangle + E_{\mathrm{xc}}[\rho]. \tag{11}$$

Here, the Hartree potential is

$$v_{\mathrm{H}}(\mathbf{r}) = \int \frac{\rho(\mathbf{r}')}{|\mathbf{r} - \mathbf{r}'|} \mathrm{d}\mathbf{r}', \tag{12}$$

which gives the larger term at the right hand side of Eq. 11, the classical Coulomb repulsion energy of the charge density $\rho(\mathbf{r})$, itself. The smaller contribution, the exchange and correlation energy, can formally be decomposed to the respective terms as

$$E_{\text{xc}}[\rho] = E_{\text{x}}[\rho] + E_{\text{c}}[\rho].\tag{13}$$

These terms contain all of the remaining electron–electron interactions beyond the Hartree energy.

Now, using the variational principle to the energy expression Eq. 5 written in terms of the orbitals ψ_i, one can derive[12] one-electron equations of the form of Eq. 4,

$$h_i(\mathbf{r}_i)\,\psi_i(\mathbf{r}_i) = \varepsilon_i\psi_i(\mathbf{r}_i).\tag{14}$$

These are called Kohn–Sham equations,[12,13] which are in the same role in DFT as the Hartree–Fock equations are in the wavefunction theory.

From Eqs. 9–11 it is relatively easy to inspect what the one-electron hamiltonian of the Kohn–Sham equations becomes to. It can be written as

$$h_i(\mathbf{r}_i) = -\frac{1}{2}\nabla_i^2 + v_{\text{eff}}(\mathbf{r}_i),\tag{15}$$

where the effective one-electron potential is

$$v_{\text{eff}}(\mathbf{r}) = v_{\text{ne}}(\mathbf{r}) + v_{\text{H}}(\mathbf{r}) + v_{\text{xc}}(\mathbf{r})\tag{16}$$

and further,

$$v_{\text{xc}}(\mathbf{r}) = \frac{\delta E_{\text{xc}}[\rho(\mathbf{r})]}{\delta\rho}\tag{17}$$

is sc. exchange–correlation potential. Solutions ψ_i to the Eq. 14 should be self-consistent, because the one-electron hamiltonian depends on the potential $v_{\text{ee}}(\mathbf{r}) = v_{\text{H}}(\mathbf{r}) + v_{\text{xc}}(\mathbf{r})$, which depends on $\rho(\mathbf{r})$ written in terms of $\{\psi_i\}$.

Note that in writing the one-electron equations we have not done any approximations to the exact DFT, so far, and we are dealing with electrons whose all interactions are included in a functional of ρ, the effective potential $v_{\text{eff}}(\mathbf{r})$, Eq. 16. This can be contrasted with the more complex Hartree–Fock equations of the wavefunction theory, which however, are known to describe an approximation including only the exchange but excluding the correlation interactions. In fact, the correlation interactions of electrons are usually defined to be those which are not described within the Hartree–Fock theory. Of course, things are not that simple with the DFT, either. The problems are just swept under the carpet for the present, i.e., the more complex interactions are gathered into the exchange–correlation potential $v_{xc}(\mathbf{r})$. It looks simple, it is just a function of $\mathbf{r}$ (for a fixed ρ) and even known exactly for the uniform electron density (for practical purposes). However, for the general case, non-uniform densities of atoms, molecules or solids, there are just various levels approximations of v_{xc} to choose from, so far.

The simplest approximations to the exchange–correlation potential v_{xc} and the corresponding energy per electron ε_{xc} are based on the properties of uniform electron gas. As mentioned above, for practical purposes these properties are known accurately enough from the Monte Carlo simulations of Ceperley and Alder.[14] In such case the constant density ρ or $r_s = (3/4\pi\rho)^{1/3}$ is the only parameter describing the whole system, if retaining to the spin-restricted case, only. Now, consider an electron gas with slow spatial variations, where we could expect the local properties of the electron gas to vary slowly, too. We could further expect that these properties depend almost entirely on the local electron density, not differing essentially from the properties of the uniform electron gas with the same density.

The slow spatial variations is the idea behind the local-density approximation (LDA), where the functional $v_{xc}[\rho]$ defined in Eq. 17 is replaced by a local function $v_{xc}^{LDA}(\rho(\mathbf{r}))$. With the same approximation to the exchange and correlation energy per electron $\varepsilon_{xc}^{LDA}(\rho(\mathbf{r}))$ the exchange and correlation energy of the density $\rho(\mathbf{r})$ can be approximated with a simple integral

$$E_{xc}^{LDA} = \int \rho(\mathbf{r})\,\varepsilon_{xc}^{LDA}(\rho(\mathbf{r}))\,\mathrm{d}\mathbf{r}. \tag{18}$$

For practical calculations there are parameterized formulae fitted to the Monte Carlo data of the homogeneous electron gas.[13–16]

The LDA works well for solid materials,[15–18] especially for metals, and it has proven to be surprisingly successful even in cases where the density variations are relatively large like in free atoms and molecules.[13] The general experience is that LDA is good in predicting bond lengths and conformations, forces on atoms and vibrational frequencies, and general trends in chemistry. On the other hand, LDA fails in predicting bond energies accurately, and nonbonding interactions, and it systematically underestimates the band gap of semiconductors.

The most popular methods to improve LDA are based on sc. generalized gradient approximation [19] (GGA), where the effects of local density variations are described by density gradients and parameterized accordingly. The GGA is known to improve calculated total energies, atomization energies, energy barriers and the band gaps of semiconductors, though not always sufficiently. There are also methods [20] that do even better with these properties, especially with the band gap problem, but are relatively heavy in practical computations.

From comparison of the results of DFT and wavefunction methods one can conclude that generally LDA does better than Hartree-Fock for the molecular properties.[13,19] However, where computationally feasible, i.e. for small molecules, the highly correlated wavefunction methods are the most accurate. For solids, on the other hand, the DFT method is the only applicable choice, in practice.

3   Computational Methods

3.1   LCAO Method

The one-electron states or orbitals of electrons are most conveniently solved as an expansion of suitable basis functions. Only spherically symmetric free atoms make an exception for which other numerical techniques are usually adopted. The set of basis functions can be chosen to suit best for the system in question. For small or disorderd structures with localized characteristics a set of atomic orbitals expanded around the nuclei may be the best choice, whereas for periodic bulk or other infinite structures plane waves may serve better. In this section, we first consider the common technique of using a set of localized basis functions, and in the next section, we consider the use of plane waves as basis functions.

The linear-combination-of-atomic-orbitals (LCAO) is the general name for methods where molecular orbitals or one-electron states ψ_i are expanded in terms of atomic one-electron orbitals $\{\varphi_{\mu k}\}$ centered around the nuclei at $\mathbf{R}_\mu$. The atomic orbitals can be written as

$$\varphi_{\mu k}(\mathbf{r}) = u_{n\ell}(r_\mu)\, Y_{\ell m_\ell}(\hat{\mathbf{r}}_\mu), \tag{19}$$

where $u_{n\ell}(r)$ is the radial part of the atomic orbital and $Y_{\ell m_\ell}(\hat{\mathbf{r}})$ is the angular part, the spherical harmonic function. Here $\mathbf{r}_\mu = \mathbf{r} - \mathbf{R}_\mu$, $r_\mu = |\mathbf{r}_\mu|$ and $\hat{\mathbf{r}} = (\theta, \phi)$, the angles of the polar coordinate representation of $\mathbf{r}$. The subscript k stands for the set of orbital quantum numbers $\{n, \ell, m_\ell\}$. Instead of the complex spherical harmonics the real combinations can be taken with the advantage of better suitability for describing bonding between atoms. For example, the combinations $Y_x = Y_{1,-1} - Y_{1,+1}$, $Y_y = Y_{1,-1} + iY_{1,+1}$ and $Y_z = Y_{1,0}$ are real functions for p-orbitals and conveniently oriented in space.

There are several forms in which the radial part of atomic type basis functions $u_{n\ell}(r_\mu)$ have conventionally been written for the computational use. The gaussian type orbitals (GTO) are composed as a sum of primitives, $u(r) = \sum_i b_i \exp(-\alpha_i r^2)$, which have convenient analytical properties. The Slater type orbitals (STO), usually scaled hydrogen like orbitals, have more realistic functional form but they are less convenient in analytical calculations. One more practical alternative is the use of numerically calculated atomic orbitals of free atoms and ions. This is the choice that can be kept in mind when reading what follows.

The one-electron molecular orbitals are expanded now as a LCAO

$$\psi_i(\mathbf{r}) = \sum_j^{N_b} c_{ij}\, \chi_j(\mathbf{r}), \tag{20}$$

where

$$\chi_j(\mathbf{r}) = \sum_{\mu k} w_{j\mu k}\, \varphi_{\mu k}(\mathbf{r}). \tag{21}$$

The N_b functions $\chi_j(\mathbf{r})$ are called symmetry adapted basis functions, which are formed with coefficients $w_{j\mu k}$ chosen to make the functions χ_j to transform according to the symmetry properties of the hamiltonian, i.e. according to the irreducible representations (irrep) of the point group of the hamiltonian. The number of the basisfunctions N_b should be much larger than the number of occupied orbitals N, that gives flexibility to the solutions and leads to better description of ψ_i as a linear combination of functions χ_j.

With substitutions of Eqs. 15 and 20, the Eq. 14 can be written in form of the matrix equation, or a set of s.c. secular equations,

$$\mathcal{H}\mathbf{c}_i = \varepsilon_i \mathcal{S}\mathbf{c}_i, \tag{22}$$

where the matrix elements of the hamiltonian $\mathcal{H}$ are

$$H_{mn} = \int \chi_m^*(\mathbf{r})\, h(\mathbf{r})\, \chi_n(\mathbf{r})\, d\mathbf{r} = \langle m|h|n \rangle \tag{23}$$

and those of the overlap matrix $\mathcal{S}$ are

$$S_{mn} = \int \chi_m^*(\mathbf{r})\, \chi_n(\mathbf{r})\, d\mathbf{r} = \langle m|n \rangle. \tag{24}$$

Note that

$$\langle m|h|n \rangle = \langle m|(-\tfrac{1}{2}\nabla_i^2)|n \rangle + \langle m|v_{\text{eff}}(\mathbf{r})|n \rangle, \tag{25}$$

that shows how the kinetic energy contribution is calculated from those of the basis functions or atomic orbitals.

The N_b solutions, eigenvalues ε_i and eigenvectors $\mathbf{c}_i = \{c_{ij}\}$, to the matrix equation 22 are obtained through diagonalization of the hamiltonian $\mathcal{H}$. As the basis functions belong to the irreducible representations of the hamiltonian, the matrices $\mathcal{H}$ and $\mathcal{S}$ are reduced to blocks, one for each irrep. This follows from the fact that $\langle m|h|n \rangle$ and $\langle m|n \rangle$ both vanish, if χ_m and χ_n belong to different irreps. Thus, the diagonalization can be carried out for each block independently, reducing the computational task considerably.

As mentioned above, the solutions $\mathbf{c}_i$ or ψ_i must be self-consistent. The charge density, obtained from Eq. 8 by summing over the N occupied orbitals, is responsible for the Hartree potential in Eq. 12, which is a large contribution in the effective potential, of the hamiltonian, Eq. 16. As the solutions are initially not known, the Hartree potential has to be approximated by an initial guess and then iterated until self-consistency. In practice, rather than evaluating $v_H(\mathbf{r})$ from Eq. 12 directly, it is usually solved from the Poisson equation

$$\nabla^2 v_H(\mathbf{r}) = -4\pi\rho(\mathbf{r}). \tag{26}$$

This usually results in a higher accuracy with less effort.

The above described LCAO procedure is natural and was initially developed for finite systems, molecules and clusters, with a finite number of atoms and electrons. It can be used, however, to evaluate the electronic structure of periodic bulk materials, too. A straightforward "large-cluster" model for infinite bulk always suffers from serious finite-size effects due to the cluster surface, and therefore, is not a good scheme, in general. Instead, dividing the bulk into identical computational unit cells or supercells, of which one is computed with LCAO under the explicit interaction of others, leads to better description of the periodic bulk. This kind of embedding of a supercell in between its identical images is what has been done in one of the commercially available computer codes.[21]

Although the embedding scheme is actually an application of the periodic boundary conditions, the periodicity or the full translational symmetry can not be used explicitly in defining the one-electron levels. The consequent drawback is that the wave vector (or wavelength) dependence of the computed one-electron eigenvalues is not known, and all the levels are assigned to the Γ-point (zero wave vector) of the supercell. This is also called Γ-point approximation. The quality of the solution can be increased by increasing the size of the supercell with respect to the primitive cell of the periodic bulk. This includes more one-electron levels from the Brillouin zone but in a reduced zone picture, or alternatively, it reduces the size of the first Brillouin zone, and thus, makes the Γ-point a better representative of the $\mathbf{k}$-points in average.

3.2   *Plane Wave Pseudopotential Method*

The one-electron levels of a solid are delocalized and adapted to the infinite and usually periodic configuration of the component atoms, except for the deep core levels. The core levels are of less importance in cohesion of atoms and solid phase formation. Therefore, the delocalized plane waves serve as a natural and convenient basis set to expand the essential one-electron wavefunctions. The localized core levels, on the other hand, could be described only with relatively large plane wave expansions. As the main role of core electrons in this context remains their "background interaction" with the valence and conduction electrons, the ion cores can be replaced with suitable pseudopotentials with only a minor loss of accuracy in the description of delocalized levels outside the core region. A general acronym for these type of plane wave pseudopotential methods is PWPP.

Transferable core pseudopotentials are ideally constructed so, that the "scattering" properties of valence and conduction electrons, i.e. the form of of their wavefunctions, remain the same outside and become smooth and radially nodeless inside the core region. Because the scattering depends on the angular momentum of the incoming wave with respect to the nucleus, different pseudopotentials may be used for different angular momentum related quantum numbers ℓ and m_ℓ. Interactions between different ℓ channels can be included. Such a pseudopotential is called non-local and can be written in the form

$$v_{\rm pp}(\mathbf{r}, \mathbf{r}') = \sum_{\mu\ell m_\ell} |\mu\ell m_\ell\rangle \, E_{\mu\ell} \, \langle\mu\ell m_\ell|', \tag{27}$$

where

$$|\mu\ell m_\ell\rangle = u_{\mu\ell}(\mathbf{r}_\mu) \, Y_{\ell m_\ell}(\hat{\mathbf{r}}_\mu), \tag{28}$$

$u_{\mu\ell}(\mathbf{r}_\mu)$ is the radial part, $Y_{\ell m_\ell}(\hat{\mathbf{r}}_\mu)$ the spherical harmonic function, $E_{\mu\ell}$ a number (energy) and μ labels the ion cores. The radial functions $u_{\mu\ell}$ and energy parameters $E_{\mu\ell}$ of pseudopotentials are fitted to suitable reference systems using various procedures.[22,27] Relativistic corrections, if essential, can easily be included into pseudopotentials, too. For more details of pseudopotentials of ion cores see Ref. 22 and references therein. Finally, the pseudopotential $v_{\rm pp}$ is added to the effective potential $v_{\rm eff}$, Eq. 16, of valence and conduction electrons.

We now consider the effects of the periodicity of the bulk lattice of atoms on the one-electron wavefunctions. Let the Bravais translation vectors defining the unit cell be $\mathbf{R}_1$, $\mathbf{R}_2$ and $\mathbf{R}_3$. This implies the periodicity of $\mathbf{R} = n_1\mathbf{R}_1 + n_2\mathbf{R}_2 + n_3\mathbf{R}_3$, where n_i are integers, in the effective potential as

$$v_{\rm eff}(\mathbf{r} + \mathbf{R}) = v_{\rm eff}(\mathbf{r}). \tag{29}$$

Now, Bloch's theorem [10] states that the one-electron eigenstates of the hamiltonian in Eq. 15 are of the form

$$\psi_{n\mathbf{k}}(\mathbf{r}) = e^{i\mathbf{k}\cdot\mathbf{r}}\, u_{n\mathbf{k}}(\mathbf{r}), \tag{30}$$

where $\mathbf{k}$ is sc. wavevector, a point in the first Brillouin zone. The radius vector $\mathbf{r}$ and the wavevector $\mathbf{k}$ form a Fourier pair, the variables in space (unit cell) and in the reciprocal space (the first Brillouin zone). Thus, $|\mathbf{k}| = 2\pi/\lambda$ is proportional to the momentum of the electron wave and λ is its wavelength. Another consequence of the Fourier transform relation is the periodicity of reciprocal space, for which reason it is sufficient to consider the first Brillouin zone, only.

In Eq. 30, the one-electron eigenfunction is a product of a plane wave and s.c. cell-periodic part. The latter can be expressed further in terms of plane waves as

$$u_{n\mathbf{k}}(\mathbf{r}) = \sum_{\mathbf{G}} c_{n,\mathbf{k}+\mathbf{G}}\, e^{i\mathbf{G}\cdot\mathbf{r}}, \tag{31}$$

where $\mathbf{G}$ are the reciprocal lattice vectors defined by the condition $\mathbf{G}\cdot\mathbf{R} = 2\pi m$, where m is an integer. Now, combining the two previous relations we obtain

$$\psi_{n\mathbf{k}}(\mathbf{r}) = \sum_{\mathbf{G}} c_{n,\mathbf{k}+\mathbf{G}}\, \exp[i(\mathbf{k}+\mathbf{G})\cdot\mathbf{r}] \tag{32}$$

for the plane wave expansion of one-electron eigenfunctions.

In an infinite solid there is an infinite number of electrons labeled with the wave vector $\mathbf{k}$. This makes the wave vector a continuous variable. On the other hand, there are more than one electronic eigenstate for each $\mathbf{k}$, labeled with the band index n above. As the eigenenergy depends on both n and $\mathbf{k}$, this results in the band structure of solids. The bands are occupied up to the Fermi energy ε_F.

It is not possible to compute an infinite number of solutions with any numerical technique. However, the eigenstates vary continuously as a function of $\mathbf{k}$, and therefore, a representative but finite set of $\mathbf{k}$-points is sufficient for an accurate description of the electronic structure. Also, the number of plane waves $|\mathbf{k}+\mathbf{G}\rangle = \exp[i(\mathbf{k}+\mathbf{G})\cdot\mathbf{r}]$ in Eq. 32 should be finite for numerical solution procedures. It is the lowest energy plane waves that are the most important in the basis set, and therefore, the basis set can be truncated to include only those plane waves who have kinetic energies $\frac{1}{2}|\mathbf{k}+\mathbf{G}|^2$ less than some particular cutoff energy. Thus, the cutoff energy describes the size of plane wave basis set and becomes a parameter that can be increased to increase the accuracy, if the available computer capacity allows.

There are a couple of more "simplifications" which we take advantage of before writing the secular equations. One is the orthogonality of plane waves, $\langle \mathbf{k} | \mathbf{k}' \rangle = \delta(\mathbf{k} - \mathbf{k}')$, that can be thought of emerging from the different translational symmetries the different plane waves represent. This is analogous with the symmetry adapted basis functions χ_j in Eq. 21, that simplified the matrix diagonalization procedure. Here, the symmetry does much better work, the hamiltonian becomes readily diagonal with respect to $\mathbf{k}$ (or kinetic energy).

The other simplification is due to the lattice symmetry, other than translational, encountered in the reciprocal lattice, too. For this reason, only one of the rotation and/or reflection related $\mathbf{k}$-points need to be considered and the solutions at the others are found using the relevant point group symmetry operations. There are methods to choose the minimal set of $\mathbf{k}$-points to represent the reduced or the whole Brillouin zone.[23]

Now, substitution of the plane wave expansion, Eq. 32, to Eq. 14 together with Eqs. 15–17 and 27 yields the matrix equation, or the s.c. secular equations,

$$\mathcal{H} c_{n,\mathbf{k}+\mathbf{G}} = \varepsilon_{n,\mathbf{k}} \, c_{n,\mathbf{k}+\mathbf{G}}. \tag{33}$$

The overlap matrix is an identity now, but for a certain $\mathbf{k}$

$$H_{\mathbf{G},\mathbf{G}'} = \frac{1}{2} |\mathbf{k} + \mathbf{G}|^2 \, \delta(\mathbf{G} - \mathbf{G}') + v_{\mathrm{eff}}(\mathbf{G} - \mathbf{G}') \tag{34}$$

and $v_{\mathrm{eff}}(\mathbf{G})$ is defined by

$$v_{\mathrm{eff}}(\mathbf{r}) = \sum_{\mathbf{G}} v_{\mathrm{eff}}(\mathbf{G}) \, e^{i\mathbf{G}\cdot\mathbf{r}}. \tag{35}$$

The diagonalization can be carried out for each $\mathbf{k}$-point in a chosen set independently. In each case the matrix size depends on the number of included reciprocal lattice vectors $\mathbf{G}$, which further depends on the cutoff energy parameter. Here, too, the solutions have to be iterated until self-consistency.

Finally, it should be mentioned that solving the secular equations is equivalent with minimization of the total energy of the electrons, Eq. 5. Based on this there are other alternative ways of solving the electronic structure of matter, see e.g. Ref. 22. There are even straightforward methods to calculate the "molecular dynamics" of atoms simultaneously with the solution of the electronic structure and forces between the atoms.[37]

4   Surface Relaxation

The cleavage faces of binary compounds may be terminated by either of the two components. Due to the charge transfer in binding of two different elements such a surface exposes charged surface layers and forms s.c. dipole bilayer. Such atomic scale charges effect dramatically on the physical and chemical properties of the surface. First, they create Coulomb forces between the surfacemost atomic layers that may lead to a reconstruction of the atomic geometry, or at least, to relaxation though retaining the original symmetry of the surface unit cell. Secondly, the change in chemical properties may change the nature and strength of the interactions between the surface and adsorbates, and thus, the consequent surface chemistry. Moreover, all these phenomena may further effect on the bulk properties of the material, such as the electrical conductivity, for example.

Comprehension of these phenomena motivates the detailed studies of the origins, mechanisms and consequences of the surface relaxation. In what follows, however, we concentrate more on the computational studies of these phenomena than the analysis and significance of the results and their consequences. We illustrate how the above presented methods can be applied to such investigations and compare the two different computational approaches, LCAO and PWPP.

4.1 $(10\bar{1}0)$ Face of CdS

Tetrahedrally coordinated compound semiconductors occur in two crystallographic allotropes: zincblende and wurtzite. Zincblende materials exhibit only (110) cleavage face but the wurtzite exhibit two cleavage surfaces: $(10\bar{1}0)$ and $(11\bar{2}0)$. All these three surfaces relax strongly resulting in deviations up the order of one Ångström from their positions in the truncated bulk geometry. There are recent reviews [4,5] of general trends and other investigations of specific cases, see Ref. 24 and references therein. The general trend in surface relaxation is anions outwards from the surface, which relates to the covalent bonding and surface electronic states. Furthermore, the surface atomic geometry is "universal" among the II–VI wurtzite semiconductors but the extent of relaxation scales linearly with the bulk lattice constant.[5]

We have chosen to consider the more general hexagonal wurtzite and the more stable of its surfaces, the $(10\bar{1}0)$. We will find that our results, like the other recent *ab initio* results [25] are in line with the general trends.

The bulk lattice parameters of the hexagonal wurtzite CdS are $a = 4.14$ Å and $c = 6.72$ Å with the c/a ratio 1.623. To be able to conveniently treat the $(10\bar{1}0)$ surface we use an orthorhombic supercell with eight atoms (two primitive cells), shown in Fig. 1. An optimization of the bulk lattice parameters with a fixed c/a ratio using total energy minimization resulted in lattice constants which deviated less than 0.1 % from the experimental ones and a very

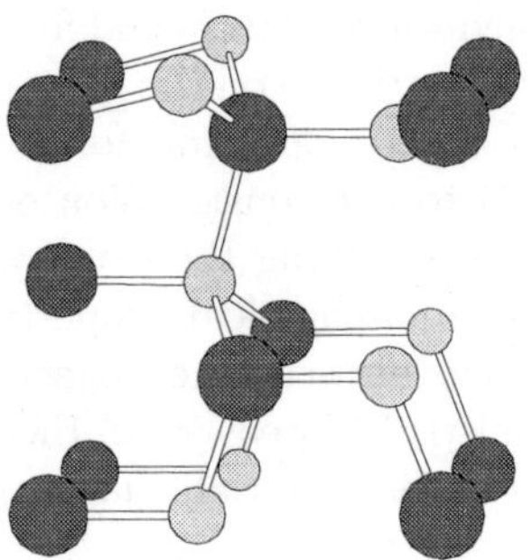

Figure 1: The orthorhombic supercell of wurtzite CdS. Smaller spheres describe anions (S) and larger cations (Cd). The upper half of the supercell illustrates the relaxed $(10\bar{1}0)$ surface but the lower half is fixed to the bulk geometry. There are eight inequivalent atoms (two primitive cells) in the corresponding bulk supercell (6.72 Å × 4.14 Å × 7.17 Å) and ten atoms in the supercell of the thinnest slab model.

small relative shift in the Cd and S lattices, with both of the computational methods. Therefore, we chose to use the experimental bulk lattice constants for the bulk and surface unit cells.

For the LCAO calculations we used a commercial software DSolid,[21] It allows self-consistent *ab initio* calculation of periodic structures with numerical atomic orbitals as a basis set, as described in Sec. 3.1. The typical extent of radial functions is 5–6 Å. For sulfur we used the following basis: $(1s^2)$, $(2s^2)$, $(2p^6)$, $3s^2$, $3p^4$, $3s^{0*}$, $3p^{0*}$, $3d^{0*}$, where those in parentheses form an atomic "frozen core" and those denoted with an asterisk have been generated in ions rather than in neutral atoms. With the same notations the basis set for cadmium is $(1s^2)$, $(2s^2)$, $(2p^6)$, $(3s^2)$, $(3p^6)$, $(3d^{10})$, $(4s^2)$, $(4p^6)$, $4d^{10}$, $5s^2$, $4d^{0*}$, $5s^{0*}$, $5p^{0*}$. Thus, the highest in energy of the inactive (frozen) set of basis functions is Cd 4p at about -65 eV (the free atom eigenvalue). The Cd 4d functions were kept active to see, if their role is important in energetics of geometric relaxation. We used the LDA parametrized by Vosko, Wilk and Nusair.[15]

For the PWPP calculations the "fhi94md" code package[26] was used. Essentially the same LDA as for LCAO was taken, but now with the Perdew-Zunger parametrization.[16] The pseudopotentials with fhi94md are of generalized norm-conserving type.[27] They consist of 1s-, 2s- and 2p-potentials for sulfur, with s being treated as a nonlocal component. The cadmium PP contains s, p and d-type functions (1s–4d) with sp-nonlocality. The number of plane waves varied from 4000 to 16000 depending on the slab thickness, as the cutoff of 10 Hartree was adopted after careful testing. For integrations over

Table 1: The perpendicular relaxation parameters of the topmost layer of wurtzite CdS $(10\bar{1}0)$ surface. Δ_{12} is the splitting of Cd and S atoms to the S and Cd layers (in Å) and ω is the corresponding rotation of the Cd–S bond from the initially horizontal direction. TB denotes tight–binding method.

CdS	Unrelaxed	LCAO	PWPP	PWPP [25]	TB [28]
Δ_{12}	0.0	0.7	0.6	0.7	0.7
ω	0°	18°	14°	16°	18°

the reciprocal space the Γ-point ($\mathbf{k} = 0$) value was taken as the representative average of the first Brillouin zone (Γ-point approximation).

The bulk supercell in Fig. 1, contains eight atoms (4 Cd + 4 S) and increase of the cell size was found to drastically increase the computational task with LCAO. With PWPP, a 16-atom (8 Cd + 8 S) supercell (two adjacent supercells of Fig. 1) was used. This was advantageous in bringing the orthorhombic cell geometry closer to cubic that was found to stabilize the internal bulk geometry very close to the experimental one. Furthermore, it increases the accuracy of the Γ-point approximation. Relaxation energies for bulk CdS with the experimental lattice constants, from the initial configuration, were 0.24 eV and 0.28 eV per primitive cell with LCAO and PWPP, respectively. The bulk relaxation results in a small relative shift of the Cd and S lattices, only.

The surface calculations were carried out with various supercell geometries. A supercell structure with four vacuum layers and four atomic layers was found to be relatively good for a quantitative description of the relaxation of the first two surface layers. The relaxed geometry is given in Table 1 together with that from the tight-binding calculations of Wang and Duke [28] and *ab initio* PWPP calculations of Schröer, Krüger and Pollmann.[25] The results from all of the calculations show a bond-length-conserving relaxation in which the top-layer cations move towards bulk and the anions move outwards from the surface plane. The rotation angle ω of the bond direction between the surface layer atoms is given in Table 1, too. The calculated relaxation is driven by the energy gain of 0.56 eV per surface unit cell according to both of our methods.

Surface densities-of-states, relevant to the chemical activity, were considered and slab models of various thicknesses were tested, too. For more details see Ref. 24.

4.2 (110) Face of SnO_2

The rutile crystal structure of SnO_2 is 6:3 coordinated and the bonding between atoms has a relatively strong ionic character. The (110) face is the most stable,

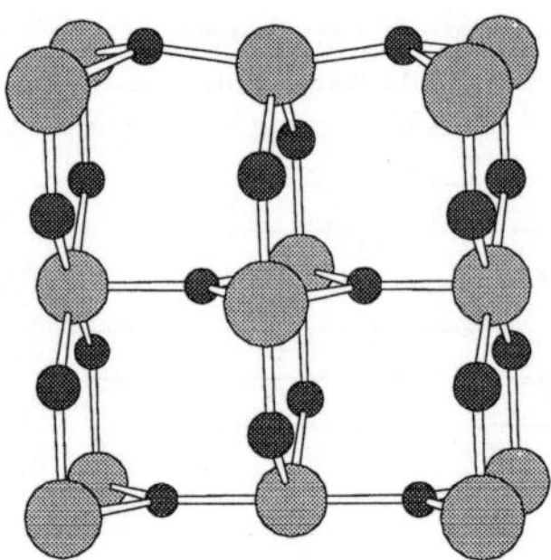

Figure 2: The orthorhombic supercell of rutile SnO_2. Smaller spheres describe anions (O) and larger cations (Sn). The upper half of the supercell illustrates the relaxed (110) surface but the lower half is fixed to the bulk geometry. There are 12 inequivalent atoms (two primitive cells) in the corresponding bulk supercell (6.67 Å × 3.19 Å × 6.67 Å) and 15 atoms in the supercell of the thinnest slab model.

and therefore, it is the most dominant surface of the crystallites of the usually porous SnO_2 material. The ideal (110) cleavage face is non-polar, but again, becomes a dipole bilayer due to the relaxation. Furthermore, the surface tends to oxidize easily, which effects on the relaxation or reconstruction, surface charge and potential, and thus, the chemical properties. Also, the oxygen vacancies present in the bulk, close to the surface and even mobile at higher temperatures interfere with the above mentioned properties.

Extensive amount of research has been done on the SnO_2 material and its surfaces both experimentally and theoretically, see Refs. 1–3, 29–31 and further references therein. Earlier, we have carried out LCAO calculations with a finite cluster model [29,30] and found out the necessity of proper embedding or inclusion of periodicity as described in Secs. 3.1 and 3.2, above. Here, we consider relaxation of the ideal (110) cleavage surface of SnO_2. This strongly relaxing face is called "reduced" in contrast to the "oxidized" surface which relaxes less.

The tetragonal unit cell contains two tin and four oxygen atoms with lattice parameters $a = 4.74$ Å and $c = 3.19$ Å. Here again we used an orthorhombic supercell with 12 atoms, shown in Fig. 2, to make calculations simpler. The same experimental lattice parameters were used throughout the study to retain simple comparability of the results from the two methods.

For the LCAO calculations the DSolid [21] software and the LDA parametrized by Vosko, Wilk and Nusair [15] were used again. For oxygen and tin we used the basis sets: $1s^2$, $2s^2$, $2p^4$, $2s^0$, $2p^0$, $3d^0$, $3d^{0*}$, and $(1s^2)$, $(2s^2)$, $(2p^6)$, $(3s^2)$, $(3p^6)$, $(3d^{10})$, $(4s^2)$, $(4p^6)$, $4d^{10}$, $5s^2$, $5p^2$, $4d^{0*}$, $5s^{0*}$, $5p^{0*}$, $5d^{0*}$, respec-

Table 2: The perpendicular relaxation parameters of the topmost layer of rutile SnO_2 (110) surface. Δ_{12} is the splitting of O and Sn atoms (average) and Δ_{11} the splitting between the two inequivalent Sn atoms (both in Å).

SnO_2	Unrelaxed	LCAO	PWPP	TB [36]
Δ_{12}	0.00	0.40	0.35	0.25
Δ_{11}	0.00	0.07	0.11	0.05

tively. The basis set notation is the same as above. The PWPP calculations were carried out with the PlaneWave software.[21,33]

Tindioxide appeared to be calculationally more demanding than CdS. The total energy minimization with PWPP led close to the experimental lattice constants only with the "extended-norm-and-hardness-conserving" pseudopotentials,[34] which required the cutoff energy of 40 Hartree (80 Ry !). The quality of the calculation was increased also by including three **k**-points for the evaluations of Brillouin zone averages. The LCAO calculations were done roughly at the same level as those for CdS. Increase of the basis set and possibly a larger supercell would have been necessary to obtain the experimental lattice constants to 1 % accuracy with LCAO. The higher demands for the calculations of SnO_2, as compared with CdS, are obviously brought by the element oxygen which is known to require not only numerically higher level calculations but also non-local corrections to LDA.

The parameters given in Table 2 describe the relaxation within the first layer of (110)-1×1 surface. They are seen to be less than those in Table 1 for CdS (roughly half). The common feature in both is, however, the anion (S or O) relaxation outwards. In case of SnO_2 the two inequivalent Sn cations relax slightly differently, too. The two methods, LCAO and PWPP, are seen to lead relatively similar relaxation, whereas the tight-binding method[36] seems to yield smaller relaxation. For more details, including supercell size effects and different slab models, see the original work.[32]

5 Conclusions

The density-functional theory is the proper framework for the *ab initio* calculation methods for solids and surfaces. It offers a straightforward formalism for numerical approaches, which become most efficient in case of large and infinite systems. Furthermore, the results are relatively accurate already at the lowest level of approximation to the exchange and correlation of electrons, the LDA, as shown in this text above. This formalism also defines the relevant concepts for the solid state electronic structure in a simple way, and that may be more suitable even for use in qualitative molecular orbital theory of small molecules than the conventional Hartree–Fock theory, as argued in a very recent paper.[38]

Evaluation of the two different computational approaches to DFT, namely LCAO and PWPP, shows that both of them are able to describe the surface structure and properties of compound semiconductors. A reasonable accuracy can be achieved with reasonable computational efforts due to the LDA. In comparison of LCAO and PWPP the largest differences, therefore, remain in the concepts which these methods use to describe the electronic structure of periodic solids. The LCAO accounts for bonding and other phenomena in terms of atomic orbitals, whereas the PWPP emphasizes the band structure in the description of valence and conduction electrons.

As expected, the relaxation of $(10\bar{1}0)$ surface of CdS and (110) surface of SnO_2 occur essentially in the perpendicular direction, only. Lateral relaxation is negligible. Symmetry breaking reconstruction was not found, in either case. The most prominent common feature for both surfaces is the anion relaxation outwards from the surface. Various *ab initio* results for the relaxation parameters seem to mutually agree, but differ somewhat from those of tight-binding results. The extent of relaxation and the consequent effects on the surface properties are definitely essential for the adsorption and surface chemistry of these materials.

Acknowledgments

The author owes thanks his coauthors Tuomo Rantala, Vilho Lantto and Juha Vaara of the original research papers related to the work reviewed in Sec. 4, above. The Center for Scientific Computing (CSC), Espoo, Finland, is acknowledged for the computational resources and software, which have been essential for carrying out the calculations in the original work.

References

1. Vilho Lantto in *Gas Sensors*, ed. G. Sberveglieri (Kluwer Academic Publishers, Dordrecht, 1993), pp. 43–88.
2. W. Göpel, *Prog. Surf. Sci.* **20**, 9 (1985); and W. Göpel, *Microel. Eng.* **32**, 75 (1996).
3. M.J. Gillan *et al.*, *Current Opinion in Solid State & Materials Science* **1**, 820 (1996); I. Manassidis and M.J. Gillan, *Surf. Rev. Lett.* **1**, 491 (1994); I. Manassidis *et al.*, *Surf. Sci.* **339**, 258 (1995); and P.J.D. Lindan *et al.*, *Surf. Sci.* **364**, 431 (1996).

4. C.B. Duke in *Surface Properties of Electronic Materials*, eds. D.A. King and D.P. Woodruff (Elsevier, Amsterdam, 1987), Chap. 3, pp. 69–118.

5. C.B. Duke in *Reconstruction of Solid Surfaces*, eds. K. Christman and K. Heinz (Springer–Verlag, Berlin, 1990).

6. Tuomo S. Rantala, Vilho Lantto and Tapio T. Rantala, CSC News **8**, 23 (1996).

7. H.-J. Freund, *Angew. Chem. Int. Ed. Engl.* **36**, 453 (1997); and M. Bäumer, J. Libuda and H.-J. Freund in *Chemisorption and Reactivity on Supported Clusters and Thin Films*, eds. R.M. Lambert and G. Pacchioni (Kluwer Academic Publishers, Dordrecht, 1997), pp. 61–104.

8. J.C. Slater *Quantum Theory of Atomic Structure, Vols. I and II; Quantum Theory of Molecules and Solids, Vols. I and II* (McGraw–Hill, New York, 1960–1965).

9. R. McWeeny, *Methods of Molecular Quantum Mechanics*, (Academic Press, London, 1978).

10. N.W. Ashcroft and N.D. Mermin, *Solid State Physics*, (Holt–Saunders, Philadelphia, 1976).

11. P. Hohenberg and W. Kohn, *Phys. Rev.* **136**, B864 (1964).

12. W. Kohn and L.J. Sham, *Phys. Rev.* **140**, A1133 (1965).

13. R.G. Parr and W. Yang, *Density-Functional Theory of Atoms and Molecules*, (Oxford University Press, New York, 1989).

14. D.M. Ceperley and B.J. Alder, *Phys. Rev. Lett.* **45**, 566 (1980); D.M. Ceperley, *Phys. Rev.* B **18**, 3126 (1978).

15. S.H. Vosko, L. Wilk and M. Nusair, *Can. J. Phys.* **58**, 1200 (1980).

16. J.P. Perdew and A. Zunger, *Phys. Rev.* B **23**, 5048 (1981).

17. R.O. Jones and O. Gunnarsson, *Rev. Mod. Phys.* **61**, 689 (1989).

18. Tapio T. Rantala, *Acta Universitatis Ouluensis A* **184**, (1987).

19. J.P. Perdew *et. al.*, *Phys. Rev.* B **46**, 6671 (1992); and *Phys. Rev. Lett.* **77**, 3865 (1996).

20. A. Seidl *et. al.*, *Phys. Rev.* B **53**, 3764 (1996).

21. User Guides of DSolid and PlaneWave (Molecular Simulations, San Diego, 1996).

22. M.C. Payne, M.P. Teter, D.C. Allan, T.A. Arias and J.D. Joannopoulos, *Rev. Mod. Phys.* **64**, 1045 (1992).

23. H.J. Monkhorst and J.D. Pack, *Phys. Rev.* B **13**, 5188 (1976).

24. Tapio T. Rantala, Tuomo S. Rantala, Vilho Lantto and Juha Vaara, *Surf. Sci.* **77**, 352 (1996).

25. P. Schröer, P.K. Krüger and J.Pollmann, *Phys. Rev.* B **49**, 17092 (1994).

26. R. Stumpf and M. Scheffler, *Comp. Phys. Comm.* **79**, 447 (1994).

27. D.R. Hamann, *Phys. Rev.* B **40**, 2980 (1989).
28. Y.R. Wang and C.B. Duke, *Phys. Rev.* B **37**, 6417 (1988).
29. Tuomo S. Rantala, Vilho Lantto and Tapio T. Rantala, *Physica Scripta* **T54**, 252 (1994).
30. Tuomo S. Rantala, Vilho Lantto and Tapio T. Rantala, *Sensors and Actuators B* **18–19**, 716 (1994).
31. Tuomo Rantala, *Acta Universitatis Ouluensis Technica C* **96**, (1997).
32. Tapio T. Rantala, Tuomo S. Rantala and Vilho Lantto (Fourth Nordic Conference on Surface Science, May 29 – June 1, 1997 in Ålesund, Norway, Book of Extended Abstracts), eds. S. Raaen and J. Bremer, PoB-9, p. 107; and Tapio T. Rantala, Tuomo S. Rantala and Vilho Lantto, *manuscript, to be published.*
33. LDA parametrization of M.P. Teter, unpublished (1990). Similar to the parametrization of Ref. 15.
34. M.P. Teter, *Phys. Rev.* B **48**, 5031 (1993).
35. Tuomo S. Rantala, Vilho Lantto and Tapio T. Rantala, *Sensors and Actuators B* **13–14**, 234 (1993).
36. T.J. Godin, and J.P. LaFemina, *Phys. Rev.* B **47**, 6518 (1993).
37. R. Car and M. Parrinello, *Phys. Rev. Lett.* **55**, 2471 (1985).
38. E.J. Baerends and O.V. Gritsenko, *J. Phys. Chem.* A **101**, 5383 (1997).

MOLECULAR-DYNAMICS STUDIES OF DEFECTS AND IMPURITIES IN BULK SEMICONDUCTORS

Stefan K. ESTREICHER
Physics Department
Texas Tech University
Lubbock, TX 79409-1051, USA
e-mail: bzske@ttacs.ttu.edu

Peter A. FEDDERS
Physics Department
Washington University
St. Louis, MO 63130-4899, USA
e-mail: paf@howdy.wustl.edu

Molecular-dynamics simulations are rapidly becoming a dominant tool in theoretical semiconductor physics. The problems addressed include the properties of bulk semiconductors (crystalline, amorphous, or liquid) and of many defects and impurities. The calculations predict equilibrium structures, vibrational modes (including local modes), diffusion paths and diffusivities of intrinsic defects and impurities, binding energies, defect reactions, a range of thermodynamic properties, and other quantities that are essential to the understanding of the fundamental processes taking place. In this chapter, we review the key aspects of the methodologies and typical results.

A great number of molecular-dynamics (MD) studies related to defects in semiconductors have been published in the past fifteen years. These studies cover a wide range of methodologies (from empirical functionals to true *ab-initio* techniques), a wide range of issues (from phonon spectra to laser melting to the diffusion of point defects), and a wide range of materials (from amorphous carbon to III-V nitride superlattices). A review of the topic is therefore necessarily just an overview. Our purpose is to give the reader an idea of how the field has evolved and what is the current state-of-the-art.

This paper consists of four sections. The first contains an overview of the methodologies, which range from highly approximate (but very fast) to highly sophisticated (but quite a bit slower). This section also discusses some of the practical aspects of the actual computations and the tricks used. The second section contains many examples of the work done in crystalline semiconductors, from the modeling of the perfect material (at low and high temperatures) to the study of localized and extended defects. The third section deals with amorphous semiconductors, with an extensive discussion of the construction of supercells. The paper concludes with some personal remarks about the present and future of MD simulations applied to the study of defects in semiconductors.

1   METHODS

A number of methods are available to perform MD simulations. Each method has its own strengths and weaknesses.[1] In general, the more sophisticated methods do a better job of getting the "chemistry" right, while the less sophisticated ones are considerably faster. Thus, if one is interested in getting the detailed properties of localized defect states correctly, including the bonding, one should use an accurate *ab-initio* method. On the other hand, if one needs to perform long simulations and/or study very large systems, a faster but less rigorous method is appropriate.

A true many-body Hamiltonian treats both the ions and electrons on a quantum mechanical basis. However, because of the large mass difference between the electrons and the nuclei, it is standard to decouple the nuclear and electronic degrees of freedom with the adiabatic or Born-Oppenheimer approximation,[2] in which the electrons are taken to respond instantly to the motion of the ions. In classical MD simulations, the nuclei or ions are treated as classical particles which move in the potential determined by the electrons and computed for the given ionic coordinates. We are unaware of MD calculations that do not follow the above procedure. Note that any quantum mechanical aspects of atomic motion is lost. For a quantum solid such as He, treating the ions classically leads to drastic errors. However, for semiconductors, this is usually unimportant (for a discussion of quantum diffusion in solids and key references, see the recent paper by Stoneham[3]).

If one is studying phonons or vibrational modes of a system via MD, then one will always be in the classical or high temperature limit, which means $k_B T > \hbar\omega$, where ω is a relevant phonon frequency. A little thought will reveal that this condition is by no means always satisfied. Thus, MD is really an intrinsically high-temperature technique.

We shall divide the methods into four distinct groupings. As always when forming groups, some methods will occur on the seams. In order of increasing complexity and accuracy (and to some degree in order of decreasing speed), we have semiempirical potentials, empirical tight binding (ETB), and density functional (DF) theory. Recent developments in order N methods shall be discussed briefly. The section concludes with a discussion of practical aspects of the computations.

1.1 Semiempirical Potentials

Semiempirical potentials give rise to the simplest and fastest MD codes and have been in use longer than any of the other methods. Essentially, one decides on functional forms for the total energy of a system of particles and the forces between them. The functional can have anywhere from a few to hundreds of parameters which one tries to determine in order to fit accurate total energy calculations and/or experimental data.

The simplest codes consist of only a 'pair potential'. The energy and forces between a pair of atoms are a function of the distance between them. With rare-gas elements and some metals, such a pair potential description gives a credible starting point. However, for semiconductors this yields terrible results and cannot reproduce tetrahedral coordination. As a minimum, one must also consider angular-dependent forces and energy terms which contain three or more atoms at a time in order to describe solid group IV materials.

More sophisticated semiempirical methods involve calculating for a given atom the equivalents of an effective field that includes a number of surrounding atoms. The embedded atom method falls into this category. Ideally, the energy functional is chosen with a reasonable dose of 'underlying physics' and this minimizes the number of parameters needed to obtain an accurate and transferable description of the system.

The enormous advantages of semiempirical potentials are the simplicity and speed of the computer codes. The energy of each atom and the forces acting on it are usually assumed to arise from the few nearest neighbors (NNs) that control the pair potential, angular dependent forces, or effective field.

Thus the number of calculations for a system of N atoms is itself of order N. The problem with semiempirical methods comes in trying to mimic global quantum effects and in transferability. For example, suppose that the forces on a mid-gap defect state depend on the number of electrons in that state – a common occurrence. Since semiempirical methods are entirely local while the Fermi level is a global property of a system, there is no way a semiempirical code can correctly account for different forces on configurations in different charge states.

Transferability, the ability of the semiempirical potential to be used in a wide variety of local bonding environments, is a critical issue. Obviously, a given semiempirical method always gives a good description of the data that are used to fit its parameters. But how will it do for bonding configurations that it is not fit to? Just giving a good description of liquid (metallic) Si or C and the tetrahedral crystalline states is a substantial accomplishment. However, there are numerous solid configurations that the potentials are usually not fit to and which have analogs in amorphous semiconductors, or distorted configurations around defect centers in crystalline semiconductors. One might think that using an energy functional with a large number of parameters and fitting them to a huge database would be a good method. Unfortunately, this is not the case. There is an annoying tendency for computer codes with more and more parameters to do a poorer and poorer job in predicting the energy of configurations that they were not fit to. For example, the number of possible defects or odd configurations in a-Si is virtually limitless and the chemistry of these configurations can be highly non-transferable.

The success of empirical MD approaches has been mixed. If the potentials are used with care, they can answer questions about model systems that are extremely large, much larger than can be described with *ab-initio* methods. However, the reliability is closely connected with limiting the application of the method to systems that exhibit bonding close to conformations for which the potential was originally fit. For amorphous, glassy, and/or liquid-like structures, it seems unlikely that empirical potentials will get the detailed chemistry right. Interesting dynamic effects have been seen with such methods but it is questionable whether the results are independent of the parameters chosen.

Probably the best known empirical method for Si is that of Stillinger and Weber[4] and, for C, the ones by Tersoff.[5] These are both rather early attempts and more recent methods may (or may not) be better. Carlsson *et al.*[6] have developed a method involving the second moment of interactions that does well in bridging the gap between tetrahedrally coordinated Si and liquid-like configurations.

An excellent review of semiempirical methods is by Carlsson.[7] A particularly intriguing approach has recently been proposed by Ercolessi and Adams[8] who have developed an efficient method for computing empirical potentials that is based upon "force matching". The idea is to take an enormous number of configurations for which the forces are accurately known (from *ab-initio* work) then minimize the error in the predicted forces relative to the *ab-initio* database. Such an approach has the appealing property that one can partly gauge its accuracy by considering the deviation of the potential from the *ab-initio* results.

1.2 Empirical Tight Binding

The simplest explicit method of electronic structure calculations is empirical tight binding (ETB). It enables the calculation of an approximate one-body Hamiltonian matrix whose eigenvalues and eigenvectors are taken to approximate the allowed electronic energies and eigenvectors of the states. Imagine that the electronic eigenstates can be represented by a linear combination of atomic orbitals

$$|\psi_i\rangle = \sum_\mu a_{i\mu}|\mu\rangle \tag{1}$$

where μ is the site-orbital index and i labels the state or eigenvalue. Although it is not necessary, the basis is almost always taken to be orthonormal $\langle\mu|\nu\rangle = \delta_{\mu\nu}$. This is actually a rather drastic approximation but leads to an enormous simplification. Further, most ETB Hamiltonians include interaction only with NNs and only two-center contributions. The sum of the occupied eigenvalues is the (attractive) electronic contribution to the total energy. One must add to this energy a repulsive interaction. This is obtained from some fitting procedure similar to that described for pair potentials.

In spite of its many shortcomings, ETB has numerous advantages. Perhaps the most salient one is that it is very easy to use and is intrinsically quantum mechanical (and defect states do depend on their charge). Consider a supercell model of a solid with N atoms that consists of a large unit cell with periodic boundary conditions. We set up the Hamiltonian matrix

$$\mathcal{H}_{\mu\nu} = \langle\mu|\hat{\mathcal{H}}|\nu\rangle \tag{2}$$

where $\hat{\mathcal{H}}$ is the Hamiltonian operator and the μ's are orbitals (typically s, p_x, p_y, and p_z) centered on each atom. The matrix elements in this representation depend on the details of the network topology and it is convenient to work with the molecular coordinates specifying the interatomic hopping. These are V_{ss}, $V_{sp-\sigma}$, $V_{pp-\sigma}$, $V_{pp-\pi}$ in the usual chemistry nomenclature for an sp^3 model.[9] Sometimes an s^* orbital is also employed [10] since it is impossible to create a band structure for Si with the valence band minimum in the correct place without this addition.

Some thought must also be given to the radial dependence of the orbitals. If it falls off too slowly, every atom will interact with all the other atoms and the computational time will increase enormously. If the interaction distance is too short, one obtains anomalous results where bandtailing virtually disappears for entirely erroneous reasons.[11] A cutoff distance of over 3 Å is necessary in a-Si in order to avoid this pitfall. Finally, the radial part of the orbitals must die off smoothly to zero so that the forces (derivatives of the energy) do not contain unphysical delta functions! In any case the matrix eigenvalue problem then reads

$$H|\psi_i\rangle = \epsilon_i|\psi_i\rangle \, . \tag{3}$$

This is the usual orthogonal eigenvalue problem where the electronic eigenvalues are represented by the ϵ_i. For electronic state density calculations related to the spectral signature of a defect, an exact diagonalization of H is sufficient. The calculation of the forces is easy once the exact eigenvectors and eigenvalues have been obtained. The Hellman-Feynman theorem[12] can be employed in the form:

$$F_\alpha = 2 \sum_{i,\text{occ}} \langle\psi_i| - \frac{\partial H}{\partial R_\alpha}|\psi_i\rangle \tag{4}$$

where the sum on i is restricted to occupied states.

The problems associated with the ETB method are as follows. The method is empirical and does well only for the configurations similar to those one fits to. If the parameters are fit to crystalline Si (as is often the case), a number of configurations in a-Si may not be correctly described. The most telling criticism is that it involves a massive $N \times N$ matrix diagonalization and thus requires about as much computer time as an *ab-initio* local orbital method – but without its accuracy and reliability.

However, there are a number of positive aspects to ETB. The programs are easily written and easy to understand. One gets reasonable forces and can investigate the localization of states and the effect of the charge state on a defect. Further, ETB has two attributes that more accurate *ab-initio* methods do not have. The energy gap is fit and thus presumably comes out correctly, while the gap obtained with virtually all DF methods is wrong. This is especially serious if one needs a comparison of energies of systems with states above and below the gap. Then, the Bethe lattice is often used profitably with a-Si.[13,14] In this approach, a small cluster is investigated that is tied off with a Bethe lattice. This is very useful when studying bandtails of defect states since the rest of the network is a defect-free amorphous system. ETB methods have been very successful in the study of dangling/floating bond defects[15] and of bandtails.[16]

There are many good ETB methods in the literature and more seem to appear almost monthly. All of them have transferability problems. In particular, some are designed for producing good band structures[10] but are quite inappropriate for total energies and forces. As with empirical potential methods, these Hamiltonians are most reliable for conformations near those the Hamiltonian was fit to in the first place.

The basic idea of an orthogonal ETB model is the notion that there exists some underlying set of generalized Wannier functions with the property that these are mutually orthonormal in real space. This is justified only for a particular topology (for a different topology, another set of Wannier functions should be used). Of course, this destroys the usefulness of the method and one hopes that results are at least approximately transferable.

The most widely used ETB Hamiltonians for force calculations are those of the Goodwin-Skinner-Pettifor type.[17] The original was devised for Si but has been adapted by Xu *et al.*[18] to carbon systems. A partially self-consistent version has been adapted by Biswas *et al.*[19] to Si-H systems.

More consistently reliable results can be obtained from a variety of non-orthogonal Hamiltonians. Menon [20] and Canel [21] have proposed empirical non-orthogonal ETB Hamiltonians which have been applied to a variety of systems. These methods appear to be more transferable than orthogonal ETB methods but still suffer from some of the same problems.

General Comments on Empirical Methods

The strengths of semiempirical methods are their simplicity and ease of interpretation. The chemical trends are usually correct, but the details of the chemistry are not accurate. As one would expect, they do very well on problems that are very close to the data that they were fit to, but do increasingly poorly for configurations that are further and further removed from the fitting base.

The reader should not be too impressed with either empirical potentials or tight binding models that have a very large number of fitting parameters. We note that some comparatively simple potentials (such as the Stillinger-Weber potential) have been quite successful while we know of some empirical potentials for Si with many dozens of free parameters which have difficulty with even some of the smallest clusters. The same is true for ETB models.

1.3   Density-Functional and Hartree-Fock Methods

The complexity of electronic structure and force calculations arises from the many-body nature of the interactions between electrons. A direct attack on the many electron problem is too difficult and computer intensive to have an impact on supercell calculations of an appreciable size. Thus many of the successful calculations relevant to semiconductors have involved a mapping of the many-body problem into an effective one-electron problem.

Note that the Hartree-Fock (HF) method was the first successful approach in this direction and is still widely used today, but not in conjunction with MD simulations. Large basis-set *ab-initio* HF calculations provide much more chemical detail than other methods and are often used to study defects in semiconductors.[22] However, the number of two-electron integrals scales as N^4, where N is the number of orbitals. This makes *ab-initio* HF very impractical for MD simulations since competitive methods scale as N^3 at worst. However, semiempirical HF techniques can be used for that purpose (see below). They require the calculation of only $\sim N^2$ two-electron integrals, but scale as N^3 because of matrix diagonalizations.

Because *ab-initio* methods are terribly complex, we will make no attempt to present any of the details of such methods. Instead, we will try to explain the difference between the methods and some of their strengths and weaknesses.

All *ab-initio* methods fail to calculate correctly the energy gap of semiconductors, or any other class of materials. The position of energy levels in the gap is also very hard to predict, especially unoccupied ones. The key reason is that the excited (unoccupied) states are never correctly described. The HF approximation produces gaps that are considerably too large. The energy difference between the lowest-unoccupied and highest-occupied levels is much overestimated since the unoccupied levels are not energy-optimized. The Koopman's theorem ionization energy only approximates the direct band gap. The DF method with the local density approximation produces gaps that are considerably too small. In most cases, the large error in the gap does not matter. However, it can introduce a substantial error in the total energy if there are occupied states in the conduction band. In many cases, this error can be removed by considering only the appropriate *differences* in total energies.

An additional warning about energy eigenvalues is in order. In the methods considered here, these eigenvalues are computed as a contribution to the total energy, but they are not the only contribution. Thus, the meaning of the energy eigenvalues is not exactly the same as in a simple single-particle energy picture. Ultimately the difference in energy between two (global) states of a system must be computed by comparing their total energies and not energy eigenvalues. In fact, the energy eigenvalues themselves change as the occupation of various single particle states changes. Finally, since all methods depend on an energy variational principle, and since unoccupied states do not contribute to the total energy, special care should be taken not to over-interpret the significance of unoccupied energy eigenvalues.

For atomistic MD calculations in solids, the current method of choice is DF theory.[23] The name arises from the connection between the total ground state electronic energy of a system and its electronic charge density. It can be rigorously shown that the ground state energy of a many electron system is a functional of the electron density $n(\vec{r})$. The trouble is that this functional is not exactly known, although there is continuing work to determine it. In practice one uses a functional with the electron kinetic energy, electron-ion potential energy, the electrostatic (Hartree) interaction of the electrons, and an unknown exchange-correlation energy. Various approximations are used for the latter term.[24] One usually makes the local density approximation (LDA), which means taking the electron density to be locally uniform and using a uniform electron gas correlation energy. This LDA leads to a nonlinear set of coupled integro-differential equations that must be solved.

Consistent with chemical intuition, a minimal basis set of s, p, and possibly d electrons is primarily responsible for the bonding and ground state electronic properties. Pseudopotentials are introduced so that only the valence electrons are treated explicitly. This is an excellent approximation and the modern classic paper on the subject is by Bachelet, Hamann, and Schlüter.[25] The calculations themselves can be linear or self-consistent. The choice of pseudopotential is partially governed by the basis used for the valence electrons. Two commonly used bases are a local basis of orbitals centered on the ions or a plane wave basis. The latter requires the use of soft pseudopotentials.[26] Both types of basis have advantages and disadvantages. We shall briefly discuss them in the next two sections.

DF with a localized basis

Because of the formal similarity to ETB, this method is sometimes called '*ab-initio* tight binding'. The basic idea, first developed by Sankey and co-workers,[27] is to construct a spatially local basis approach to the DF orbitals. The method was originally characterized by a limited number of functions (s and p orbitals on each atom) which made the method very fast. Recently, more orbitals have been added, but with a huge penalty in speed. These basis functions are not true atomic orbitals but instead are slightly excited (confined) pseudoatomic orbitals (PAO). The PAOs are calculated self-consistently within the LDA using pseudopotentials to eliminate the core. In the case of Si, for example, they can be thought of as the $3s$ and $3p$ level orbitals.

The use of PAO's is appealing because the chemistry is naturally built into the basis functions. For a supercell with N atoms, the overlap and Hamiltonian matrices have dimensionality nN where n is the number of orbitals per site. For group IV materials, a minimal basis of four PAOs per site can be used, and an N atoms calculation involves matrices of dimensionality $4N$. The corresponding plane wave calculations requires 100 to 1000 (or more) plane waves per atom. In the case of states that are associated with strong potentials — such as the $1s$ state of H, the $2p$ states of O or F, or the $3d$ states of fourth-row transition metals — plane-wave basis calculations run into difficulties.

The eigenvalues and eigenvectors can easily be expressed in terms of the local PAO's, which is very helpful in interpreting the results. Finally, since the method is a real-space method, periodicity is natural and is very amenable to cluster and surface calculations.

The original work of Sankey *et al.* made use of a linearized (non self-consistent) version of DF theory, the Harris functional.[28] The calculations are made without self-consistent iterations. This approximation is well discussed by Foulkes and Haydock [29] and is remarkably good in almost all situations. The exceptions seem to be in systems like SiO_2 where there is a lot of charge transfer. The method is also suspect in investigation of defects in a-Si which depend heavily on the charge state or of the relationship of the Fermi level to the charge states of the defect. Again, this is a problem where there is a large charge transfer. However, this limitation of non self-consistency has recently been lifted by Demkov *et al.*[30] and Ordejón *et al.*[31]

DF with plane-waves: Car-Parrinello methods

In 1985, Car and Parrinello (CP)[32] revolutionized *ab-initio* MD with what has become known as the CP method. To some degree, CP means slightly different things to different people. Usually it is taken to mean a MD simulation using a plane wave basis and an iterative minimization scheme to solve the electronic structure problem with the self consistent LDA equations. More properly, CP refers only to a method for coupling the approximation of stationary states of a huge basis eigenvalue problem with associated ion dynamics, even for a basis other than plane wave. However, in many peoples minds, CP will always be associated with the plane wave basis. The reason why the plane wave basis works so well is the great speed of fast Fourier transforms.

A plane-wave basis restricts the studies to periodic systems and makes problems with clusters and surfaces awkward, although not impossible. After all one can put a cluster in a box within a periodic array. Of course, one must make the box big enough so that neighboring clusters do not interact. But then, one has a very large empty volume that the plane wave basis must fill.

In order to calculate the eigenstates with a plane wave basis, a very large matrix must be diagonalized. The key contribution of CP was to note that an actual diagonalization was not necessary in order to find the LDA energy. They used a fictitious Lagrangian.[32] More recently other methods have been use to avoid the complete diagonalization of the full Hamiltonian matrix. This is a complicated subject with numerous pitfalls and will not be discussed here. A faster version of the CP approach was developed by Scheffler and co-workers.[33,34] Their energies appear to be as accurate than the CP ones but the method allows for the use of longer time steps. We close this section by noting that the methods still all involve N^3 computations per iteration, where N is the number of atoms times the number of plane waves per atom.

Beyond the local density approximation

At present, the LDA is the almost universal choice for *ab-initio* calculations on crystalline and amorphous systems. As alluded to earlier, there are a number of well-known problems associated with this approximation. (1) The optical gap is poorly estimated, almost always underestimated. This is a serious problem when calculations of conduction band states are important. Of course, one can artificially add (or subtract) energy to the gap, and this is often done. However, this does not help in comparing different materials, and casts some doubt on the calculations of mid-gap defect states. (2) In systems that are strongly electronically inhomogeneous (like SiO_2) as well as for elements with strong potentials (like O, F, or $3d$ transition metals), the basic LDA assumption of weak spatial variation of the charge density is not well fulfilled and the LDA has difficulties. (3) The LDA implicitly assumes that any system is diamagnetic. The so-called local spin density approximation,[35] in which separate spin up and spin down density functionals are used, can overcome this problem. (4) Hydrogen bonding (as in H_2) is not well described by the LDA and neither is the fluctuating dipole (Van der Waals) bonding.

Progress has been made on most of these problems. Perdew[36] and others have added the gradient of charge density to the LDA and this has led to significant improvements for SiO_2.[37] Another approach is to abandon the LDA and try something else. One such approach is the Quantum Monte Carlo (QMC) method.[38] At present, systems of up to about 100 particles have been handled.

Semiempirical HF

The only MD simulations we are aware of in which the electronic problem is solved with a semiempirical HF technique uses the method of Complete Neglect of Differential Overlap (CNDO). The problem is solved in molecular clusters rather than periodic supercells. While fully self-consistent, the CNDO method neglects many interactions but introduces parameters which are fit to atomic properties.[39,40]

The method suffers from the drawbacks that are common to all semiempirical techniques (see above) and uses a minimal basis set. However, it is computationally efficient and has been used successfully to discuss various defect problems in crystalline diamond and silicon (see below).

1.4   Order N Methods

A number of physical problems require very large supercells. Further, in amorphous or crystalline systems, an impurity can interact with an image of itself in the next unit cell, unless the unit cell is very big. We have found this to be the case for dangling bonds in a-Si with cubic cells of side 11 Å (64 Si-atoms cell), and the problem is still evident in cells eight times as large (512 Si-atoms cells). The effect is not limited to the electronic structure, but also shows up in the vibrational spectrum. This can decrease the localization considerably, and this delocalization is unphysical because in real materials the defect density is of order parts per million instead of parts per hundred. Further, for a problem like band tailing in amorphous semiconductors, bandtail states are of order one per thousand, so that one needs a supercell containing thousands of atoms in order to investigate the effect properly.

As mentioned above, *ab-initio* HF is an order N^4 method (because one must evaluate $\sim N^4$ four-center integrals) while LDA or ETB are N^3 methods (because the diagonalization of an $N \times N$ matrix requires $\sim N^3$ operations). For a local orbital code, N varies from 4 to 9 times the number of atoms, while the multiplier for plane wave methods is much greater. In any case, it is clear that computations for really large systems using these methods are prohibitively expensive. Thus, there is presently a vigorous quest for quantum mechanical methods which scale linearly with the size of the problem, or $O(N)$. Since this is a competitive, fast-moving field, we will not attempt to explain the mathematics of the various methods (see e.g., the detailed discussions in Refs. [41,42]). Instead, will try to define the issues and give the reader some feel for the subject.

For large systems, the matrix must be sparse because any given atom is within interaction range of only a finite number of other atoms, and this number does not increase as the size of the supercell increases. One would like to obtain the eigenvalues ϵ_i where

$$H|\psi_i\rangle = \epsilon_i|\psi_i\rangle \tag{5}$$

where H is the Hamiltonian matrix. We note that obtaining the eigenvalue spectrum, or parts of it, is a problem that has been studied for many years, and Lanczos [43] understood this as early as 1956. The Hamiltonian has moments μ_n where

$$\mu_n = \int_{-\infty}^{\infty} dE \; E^n \; \rho(E) = \frac{1}{n} \, \mathrm{Tr}(H^n) \tag{6}$$

where Tr denotes the trace of a matrix and $\rho(E)$ is the density of states. The first moment, of course, is the total energy.

Reconstructing the density of states from the moments is a problem that has been extensively worked on. Skilling [44] noted the possibility of extracting moment data from the operation of sparse matrices on random vectors and derived useful error estimates for integrals over the density of states. That is, approximate moments of a large Hamiltonian matrix can be obtained by taking the expectation value of H (or powers of H) with random vectors. The importance of this is that an expectation value is an $O(N)$ operation but the trace is not. The above method allows one to get energies rather easily as well as the density of states itself. However, while this is useful for analyzing a large system, it does not yield forces.

The above methods give only global properties of a system. It is far more challenging to extract local properties in an order N fashion, including forces. This is highly mathematical, and complete discussions are given in Refs. [41,42,45,46]. The various methods all use localized Wannier-like functions that are truncated beyond a certain radius. These methods work quite well in systems where the Fermi level falls in the middle of a substantial gap — of order one eV or so. There are problems if the system has localized states in the middle of the gap. Finally, while the required computer time scales linearly with the size of the system, the prefactor is huge, and these methods become truly efficient in very large systems only.

1.5 *Practical Considerations*

The range of problems to which MD simulations can be applied is continuously expanding because of advances on several fronts. First, computer hardware is continuing to improve. In workstations, the CPUs are more powerful and the memory and disks are faster and bigger. Second, internet access to supercomputers allows easy use of remote sites. Third, the software is getting

better: more efficient routines and compilers, including the use of paralleliza-
tion. Fourth, the theory itself is evolving, in particular in the area of TB and
$O(N)$ methods.

Typical uses of MD simulations include crystal growth and melting, ion
implantation, as well as the calculation of vibrational frequencies, the search for
global or local minima of the potential energy surface, the simulations of defect
reactions and diffusivities. Searches for minima involve simulated quenching
from an initial high temperature down to (almost) $0\,K$. In other simulations,
the temperature is kept constant at some selected value and the system evolves
on its own for as long periods of time as computationally tractable.

Simulated quenching consists of removing a fraction of the ionic kinetic
energy every N time steps. It mimics the cooling down of the material. Of
course, computer quenching occurs much faster (real time) than any actual
quenching, since the temperature drops from a very high temperature to near
zero within nanoseconds or less. This method allows a search for local (fast
quench) and global (slow quench) minima of the potential energy surface. Note
that one really wants the minima of the free energy, but the entropy term drops
out at zero temperature. The system must be allowed to equilibrate (at least
a few hundred time steps) before the actual quenching is performed. If one
deals with a defect complex with many degrees of freedom, the number of local
minima increases exponentially with problem size. Therefore, in order to find
the global minimum of the energy, one must quench many times, starting with
a large variety of possible initial configurations, in order to make sure that the
global minimum has been reached. Note that the reverse process (**simulated
annealing**) is also possible.

Constant-temperature simulations are limited to reactions occurring
very fast. The total energy must be calculated at each time step, with Δt
small enough so that the nuclei move only short distances. In the case of
relatively heavy atoms, such as the Si, typical time steps are of the order of
$2 \times 10^{-15}\,s = 2\,fs$ or less. For light impurities, such as interstitial H, Δt may
have to be as short as $0.2\,fs$. In supercells containing 60 to 200 atoms (these
are typical sizes), simulations lasting several thousand Δt's are common, those
lasting a couple of hundred thousand Δt's are rare and require highly efficient
and/or approximated methods. Thus, in the most favorable situations, it is
possible to simulate events lasting a fraction of a nanosecond (real time). More
commonly, real times of two to ten picoseconds are a practical limit. Note that
the forces are calculated and Newton's equations solved at each time step.
Therefore, the accumulation of systematic errors is not a serious problem,
even in very long simulations.

Finally, a number of tricks are used to accelerate a reaction too slow to be calculated under normal conditions, simulate the capture of a phonon or photon, force a diffusion jump, or examine an assumed diffusion path. Several examples are mentioned below.

Kicks of one (or more) atom(s) are used to overcome a potential barrier or simulate the capture of a phonon. A kick may consist in assigning a large initial velocity to one or more nuclei, that is assign some kinetic energy. Alternatively, one can underestimate the initial lattice relaxation around a defect center, that is assign additional potential energy. The magnitude of a kick must be sufficient to overcome most of a potential barrier and force the reaction. The kick must be carefully controlled and many trials may be needed to make sure that all one does is speed up a slow reaction, not shoot atoms in random directions. This control implies a good understanding of the processes that are likely to take place. The extra energy associated with a kick is very rapidly absorbed by the phonon bath, which results in a change in temperature. The most extreme case involves assigning a kinetic energy of several keV's to one surface atom in order to study the effect of ion implantation.

In order to find a diffusion path, one can **freeze some degrees of freedom** by assigning an arbitrarily large mass to some nuclei, thus effectively prohibiting their motion. In the case of a slow diffusing impurity, the mass of some atom(s) can be set to a large value and an assumed diffusion path be studied by pushing the impurity along while allowing only a part of the crystal to relax at intermediate steps.

Alternatively, one can **force the diffusion** of an atom by pushing one or more nuclei in the general desired direction, for example by cancelling any component of the net force pointing in the opposite direction. In the 'adiabatic trajectory' approach,[47] the diffusing atom is forced to move at a small constant speed along a predetermined path while the other atoms in the supercell continuously relax in response to its motion. Such treatments provide much needed insight into a problem that would otherwise not be tractable within realistic computer times.

Very long simulations are needed to study phenomena such as epitaxial growth, defects at a liquid/solid interface (crystal growth), propagation of dislocations or other extended defects, wafer bonding, calculation of diffusivities at various temperatures, etc. On the other hand, relatively short simulations (several ps) at some $1000\,K$ suffice to observe one or more diffusion jumps of impurities such as interstitial H or simple defects such as the monovacancy in c-Si. However, the vast majority of processes are (and likely will remain) out of the reach of MD simulations. One example is the aggregation of oxygen interstitials in Si and the formation of O-related thermal donors, since the diffusivity of interstitial oxygen, $0.13\,\exp\{-2.53\,eV/k_BT\}\,cm^2/s$, is many orders of magnitude too low for direct simulation at any temperature. In such situations, one must play tricks, and these work when one already has a rather precise idea of the microscopic processes taking place.

As mentioned above, MD simulations are unable to describe *quantum tunneling*. Tunneling is important when dealing with light impurities such as H, which has been reported to tunnel around shallow acceptors [3,48,49] in Si.

2   CRYSTALLINE MATERIALS

The construction of periodic supercells which describe the perfect crystal is not a problem. Cell sizes ranging from 16 to 216 and up are commonly used, with a small number of k-points in the first Brillouin zone included (typically 4 k-points for a 64-atoms cell, and only $k = 0$ for the largest cells). Many calculated properties of the perfect crystal are well reproduced by most methods. The energies per atom of various crystalline structures can be obtained, as well as band structures, bulk moduli, phonon frequencies, lattice constants, cohesive energies, etc. Such quantities are largely independent of cell size, because the periodic boundary conditions simulate an infinite perfect crystal in any case.

The study of impurities and defects in periodic supercells brings in one major approximation since the defect center itself is periodic. Suddenly, one studies not an isolated defect but a periodic arrangement of defects, often corresponding to unrealistically high concentrations. Interactions between defects in neighboring cells result in defect energy levels becoming defect bands, the width of which varies considerably with cell size (see Ref. [22] for examples). Repeating the same calculations in cells of increasing size becomes necessary to convince oneself (and a referee) that the calculated results describe an isolated defect center.

If one studies defect centers in a compound semiconductor, the long-range Madelung energy contribution is not that of an isolated defect in an otherwise perfect material but that of a periodic arrangement of defects. Dipole-dipole interactions between defects in adjacent cells give an artificial contribution to the energy. This affects the calculated equilibrium geometry as well as activation energies and other properties. A recent study [50] of Madelung energy corrections for interstitial hydrogen in clusters of cubic GaN (a fairly strongly ionic compound) shows that corrections range from very small up to a few tenths of an eV, depending on the configuration and charge state of the impurity.

2.1   Elemental Semiconductors

Most of the work done for group IV crystalline materials involves silicon. However, a number of issues dealing with diamond and germanium have been addressed as well.

Diamond

There are few constant-temperature MD calculations involving defects in semiconductor diamond. The reasons are that C forms strong and short bonds and a defect or impurity is trapped in a small and stiff environment. As a result, activation energies for diffusion are high for both intrinsic defects and impurities, making direct simulations computationally expensive. MD simulations have been used to study structural properties, the liquid states, small C clusters, and properties of a few impurities and defects.

Drabold *et al.*[51] used a tight-binding MD approach with a small basis set of pseudo-atomic localized functions. The approach is very similar to that developed by Sankey *et al.*[27] for Si. The method was tested for small carbon clusters, as well as bulk diamond and graphite. The calculations predicted vibrational spectra, bond lengths, and binding energies.

Galli *et al.*[52] used the first-principles CP method to study of liquid carbon in the $2500 - 5000\,K$ temperature range. They discussed the problem of energy transfer to electronic degrees of freedom. Most simulations were done in a 54 C-atoms fcc periodic supercell with a 12,000 plane-waves basis set ($32\,Ry$ energy cutoff). The simulations used a time step of $0.1\,fs$ with over 10,000 time steps ($1\,ps$ real time). The authors predicted a metallic liquid state with two-, three-, and four-fold coordinated C-atoms. They observed N-membered rings with N greater than 9.

Goedecker and Colombo[53] developed at linear scaling algorithm for tight-binding MD simulations optimized for a parallel supercomputer. The method[42] allows the calculation of very large systems (thousand of atoms) and/or very long simulation times (hundreds of thousands of times steps). They applied the method to the problem of liquid and solid carbon in the high- and low-density phases. Most calculations were done with 512-atom supercells and a time step of $0.5\,fs$, although larger cells were also tested. The results and predictions for *l*-C are quite similar to those obtained at the first-principles level by Galli et al.[52] with a slightly larger diffusion coefficient.

Phonon anharmonic effects in diamond were studied[54] by Wang *et al.* at the ETB level in 64-atom cells at $100\,K$.

Several *a priori* possible *n*-type dopants in diamond were studied with the CP method by Kajihara *et al.*[55] Most of the reported results are static ($0\,K$)

ones, such as formation energies, solubilities, equilibrium configurations, and position of energy levels in the gap.

Finally, MD simulations in $\sim$ 70 C-atoms clusters were performed using the semiempirical HF method of CNDO. The parameters were used to reproduce the band structure of diamond and various C–N configurations. The calculations deal with the core exciton,[56] nitrogen,[56,57,58] and vacancy-nitrogen complexes.[57]

Silicon

Although the focus of this chapter is on defect issues, it is worth mentioning that MD simulations are being used extensively to study the properties of defect- and impurity-free semiconductors as well, including the liquid state. These issues are closely related to defect problems. In particular, a number of crucial and poorly understood processes and taking place during crystal growth at the liquid/solid interface. Depending on the ratio of the pull rate and the temperature gradient, the crystal may be vacancy-rich (with D-defects), interstitial-rich (with A-defects), show oxygen-rich precipitates (OSF ring), or have other types of problems. Clearly, theory is far from being capable of simulating realistically these defects. However, one should welcome steps in this direction.

Liquid Si (ℓ-Si) has been studied at various levels of MD simulations. These studies often include the calculations of a number of thermodynamic properties, such as the entropy, specific volume, heat capacity, etc. The properties of ℓ-Si were calculated using Tersoff's empirical potential [59] in a large supercell (512 atoms) for long times (30,000 time steps of 2 fs). This study finds that in the liquid state, the Si atoms exhibit a range of coordination numbers, with a maximum at 6. The Stillinger-Weber potential was used [60] to study the properties of c-Si, a-Si, and ℓ-Si in large supercells. While the calculated properties near the triple point are reasonable, the melting temperature of a-Si was found to be below room temperature.

In a recent study at the *ab-initio* level with the CP method,[61] a 64-atoms supercell at 1800 K was used for a simulation lasting 1.2 ps. The results show that Si is metallic, and that covalent bonding persist above the melting point despite the loss of the tetrahedral network. The Nosé thermostat [62] kept the ions at constant temperature despite the transfer of energy to electronic degrees of freedom. The results proved to be sensitive to the (plane wave) basis set size. The CP method was also used [63] to calculate the thermal expansion coefficient of Si above the Debye temperature.

The CP method was also employed to calculate the solid-liquid phase boundary [64] and how various thermodynamic quantities are affected by the melting. The melting temperature was found to be rather close to the experimental value. In this study, a 64-atoms cell was used with 4 k-points as well as a 216-atoms cell but at the Γ-point only. The physics of laser-melting was also investigated.[65] This study included the finite-temperature version of DF theory, including the effects of electron excitations and fractionally occupied states. The results show the existence of a liquid state distinct from the usual one discussed above.

Phonon anharmonic effects in silicon were studied[54] by Wang *et al.* at the ETB level in 64-atom cells at $100\,K$.

The effects of ion implantation have been studied by a number of groups at the ETB level [66,67] and with $O(N)$ methods.[68,69] The latter calculations involved a 160,000 atoms cell and simulations times of several hundred ps. A 5 keV Si ion is shown to impact as many as 1,000 Si atoms. Finally, the interactions and annihilation of two screw dislocations was discussed [53] using an $O(N)$ method. This work also includes a discussion of ℓ-Si. Stillinger-Weber and Tersoff potentials are used to study wafer bonding in large supercells,[70,71,72] and the 90° twist boundary.[73]

Point defects in silicon

The calculation of properties of defects and impurities in Si and other semiconductors require methods with total energy capabilities, because lattice relaxations and distortions play a dominant role. Further, one must compare the energies of inequivalent (relaxed) configurations, especially in the case of defect complexes. MD simulations are very useful to obtain the various minima of the potential energy surface by simulated quenching, and are routinely used for that purpose. However, in a few cases, dynamics aspects of the problem can also be calculated. These include vibrational modes, defect reactions, and diffusivities.

Systematic studies of point defects in silicon using MD simulations are a relatively recent development. Static *ab-initio* calculations at the HF or MD level have focused on the equilibrium configurations and electronic structures, the metastable states and transition points, the formation energies, and other quantities requiring high-level calculations. Even for rather simple defects such as the monovacancy or interstitial H, there is disagreement among theorists (see for example Ref. [22]). The details depend on how well the methodologies used are able to describe for example the reconstruction of vacancies, the three-center two-electron bonds for H, the hybridization of the self-interstitial, and other chemically unusual features.

Many MD simulations approximate the solution to the electronic problem and are therefore likely to describe static properties (quenched to 0 K) imperfectly or poorly. However, at high temperatures, the finer details of the potential energy surface become less important and very different MD simulations may predict very similar behavior. The following examples deal with interstitial H, the vacancy and related complexes, the self-interstitial, and H-enhanced diffusion of interstitial O.

Buda *et al.*[74] performed the first *ab-initio* MD study of the diffusion of H^+ in Si with the CP method. They used a 128-atoms supercell, with plane-wave up to 6 Ry. The simulations lasted 4 ps with a time step of 0.12 fs, with temperatures ranging from 1,000 to 1,900 K. The lowest (potential) energy site for H^+ is found to be at the relaxed bond-centered (BC) site, in agreement with other calculations. The proton diffuses by jumping from one BC site to another, spending most of its time around BC sites. The calculated diffusivities at high temperature agree very well with the high-temperature experimental data, with a calculated activation energy of $0.33\pm0.25eV$ (expt: $0.48\pm0.05eV$).

The same problem was discussed with ETB simulations by Boucher and DeLeo[75] as well as Panzarini and Colombo.[76] They studied the diffusion of H^0 at high temperatures. Both groups report jump-like diffusion among BC sites, in good agreement with the *ab-initio* results. Figure 1 shows the result from Ref.[76].

However, the diffusivities calculated in Ref.[76] substantially drop below the extrapolated experimental value for temperatures lower than 1000 K or so. While such a shift toward low diffusivities is indeed experimentally observed, it is caused by trap-limited diffusion, not by some change in the BC-to-BC diffusion process which should follow an Arrhenius behavior. When all the calculated points are taken into account, the diffusivity predicted in Ref.[76] is substantially less than expected. Further, the static potential surfaces for H calculated with the ETB methods[75,76] predict minima at the incorrect sites.[22] Neutral H is found to prefer the tetrahedral interstitial (T) site to the BC site in Ref.[75] and the hexagonal interstitial (H) site to be degenerate with the H site in Ref.[76].

Thus, while all the MD simulations (CP method,[74] ETB,[75,76] and *ab-initio* TB[77]) predict a qualitatively similar diffusion of bond-centered hydrogen along BC-to-BC paths at high-temperatures, the approximate calculations show substantial problems at lower temperatures. This is to be expected since these methods heavily approximate the quantum mechanical part of the problem.

Three more comments regarding the diffusion of hydrogen. First, two authors[75,77] report that if the initial position of H is at its metastable site (the T site), it diffuses along a T–H–T path faster than it does along BC-to-BC

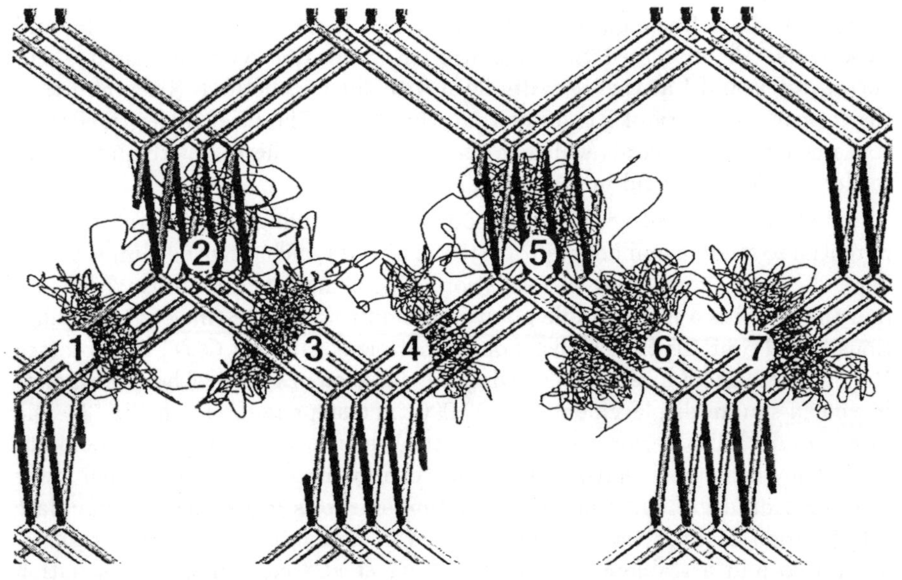

Figure 1: H trajectory at $1,200\,K$ during 25 ps. The Si atoms are left at unrelaxed positions in this picture, but H hops from relaxed BC to BC sites. The time evolution of the position of H is 6-7-6-5-2-1-3-4. This figure is from Ref.[76].

paths. The metastable (T) to stable (BC) conversion occurs at distorted sites in the Si crystal, such a near interstitial O.[77] Second, at lower temperatures, the diffusion of H becomes trap-limited. This cannot be simulated in supercells which are free of traps (such as vacancies or other defects). Third, anomalous diffusion of hydrogen at low temperature has been reported,[78] a process that cannot be reproduced by classical MD simulations.

The vacancy (V) and the Si self-interstitial (I) have also been studied at various levels of MD theory. Wang *et al.*[79] used an ETB approach to calculate the formation energies and configurations of V and I, assuming that the latter is at the T or the H site. The simulations involved a 512-atom supercell and times of the order of 8 to 9 ps. The formation energies ($\sim 4.1\,eV$ for V and $\sim 4.4\,eV$ for I at the T site) are rather close to those of other authors, but the geometries differ (see e.g., Ref.[80] for the V and Ref.[81] for I). Car, Blöchl, and co-workers [82,83] discussed the configuration, migration path, diffusivity, and entropy of V and I at the *ab-initio* level with the CP method. They find that the formation energy of I is lower than that of V. The brief review in Ref.[82] contains numerous details which will not be discussed here, but the interested reader will find it useful in many respects.

Sinno *et al.*[81] used an ETB method with the Stillinger-Weber interatomic potential to study a number of properties of V and I, including the temperature-dependence of the entropy of formation. They calculate the diffusivities and find the T site is a likely saddle point in the diffusion of I. Some of the calculations lasted over 10^6 time steps. Long simulations using an $O(N)$ method (see Ref.[42]) produced a nice video showing the diffusion of I in Si. The corresponding mean-square displacement for 200,000 time steps is shown in Fig. 2. The same group [84] also performed extensive studies of V and I and V-I recombination. They concluded that the activation energy for V diffusion is much lower than for I diffusion, but that the formation entropies imply that I's dominate self-diffusion at high temperature while V's dominate at low temperatures. They also find a stable V-I complex. Tang *et al.*[85] compared the formation energies for V and I (in various configurations) calculated with several empirical potentials and the LDA method. The range of calculated formation energies strongly suggests that empirical potentials should be used with great care when predicting formation energies and static properties.

Using semiempirical HF-based MD simulations (CNDO), Mainwood[86,87,88] studied a range of issues related to carbon and the self-interstitial in Si. Their migration,[86] interactions with substitutional B,[87] and C-H interactions.[88] In contrast to all other MD simulations, these calculations are performed in clusters rather than periodic supercells, and the electronic problem is solved at the HF rather than DF level. It would be of considerable interest to study the

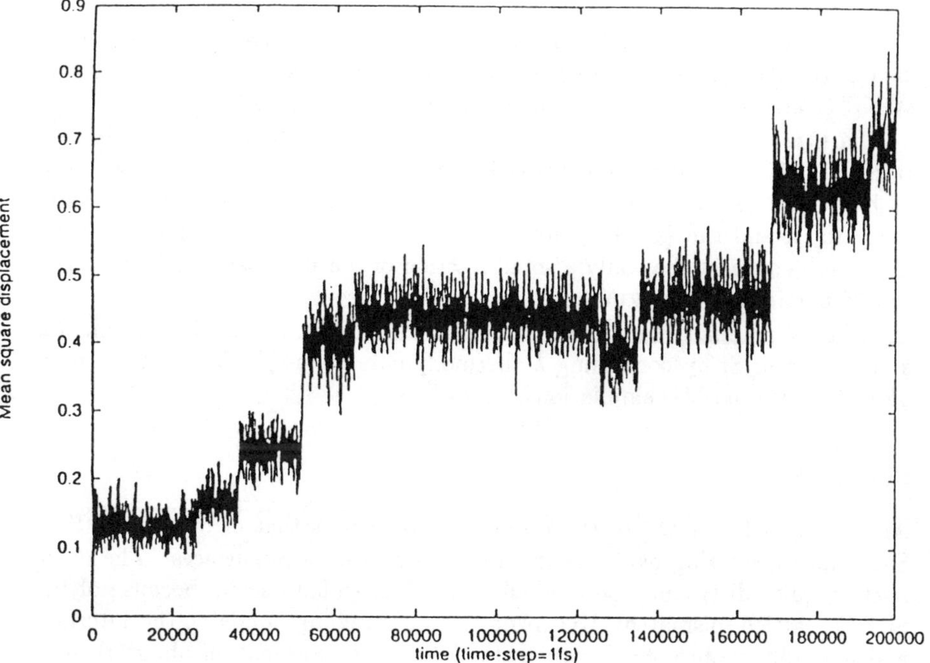

Figure 2: Mean-square displacement ($Å^2$) of the Si self-interstitial for 200,000 times steps. This figure is from Ref.[42].

same problems within a DF/supercell scheme in order to assess the strengths and weaknesses of the two approaches.

Vacancy-hydrogen complex formation and dynamic properties have been studied by Park *et al.*[89] using the *ab-initio* TB method.[27] In addition to the equilibrium structures and the capture of H by the vacancy, they calculated the stretching frequencies of the four (neutral) V-H complexes (with 1, 2, 3, and 4 H's in the vacancy). The vibrational spectral densities calculated at room temperature are shown in Fig. 3. The calculated frequencies agree quite well with the experimental values (2068, 2144, 2191, and $2222 \, cm^{-1}$, respectively). The agreement is even much better when D substitutes for H: the theoretical predictions for all four stretch frequencies equal the experimental value plus about $30 \, cm^{-1}$.

Estreicher *et al.*[90] calculated the stability of all the possible clusters of vacancies (up to 7) at the *ab-initio* TB and HF levels. They report a particularly stable, electrically and optically inactive, ring-hexavacancy.

Park *et al.*[77] studied the H-enhanced diffusion of interstitial O in Si. *Ab-initio* TB MD simulations show that H is attracted to O, but tends to self-trap at BC sites in its vicinity. However, if H reaches a T site near O, it rapidly binds to it, and the $\{O, H\}$ pairs becomes much more mobile than O alone. The results suggest that only a small fraction of the H present in the material should participate in the process.

Finally, Milman *et al.*[91] calculated the free energy of diffusion for Li, Na, and K ions in Si by combining a thermodynamical integration method with *ab-initio* CP-type MD simulations.

Silicon oxides

An issue closely related to the problem of H in Si is that of interstitial H in SiO_2 since insulating oxide layers are often grown on Si surfaces. The oxide itself is quite dirty and contains all sorts of extended defects because of the Si–SiO_2 lattice mismatch. Hydrogen ties up dangling bonds at the interface and may diffuse into the Si substrate. Therefore, information about H in Si oxides is important. A recent semiempirical tight-binding MD simulation[92] of interstitial H in crystalline SiO_2 has addressed some of these issues.

This study focused on the diffusivity of neutral interstitial H in the range $700 - 1800 \, K$ in α-quartz, β-quartz, β-tridymite, and β-cristobalite. Empirical potentials were used to model the Si–O, Si–H, and O–H interactions. The supercells for quartz, tridymite, and cristobalite contained 72, 96, and 192 atoms, respectively, and trajectories were computed for 10^6 times steps. In quartz, H is reported to diffuse in channels along the c-axis, while isotropic

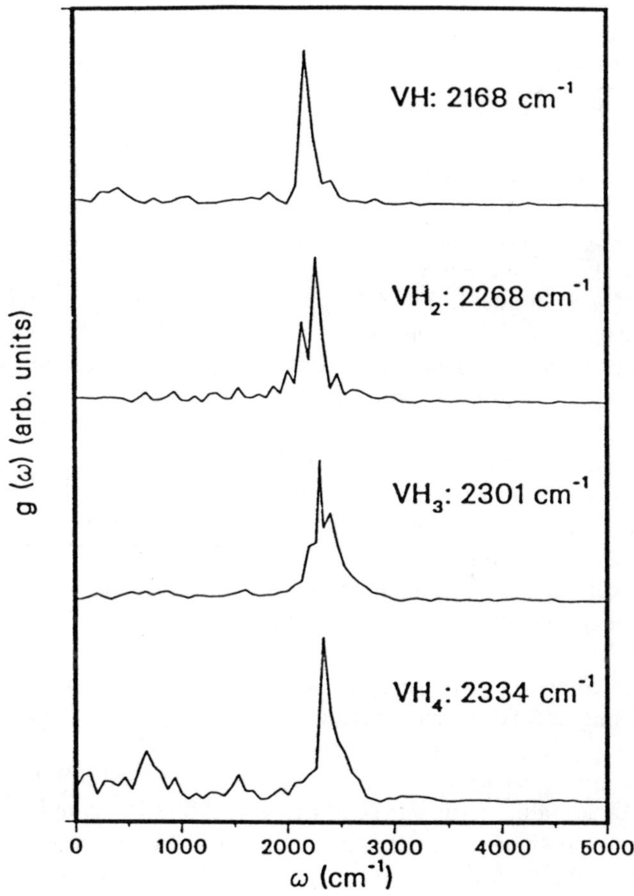

Figure 3: Si–H stretching modes of the four vacancy-hydrogen complexes calculated at room temperature (see text). This figure is from Ref.[89].

diffusion occurs in the other structures. The calculated diffusivities $D(T) = D_0 \exp\{-E_a/k_B T\}$ for H in α-quartz, β-quartz, β-tridymite, and β-cristobalite are $D_0 = 417$, 2.1×10^{-2}, 2.0×10^{-3}, and $7.3 \times 10^{-2}\, cm^2/s$, respectively, and $E_a = 1.29$, 0.62, 0.71, and $1.19\, eV$, respectively. *Ab-initio* calculations of activation energies for diffusion are needed to confirm these values.

Germanium

Payne *et al.*[93] predicted the geometrical structure of twist grain boundaries in germanium using an simulated annealing approach. The supercell represented two six-layer crystal slabs rotated relative to each other to simulate various grain boundaries. All the atoms in the cell were allowed to relax except those in the central two layers which were fixed. Further, the distance between the layers was allowed to vary to account for volume changes resulting from the presence of the boundary. The CP approach was used, with 4,300 plane-waves (10 Ry cutoff). Two k-points were used for Brillouin-zone averaging. The authors report that quenching from different initial coordinates for the atoms results in a number of metastable configurations being realized. They also report structures unusual for Ge, such as four-fold rings.

Several groups have studied the properties of liquid Ge.[94,95,96] The most recent calculations by Kulkarni *et al.*[97] included five temperatures between 1250 and 2000 K. These simulations use the finite-temperature version of DF theory, with an electronic temperature such that $k_B T^{el.} = 0.1\, eV$. Germanium was described with a 64-atoms supercell (Γ-point only) with volumes adjusted to match the experimental density data. Planes waves up to 10 Ry were included and a 3 fs time step was used up to 1.5 ps.

The calculations yield the pair correlation function, static structure factor, bond-angle distribution function, electronic density of state, atomic self-diffusion coefficient, and the ac conductivity at various temperatures. The results are nicely consistent with a range of experimental data. Liquid Ge is found to be metallic, with a pseudo-gap separating the s and p bands. While a substantial amount of covalent-like bonding has been reported in MD simulations of liquid carbon and silicon (see above), much less of this is visible in liquid Ge.

2.2   *Compound Semiconductors*

III-V nitrides

The theory of defects in III-V nitrides has recently been reviewed.[98] These materials are quite dirty by Si standards. Typical GaN films contains many intrinsic defects, and impurities such as H, O, C, etc. have been seen in concentrations up to 10^{20} cm^{-3}! The films also contain stacking faults and other extended defects because there is no substrate on which to grown strain-free films and because of the presence of polytype grain boundaries. Virtually all the as-grown samples exhibit a yellow luminescence associated with some deep-level defect, and strong n-type behavior which is associated with some abundant shallow-donor impurity or defect. To date, there are few microscopic experimental data about point defects, such as FTIR or Raman studies of specific defect centers. As a result, theoretical studies are heavily relied upon, but the predictions have yet to be confirmed by direct observations.

Bogusławski *et al*.[99,100] and Neugebauer *et al*.[101,102] use local density-functional theory with pseudopotentials and plane wave basis sets (30 and 60 Ry cutoffs, respectively). The former use 72-atoms supercells with only the Γ-point and the latter 32-atoms cells with several special k-points.

Both groups use simulated quenching to obtain local minima of the potential energy and calculate the electronic structures and formation energies of intrinsic defects such as vacancies, antisites, and self-interstitials. Note that the formation energies depend on the chemical potentials of the atomic 'reservoir' used (e.g., fcc-Ga vs. α-Ga vs. ℓ-Ga) and the growth conditions (theory assumes equilibrium). Properties of impurities such as interstitial hydrogen[103] or substitutional group IV impurities[104] have also been obtained. The latter paper also discusses the strain and reconstruction at surfaces and various interfaces.

Stumm and Drabold[105] expanded the *ab-initio* TB scheme of Sankey and co-workers[27] for GaN. The basis set consists of four pseudo-atomic orbitals per atomic site with confinement radii of 3.8 Å (for N) and 5.4 Å (for Ga). The Ga $3d$ electrons are treated with the core. The calculated band structures of zincblende and wurtzite GaN agree with those of other authors. The method was tested for various supercell sizes (up to 216 atoms) and k-point sampling in the Brillouin zone (up to 14 k-points). The structure and key electronic properties of the various possible vacancies and divacancies are in agreement with static *ab-initio* results.

An ETB code was developed by Boucher *et al*.[106] It uses two-center approximations for the Ga–N and N–N interactions and three-body terms for Ga–Ga interactions. The basis set consists of sp^3 hybrids on each N and Ga atom, and a correction term is added to account for the Ga $3d$ orbitals. An exponential cutoff is introduced to limit the spatial extent of the interactions, and the model parameters are fit to a range of experimental data such as molecular bond lengths and vibrational frequencies. The host crystals are approximated with periodic supercells containing 64 atoms (zincblende GaN) and 96 atoms (wurtzite GaN) with k-point sampling reduced to the Γ point. The calculated structural properties are in good agreement with experimental data.

The authors focus on static properties of a number of intrinsic defects in wurtzite GaN by quenching from a high temperature. The results agree at least qualitatively with those predicted at the *ab-initio* level by other authors,[101,99] with differences in the details of the equilibrium structures and the energetics.

Serra *et al.*[107] also developed an ETB approach to study the properties of GaN at high temperatures. Their two-center potential is fitted to reproduce the band structure and phase diagram of GaN in the zincblende, wurtzite, and rocksalt structures. Their basis set consists of sp^3s^* functions, where s^* mimics the filled $3d$ orbitals of Ga. The high-T simulations are performed for several ps with a time step of $1\,fs$, with temperature varying from 300 to $4,000\,K$. The pair correlation functions are calculated and show that local ordering is preserved, with more N–N than Ga–Ga interactions. The density of states at $2,300\,K$ still exhibits a large gap, although smaller than at low temperatures. The authors calculated the formation energies of several native defects in zincblende GaN and found values close to those predicted at the *ab-initio* level by Neugebauer *et al.*[101]

GaAs and related compounds

The problem of self-diffusion in GaAs was studied at the *ab-initio* MD level by Bockstedte and Scheffler.[108] The predicted dominant migration mechanism of the Ga vacancy, its free energy of formation, and the rate constant for the Ga self-diffusion. The plane-wave basis set had an $8\,Ry$ cutoff, a 63 atom supercell was used, and the time step was about $9.5\,fs$. The simulations were performed at temperatures around 1600-1700 K for several ps.

The calculations predicts that the $\{As_{Ga} - V_{As}\}$ complex is unstable in n-type material, implying that the nearest-neighbor (NN) jump of the Ga vacancy is impossible. In this material, the Ga vacancy diffuses via second-NN jump, involving only the Ga sublattice, as shown in Fig. 4. The calculated Ga self-diffusion constant is $D_{Ga} = 208\exp\{-4\,eV/k_BT\}\,cm^2/s$.

Ab-initio MD simulations were also used to describe diffusion processes in GaAs/AlAs superlattices, first Zn-enhanced disordering,[47] then the diffusion of Si.[109] The latter work also includes the diffusion of Si in GaAs. These simulations were of the CP type at $600\,K$, with plane waves up to $14\,Ry$, and a 64 atoms supercell. In order to calculate high migration barriers, the authors propose an 'adiabatic trajectory' approach (see Sec. 1.5). The lowest-energy diffusion process for substitutional Zn is predicted to be the (100) kick-out process, assisted by group-III interstitials.[47] During the Zn in-diffusion, group-

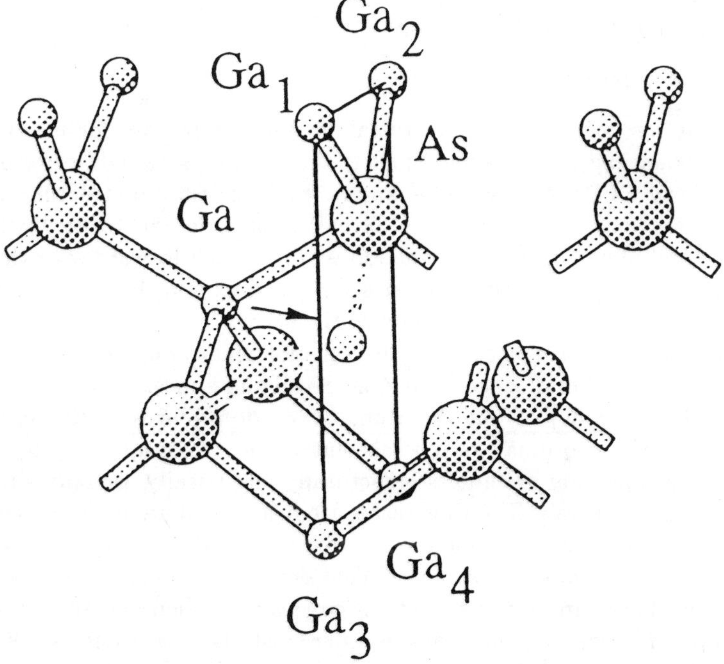

Figure 4: Second NN jump geometry predicted for the self-diffusion of Ga in n-type GaAs. The arrow indicates the direction of the hop. This figure is from Ref.[108].

III atoms become interstitials, diffuse, and finally exchange with substitutional group III atoms, thus disordering the superlattice.

A variety of sites for Si in GaAs are discussed[109] (including the Si_{Ga} − Si_{As} pair) with their formation energies and activation energies for diffusion in various charge states. The lowest-energy diffusion path for substitutional Si_{Ga} involves second-NN jumps assisted by group-III vacancies. The Si-induced disordering for GaAs/AlAs superlattices is predicted to occur with the help of group-III vacancies through the formation of Si^+_{III} − V^{3-}_{III} pairs.

3   AMORPHOUS MATERIALS

3.1   *The Construction of Supercells*

The construction of supercells of crystalline semiconductors for electronic-structure calculations or MD simulations is almost always straightforward and not particularly CPU intensive. One knows the basic crystal structure or, at the worst, the structure might be one of several crystalline structures. With impurities or small impurity clusters, the job is still straightforward even though one might have to place the impurity at a variety of sites in the supercell, then relax the coordinates in order to minimize the energy.

In contrast, the construction of supercells for amorphous materials (even group IV semiconductors) is decidedly non-trivial and almost always computer intensive. One cannot just put the atoms at random places in the unit cell and then relax the coordinates. It is absolutely out of the question to make a computer sample that resembles structurally, electrically, or optically the material grown in the lab. The time scales are just too different. The criterion that we have adopted is that to a first approximation, it doesn't really matter how one constructs a supercell sample. What does matter is its resemblance to real material. There are minor exceptions to this rule when one alters the so-called computer "growth" conditions in order to study how it affects the final supercell. This change in conditions is then suggestive as to how variations in temperature, pressure, density, annealing time, etc. affect the lab-grown material.

Amorphous silicon

The criteria and their weighting that we suggest in this section are by no means universally agreed upon. Thus, this section is more colored by personal preference than most. Further, we shall start out considering a-Si, which has a much simpler chemistry than a-C. Good amorphous Si is always hydrogenated. Otherwise, it has monstrous bandtails and a large number of dangling bond defect states located approximately midgap. By 'large', we mean from 10^{19} to 10^{20} cm^{-3}, which is of the order of 1 atomic %. This makes the material completely useless as a semiconductor. Good hydrogenated material may have from one to ten atomic percent hydrogen. In such material, the band tails are severely reduced and the number of dangling-bond-related states is of order several times 10^{16} cm^{-3}, that is about one defect state for every 10,000 to 100,000 atoms. The Si atoms in decent a-Si:H are almost all four-fold coordinated, with each Si atom bound to a combination of four Si and H atoms. The dangling bonds and three-fold coordinated Si atoms are the defects which lead to the localized states in the gap.

In a-Si:H, pair correlations between pairs of Si atoms show a peak at a Si–Si distance of 2.35Å, which is the NN distance for Si atoms in c-Si. However, in the amorphous material, the peak in the bond length distribution is several tenths of an Å wide. A broad second NN peak also exists, and at larger distances the pair correlation function becomes rather flat. The average bond angle is the tetrahedral angle (109.47 °) with an rms width of about 10 °. Finally, at temperatures below 600 K, the samples are stable, except when exposed to light. This does not mean that atoms cannot move, but that the average properties of the samples do not change with time.

Another issue is the positioning of H in supercells and in the actual material. Nuclear magnetic resonance (NMR) experiments[110] show a clustered component of H with distances between adjacent Hs of about 2.0 to 2.1Å, and a dilute phase where the H atoms are much further apart. The fraction of H atoms in each of these phases depends on the growth procedure. Further, NMR measurements on samples made with D instead of H show that some of the hydrogen is bonded with a simple Si-H bond configuration, and some is in an entirely different form, possibly molecular. The bonded H is well understood but the other component is not. For the bonded component, the Si-H bond length is 1.5 to 1.6Å.

Many a-Si and a-Si:H supercells have been generated and all of them suffer from a number of deficiencies. Very few authors point those out and one may get the impression from the literature that all is well. This is not the case, although the deficiencies do not negate the value of the work.

All the samples we have studied, read about, or heard of, suffer from one serious deficiency. They all exhibit an energy gap that is too small. This is in addition to problems with the LDA. All known supercells would exhibit a gap that is too small even if the density of states were computed exactly. An example will illustrate this. Using ETB one can get the band structure of c-Si virtually exactly, since the ETB parameters are fit to the band structure. One then can use ETB to compute the density of states of a-Si using the

Bethe lattice.[14] This lattice is almost perfectly suited to simple amorphous calculations because the bond lengths and angles are preserved perfectly, yet translational invariance and long-ranged order are lost. In any case, the ETB density of states for the Bethe lattice gives a gap of 1.9 eV.[13] This is larger than the gap in c-Si but is about the size of the gap in c-Si averaged over the Brillouin Zone, as it should be. However, we have applied the same ETB theory to a large number of amorphous supercells and always got much smaller gaps! The reasons for this are not exactly clear but we believe that it is because of excessive band tailing caused by a few abnormally large or small bond angles. This is a problem that needs solving, but it is not clear how to proceed.

Before considering actual supercell samples and their construction, a few words about defects in a-Si:H are in order. Defects can and are defined in two different ways. First, one can define a *geometrical defect* as any Si atom that is not four-fold coordinated or any H atom that is not one-fold coordinated. This involves an arbitrary definition for the broken bond distance between NNs. Usually, any distance near 2.8Å will work for Si-Si neighbors and 1.9 to 2.0Å for Si-H neighbors. Although these defects are important for supercell constructors, they are largely incidental to what most people are really interested in. What usually characterizes real materials and where the real physical interest lies are *electronic defects*, which lead to localized or quasi-localized states in the gap or to band tails. It is not always true, and by no means universally accepted,[11] that these two types of defects describe the same thing and that geometrical and electronic defects have a one-to-one correspondence with each other.

In supercells with no greatly strained bonds (either bond lengths or bond angles) and with a concentration of defects of less than one percent, geometrical and spectral defects do bear a one-to-one correspondence to each other. That is, dangling bond geometries (three-fold coordinated Si atoms) are associated with specific localized states in the gap that are reasonably well localized at the geometric defect. However, already near the one percent level, the defect states tend to hybridize with each other, and the localization deteriorates. Many supercells have geometrical defects at the several percent level. In these cases, the defect states are well hybridized and the electronic states cannot be associated with given geometrical sites.

Further, there are more geometrical defects than spectral defects in the gap because the hybridization pushes energy levels into the conduction and valence bands. It is difficult for us to believe that such supercell samples are representative of real a-Si:H. In addition, even in supercells with relatively few geometrical defects, spectral defects can and do arise from purely four-fold coordinated Si atoms. Badly stretched bonds (say Si–Si bonds of 2.6Å instead of the unstrained 2.35Å) lead to spectral states in the gap.[111] So will badly strained bonds with bond angles diverging from the tetrahedral angle by 30° or more.[111] There is no reason to believe that such geometrical anomalies exist in real a-Si:H. They may simply be artifacts of supercell construction techniques.

There is no consensus about the ranking of characteristics for constructed supercell samples. Our view is that good supercells resemble real a-Si:H as much as possible in the gross sense, even if some of the details are not perfect. Thus we feel that it is most important to have very few defects, none if possible at or above the one percent level. Very few supercell samples satisfy this. Of course, one must have at least one defect in order to study the properties of defects. This will usually ensure a reasonable density of electronic states and a good clean energy gap. With several percent defects, the gap is virtually non-existent.

It is important for the cells to have a good pair correlation function, and Si–H vibrational modes that agree reasonably well with experiment; the closer to experiment, the better. All the supercells that we have looked at show a reasonable pair correlation function. It appears almost impossible to go very wrong in this regard even for supercells that are loaded with geometric and electronic defects. Vibrational frequencies are very touchy to small changes of parameters in a computer code. This is because frequencies are proportional to the second derivative of the energy while force are only proportional to the first.

The first respectable supercells were constructed by Wooten, Winer, and Weaire [112] (WWW) and their method is well documented in the literature. WWW start with a large (216 atoms or greater) crystalline supercell, then switch atoms around by a clever algorithm that ensures that all atoms remain four-fold coordinated. However, the topology is substantially disordered in that a number of five- and seven-membered rings are made, while c-Si has only six-membered rings. The supercell sample is then relaxed using MD simulations. The cells are surprisingly good and are still used today. Their biggest drawback is that they do not contain hydrogen and that, by construction, they contain no dangling bonds. Further, as one might expect, they contain a number of stretched and strained bonds that lead to spectral defects. However, these defects are not particularly numerous. Since the method does not employ a band structure code, the WWW models do not have a density of electronic states associated with them. However, using TB codes, they produce reasonable densities of states. Of course the gaps are too small and the band tailing too severe, as in all supercells.

Some of the biggest drawbacks of the WWW cells can be rather easily fixed. In particular, the badly stretched/strained sites that produced spectral defects, the lack of hydrogen, and the lack of three-fold coordinated Si atoms that are necessary for studying dangling bonds, can all be fixed at once.[113] One can simply remove the badly stretched/strained Si atoms, tie up the groups of four dangling bonds with H, and relax or anneal the result. It has been found that besides the incorporation of H, this removes spectral defects and band tailing, and widens the gap. Further, one can create dangling bonds at will by removing one or more H atoms.

With the advent of faster computers and more efficient MD algorithms, investigators started to construct a-Si:H supercells in more conventional ways. The widespread and typical method is to start with a crystal or collection of Si atoms, heat them to a high temperature to ensure disorder, and then cool the supercell as slowly as one has patience or computer time for. Hydrogen can easily be incorporated into the soup that one starts with. Supercells have been constructed by this method - or variation of it - by numerous investigators using empirical potential, ETB, and a variety of *ab-initio* methods. These include efforts by Guttman and Fong,[114] Mousseau and Lewis,[115] Buda, Chiarotti, Car, and Parrinello,[116] Holender, Morgan, and Jones (HMJ),[117] Fedders and Drabold (FD),[113] and Tuttle and Adams.[118] The above selection is far from exhaustive but does include different types of codes.

A number of reviews of amorphous supercell samples are available.[118] We shall briefly summarize some of the results from these papers. Most of the samples cited above have about a 10% H concentration, which is at the high end of the normal range for good lab-prepared material. Also, most of the samples were formed from a crystal or liquid of Si and H atoms, and then cooled. Of these cells, only FD and HMJ produced cells with no defects. The other ones had of order 10% or more geometrical defects and virtually no gap. Because of these enormous defect densities, we believe that they are a poor representation of a-Si:H for most (but not necessarily all) purposes. A price had to be paid for the two cells with no defects. The HMJ cell has 23% H content, which is far too much to represent lab-prepared material. The other cell with no defects, the FD cell, was prepared as follows. After the usual quench, some of the strained bonds were healed by removing one Si atom from a strained portion of the supercell and four H's were inserted to remove the dangling bonds. The supercell was then annealed. Further, naturally occurring dangling bonds were hydrogenated or not, depending on whether one wished to study dangling bonds or a defect-free supercell. However, the FD cell produced a pair correlation function that is not quite right at distances greater than the second NN distance. This flaw is probably due to a remnant of crystallinity.

In device-quality lab-grown a-Si:H, hydrogen occurs in both clustered and unclustered forms. In the clustered regions, NMR experiments[110] show that the H atoms are about 2.0 to 2.1Å apart. This is very close, but the evidence is quite solid and not in dispute. The fraction of H that is in clustered regions varies from sample to sample, but can be considerably greater than 50%. The clustered nature of the hydrogen in a-Si:H supercells presents another problem for the MD supercell maker. The usual quenched liquid approach does not produce supercells with H clustered anywhere near 2.0 to 2.1Å apart. Again, the FD model does produce such clusters by construction. That is, by removing Si atoms and replacing each by four H atoms and then annealing, it is almost guaranteed that the H's will remain close together. This produces H clusters that produce a second NMR lineshape moment that is virtually identical to the second moments observed in experiments. Since the lineshape scales as the inverse third power of H spacing, this a quite a stringent test. However, the spread of frequencies of the Si-H vibrational modes is too broad in the FD samples. For further details on the structure and vibrational modes, see Ref. [118].

There is one important property of supercells that we have not yet discussed. That is stability. A supercell prepared by one MD method may well not be stable when investigated with another MD method. This is hardly surprising but can be disconcerting. All the supercells described above are stable within their own MD methods (the given configuration is a local minimum of the energy). However, many of them are not stable as a function of time! That is, many supercell samples that are at a local potential energy minimum change drastically when annealed for several picoseconds at temperatures as low as 600 K. This is unphysical since a-Si is known to be stable at these temperatures. In fact, the only published results on stability was one of the FD samples where it was found to be stable over long times at 600 K.[119] Many other samples have been found unstable in the above sense, but these results were not published.

In concluding this section, we note that FD have a somewhat different philosophy for supercell creation than the other investigators. They believe that it is impossible to create supercells in a way close to the way they are created in the lab, and that they should do all they can to make their supercells resemble real material regardless how this is done. They have recently constructed [120] large (over 500 atoms) supercell samples hat have a better gap (although still too small) than previously reported, as well as an excellent pair correlation function and good vibrational modes. The started with a WWW supercell, removed a few Si atoms from strained regions, hydrogenated the dangling bonds, and re-annealed. This produced the best supercells to date.

Amorphous carbon

Carbon is similar to silicon, except that its chemistry is much richer. This makes the study of a-C more demanding but possibly also more rewarding. Carbon comes as diamond (tetrahedral or sp^3 bonding), graphite (trigonal or sp^2 bonding and some π bonding), buckyball-like (with more complex hybridizations[121]), not to mention the various distinct tetrahedral amorphous phases. In all known samples of a-C, both types of bonding occur.[122] McKenzie et al.[123] and others have grown 'tetrahedral amorphous carbon' (ta-C). This material has about 80% sp^3 bonding and 20% sp^2 bonding, a gap of about 2 eV, and is very hard. The study of this material is an exciting new field with possibly very important applications.

The mix of sp^3 and sp^2 configurations is of considerable interest both geometrically and spectrally. In early pioneering work, Robertson[124] suggested some possibilities and, recently, MD simulations have partially answered a number of questions. Reliable MD simulations for a-C are much more difficult to produce than for a-Si. But the actual mechanics of creating a-C supercells on the computer is almost identical to the creation of a-Si ones discussed above. However, the creation of ta-C supercells is much more difficult than that of other forms of carbon. A failure to produce ta-C does not mean that a method is not perfectly good for other forms of amorphous carbon or even for analyzing supercells of ta-C.

Tersoff [125] has performed calculations with an empirical potential. His potential has worked well for a number of applications. For ta-C, his work suggests that an sp^3 bonded network is badly strained, because the sp^3 fraction is too large. It now appears that his structure is not representative of lab-grown ta-C, and we believe that empirical potentials are not up to describing the variety of types of bonding in that material.

The WWW bond switching scheme has worked well in creating a-Si su-percells. Recently Djordjevic et al.[126] have constructed cells using the WWW methods, but suitable for C. These cells are useful for studying a-C, but contain only four-fold coordinated C atoms and so cannot be used for studying the interplay of sp^3 and sp^2 bonding. However, we suspect that by some clever manipulations and re-annealing, mixed cells could be made from them.

Since diamond is denser than graphite, one expects a higher fraction of sp^3 bonds at the expense of sp^2 bonds as the density of a-C is increased. This has been observed experimentally and was first described in MD simulations by Fraunheim and co-workers.[127] They were concerned primarily with the trends in sp^3 concentration as a function of density and H concentration. Their supercell samples do a good job of describing this. Their computational scheme is an interesting hybrid involving elements of DF and non-orthogonal TB theories, and appears to correctly describe the challenging mixtures of different kinds of C bonding. Their work is in reasonable agreement with experiment. We believe that the non-orthogonality of their method is vital. Indeed, Menon and Subbaswamy [20] have demonstrated that the assumption of orthogonality reduces the transferability for C microclusters. However, the electronic density of states obtained by Fraunheim *et al.* for ta-C is poor since there are many midgap defect states.

Orthogonal ETB has also been used to construct and study a-C supercell samples. Wang and Ho [128] have presented a 216 atom structure of a-C. As expected of methods fit to a class of substances, the method describes those substances well. However, they can have difficulty with an amorphous structure of something with as rich a chemistry as C. In fact, the (orthogonal) ETB sample is not representative of lab grown ta-C. There is no clean energy gap in the density of states and there is no tendency for the defects to arrange themselves to form a gap as has been observed with more sophisticated methods.

The first major *ab-initio* simulation of an a-C supercell was performed by Galli *et al.*[129] This sample was constructed with the CP code. This work is a major study of liquid and a-C including energies as a function of the number of k-points, radial distribution functions, and microstructure in the mixture of sp^3 and sp^2 bonding. However, the study is limited to low density supercells. They find a clustering of the four-fold coordinated atoms and thick planes of mostly three-fold coordinated C atoms. Their findings are consistent with experimental observations and there is every reason to believe that the cells are good ones. However, since the densities of their supercells are too low to correspond to ta-C, little about the latter can be learned from them.

Recently, Drabold *et al.*[130] have used an *ab-initio* TB code [27] to construct high-density supercells of ta-C. Using a very slow quench from a liquid, they were able to construct supercells that have no defect states in the gap, thus reproducing the most impressive experimental feature of ta-C. Further, a mechanism for creating the clean gap was partially explained. All of the three-fold coordinated C atoms (those with sp^2-type bonding) are paired or in groups of an even number of three-fold atoms. In two samples, there were no isolated three-fold coordinated atoms (or dangling bonds) that would lead to a state in the gap. Further, other strained configurations of atoms that should have produced states in the gap hybridized with the sp^2 pairs and thus were swept from the gap. Finally, it had been believed that the conduction and valence band tails in ta-C were almost entirely due to the three-fold coordinated atoms. The above study showed that this was not true, although the band tail states were heavily weighted toward the three-fold coordinated sites.

3.2 Defects and Impurities

As noted above, much of the MD effort has been concentrated on creating supercell samples of good quality and large size. Many of these supercells have been used to study defects, doping, and other properties of amorphous materials but these applications are not within the scope of this review. This section will be limited to a few MD studies of the actual dynamical properties of amorphous systems.

In reading these studies, one must be slightly suspicious. It is entirely possible that some phenomena may simply be artifacts of either an inadequate method or of a supercell that is not stable.

Starting from some of the earliest MD studies,[61,11,131] there have been reports of five-fold coordinated Si atoms in a-Si supercells both with and without hydrogenation. These defects, called floating bonds,[132] were at one time seriously thought to be responsible for the dominant defect state in a-Si:H.[133] However, further studies [13] proved that these defects showed localization and gap position properties that did not agree with experiments. Further, in well annealed supercells with no or very few defects, such floating bond defects do not occur. Nevertheless, they are still observed by virtually all investigators in dynamical MD runs! Usually the five-fold coordinated atoms exist for rather short times, far less than a picosecond. Any quench to a metastable configuration makes them disappear in good supercell samples. Further, pairs of dangling bonds keep forming and disappearing in dynamical MD runs.[61,131] These are merely dynamical fluctuations and do not lead to either localized states or states in the energy gap even for the short times that the dangling bond pairs exist.

Fedders and Drabold [119] performed the only dynamical MD run that actually measured the dynamical stability of a-Si:H supercell samples over time. They showed that their supercell samples were stable over time on the average, but that individual bond lengths and angles did change. This was done in the following way. A statistically stable supercell was allowed to vibrate of evolve freely at a given temperature. From time to time a snapshot of the coordinates was saved and was quenched. The energy of the quenched configuration was saved and the bond angle and bond lengths were stored. In runs lasting many picoseconds, the energies of the system did not change by more that a few hundredths of an eV and the bond angle and length distributions were essentially identical. This is as it should be since the average properties of real a-Si:H do not change over short time scales. However, individual bond angles change by as much as several degrees, and individual bond lengths change by a few hundredths of an Å. Thus, over time, the average properties of the system remain static while the properties of individual atoms changed. This is exactly what one would expect from a glassy or amorphous system. This study was performed only on defect-free supercells and, to our knowledge, there have been no dynamical stability studies performed on other supercells. However, in unpublished work we have seen many supercells (both some of ours and others) change drastically. This makes the dynamical stability of most supercell samples slightly suspect.

In a series of notable papers,[134] Li and Biswas have investigated metastable defects, the motion of hydrogen, hydrogen rebonding, and the annealing of defects using a ETB code developed by Biswas and collaborators.[19] The investigators found that defect densities and formation energies are controlled by the bond-length disorder of the material. From their study, they suggest that bond-length disorder plays an important role in metastable defect formation.

Fedders[135] has also studied the migration of H and defects but using an *ab-initio* method. He found that dangling bonds and bond centered H move very quickly when there are dangling bond defects around and when H is around as well. Hydrogen allows the network to distort easier, and H and dangling bond move in concert. In fact, the motion of the defect-hydrogen complex involves the substantial rearrangement of up to 20 or more atoms. However, the above mechanism works only where there are a number of hydrogen atoms around and is thus not a mechanism for global H diffusion.

Lanzavecchia and Colombo[136] also have investigated H bonding and migration in a-Si:H. Among other things, they obtain a value for the activation energy for diffusion of H among bond-centered sites of about 0.5 eV, which is in agreement with experiment.

Colombo and a number of co-workers have used a TB code to investigate the dynamical properties of a number of liquid and amorphous systems. This includes the investigation of defect-induced amorphization in Si.[137] They also have investigated amorphous and liquid GaAs.[138,139] This latter work is virtually the only work on this subject.

Finally, Stumm and Drabold[140] proposed several interesting structures for a-GaN, on the basis of *ab-initio* TB simulations.

4   TRENDS AND OUTLOOK

When we accepted to contribute this chapter, we had no clear idea of the wide variety of MD methods and the range of applications, from perfect crystal properties to the interactions between dislocations or wafer bonding. Clearly, this chapter does not give credit everywhere credit is due and we wish to apologize to those whose work was not even mentioned. We hope to have covered the key methodologies and many examples of applications. This should give the reader the tools needed to pursue a literature search in a given area with a critical eye.

It is very obvious that the field of MD simulations is rapidly expanding, in terms of the range of methods used, of materials described, and of applications. There is no doubt in our minds that this powerful and appealing tool will continue to be used to study an ever-wider range of issues.

Among the capabilities of MD simulations and the search for global as well as local minima of any (or almost any) potential energy surface, the calculation of diffusion paths and diffusivities, of vibrational modes (local or non-local), of defect reactions, etc. Further, thermodynamic quantities such as entropies are being calculated. Most MD methods are highly versatile, can handle finite electronic temperatures, and large numbers of degrees of freedom.

Linear scaling methods allow studies requiring very large numbers of time steps and/or very large systems. Maybe as many as one million host atoms or one million time steps can be handled.

However, the euphoria should be accompanied by a strong dose of realism. The results of MD simulations are extremely visual, intuitive, and appealing. Often, one sees an actual 'movie' showing atoms vibrating around, impurities diffusing, molecules bouncing off of surfaces, and it is tempting to believe more than one should. All the methods have approximations, most have many, important, approximations. Regardless of how beautiful, some results may be wrong - simply because the heavy approximations used to solve efficiently the electronic part of the problem do not allow for an accurate description of the interactions taking place.

Some example are as follows. Empirical potentials are suspect, especially when applied to situations for which they were not parameterized. Very efficient methods can handle huge system, but the quantum mechanics is so heavily approximated that only the high-temperature behavior is approximately correct. On the other hand, *ab-initio* codes are much more trustworthy when it comes to the details of a potential surface, but suffer from the restriction to small cells and/or insufficient k-point sampling in the first Brillouin zone. In fact, it is not unusual to see various authors using the same *ab-initio* method to predict different potential surfaces for an impurity [22] (to get the details of the chemistry right, large basis-set ab-initio HF are still is the best, because the chemical analysis is based on one-electron atomic-like orbitals, not on total densities). Another issue is that classical MD simulations are not capable of describing any quantum aspect of atomic (or ionic) motion.

Another fundamental limitation is the amount of real time that can be described. It is likely that the fastest techniques will achieve one nanosecond, but it is not likely that this will be within the reach of *ab-initio* techniques any time soon. Tricks will be needed, such as assumed adiabatic trajectories for diffusion, kicks of various types, or other inventive ways to speed up reactions too slow to be studied.

However, who can tell if, some day, the growth of a realistic chunk of silicon will actually be simulated at the *ab-initio* level, showing the precipitation of 'A' or 'D' defects, and the formation of an OSF ring? Who knows if the old issue of O-related thermal donors in Si will not be resolved once and for all by MD simulations?
Nah, these are just dreams...

Acknowledgments

The work is SKE is supported by the grant D-1126 from the R.A. Welch Foundation and the contract RAD-7-17652-01 from the National Renewable Energy Laboratory. The work of PAF is supported in part by the NSF under the grant DMR 93-05344. The authors thank D.A. Drabold for help with Sec. I.

References

1. This section relies on D.A. Dabold, *Electronic Structure Methods with Applications to Amorphous Semiconductors*, in *Amorphous Insulators and Semiconductors*, ed. M.F. Thorpe and M.I. Mitkova (Kluwer, Dordrecht, 1997).
2. M. Born and K. Huang, *Dynamical Theory of Crystal Lattices*, (Oxford Univ. Press, Clarendon, 1954).
3. A.M. Stoneham, *J. Chem. Soc. Faraday Trans.* **86**, 1215 (1990).
4. F.H. Stillinger and T.A. Weber, *Phys. Rev.* B **31**, 5262 (1985).
5. J. Tersoff, *Phys. Rev. Lett.* **56**, 632 (1986) and *Phys. Rev.* B **37**, 6991 (1988).
6. A.E. Carlsson, P.A. Fedders, and C.W. Myles, *Phys. Rev.* B **41**, 1696 (1990).
7. A.E. Carlsson, *Sol. St. Phys.* **43**, ed. H. Ehrenreich and D. Turnbull (Academic, New York, 1990) p1.
8. F. Ercolessi and J.B. Adams, *Europhys. Lett.* **26**, 583 (1994).
9. W. Harrison, *Electronic Structure*, (Freeman, San Fransisco, 1980).
10. P. Vogel, H.P. Hjalmarson, and J.D. Dow, *J. Phys. Chem. Sol.* **44**, 365 (1983).
11. P.A. Fedders, D.A. Drabold, and S. Klemm, *Phys. Rev.* B **45**, 4048 (1992).

12. R.P. Feynman, *Phys. Rev.* **56**, 340 (1939).
13. P.A. Fedders and A.E. Carlsson, *Phys. Rev.* B **39**, 1134 (1989).
14. D. Allen and J.D. Joannopoulos, in *Hydrogenated Amorphous Silicon II*, ed. J.D. Joannopoulous and G. Lucovsky (Springer-Verlag, Berlin, 1984) p.5.
15. P.A. Fedders and A.E. Carlsson, *Phys. Rev.* B **37**, 8506 (1988).
16. B.N. Davidson, PhD thesis, North Carolina State University, 1992.
17. L. Goodwin, A.J. Skinner, and D.G. Pettifor, *Europhys. Lett.* **9**, 701 (1989); J. Mercer and M.Y. Chou, *Phys. Rev.* B **47**, 9366 (1993).
18. C.H. Xu, C.Z. Wang, C.T. Chan, and K.M. Ho, *J. Phys. Cond. Mat.* **4**, 6047 (1992).
19. Q. Li and R. Biswas, *Phys. Rev.* B **50**, 18090 (1994).
20. M. Menon and K.R. Subbaswamy, *Phys. Rev.* B **47**, 12754 (1993) and **48**, 8398 (1993).
21. L.M. Canel, A.E. Carlsson, and P.A. Fedders, *Phys. Rev.* B **48**, 10739 (1993).
22. S.K. Estreicher, *Mat. Sci. Engr.* R **14**, 319 (1995).
23. P. Hohenberg and W. Kohn, *Phys. Rev.* B **136**, 864 (1964); W. Kohn and L.J. Sham, *Phys. Rev.* A **140**, 1133 (1965).
24. D.M. Ceperly and G.J. Adler, *Phys. Rev. Lett.* **45**, 566 (1980).
25. G.B. Bachelet. D.R. Hamann, and M. Schlüter, *Phys. Rev.* B **26**, 4199 (1982).
26. D. Vanderbuilt, *Phys. Rev.* B **41**, 7892 (1990).
27. O.F. Sankey and D.J. Niklewski, *Phys. Rev.* B **40**, 3979 (1989).
28. J. Harris, *Phys. Rev.* B **40**, 1770 (1985).
29. W.M.C. Foulkes and R. Haydock, *Phys. Rev.* B **39**, 12520 (1989).
30. A.A. Demkov, J. Ortega, O.F. Sankey, and M.P. Grumbach, *Phys. Rev.* B **52**, 1618 (1995).
31. P. Ordejón, E. Artacho, and J.M. Soler, *Phys. Rev.* B **53**, 10441 (1996).
32. R. Car and M. Parrinello, *Phys. Rev. Lett.* **55**, 2471 (1985).
33. R. Stumpf and M. Scheffler, *Comp. Phys. Comm.* **79**, 447 (1994).
34. M. Brockstedte, A. Kley, and M. Scheffler, *Comp. Phys. Comm.* (to be published).
35. P. Fulde, *Electron Correlations in Molecules and Solids*, (Springer-Verlag, Berlin, 1993).
36. J. Perdew, *Int. J. Quant. Chem.* **57**, 309 (1996).
37. D.R. Hamann, *Phys. Rev. Lett.* **76**, 660 (1996).
38. L. Mitas in *Electronic Properties of Solids Using Cluster Methods*, ed. T.A. Kaplan and S.D. Mahanti, (Plenum, New York, 1995), p. 151.

39. D.S. Wallace, A.M. Stoneham, W. Hayes, A.J. Fisher, and A.H. Harker, *J. Phys. Cond. Matter* **3**, 3879 (1991).
40. A. Mainwood, *Mat. Sci. Forum* **143-147**, 51 (1994).
41. S. Goedecker and M. Teter, *Phys. Rev. B* **51**, 9455 (1995).
42. L. Colombo, *Ann. Rev. Comp. Phys.* **IV**, 147 (1996).
43. C. Lanczos, *Applied Analysis* (Prentice Hall, New York, 1956).
44. J. Skilling, *Maximum Entropy and Bayesian Methods*, (Kluwer, Dordrecht, 1989), p. 455.
45. B. Strohmaier, S.M. Grimes, and S.D. Bloom, *Phys. Rev.* C **32**, 1397 (1985).
46. P. Ordejón, D.A. Drabold, R.M. Martin, and M.P. Grumbach, *Phys. Rev. B* **51**, 1456 (1995) and references therein.
47. C. Wang, Q.-M. Zhang, and J. Bernholc, *Phys. Rev. Lett.* **69**, 3789 (1992).
48. K. Muro and A.J. Sievers, *Proc.* 18th *ICPS*, ed. O. Engström (World Scientific, Singapore, 1986), p. 891.
49. Y.M. Cheng and M. Stavola, *Phys. Rev. Lett.* **73**, 3419 (1994) and *Sol. St. Com.* **93**, 431 (1995).
50. C. He and S.K. Estreicher, unpublished.
51. D.A. Drabold, R. Wang, S. Klemm, O.F. Sankey, and J.D. Dow, *Phys. Rev. B* **43**, 5132 (1991).
52. G. Galli, R.M. Martin, R. Car, and M. Parrinello, *Phys. Rev. Lett.* **63**, 988 (1989).
53. S. Goedecker and L. Colombo, *Phys. Rev. Lett.* **73**, 122 (1994).
54. C.Z. Wang, C.T. Chan, and K.M. Ho, *Phys. Rev. B* **42**, 11276 (1990).
55. S.A. Kajihara, A. Antonelli, J. Bernholc, and R. Car, *Phys. Rev. Lett.* **66**, 2010 (1991).
56. A. Mainwood and A.M. Stoneham, *J. Phys. Cond. Mat.* **6**, 4917 (1994).
57. A. Mainwood, *Phys. Rev. B* **49**, 7934 (1994).
58. I. Kiflawi, A. Mainwood, H. Kanda, and D. Fisher, *Phys. Rev. B* **54**, 16719 (1996).
59. M. Ishimaru, K. Yoshida, T. Kumamoto, and T. Motooka, *Phys. Rev. B* **54**, 4638 (1996).
60. J.Q. Broughton and X.P. Li, *Phys. Rev. B* **35**, 9120 (1987).
61. I. Stich, R. Car, and M. Parrinello, *Phys. Rev. Lett.* **63**, 2240 (1989).
62. S. Nosé, *Mol. Phys.* **52**, 255 (1984) and *J. Chem. Phys.* **81**, 511 (1984).
63. F. Buda, R. Car, and M. Parrinello, *Phys. Rev. B* **41**, 1680 (1990).
64. O. Sugino and R. Car, *Phys. Rev. Lett.* **74**, 1823 (1995).
65. P.L. Silvestrelli, A. Alavi, M. Parrinello, and D. Frenkel, *Phys. Rev. Lett.* **77**, 3149 (1996).

66. D. Stock, M. Nitschke, K. Gärtner, and T. Kandler, *Rad. Eff. Def. Sol.* **130**, 67 (1994).
67. R. Smith, D.E. Harrison, Jr., and B.J. Garrison, *Phys. Rev.* B **40**, 93 (1989).
68. T. Diaz de la Rubia and G.H. Gilmer, *Phys. Rev. Lett.* **74**, 2507 (1995).
69. M.J. Cartula, T. Diaz de la Rubia, and G.H. Gilmer, *Nucl. Instr. Meth. Phys. Res.* B **106**, 1 (1995).
70. K. Scheerschmidt, D. Conrad, and U. Gösele, *Comp. Mat. Sci.* **7**, 40 (1996).
71. K. Scheerschmidt, S. Ruvimov, P. Werner, A. Höpner, and J. Heydenreich, *J. Microsc.* **179**, 214 (1995).
72. D. Conrad, K. Scheerschmidt, and U. Gösele, *Appl. Phys.* A **62**, 7 (1996).
73. A.Yu. Belov, D. Conrad, K. Scheerschmidt, and U. Gösele, *Phil. Mag.* B (in print).
74. F. Buda, G.L. Chiarotti, R. Car, and M. Parrinello, *Phys. Rev. Lett.* **63**,4294 (1989).
75. D.E. Boucher and G.G. DeLeo, *Phys. Rev.* B **50**, 5247 (1994).
76. G. Panzarini and L. Colombo, *Phase Trans.* **52**, 137 (1994) and *Phys. Rev. Lett.* **73**, 1636 (1994).
77. Y.K. Park, S.K. Estreicher, and P.A. Fedders, in *Early Stages of Oxygen Precipitation in Silicon*, ed. R. Jones (Kluwer, Dordrecht, 1996), p.179.
78. Ch. Langpape, S. Fabian, Ch. Klatt, and S. Kalbitzer, *Appl. Phys.* A **64**, 207 (1997).
79. C.Z. Wang, C.T. Chan, and K.M. Ho, *Phys. Rev. Lett.* **66**, 189 (1991).
80. M.A. Roberson and S.K. Estreicher, *Phys. Rev.* B **49**, 17040 (1994).
81. T. Sinno, Z.K. Jiang, and R.A. Brown, *Appl. Phys. Lett.* **68**, 3028 (1996).
82. R. Car, P.E. Blöchl, and E. Smargiassi, *Mat. Sci. Forum* **83-87**, 433 (1992).
83. P.E. Blöchl, E. Smargiassi, R. Car, D.B. Laks, W. Andreoni, and S.T. Pantelides, *Phys. Rev. Lett.* **70**, 2435 (1993).
84. M. Tang, L. Colombo, J. Zhu, and T. Diaz de la Rubia, *Phys. Rev.* B (in print).
85. M. Tang, L. Colombo, and T. Diaz de la Rubia, *MRS Proc.* **396**, 33 (1996).
86. A. Mainwood, *Mat. Sci. Forum* **196-201**, 1589 (1995).
87. A. Mainwood, *Proc. 7th Int. Conf. Shallow-Level Defects in Semic.*, ed. C.A.J. Ammerlaan and B. Pajot (World Scientific, Singapore, 1997), p.523.

88. A. Mainwood, *Proc. 19th Int. Conf. Defects Semic.*, ed. M.H. Nazare, Mat. Sci. Forum (in print).

89. Y.K. Park, S.K. Estreicher, C.W. Myles, and P.A. Fedders, *Phys. Rev.* B **52**, 1718 (1995).

90. S.K. Estreicher, J.L. Hastings, and P.A. Fedders, *Appl. Phys. Lett.* **70**, 432 (1997); J.L. Hastings, S.K. Estreicher, and P.A. Fedders, *Phys. Rev.* B (in print).

91. V. Milman, M.C. Payne, V. Heine, R.J. Needs, J.S. Lin, and M.H. Lee, *Phys. Rev. Lett.* **70**, 2928 (1993).

92. A. Bongiorno, L. Colombo, and F. Cargnoni, *Chem. Phys. Lett.* (in print).

93. M.C. Payne, P.D. Bristowe, and J.D. Joannopoulos, *Phys. Rev. Lett.* **58**, 1348 (1987).

94. G. Kresse and J. Hafner, *Phys. Rev.* B **49**, 14251 (1994).

95. N. Takeuchi and I.L. Garzón, *Phys. Rev.* B **50**, 8342 (1994).

96. V. Godlevsky, J.R. Chelikowski, and N. Troullier, *Phys. Rev.* B **52**, 13281 (1995).

97. R.V. Kulkarni, W.G. Aulbur, and D. Stroud, *Phys. Rev.* B **55**, 6896 (1997).

98. S.K. Estreicher and D.E. Boucher in *GaN and Related Materials*, ed. S.J. Pearton (Gordon and Breach, New York, 1997), p.171.

99. P. Bogusławski, E.L. Briggs, and J. Bernholc, *Phys. Rev.* B **51**, 17255 (1995).

100. P. Bugosłwaski, E.L. Briggs, T.A. White, M.G. Wensell, and J. Bernholc, *MRS Proc.* **339**, 693 (1994).

101. J. Neugebauer and C.G. Van de Walle, *Phys. Rev.* B **50**, 8067 (1994).

102. J. Neugebauer and C.G. Van de Walle, *Phys. Rev. Lett.* **75**, 4452 (1995); *MRS Proc.* **339**, (1994); *MRS Proc.* **378**, 503 (1995).

103. J. Neugebauer and C.G. Van de Walle, *MRS Proc.* **423**, 619 (1996).

104. P. Bogusławski, E.L. Briggs, and J. Bernholc, *Appl. Phys. Lett.* **69**, 233 (1996); J. Bernholc, P. Bogusławski, E.L. Briggs, M. Bongiorno Nardelli, B. Chen, K. Rapcewicz, and Z. Zhang, *MRS Proc.* **423**, 465 (1996).

105. P. Stumm and D.A. Drabold, *MRS Proc.* **449**, 941 (1997).

106. D.E. Boucher G.G. DeLeo, and W.B. Fowler, *Phys. Rev.* B (in print).

107. S. Serra, L. Miglio, and V. Fiorentini, *MRS Proc.* **395**, 435 (1996).

108. M. Bockstedte and M. Scheffler, *Z. Phys. Chemie* (1997). **69**, 3789 (1992).

109. B. Chen, Q.-M. Zhang, and J. Bernholc, *Phys. Rev.* B **49**, 2985 (1994).

110. K.K. Gleason, M.A. Petrich, and J.A. Reimer, *Phys. Rev.* B **36**, 3259 (1987).

111. P.A. Fedders, *J. Non-Cryst. Sol.* **137-138**, 141 (1991).
112. F. Wooten, K. Winer, and D. Weaire, *Phys. Rev. Lett.* **54**, 1392 (1985); F.Wooten and D. Weaire, *Sol. St. Phys.* **40**, 2 (1991).
113. P.A. Fedders and D.A. Drabold, *Phys. Rev.* B **47**, 13277 (1993).
114. L. Guttman, *Phys. Rev.* B **23**, 1866 (1981); L. Guttman and C.F. Fong, *ibid.* **26**, 6756 (1982).
115. N. Mousseau and L.J. Lewis, *Phys. Rev.* B **43**, 9810 (1991) and **41**, 3702 (1990).
116. F. Buda, G.L. Chiarotti, R. Car, and M. Parrinello, *Phys. Rev.* B **44**, 5908 (1991).
117. J.M. Holender, G.J. Morgan, and R. Jones, *Phys. Rev.* B **47**, 3991 (1993).
118. B. Tuttle and J.B. Adams, *Phys. Rev.* B **53**, 16265 (1996).
119. P.A. Fedders and D.A. Drabold, *Phys. Rev.* B **53**, 3841 (1996).
120. P.A. Fedders and D.A. Drabold, unpublished.
121. D.S. Marynick and S.K. Estreicher, *Chem. Phys. Lett.* **132**, 383 (1986).
122. J.R. Dennison, M.W. Holtz, and G.M. Swain, *Spectroscopy* **11**, 38 (1996).
123. D.R. McKenzie, D.A. Muller, and B.A. Pailthorpe, *Phys. Rev. Lett.* **67**, 773 (1991).
124. J. Robertson, *Phil. Mag.* B **47**, L33 (1983) and *Adv. Phys.* **35**, 317 (1986).
125. J. Tersoff, *Phys. Rev.* B **44**, 12039 (1991).
126. B.R. Djordjevic, M.F. Thorpe, and F. Wooten, *Phys. Rev.* B **52**, 5685 (1995).
127. Th. Fraunheim, P. Blaudeck, U. Stephen, and G. Jungnickel, *Phys. Rev.* B **48**, 4823 (1993); U. Stephan, Th. Fraunheim, P. Blaudeck, and G. Jungnickel, *ibid.* **49**, 1489 (1994).
128. C.Z. Wang and K.M. Ho, *Phys. Rev. Lett.* **71**, 1184 (1993).
129. G. Galli, R.M. Martin, R. Car, and M. Parrinello, *Phys. Rev.* B **42**, 7470 (1990).
130. D.A. Drabold, P.A. Fedders, and P. Stumm, *Phys. Rev.* B **49**, 16415 (1994); D.A Drabold, P.A. Fedders, and M.P. Grumbach, *Phys. Rev.* B **54**, 5480 (1996).
131. D.A. Drabold, P.A. Fedders, S. Klemm, and O.F. Sankey, *Phys. Rev. Lett.* **67**, 2179 (1991).
132. S.T. Pantelides, *Phys. Rev. Lett.* **57**, 2979 (1986).
133. S.T. Pantelides, *Phys. Rev. Lett.* **58**, 1344 (1987); *Phys. Rev.* B **36**, 3479 (1987).
134. Q. Li and R. Biswas, *Phys. Rev.* B **52**, 10705 (1995); *Appl. Phys. Lett.* **68**, 2261 (1996).

135. P.A. Fedders, *J. Non-Cryst. Sol.* **198-200**, 56 (1996).

136. S. Lanzavecchia and L. Colombo, *Mat. Sci. Engr.* B **36**, 264 (1996).

137. S. Serra, M. Manfredini, P. Milani, and L. Colombo, *Chem. Phys. Lett.* **238**, 281 (1995).

138. C. Moleni, L. Colombo, and L. Miglio, *Phys. Rev.* B **50**, 4371 (1994).

139. C. Molteni, L. Colombo, and L. Miglio, *Europhys. Lett.* **24**, 659 (1993).

140. P. Stumm and D.A. Drabold, *Phys. Rev. Lett.* **79**, 677 (1997).

Tight-Binding Molecular Dynamics Study of Structures and Dynamics of Carbon Fullerenes

C. Z. Wang, B. L. Zhang, and K. M. Ho

Ames Laboratory and Department of Physics and Astronomy,
Iowa State University, Ames, IA 50011

The discovery of buckminsterfullerene has stimulated considerable interest in carbon fullerene research in the past eight years. As a unique computational method, tight-binding molecular dynamics has been widely used to investigate the structures and properties of this new phase of carbon. In this paper, we will review some results and understanding on the structures and dynamics of carbon fullerenes obtained from the tight-binding molecular dynamics simulations.

1 Introduction

Emerging as a new phase of carbon that is distinct from diamond and graphite, fullerene has attracted considerable interest from scientists in the fields of chemistry, physics and materials science. The term fullerene describes a closed network of carbon atoms that forms a cage structure consisting of 12 five-membered rings and $(N/2 - 10)$ six-membered rings, where N is the number of carbon atoms in the cluster. The prototypical fullerene, the "Buckminster-fullerene" or C_{60} in which the carbon atoms are arranged at the 60 vertices of a truncated icosahedron, was first introduced in 1985 by Kroto, Heath, O'Brien, Curl, and Smalley [1] to account for the exceptional stability of the C_{60} species in the carbon cluster beams produced by laser vaporization of graphite. This elegant soccerball-like structure was verified in 1990 [2,3] following the breakthrough of Krätschmer, Lamb, Fostiropoulos and Huffman (KLFH) [4] in producing macroscopic quantities of C_{60} and C_{70}. Besides C_{60} and C_{70}, the mass spectrum of carbon cluster beams also indicate presence of even-numbered carbon clusters in the range from thirty to several hundred atoms in carbon vapor [1,5,6,7,8]. It was conjectured that these clusters are also in the form of fullerenes [6,7,8,9,10]. Following the method of KLFH, Diederich and Whetten and their collaborators [11] in 1991 obtained clear evidence for the presence of fullerenes larger than C_{70}. Since then, the chromatographic separation of the toluene-soluble soot have resulted in the isolation of higher fullerenes such as C_{76}, C_{78}, C_{82}, C_{84}, C_{90}, C_{94}, and C_{96} [11,12,13,14,15,16]. The success in the discovery and isolation of large fullerenes received a lot of experimental and theoretical attention. Considerable efforts have been devoted to understanding the structural and properties of this new form of material.

In the past several years, we have performed extensive tight-binding molecular dynamics calculations and simulations to study the structures and properties of carbon fullerenes. We have successfully determined the fullerene structures of every even-number carbon clusters ranging from C_{20} to C_{102}. We have also investigated the dynamical behavior of the carbon fullerenes as a function of temperature or under different collision conditions. Our studies have provided some useful insight into the structure and stability of this novel form of carbon.

In this article, we will review our studies on carbon fullerenes using the tight-binding molecular dynamics method. The paper is organized as follows: In section 2, we present the tight-binding model for describing the interatomic interaction of carbon systems. The methods employed to determine the structures of fullerenes are discussed in section 3. The optimized fullerene structures and their stabilities in the range C_{20} to C_{102} are presented and discussed in sections 4 and 5 respectively. In section 6, the dynamical behavior of the fullerenes, such as the formation and disintegration process and the behavior under collision are discussed. The study of solid C_{60} under high pressure compression is reviewed in Section 7. Finally, concluding remarks are given in Section 8.

2   Tight-Binding Potential for Carbon

Tight-binding molecular dynamics (TBMD) is a recent developed simulation method which incorporates electronic structure effects into molecular dynamics through an empirical tight-binding Hamiltonian H_{TB}. The quantum mechanical many-body nature of the interatomic forces is taken into account naturally through the Hellmann-Feynman theorem. Since the scheme usually uses a minimal basis set for the electronic structure calculation and the Hamiltonian matrix elements are parametrized, larger numbers of atoms can be tackled within the present computer capabilities. One of the distinctive features of this scheme in comparison with other empirical schemes is that all the parameters in the model can be obtained theoretically. It is therefore very useful for studying novel materials where experimental data are not readily available. The scheme has been demonstrated to be a powerful method for studying various structural, dynamical and electronic properties of covalent systems [17].

In our study of carbon fullerene, the tight-binding potential describing the interactions between the carbon atoms is taken from our previous work [18]. The total energy in this model is expressed as:

$$E(\{\vec{r}_i\}) = \sum_n^{occupied} < \psi_n | H_{TB}(\{\vec{r}_i\}) | \psi_n > + E_{rep}(\{\vec{r}_i\}) \tag{1}$$

The first term in (1) is the electronic energy calculated by a parametrized tight-binding Hamiltonian $H_{TB}(\{\vec{r}_i\})$, and the second term is a short-ranged repulsive energy representing the ion-ion repulsion and the correction to the double counting of the electron-electron interaction in the first term. The orthogonal sp^3 basis tight-binding Hamiltonian $H_{TB}(\vec{r}_i)$ is described by two on-site atomic energies ε_s and ε_p and four hopping parameters $v_{ss\sigma}(r_{i,j}), v_{sp\sigma}(r_{i,j}), v_{pp\sigma}(r_{i,j})$ and $v_{pp\pi}(r_{i,j})$. The repulsive energy is in the form of $E_{rep} = \sum_i f[\sum_j \phi(r_{i,j})]$ where $\phi(r_{i,j})$ is a pairwise interaction and f is a function with argument $\sum_j \phi(r_{i,j})$. The dependence of the TB hopping parameters and the pairwise potential on the interatomic separation is given by:

$$h_\alpha(r) = h_\alpha(r_0)(\frac{r_0}{r})^n exp[n\{-(\frac{r}{r_c})^{n_c} + (\frac{r_0}{r_c})^{n_c}\}] \tag{2}$$

and

$$\phi(r) = \phi_0(\frac{d_0}{r})^m exp[m\{-(\frac{r}{d_c})^{m_c} + (\frac{d_0}{d_c})^{m_c}\}] \tag{3}$$

where r_0 denotes the nearest-neighbor atomic separations in diamond and $h_\alpha(r_0), n, n_c, r_c, \phi_0, m, d_0, d_c,$ and m_c are parameters determined by fitting to first-principles local-density-functional calculations results of band structures and cohesive energies of crystalline carbon polytypes. Details of the parameters can be found in Ref[18].

This tight-binding model not only reproduces the binding energies and bond lengths of crystalline carbon with different coordination numbers (see Fig. 1), but also describes well the properties of the carbon in disordered amorphous and liquid states[18,19,20]. The reliability of this potential for cluster calculations is further tested by studying the ground-state geometries of small clusters and C_{60}. The cohesive energy as a function of carbon cluster size for linear-chain, ring, and graphite fragments in the range of 2 to 20 atoms obtained from our tight-binding model are shown in Fig. 2. In the range $5 \leq n \leq 11$, we found that odd-numbered clusters prefer a linear structure, while even-numbered clusters prefer a ring structure, in good agreement with ab-initio quantum mechanical calculations [18,21]. The fully relaxed C_{60} molecule obtained by the present tight-binding potential has iscosahedral symmetry and bond lengths of 1.40 Å and 1.46 Å respectively for the double and single bonds, which agree very well with the first-principles density functional calculation results of 1.40 Å and 1.45 Å. Our tight-binding calculation also yields a cohesive energy of 0.41 eV per atom relative to graphite and HOMO-LUMO energy separation of 1.61 eV. These results agree very well with the LDA results of 0.40 eV per atom and 1.71 eV respectively. Typical accuracy of the present tight-binding

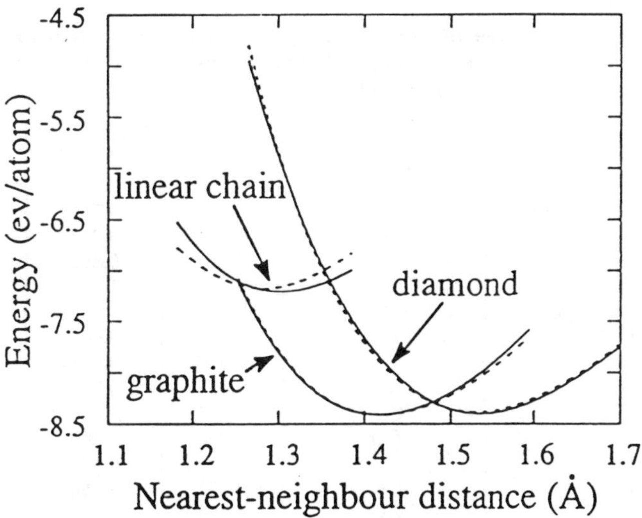

Figure 1: Cohesive energies as a function of nearest-neighbor distance for carbon in diamond, graphite and linear chain structures. Solid and dashed lines show the results from tight-binding and LDA calculations respectively

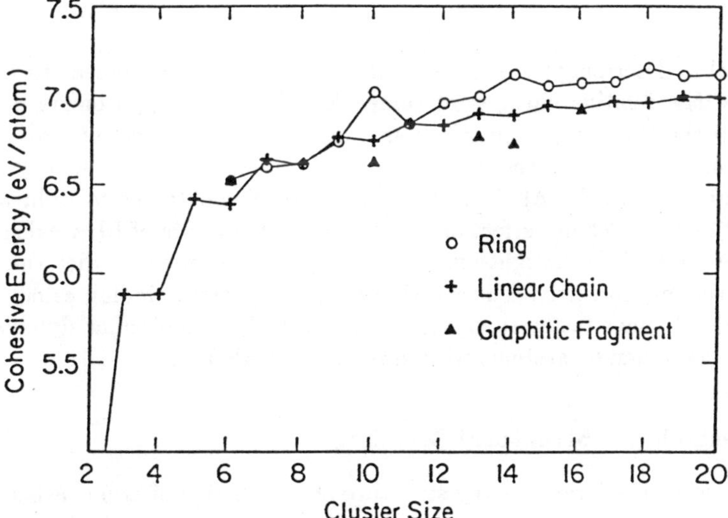

Figure 2: Cohesive energy of carbon linear-chain, carbon ring and graphite fragment structures as a function of cluster size from n=2 to 20.

Table 1: The relative energies of some C_{78}, C_{82}, and C_{84} fullerene isomers obtained by present tight-binding potentials are compared with the results from the LDA calculations. The energies are in the unit of eV/molecule

Cluster Size	Isomer Symmetry	ΔE (TB)	ΔE (LDA)
C_{78}			
	$C_{2v'}$	0.000	0.000
	$D_{3h'}$	0.087	0.160
	C_{2v}	0.284	0.210
	D_3	0.324	0.361
	D_{3h}	0.913	1.200
C_{82}			
	C_2	0.000	0.000
	C_s	0.076	0.123
	$C_{2'}$	0.166	0.312
	$C_{2''}$	0.191	0.418
	$C_{s'}$	0.226	0.385
	$C_{s'}$	0.229	0.467
	C_{2v}	0.285	0.656
	C_{3v}	0.566	1.287
	$C_{3v'}$	0.731	1.025
C_{84}			
	D_2-22	0.000	0.000
	D_{2d}-23	0.033	0.025
	C_2-11	0.277	0.370
	D_2-1	1.915	2.150

potential in describing the energetics of large fullerenes can be seen from Table 1 where the relative energies of some C_{78}, C_{82}, and C_{84} isomers obtained from our tight-binding calculations are compared with the results from the ab initio calculations using the density functional method within the local density approximation (LDA). Although the energy ordering obtained by tight-binding calculations are not in perfect agreement with the results of LDA calculations (see the case of C_{82}), tight-binding calculations seem to predict the lower-energy isomers correctly. The speed of the present tight-binding calculation is hunderds of times faster than regular first-principles molecular dynamics, so we can easily handle systems with several hundreds atoms.

3   Methods for Structural Searches

To determine the lowest-energy structure of a collection of atoms using molecular dynamics or Monte Carlo computer simulations is a very challenging problem. Because the number of local energy minima grows exponentially with the number of atoms, the computational effort to obtain lowest-energy structure of a cluster also increases exponentially.

Traditionally, simulated annealing techniques are used to search for ground-state structures by minimizing the total potential energy of N atoms with respect to the $3N$ coordinate degrees of freedom of the cluster. This approach for carbon fullerenes requires an accurate and efficient interatomic potential for carbon that allows sampling of many possible configurations. There exists, however, a serious problem in the implementation of this approach due to the strong directional bonding between carbon atoms which creates large energy barriers between various metastable structures. Although this method has been used to search for the ground-state structures of carbon fullerenes by several groups [21,22,23,24], it was found that for cluster sizes larger than thirty, the system is invariably trapped in metastable structures [21]. Without invoking additional "constraints", this approach is not capable of producing ground-state structures for fullerenes larger than C_{30}.

To resolve this problem, we have devised a new simulated annealing scheme for generating energetically favorable structures for large fullerenes [25,26]. Instead of looking at the network connecting individual atoms, our new scheme focuses on a "face-dual network" which links the centers of polygonal faces of the cage structure. Since each carbon atom in a fullerene molecule is three-fold coordinated, the face-dual network is a triangular mesh. This face-dual system can be easily generated by the method of simulated annealing. The search for the ground-state structure of the fullerenes is therefore performed in three steps. Firstly, plausible face-dual networks are generated using a simulated annealing method; Secondly, the face-dual networks are inverted to obtain the structures of the candidate fullerenes; Finally, the ground-state structures are selected by total energy optimization using tight-binding molecular dynamics, we call this scheme the "face-dual"-TBMD scheme.

According to Euler's theorem, the number of polygonal faces is $(N/2+2)$[27] for a fullerene molecule of N atoms. To generate the face-dual network for carbon clusters of a given size, we represent each face of the molecule by a point lying on a fixed ellipsoidal surface. These points are initially placed in random positions and allowed to interact with each other via a two-body repulsive potential. The energetically favorable face-dual network is found by simulated annealing using molecular dynamics simulation. In such a scheme, whether a face is pentagonal or hexagonal is determined by the number of neighboring faces. Due to the nature of the two-body repulsive interaction, the energy barriers for changing hexagonal faces to pentagonal or vice versa are very small, and thus even for quite large number of faces, the system does not get trapped in metastable states. The shapes of the ellipsoidal constraining surface are varied to obtain various candidate networks for energetically favorable fullerene

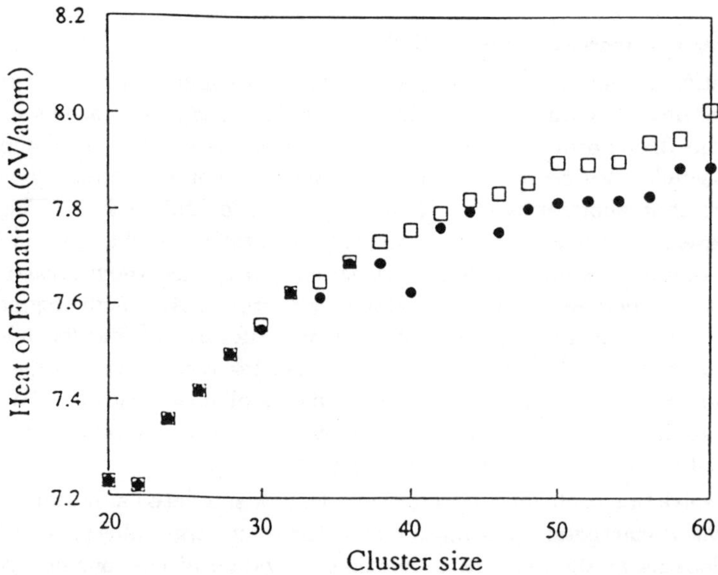

Figure 3: Formation energies of carbon fullerenes obtained from the use of the "face-dual" network scheme (squares) are compare to those obtained by simulated annealing using the atomic network (solid circles). It is evident that the "face-dual" network scheme is much better in locating the lowest energy structures when cluster size larger than 30 atoms.

structures. Because of the repulsion between faces, the ground-state structures tend to distribute the faces as evenly as possible on the ellipsoidal surface and the resultant networks tend to separate the pentagonal faces far apart from one another. Fig. 3 compares the results of this new scheme with those obtained by the atomic network scheme [21] for C_{20} to C_{60}. It is evident that the new scheme is doing much better in locating the lowest energy structures.

The parameters that need to be adjusted in the face-dual network scheme are the shape of the predetermined surface and the range of the two-body repulsive potential. Although in principle the search should be performed for every possible constraining surface, we found that for fullerenes in the range from C_{20} to C_{102} as discussed in the following sections, confining the face-dual network to ellipsoidal surfaces is already quite sufficient. Further symmetry breaking of the surface, if it is energetically favorable, can be realized in the subsequent unconstrained tight-binding molecular-dynamics optimization process.

4 Search for the Ground State Structure of C_{84}

Our first application of the "face-dual"-TBMD scheme is to search for the ground state structure of C_{84}. C_{84} is the most abundant fullerene molecule after C_{60} and C_{70}. The structure of this fullerene had been a subject of extensive debate. C_{84} was isolated by the UCLA group in early 1991 and subsequently by Achiba's group in Japan. According to the "ring-spiral" algorithm proposed by Manolopoulos, C_{84} has 24 topologically distinct isomers satisfying the isolated-pentagon rule[28]. We have also generated the 24 distinct isolated-pentagon isomers using the face-dual network simulated annealing method described in the previous section. These isomers are subsequently optimized using the tight-binding energy model described in section II. The resultant structures are plotted in Fig. 4.

Among these 24 isomers, three candidates for the structure of C_{84} were first proposed by Fowler[29] based on the "leapfrog" method. These isomers have T_d, D_{6h} and helical-D_2 symmetries, respectively (number 20, 24, and 1 of Fig 4). The helical-D_2 isomer was favored because of its closed-shell electronic structure[29] and its similarity to the known helical structure of C_{76}[12].

In late 1991, we found that the helical-D_2 isomer is in fact energetically very unfavorable in comparison to other isomers[25,30]. The relative energies of the 24 isolated-pentagon isomers obtained from our tight-binding calculation are shown in Fig. 5 and Table 2. Our calculation predicted that two isomers, one with D_2, and the other with D_{2d} symmetry should be the most stable structures of C_{84}[25,30]. These predictions were further supported by subsequent first-principles density-functional calculations[31]. In May 1992, a detailed NMR spectrum was published by Kikuchi *et al.*[15]. They observed 32 NMR lines, which is consistent with a mixture of D_2 and D_{2d} isomers forming in a ratio of 2:1. Subsequent NMR studies by the UCLA[32,33] group and by Manolopoulos *at el.* in U. K.[16] also obtained very similar results. These experimental results confirm the prediction from our tight-binding calculation.

5 Structure of Fullerenes from C_{20} to C_{102}

Combining the face-dual network generating scheme with the tight-binding molecular dynamics scheme, we have also systematically studied the structure of every even-number carbon clusters ranging from C_{20} to C_{102}. The results are summarized in the following two subsections.

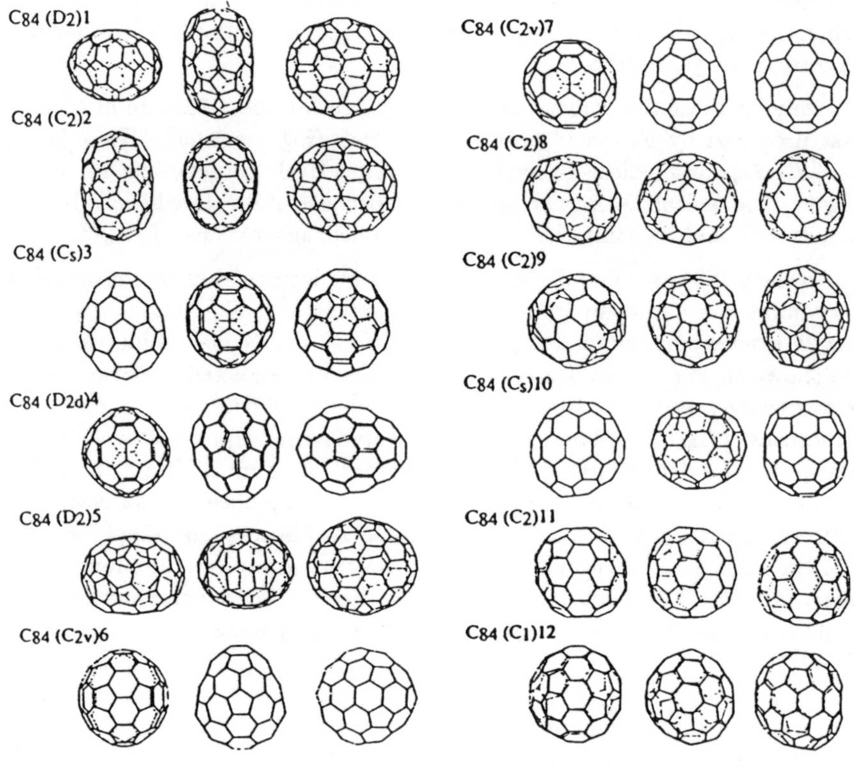

Figure 4: Fully optimized structures of the 24 topologically distinct isolated-pentagon isomer of C_{84}, viewing along the three orthogonal axes.

C84 (C2)13

C84 (Cs)14

C84 (Cs)15

C84 (Cs)16

C84 (C2v)17

C84 (C2v)18

C84 (D3d)19

C84 (Td)20

C84 (D2)21

C84 (D2)22

C84 (D2d)23

C84 (D6h)24

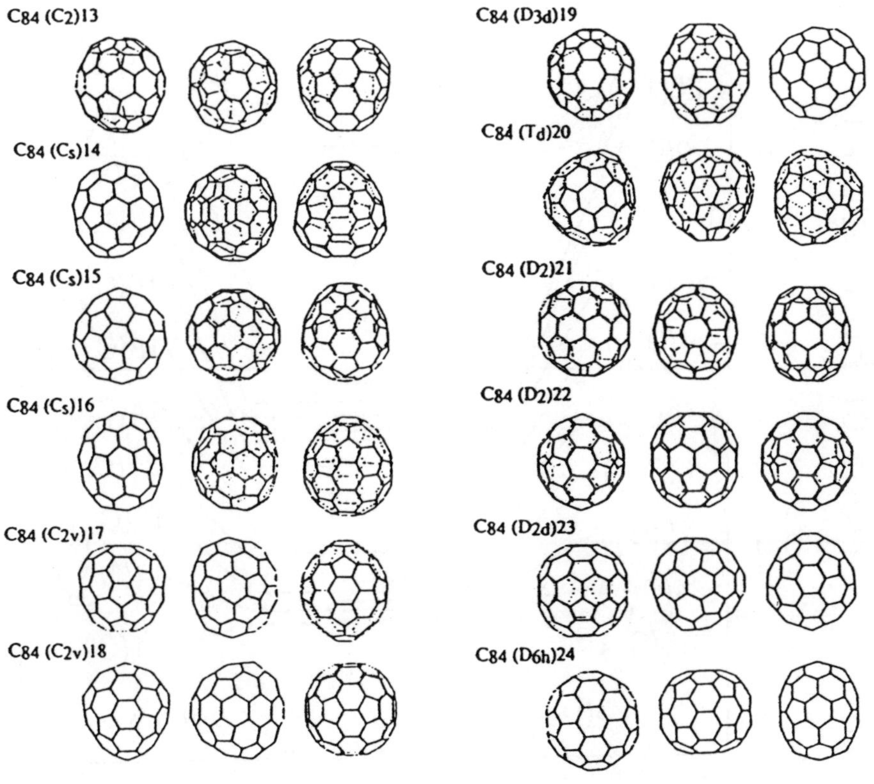

Figure 4: (*continued*)

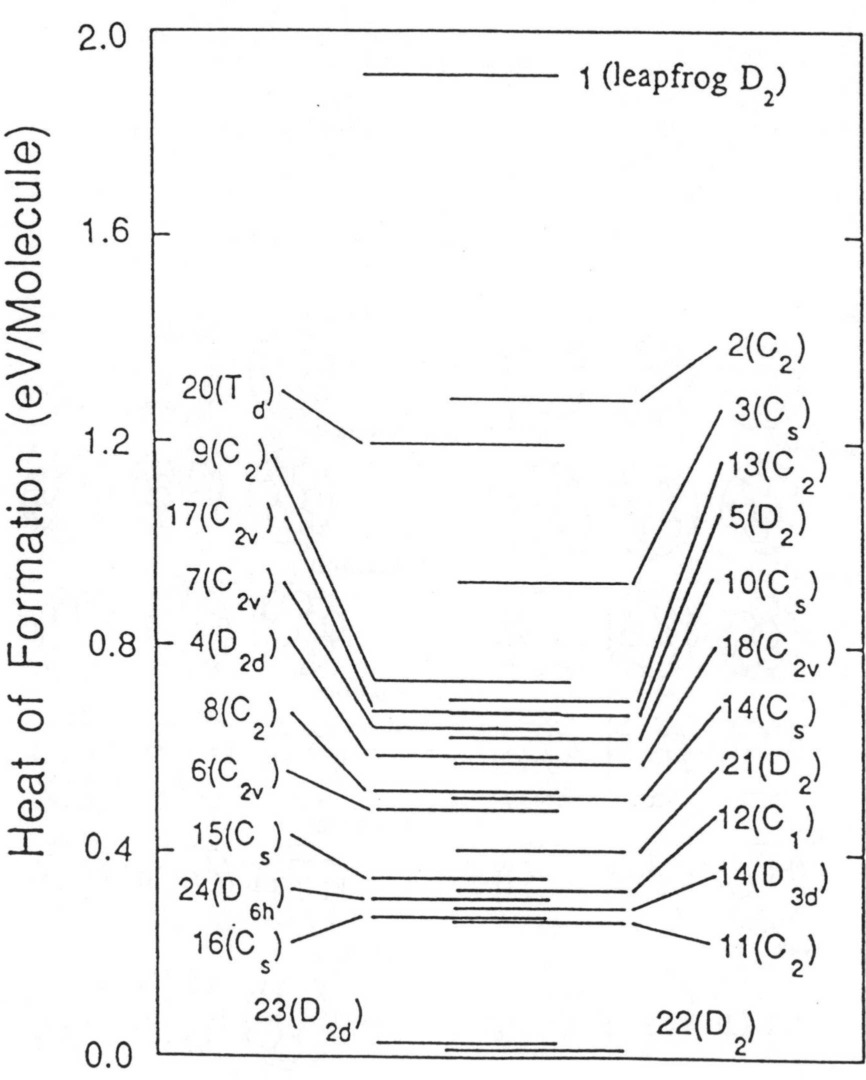

Figure 5: Heat of formation of the C$_{84}$ isomers relative to the ground-state isomer (in units of eV per molecule).

Table 2: The structures and energies of the isolated-pentagon isomers for C_{84} obtained from the tight-binding calculations. The values listed under Energy are the heat of formation of isomers relative to that of ground-state isomer in the units of eV per molecule. The HOMO-LUMO energy separations (in eV) are listed in the column HOMO-LUMO and the values in the column NMR are the numbers of distinct NMR lines expected. The order numbers follow the notation of Manolopoulos's work.

Order number	Symmetry	Energy	HOMO-LUMO	NMR
1	D_2	1.908	1.28	21
2	C_2	1.284	0.79	42
3	C_s	0.922	0.14	44
4	D_{2d}	0.642	0.82	12
5	D_2	0.677	0.61	21
6	C_{2v}	0.525	0.53	22
7	C_{2v}	0.680	0.49	23
8	C_2	0.594	0.34	42
9	C_2	0.743	0.17	42
10	C_s	0.656	0.22	46
11	C_2	0.273	0.66	42
12	C_1	0.332	0.53	84
13	C_2	0.689	0.31	42
14	C_s	0.534	0.89	43
15	C_s	0.358	0.55	44
16	C_s	0.290	0.81	43
17	C_{2v}	0.688	0.49	24
18	C_{2v}	0.606	0.82	22
19	D_{3d}	0.299	0.55	8
20	T_d	1.181	1.49	4
21	D_2	0.429	0.40	21
22	D_2	0.000	0.82	21
23	D_{2d}	0.033	0.84	11
24	D_{6h}	0.307	1.14	5

Table 3: Properties of C_{20} - C_{70}. The values listed under Energy are the heat of formation of cluster relative to bulk graphite in the units of eV per atom. If there are two entries for the symmetry, the first one is topological symmetry and the second one is real symmetry due to Jahn-Teller distortion. The HOMO-LUMO energy separations are listed in the column HOMO-LUMO and the values in the column NMR are the numbers of distinct NMR lines expected.

Cluster Size	Energy	Symmetry	HOMO-LUMO	NMR
20	1.180	$I_h(C_2$	0.015	10
24	1.050	$D_{6d}(C_2$	0.448	12
26	0.989	$C_{2v}(C_2)$	0.292	13
28	0.912	$T_d(C_1)$	0.243	28
30	0.850	C_{2v}	0.333	10
32	0.781	D_3	0.881	6
34	0.758	C_1	0.239	34
36	0.706	C_{2v}	0.526	10
38	0.673	C_2	0.708	19
40	0.641	D_2	0.950	10
42	0.614	D_3	0.787	7
44	0.589	D_2	0.732	11
46	0.573	C_2	0.577	23
48	0.552	C_2	0.195	24
50	0.509	D_{5h}	0.503	4
52	0.502	C_2	0.657	26
54	0.482	C_{2v}	0.412	16
56	0.467	D_2	0.665	14
58	0.453	$C_s(C_1)$	0.111	58
60	0.401	I_h	1.61	1
62	0.434	C_2	0.378	31
64	0.410	D_2	0.952	16
66	0.409	C_2	0.471	33
68	0.398	C_2	0.725	34
70	0.365	D_{5h}	1.103	5

5.1   Small fullerenes C_{20} to C_{70}

The ground-state structures of small fullerenes C_{20} to C_{70} obtained from our simulation are plotted in Fig. 6. A summary of our results on the cohesive energy, symmetry, HOMO-LUMO energy separation and number of distinct NMR lines for these fullerenes is given in Table 3. We found that except for C_{22}, all the ground-state structures consist of six-membered rings and twelve five-membered rings. Among all these structures, only C_{60} and C_{70} satisfy the isolated-pentagon rule. Most fullerenes, except C_{60}, C_{70} and C_{50}, have low symmetries and some of those have no symmetry at all.

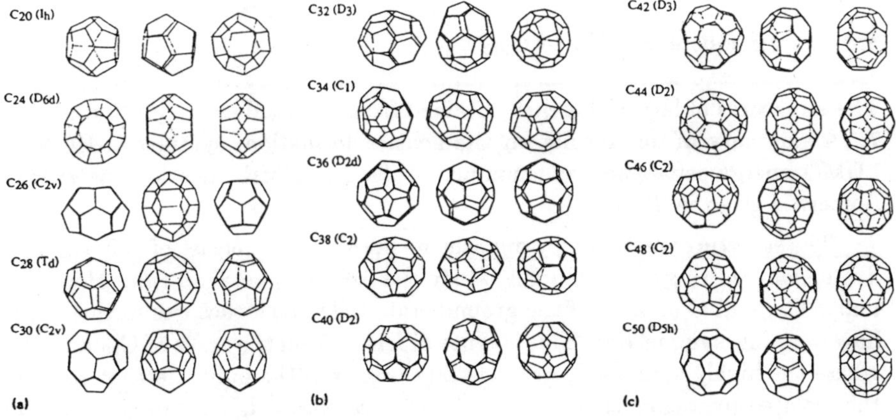

Figure 6: The line drawings of fully relaxed ground-state structures of fullerenes from C_{20} to C_{70}. The views along three orthogonal axes are provided to appreciate the symmetries and the shapes of the cages.

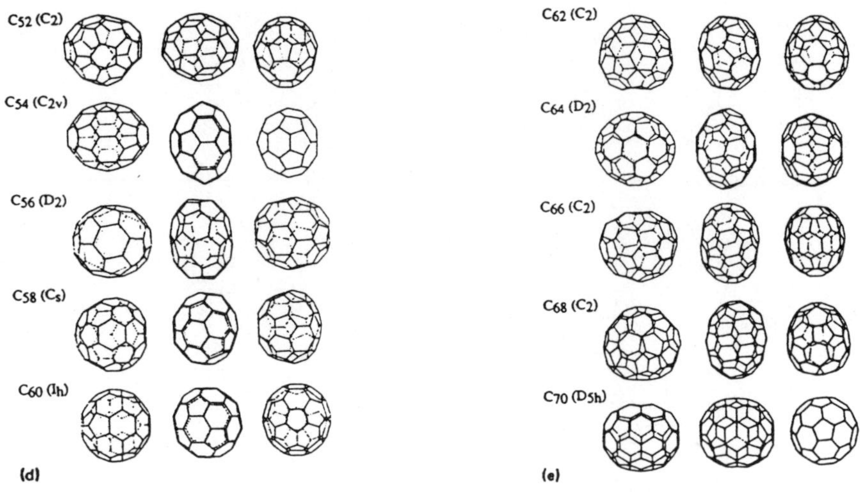

Figure 6: (*contined*)

5.2   Larger fullerenes: C_{72} to C_{102}

The structures of large fullerenes obtained from our studies are plotted in Fig. 7. A summary of our results on the heat of formation, symmetry, HOMO-LUMO energy separation and number of distinct NMR lines for low energy isomers is given in Table 4.

The structures obtained by our scheme in this range consist of only six- and five-membered rings and they all satisfy the isolated-pentagon rule. However, unlike C_{60} and C_{70}, most of the ground-state fullerene structures in this range have very low symmetries. This is quite different from the cases of C_{60} and C_{70} where the highly symmetric ground-state isomer is the only one that satisfies the isolated-pentagon rule. In the large fullerenes, high symmetry usually restricts the distribution of pentagons. This point is well demonstrated by the D_2 and D_5 isomers of C_{100}. In the D_5 isomer of C_{100}, the 12 pentagons are confined to the two end of a sausage-like cage in order to satisfy the D_5 symmetry. While in the D_2 isomer, the 12 pentagons are spread much more uniformly throughout the cage. As a result, the energy of the D_5 isomer is 0.6 eV per molecule higher than that of the D_2 isomer.

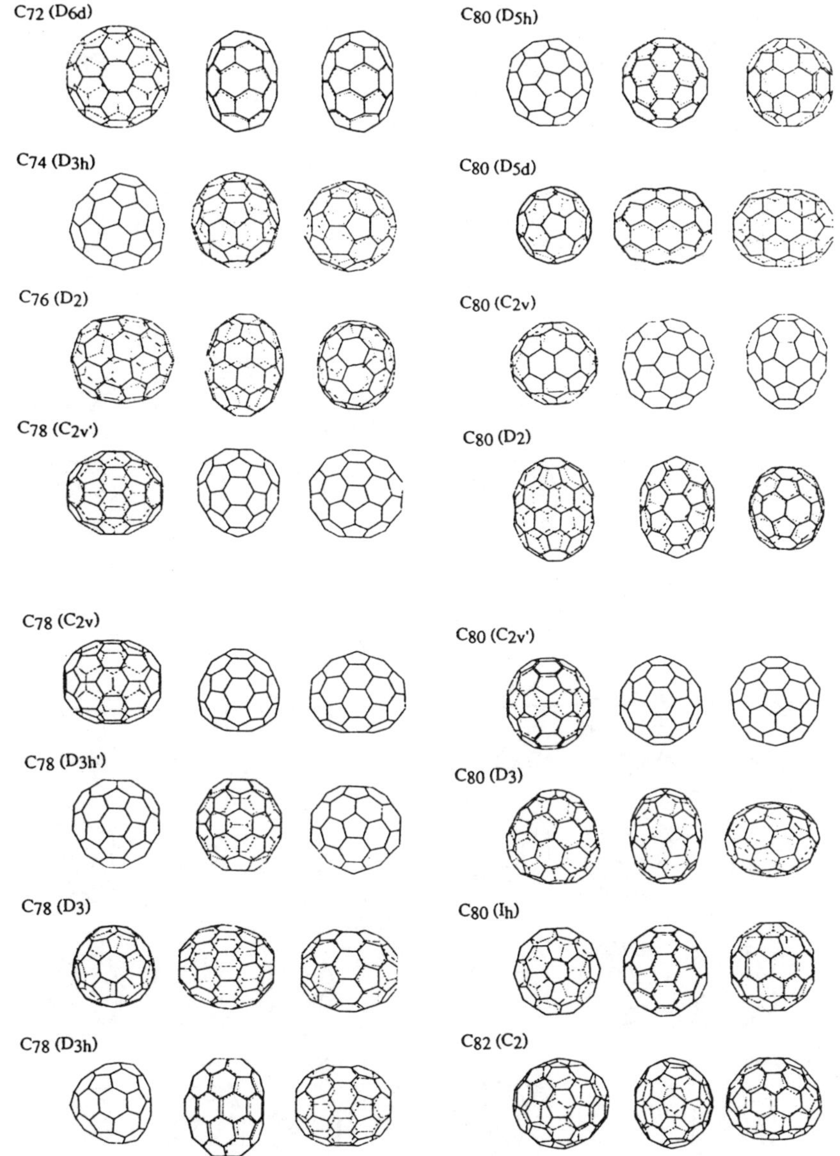

Figure 7: The line drawings of fully relaxed ground-state structures of fullerenes from C_{72} to C_{102}. The views along three orthogonal axes are provided to appreciate the symmetries and the shapes of the cages.

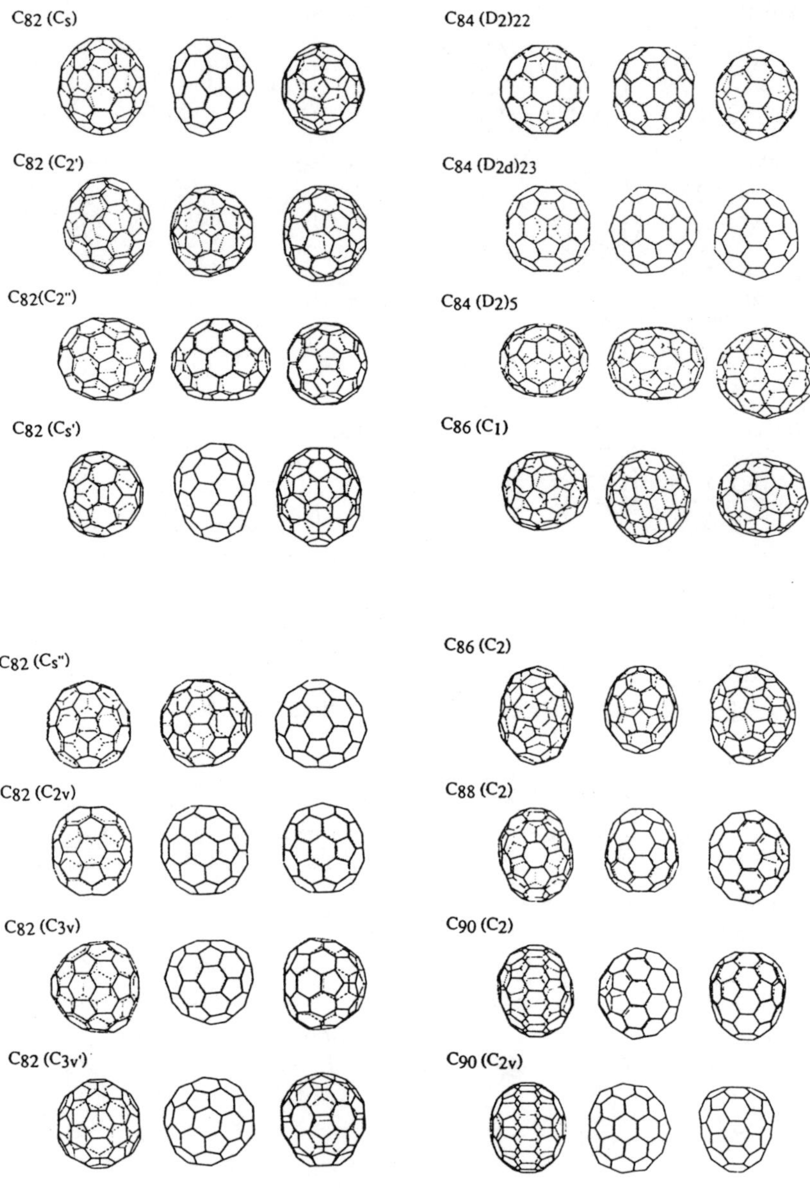

Figure 7: *(continued)*

C$_{90}$ (C$_2$')

C$_{100}$ (D$_5$)

C$_{92}$ (C$_2$)

C$_{102}$ (C$_1$)

C$_{92}$ (D$_2$)

C$_{94}$ (C$_2$)

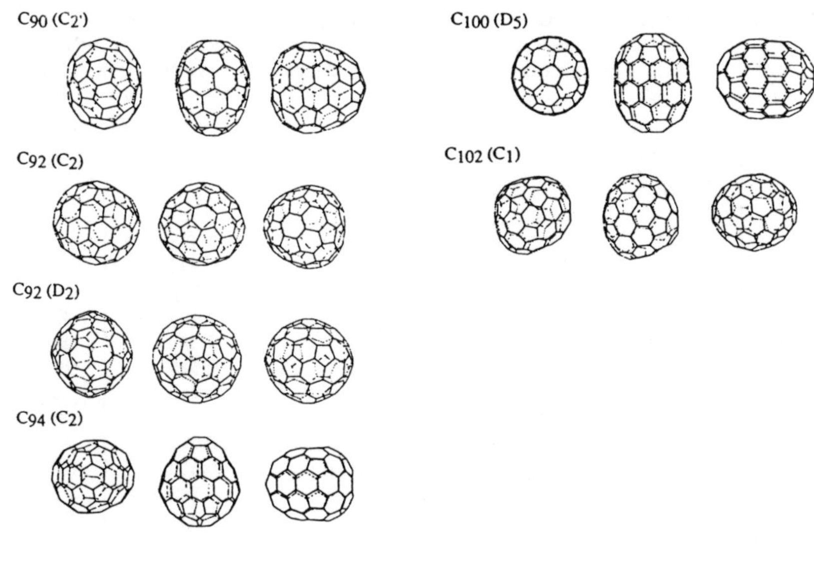

C$_{96}$ (C$_1$)

C$_{96}$ (D$_{6d}$)

C$_{98}$ (C$_1$)

C$_{100}$ (D$_2$)

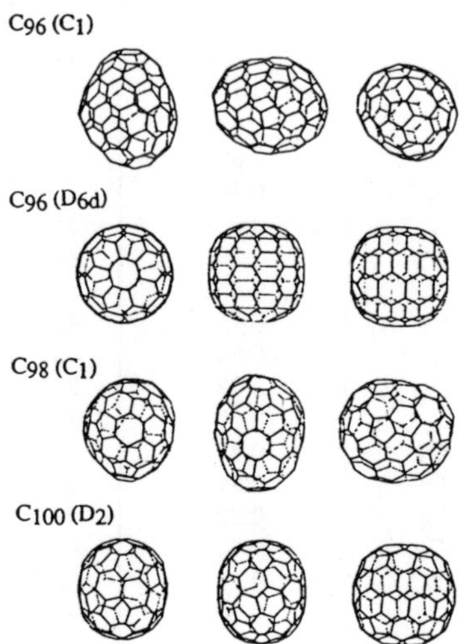

Figure 7: *(continued)*

Table 4: Properties of C_{60} - C_{102}. The energy is relative to that of bulk graphite in the units of eV/atom. ΔE is the energy difference between the isomer and the corresponding ground-state isomer of the same size in the unit of eV/molecule. The symmetry in the parenthesis is the symmetry after Jahn-Teller distortion. The HOMO-LUMO gaps are in unit of eV and the values in the column NMR are the numbers of distinct NMR lines expected.

Cluster Size	Energy	ΔE	Symmetry	HOMO-LUMO	NMR
60	0.401		I_h	1.61	1
70	0.365		D_{5h}	1.103	5
72	0.377		D_{6d}	1.388	4
74	0.357		D_{3h}	0.224	9
76	0.354		D_2	0.796	19
78	0.346		$C_{2v'}$	0.493	22
		0.087	$D_{3h'}$	0.353	8
		0.284	C_{2v}	0.545	21
		0.324	D_3	0.443	13
		0.913	D_{3h}	1.373	8
80	0.344		$D_{5h}(C_s)$	0.076	40
		0.015	D_{5d}	0.350	5
		0.030	C_{2v}	0.198	22
		0.090	D_2	0.518	20
		0.165	$C_{2v'}$	0.141	23
		0.230	D_3	0.280	14
		0.381	$I_h(C_{2v})$	0.086	23
82	0.335		C_2	0.649	41
		0.076	C_s	0.568	44
		0.166	$C_{2'}$	0.362	41
		0.191	$C_{2''}$	0.477	41
		0.226	$C_{s'}$	0.773	44
		0.229	$C_{s'}$	0.234	44
		0.285	$C_{2v}(C_2)$	0.081	41
		0.566	C_{3v}	0.120	17
		0.731	$C_{3v'}(C_s)$	0.088	43
84	0.325		D_2-22	0.823	21
		0.033	D_{2d}-23	0.844	11
		1.231	D_2-5	0.610	21
86	0.329		C_1	0.344	86
		0.038	C_2	0.351	43
88	0.327		C_2	0.154	44
90	0.314		C_2	0.633	45
		0.180	C_{2v}	0.629	24
		0.307	$C_{2'}$	0.564	45
92	0.311		C_2	0.471	46
		0.071	D_2	0.575	23
94	0.304		C_2	0.679	47
96	0.303		C_1	0.630	96
		0.081	D_{6d}	0.318	5
98	0.300		C_1	0.202	98
		0.054	C_s	0.606	50
100	0.292		D_2	0.432	25
102	0.288		C_1	0.365	102

6 General Trend of Stability

Although all even-numbered clusters with $N \geq 30$ have been observed in the gas phase, the extraction products from evaporated graphite are dominated by C_{60} and C_{70} plus significant minor fractions of certain fullerenes (i.e., C_{76}, C_{78}, C_{82}, C_{84}, C_{90}, C_{94}, and C_{96}). Not every even-numbered cluster in this range [11,14] is present after extraction. The origin of these "magic" numbers is not yet completely understood.

The isolated-pentagon rule may be used to explain the stability of C_{60} and C_{70}, because the I_h isomer of C_{60} and the D_{5h} isomer of C_{70} are the only fullerenes in the range $20 \leq n \leq 70$ satisfying the isolated-pentagon rule. However, this rule is not sufficient to explain the stabilities of fullerenes with $n \geq 76$, since each fullerene in this range has more than one isolated-pentagon isomers.

Using a simple Hückel theory, Fowler suggested that three homologous series of fullerenes with closed-shell electronic configuration corresponding to the stoichiometries $C_{60+6k}(k = 0, 2, 3, 4, ...)$, $C_{70+30k}(k = 0, 1, 2, 3, 4, ...)$, and $C_{84+36k}(k = 0, 1, 2, 3, ...)$ would be more stable than the others [34,29,35]. This empirical rule has been used to explore the stability of C_{78} [35] and C_{84} [29]. However, it failed to explain the presence of C_{76}, C_{82}, C_{94} and the absence of C_{72} in the extraction products.

Based on the results presented in the previous section, we have systematically studied the general trends in the structures, energies and HOMO-LUMO gaps for carbon fullerenes in the range from C_{20} to C_{102} [26,36]. In Fig. 8, the formation energy as a function of fullerene size obtained from our calculations is plotted. While the energies of small cages grow rapidly as the cluster size increases, the energies of large fullerenes with no adjacent pentagons are found to increase at a slower rate. The most interesting feature as shown in Fig. 8 is that C_{60}, C_{70}, and C_{84}, which correspond to "magic" numbers in connection with the abundance peaks in the carbon cluster beam mass spectrum, are energetically more favorable than their neighbors.

Furthermore, we found that the abundance of fullerenes observed in the extraction products is strongly correlated with both the fragmentation energies and electronic HOMO-LUMO gaps of the fullerenes obtained from our tight-binding calculations.

In the fragmentation process with successive C_2-loss as suggested by O'Brien *et al.* [37], the fragmentation energy can be defined as:

$$E_{frag}(n) = E_{coh}(n - 2) + E_{coh}(2) - E_{coh}(n). \qquad (4)$$

Here $E_{coh}(n)$ is the cohesive energy of C_n and $E_{frag}(n)$ is the energy needed for

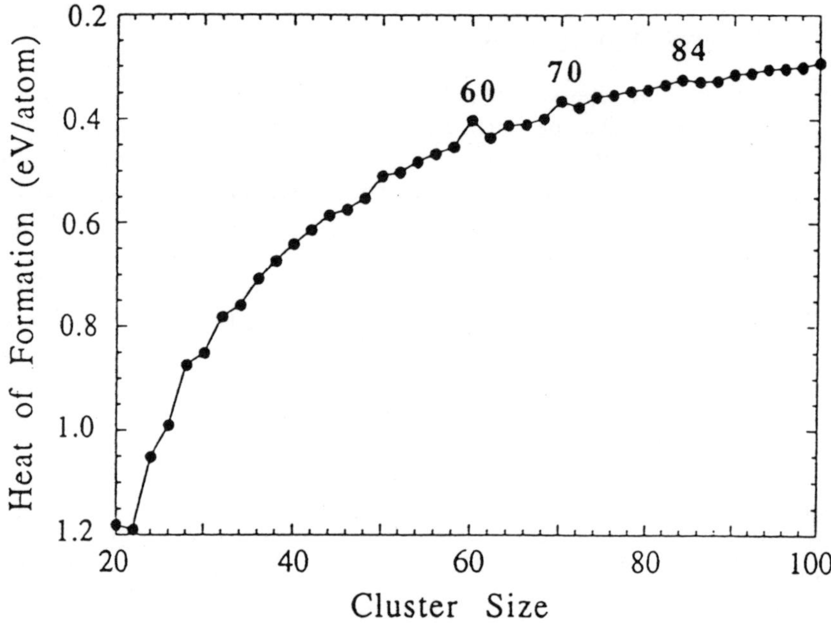

Figure 8: The heat of formation of carbon fullerene (relative to that of graphite) as a function of cluster size.

C_n to fragment into C_{n-2} and C_2. Therefore clusters with larger fragmentation energy are more stable against fragmentations.

The HOMO-LUMO gap is a measure of the stability of fullerenes towards chemical reactions. A large HOMO-LUMO energy separation makes it more difficult to extract electrons from the low-lying HOMO or to add electrons to the high-lying LUMO. It is easier for clusters with small HOMO-LUMO energy gaps to react with the solvent and other encountered chemicals. A larger HOMO-LUMO gap might also make the cluster more stable towards further accretion of extra C atoms or clusters.

Plotted in Fig. 9 are the fragmentation energy and the HOMO-LUMO gap as a function of cluster size obtained from our study. It is clearly shown that C_{60} and C_{70} not only have very large fragmentation energies, but also possess large HOMO-LUMO gaps. These results may explain why these two fullerenes are superstable. It should be noted that all fullerenes which are more abundant in the extraction products, i.e. C_{76}, C_{78}, C_{82}, C_{84}, C_{90}, C_{94}, and C_{96}, have both relatively large HOMO-LUMO gaps and fragmentation energies. On the contrary, those with small HOMO-LUMO gaps (C_{74}, C_{80}, C_{86}, C_{88}) or small fragmentation energies (C_{72}) are not observed in the extraction products.

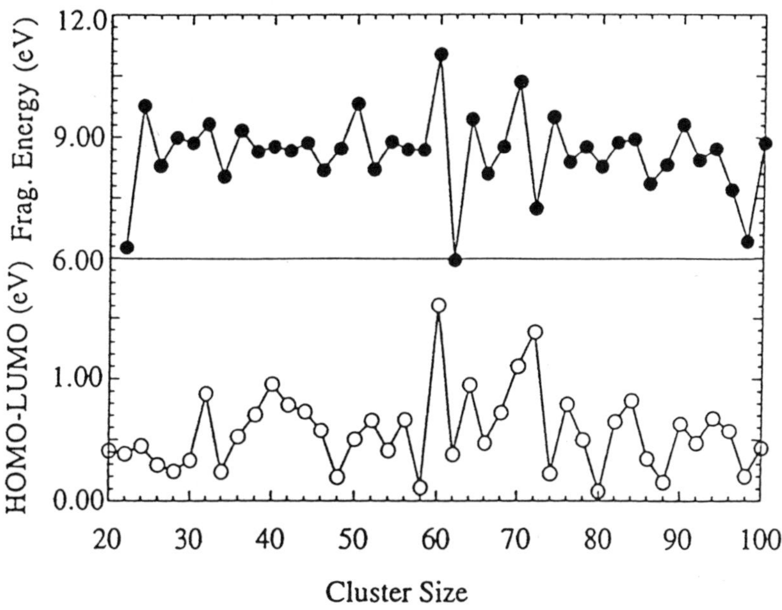

Figure 9: (a) The Fragmentation energy and (b) HOMO-LOMO gap of carbon fullerene as a function of cluster size.

7 Dynamical Behavior of Carbon Fullerenes

7.1 Simulation of fullerene formation process

In order to get some insight into the formation mechanism and the growth conditions of the C_{60}, we have performed molecular dynamics simulations using the tight-binding energy model to study the formation of C_{60} from gaseous carbon atoms [38]. We enclosed 60 carbon atoms in a hollow sphere of radius R with specular reflection imposed when the atoms hit the inner surface of the sphere. We start the simulation by heating the carbon atoms to very high temperatures (10000 K) in a larger sphere (R=9.22 Å). The carbon atoms under this condition are found to be gas like. Then we gradually reduce the temperature and the radius of the sphere. When the temperature is reduced to 6000K within a sphere of radius 5.3 Å, we found that polygonal rings nucleate rather rapidly from the originally loose linear-chained cluster (Fig. 10(a)-(c)). By reducing the sphere radius gradually from 5.3 to 3.832 Å while keeping the temperature at 6000 K, we found that a close cage structure is forming in the process of simulation as one can see from Fig. 10(d)-(f). We note that the structure of the C_{60} cage obtained from this simulation is similar to that of buckyball but with some defects due to the rapidity of the compression and cooling in the simulation. These defects include seven- and four-membered

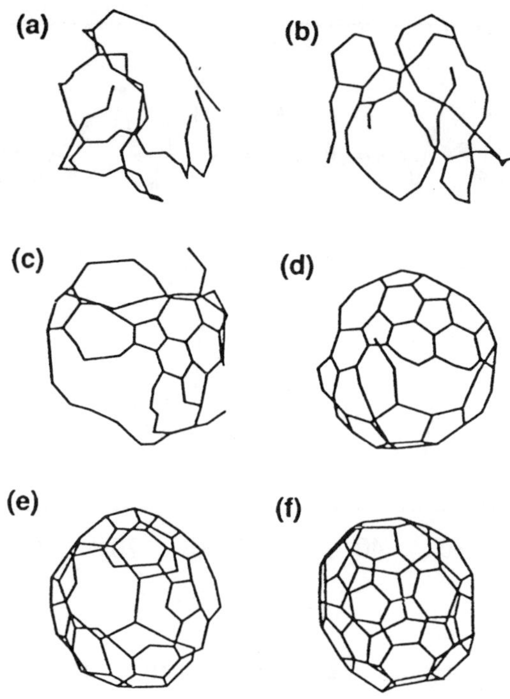

Figure 10: Perspective view of C_{60} cage formation process at T=6000 K. The carbon atoms are connected by straight lines when the interatomic distances are less than 1.8 Å. (a), (b), and (c) are snapshots at 2.8ps, 4.2ps, and 5.6ps respectively and with the sphere radius R=5.32 Å. (d), (e), and (f) are typical snapshot when the sphere radius is reduced to 4.61 Å, 3.90 Å, 3.832 Å respectively. Note a closed cage forms when R=3.832 Å. The total simulation time is around 30ps.

rings and adjacent pentagons.

While the hollow sphere used in our simulation is originally proposed to contain the gaseous atoms, the simulation results suggest that the inner curved surface of the sphere may play a role in facilitating the nucleation of the cage. When collision of hot loose filament-like clusters with cool condensed graphitic or cap-like fragments occurs the momentum of the hot cluster deforms the cool fragments and produces a curved surface with decreasing radius of curvature which we represent by the shrinking sphere in our simulations. Although our simulation didn't provide a fully understanding the formation mechanism of carbon fullerenes, the results suggests that the cooling and compression of the hot clusters in such collisions provides the nucleation mechanism for the fullerene cages.

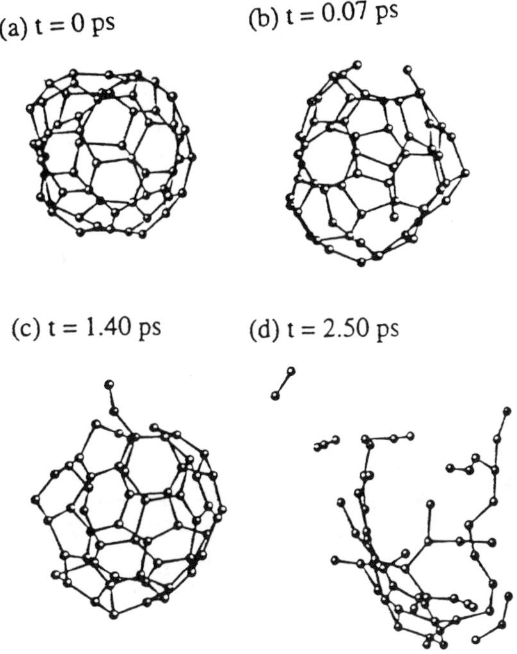

Figure 11: Snapshots of C_{60} fragmentation process at 4500 K.

7.2 Thermal disintegration of carbon fullerenes

Tight-binding molecular dynamics simulations are also performed to study the thermal disintegration of carbon fullerenes.[39] A schematic illustration of the fullerene fragmentation process is plotted in Fig. 11. Our simulation results show that after the onset of bond breaking, the atoms with dangling bonds stick out from the cage surface. These off-surface atoms vibrate much more violently than the rest of the network and the whole cage structure is finally be unravelled into a ring or multiple-ring structure with the breaking off of fragments like dimers.

In Fig. 12 the fragmentation temperatures obtained from the simulations are plotted as a function of cluster size ranging from C_{20} to C_{90}. As a general trend, we find that the fragmentation temperature increase almost linearly with the cluster size in the range C_{20} to C_{58}. This trend is quite different from that of the formation energy (see Fig. 8) in which the slope decreases rapidly as the cage size increases. For clusters with more than 60 atoms, the fragmentation temperature is almost a constant, although their bonding energy

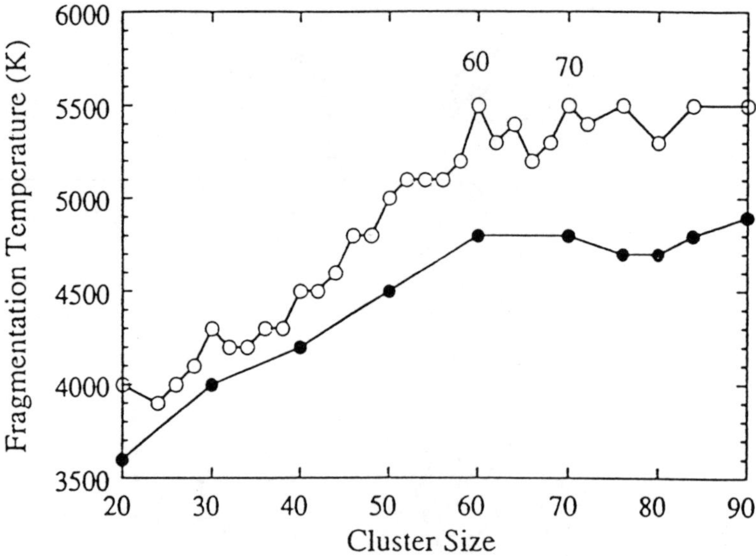

Figure 12: The fragmentation temperature of fullerene is plotted as function of cluster size. Where the solid dots are the results without F-D distribution and the open dots are the results with F-D distribution.

keeps on increasing. While both the heat of formation and the fragmentation temperature can be used to measure the stability of the fullerene structure, the different behavior in the trends for these two properties suggests that the fragmentation temperature is more closely related to the weakest bond in the cage, while the heat of formation is a measure of the average bonding in the cluster.

It should be pointed out that the criterion we used to determine the fragmentation point in the study may have overestimated the fragmentation temperature due to the rather short simulation times practicable in TBMD simulations. It should also be kept in mind that the present tight-binding model is fitted to reproduce the binding energy of first-principles density functional calculation results, which overestimate the binding energy of graphite by about 10% in comparison with experimental value. Thus there exists a corresponding systematic error in the fragmentation temperature if one wants to make a direct comparison with experimental results.

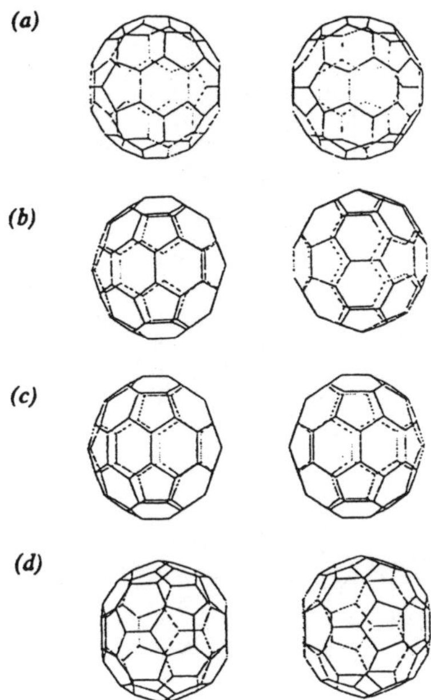

Figure 13: The schematic views of orientations of collisions: (a)The pentagon to pentagon. (b)bond to bond (perpendicular). (c) bond to bond (parallel). (d) hexagon to hexagon.

7.3 Simulation of buckyball collisions

Collisions between fullerenes play a very important role in the process of fullerene formation and thermal equilibrium, and it may be one of the key process for understanding the abundance of various fullerenes. The stability of the C_{60} molecule under high energy collisions is also an important consideration in some potential applications. Furthermore, understanding the C_{60} collision process will help us predict the behavior of solid C_{60} under sudden shock compression when cluster collisions occur in the solid state.

We have simulated the collision between C_{60} buckyballs using tight-binding molecular-dynamics. The simulations are performed at three temperatures : 0, 2000 and 3000K. For each temperature, the results of head-on collisions are observed over a wide range of collision energies. At zero temperature, various orientations of the colliding molecules as shown in Fig. 13 have been investigated: (a) pentagon to pentagon, (b) bond to bond(perpendicular),(c) bond to bond(parallel) and (d) hexagon to hexagon. Collisions between C_{30} clusters are also studied for comparison with the results of C_{60}.

For all the cases we have studied, there are qualitatively three distinct

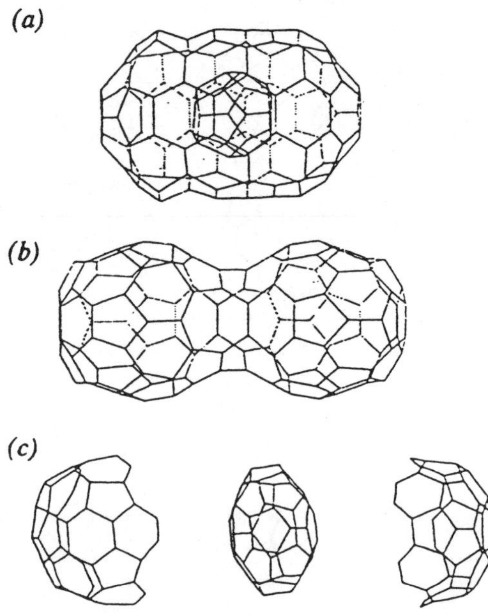

Figure 14: The line draws of the resultant typical molecules produced by collisions between C_{60}: (a) 'Russian Doll'. (b) 'baby-cage'. (c) 'peanut'.

regimes as the collision energy increases. At low collision energies, the clusters deform upon contact and bounce back with their cage structures intact. After collision, a fraction of the translational kinetic energy of the clusters is converted to intra-molecular vibrations, notably in the breathing vibrational mode. At higher collision energies, fusion occurs : the clusters stick together to form a larger cluster. In most cases, the new cluster is not a perfect fullerene and usually contains dangling bonds. These 'poor' clusters are energetically quite unstable and it will be quite easy for them to grow in subsequent collisions with other clusters or small grains existing in the carbon gas phase in the process of soot formation. However, in some cases at this collision energy range, some well ordered structures may be resulted. Fig. 14 shows some examples of these special cases where a Russian Doll, a peanut and a baby-cage have been observed as the collision products. If the kinetic energy is even larger, the collision will reach the third regime in which the clusters will break up into smaller pieces. The most commonly observed fragments are carbon dimers, trimers, linear chains and multiple rings.

The critical collision energies for the occurrence of fusion and fragmenta-

Table 5: The fusion energies of carbon clusters at different temperature and orientation. The collision energy is defined as total kinetic energy of system in the center of mass frame. Note: Since the clusters are randomly orientated at high temperature, there is only one value of collision energy at high temperature.

Survival Energy	T=0K	T=2000K	T=3000K
60(a)	102		
60(b)	84	66	60
60(c)	84		
60(d)	84		
30	51	34	12

Table 6: The fragmentation energies of carbon clusters at different temperature and orientation. The collision energy is defined as total kinetic energy of system in the center of mass frame. Note: Since the clusters are randomly orientated at high temperature, there is only one value of collision energy at high temperature.

Survival Energy	T=0K	T=2000K	T=3000K
60(a)	210		
60(b)	210	114	72
60(c)	204		
60(d)	198		
30	78	54	48

tion are listed in Table 5 and Table 6 respectively. The results show that the critical energies for fusion and fragmentation are not sensitive to the initial orientations of the colliding buckyballs. Because thermal vibrations weaken the carbon bonding of the cage, we observed a decrease in the critical energies as the temperature increase. From Table 5 and Table 6, we can see that C_{60} is much more stable against fusion and fragmentation compared with C_{30}. This is particularly evident in the high temperatures range from 2000K to 3000K which is relevant to the fullerene formation process. The high stability of the C_{60} molecule at high temperatures may be related to the high yield of C_{60} in the experimental formation process. However, a more detailed study of collisions between different carbon clusters and small carbon radicals is needed to make a final conclusion.

We have performed a detailed analysis of the energy transfer in the collision process as a function of collision energy for the case of collisions between C_{60} molecules at 2000K. The change in the translational kinetic energy of the system is defined as

$$\Delta K = K_{cm}^{initial} - K_{cm}^{final} \tag{5}$$

$$K_{cm} = N\frac{1}{2}mV_{cm}^2 \tag{6}$$

where N is the number of atoms in the system and V_{cm} is the velocity of the center of mass of one of the molecules. In Fig. 15(a) we plot the fractional decrease in the translation energy

$$R_t = \frac{\Delta K}{K_{cm}^{initial}} \tag{7}$$

as a function of the collision energy. We note that the R_t stays fairly constant at $25 \pm 3\%$ independent of the collision energy until the molecules stick together at a collision energy of 66 eV. The loss in translational energy ΔK is transferred to the internal degrees of freedom of the molecules. In the low energy regime, the cage structures of the molecules remain intact, thus all of ΔK goes to heating up the outgoing molecules. In the fusion regime, almost all the translational kinetic energy is lost after collision since the two molecules stick together after the collision, but not all of it is transformed into heat. Part of the energy is stored as deformation energy in the new cluster. In fig. 15(b), we plot the final temperature of the system after collision for both the bouncing as well as the fusion regime. The final temperature is obtained from the vibrational kinetic energy per degree of freedom. In the bouncing regime, the temperature of the molecules after the collision increases as the collision energy increases indicating that more and more heat is generated during the collision process. In the sticking regime, the final temperature of the cluster stays quite constant as the collision energy increases and even drops at the high energy end of the fusion regime. The reason is that, as the collision energy increases, many bonds are broken in the collision process and most of the translational kinetic energy goes into deformation energy of the system. In fact, when the collision is very close to the fragmentation regime, the final form of the cluster is no longer cage-like but becomes more like a sheet. This introduces a lot of dangling bonds which cannot be balanced by the increase in collision energy and the temperature of the cluster drops.

A complete picture requires further investigations of collisions between clusters of different sizes and collisions between clusters and small carbon fragments in the gas phase. Also in reality, most collisions are not head-on so collisions. A more detailed study of collision with variable impact parameters is also useful to make a reasonable comparison with experimental measurements.

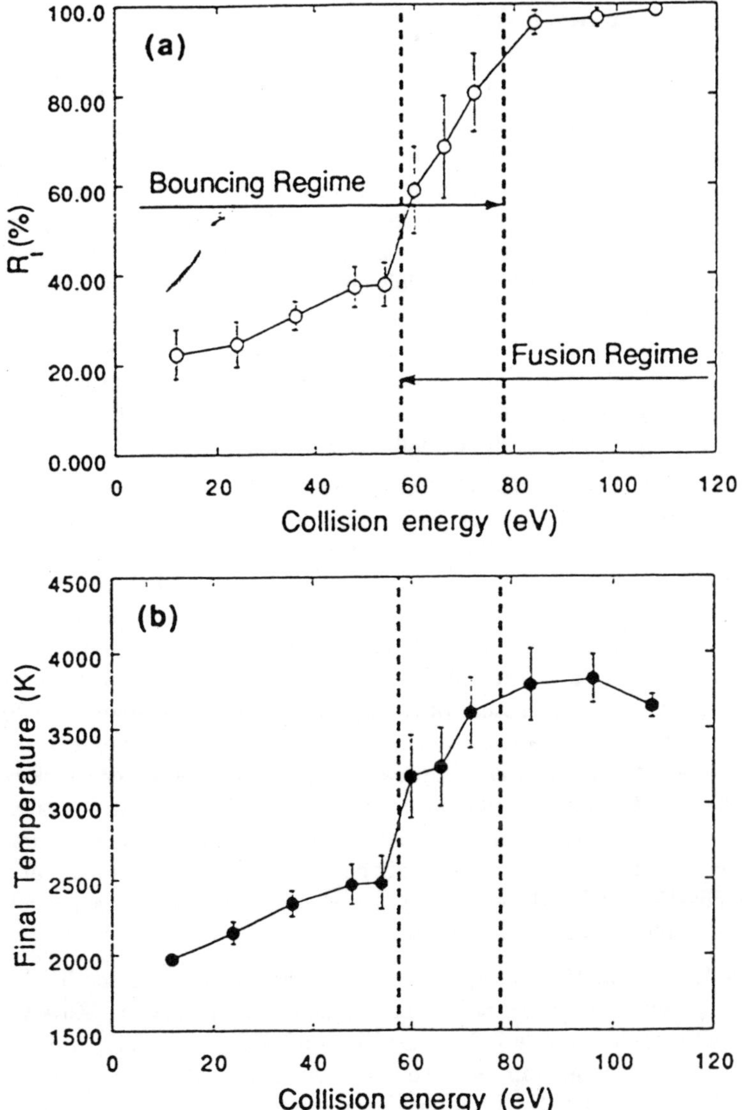

Figure 15: he energy transfer ratio R_t (a) and final temperature (b) are plotted as the function of initial kinetic energy of center of mass.

8   Solid C_{60} under high pressure

There have been considerable interest in the study of solid C_{60} under high pressure [42,43,44,45]. Experiments showed that the optical absorption edge of C_{60} films shifts to lower energy as pressure increases, reflecting a decreasing gap between the valence and conducting bands of C_{60} solid with applied pressure [45]. However, instead of transforming into a metallic phase, solid C_{60} undergoes an irreversible transition to a more insulating phase when the pressure is increased to about 20 GPa [42,43,45]. This new phase of carbon is stable even when the pressure is released. From Raman scattering measurements [45], this new phase was found to be distinct from C_{60}, diamond, and graphite-like amorphous carbon. While it is believed that this phase transformation is caused by the collapse of the C_{60} cage structure, details of the structural and electronic properties of the new phase are still not well understood.

Our simulations of solid C_{60} under pressure were performed with unit cell containing four C_{60} molecules (240 atoms) arranging in *fcc* lattice with periodic boundary conditions. We have performed two kinds of simulations which correspond to the different extremes of shock and hydrostatic compressions. For shock compression, the whole process is considered as adiabatic and the lattice constant is reduced at a rate of 0.0003 $\mathring{A}$ per step. The temperature of the sample increases as the volume decreases. When the density is increased to 4.0 g/cm^3, the temperature of the sample can reach as high as 3000 K. For the case of hydrostatic pressure, the system is kept at a temperature of 2500 K, which is close to the final temperature of the sample under shock compression. We adopt a high temperature for the simulation so that annealing of the sample, which happens over an extended period of time in the experiment, can occur within the short time frame of the molecular dynamics simulation. The lattice constant was reduced by 0.2 $\mathring{A}$ each time and the system is allowed to relax for 4000 steps at each volume.

We found that the energy gap between the valence and conduction bands of the C_{60} solid decreases as the lattice constant decreases. However, when the density of the sample is increased beyond a certain value, which is about 3.3 g/cm^3, the band gap tend to open up again, although in most of our simulations the gap in the high density phase is not 'clean', i.e. there are some residual states in the gap region.

We have calculated the optical absorption spectra of the C_{60} solid as a function of density (pressure) in the course of molecular dynamics simulation. The optical absorption intensity between two eigenstates n and m is in proportion to the oscillator strength:

$$f_{nm} = \frac{2}{m_0} \frac{|(\mathbf{p}_{nm}|^2}{E_n - E_m}.$$ (8)

where P_{nm} is the momentum matrix element, m_0 is the free electron mass, E_n and E_m are the energy of the states n and m respectively. Within the tight-binding formalism, the momentum matrix elements $\mathbf{P}_{nm}$ can be expressed as:[46,47]

$$\mathbf{p}_{nm}(\mathbf{k}) = \frac{im_0}{\hbar} \sum_{b\alpha, b'\alpha'} C^*_{nk}(b, \alpha) C_{mk}(b', \alpha') \sum_{\mathbf{R}} e^{i\mathbf{k}\cdot\mathbf{R}} \mathbf{R} E^{bb'}_{\alpha\alpha'}(\mathbf{R}).$$ (9)

Here R, b and α label the atomic positions, basis atoms in a unit cell and the atomic orbitals, respectively. $E^{b,b'}_{\alpha\alpha'}$ is the tight-binding Hamiltonian matrix element between atomic orbitals $|b, \alpha >$ and $|b', \alpha' >$.

Fig. 16(a) shows the optical absorption spectra obtained from our simulations. The spectra at four different densities (i.e. ρ_1, ρ_2, ρ_3, and ρ_4) were calculated at a temperature of 2500K and averaged over 2000 configurations. The curve labeled 'collapsed phase' is the optical spectrum calculated from a collapsed solid C_{60} sample whose structure is shown in figure 17. The experimental spectrum [45] of the collapsed phase of solid C_{60} is also plotted in the insert for the purpose of comparison. The tendency of gap reduction and reopening as a function of density can be seen from Fig. 16(a). As the density of the sample is increased, the first absorption peak broadens and the absorption edge shifts towards lower energy. At the same time, a very broad second absorption peak emerges around $30 \times 10^3 cm^{-1}$. When the density is increased to 3.38 g/cm^3, the C_{60} solid collapses and the broad peak around $30 \times 10^3 cm^{-1}$ becomes a dominant feature in the spectrum. This peak becomes more pronounced when the sample is compressed to higher densities.

It has been observed in experiment that the optical absorption intensity of the crushed C_{60} phase is much larger than that of the pure diamond sample [45], but is only about one tenth of that of graphite-like amorphous carbon. It was also observed that the optical absorption intensity of graphite-like amorphous carbon is linear as function of energy while collapsed C_{60} sample exhibits a clear maximum around $21 \times 10^3 cm^{-1}$. Our calculated spectrum of the collapsed C_{60} as shown in Fig. 16 (a) (labeled 'collapsed phase') is similar to the experimental result. In particular, both spectra increase very fast after passing about $11 \times 10^3 cm^{-1}$ although the peak absorption frequency of our calculated spectrum is higher than the experimental one.

The initial reduction of the gap with applied pressure arises from the broadening of the energy bands due to the stronger interactions between the C_{60}

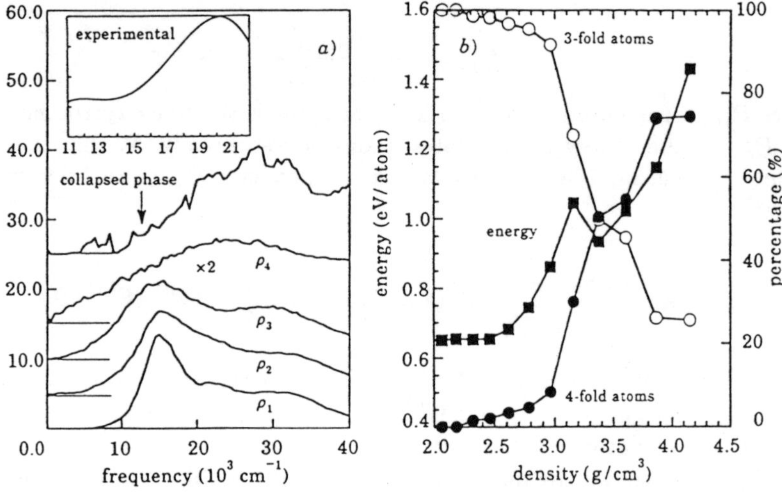

Figure 16: (a)Calculated optical absorption spectra at different densities. ($\rho_1 = 1.75\ g/cm^3$, $\rho_2 = 2.32\ g/cm^3$, $\rho_3 = 2.78\ g/cm^3$ and $\rho_4 = 3.38\ g/cm^3$). The curve labeled 'collapsed phase' is the optical spectrum calculated from the structure of figure 2 (see text for detail). Experimental spectrum of crushed C_{60} at ambient pressure intercepted from the work of Moshary *etal*. is shown in the inset. (b) The energy (relative to the graphite) and the percentages of 3-fold and 4-fold atoms as function of density.

molecules under pressure. The gap opening at high pressure are attributable to the increase of sp^3 bonding in the collapsed phase [43]. We have monitored the sp^2 and sp^3 ratios in the sample during the simulation as the volume is decreased. These results are plotted in Fig. 16(b). From Figure 16(a) and (b), it is clear that there is a strong correlation between the gap opening and the increase of sp^3 bonding in the system. The energy of the system as a function of density (Fig. 16(b)) also shows a clear kink due to the formation of sp^3 bonding, indicating a phase transition from ordered C_{60} solid to a disordered amorphous carbon phase.

In order to clarify the atomistic structure of the collapsed C_{60}, we took a highly compressed sample (density of 4.4 g/cm^3 and temperature of about 2500K) and quenched it to room temperature. To release the internal pressure of the sample, we slowly increase the volume until the system reaches the lowest energy. This procedure does not change the topological network formed at high pressure, but allows local relaxation of the network. The equilibrium density of the collapsed C_{60} sample, when the pressure is reduced to zero, is 3.35 g/cm^3 which is close to that of diamond (3.52 g/cm^3). The atomistic structure obtained by the above procedure is shown in Fig. 17. This structure

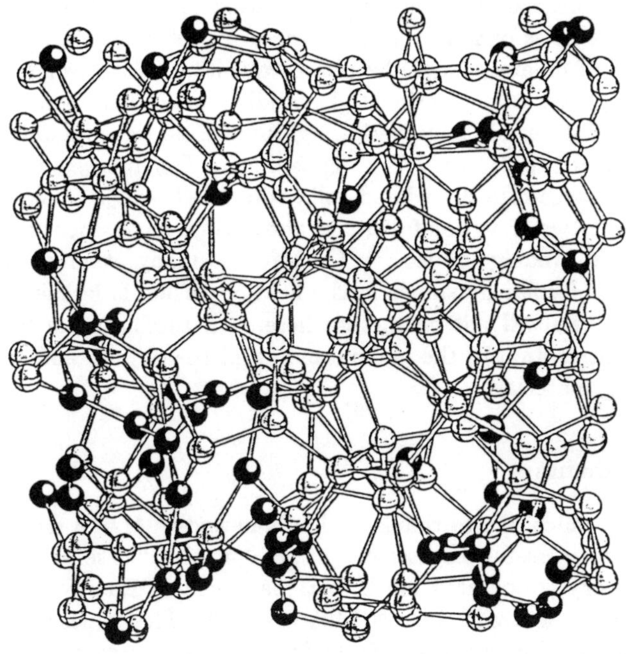

Figure 17: The atomic network of collapsed C_{60} sample in an cubic unit cell with 240 atoms. The darker and lighter balls represent the 3-fold and the 4-fold atoms respectively.

consists of 79% of sp^3 sites and 21% of sp^2 sites, which is very similar to the diamond-like amorphous carbon structure obtained by quenching from the high pressure high temperature liquid [20]. The networks is found to be very hard, with bulk modulus about 4 Mbar, as compare to 4.43 Mbar of diamond. The energy of this structure is about 0.65 eV/atom higher than that of diamond.

9 Concluding Remarks

We have shown in this paper that tight-binding molecular dynamics is a very useful and powerful scheme for studying the structural and dynamical properties of carbon fullerenes. Although the examples chosen in this paper are from our own research work, the tight-binding molecular dynamics method has been used by many other research groups to study the carbon fullerenes and other closely related materials such as carbon nanotubes. For example, tight-binding molecular dynamics has been used to study the structures and vibrational properties of solid C_{60} and carbon nanotubules. [48] Annealing and fragmentation of fullerenes and the phase transition in C_{60} solid under pressure have also been studied using tight-binding molecular dynamics at Rice University [49]. Galli et al. have performed large-scale tight-binding molecular dynamics simulations to study the deposition of C_{60} fullerenes on diamond surfaces and growth of a disordered solid composed of C_{28} fullerenes. [50] The electronic properties of carbon fullerenes and carbon nanotubes have also been studied by tight-binding models. [51,52] There are many other applications of tight-binding molecular dynamics in the field of carbon fullerenes and carbon nanotubes that we will not enumerate here. All these studies have made a substantial contribution to the understanding of the physics and chemistry of these emerging materials.

Acknowledgments

We would very much like to thank C. T. Chan and C. H. Xu for their collaboration and many useful discussions throughout the course of these studies. Ames Laboratory is operated for U.S. Department of Energy by Iowa State University under Contract No. W-7405-ENG-82. This work was supported by the Director for Energy Research, Office of Basic Energy Science, the High Performance Computing and Communications Initiative, including a grant of computer time at the NERSC.

References

1. H. W. Kroto, J. R. Heath, S. C. O'Brien, R. F. Curl, and R. E. Smalley, Nature (*London*) **318**, 162 (1985).
2. R. Taylor, J. P. Hare, A. K. Abdule-Sada, and H. W. Kroto, J. Chem. Soc., Chem. Commun., **20**, 1423 (1990).
3. R. D. Johnson, G. Meijer, and D. S. Bethune, J. Am. Chem. Soc. **112**, 8983(1990).
4. W. Krätschmer, L. D. Lamb, K. Fostiropoulos, and D. R. Huffman, Nature (London) **347**, 354 (1990).
5. E. Rohlfing, D. M. Cox, and A. Kaldor, J. Chem. Phys. **81**, 3322 (1984).
6. Q. L. Zhang, S. C. O'Brien, J. R. Heath, Y. Liu, R. F. Curl, H. W. Kroto, and R. E. Smalley, J. Phys. Chem. **90**, 525(1986).
7. S. Maruyama, L. R. Anderson, and R. E. Smalley, Rev. Sci. Instrum. **61**, 3686 (1990); S. Maruyama, M. Y. Lee, R. E. Haufler, Y. Chai, and R. E. Smalley, Z. Phys. D **19**, 409 (1991).

8. D. H. Parker, P. Wurz, K. Chatterjee, K. R. Lykke, J. E. Hunt, M. J. Pellin, J. C. Hemminger, D. M. Gruen, and L. M. Stock, J. Am. Chem. Soc. **113**, 7499 (1991).

9. H. W. Kroto, Nature (London) **329**, 529 (1987).

10. H. W. Kroto and K. Mckay, Nature (*London*) **331**, 328(1988).

11. F. Diederich, R. Ettl, Y. Rubin, R. L. Whetten, R. Beck, M. M. Alvarez, S. Anz, D. Sensharma, F. Wudl, K. C. Khemani, A. Koch, Science **252**, 548 (1991).

12. R. Ettl, I. Chao, F. Diederich, and R. L. Whetten, Nature (London) **353**, 149 (1991).

13. F. Diederich, R. L. Whetten, C. Thilgen, R. Ettl, I. Chao, M. M. Alvarez, Science **254**, 1768 (1991).

14. K. Kikuchi, N. Nakahara, T. Wakabayashi, M. Honda, H. Matsumiya, T. Moriwaki, S. Suzuki, H. Shiromaru, K. Saito, K. Yamauchi, I. Ikemoto, and Y. Achiba, Chem. Phys. Lett. **188**, 177 (1992).

15. K. Kikuchi, N. Nakahara, T. Wakabayashi, S. Suzuki, H. Shiromaru, Y. Miyake, K. Saito, I Ikemoto, M. Kainosho, Y. Achiba, Nature (London) **357**, 142 (1992).

16. D. E. Manolopoulos, P. W. Fowler, R. Taylor, H. W. Kroto, and D. R. M. Walton, J. Chem. Soc. Fraraday Trans. **88** 3117 (1992).

17. C. Z. Wang and K. M. Ho, Adv. Chem. Phys. XCIII, 651 (1996).

18. C. H. Xu, C. Z. Wang, C. T. Chan, and K. M. Ho, J. Phys. Condens. Matter, **4**, 6047 (1992).

19. C. Z. Wang, C. T. Chan, and K. M. Ho, Phys. Rev. Lett. **70**, 611 (1993).

20. C. Z. Wang and K. M. Ho, Phys. Rev. Lett. **71**, 1184 (1993).

21. C. H. Xu, C. Z. Wang, C. T. Chan, and K. M. Ho, Phys. Rev. B **47**, 9878 (1993).

22. P. Ballone and P. Milani, Phys. Rev. B **42**, 3201 (1990).

23. J. R. Chelikowsky, Phys. Rev. Lett. **67**, 2970 (1991).

24. C. Z. Wang, C. H. Xu, C. T. Chan, K. M. Ho, J. Phys. Chem. **96**, 7603 (1992).

25. B. L. Zhang, C. Z. Wang, K. M. Ho, Chem. Phys. Lett. **193**, 225 (1992).

26. B. L. Zhang, C. H. Xu, C. Z. Wang, C. T. Chan, and K. M. Ho, Phys. Rev. B **46**, 7333 (1992).

27. P. W. Fowler, J. E. Cremona, and J. I. Steer, Theor. Chim. Acta **73**, 1 (1988).

28. D. E. Manolopoulos and P. W. Fowler, J. Chem. Phys. **96**, 7603 (1992).

29. P. W. Fowler, J. Chem. Soc. Faraday Commun. **87**, 1945 (1991).

30. B. L. Zhang, C. Z. Wang, and K. M. Ho, J. Chem. Phys. **96**, 7183 (1992).

31. X. Q. Wang, C. Z. Wang, B. L. Zhang, K. M. Ho, Phys. Rev. Lett. **69**, 69 (1992).

32. F. Diederich and R. L. Whetten, Acc. Chem. Res. **25**, 119 (1992).

33. C. Thilgen, F. Diederich, R. L. Whetten, in *The Fullerenes*, ed. W. Billups, VCH Deerfield, Flordia, p 408 (1992).

34. P. W. Fowler, J. Chem. Soc., Faraday Trans., **86**, 2073(1990).

35. P. W. Fowler, R. C. Batten, and D. E. Manolopoulos, J. Chem. Soc. Faraday Commun. **87**, 3103 (1991).

36. B. L. Zhang, C. Z. Wang, K. M. Ho, C. H. Xu, and C. T. Chan, J. Chem. Phys. **97**, 5007(1992); **98**, 3095(1993).

37. S. C. O'Brien, J. R. Heath, R. F. Curl, and R. E. Smalley, J. Chem. Phys. **88**, 220 (1988).

38. C. Z. Wang, C. H. Xu, C. T. Chan, and K. M. Ho, J. Phys. Chem. **96**, 3563 (1992).

39. B. L. Zhang, C. Z. Wang, C. T. Chan, and K. M. Ho, Phys. Rev. B **48**, 11381 (1993).

40. B. L. Zhang, C. Z. Wang, C. T. Chan, and K. M. Ho, J. Phys. Chem. **97**, 3134 (1993).

41. B. L. Zhang, C. Z. Wang, K. M. Ho, and C. T. Chan, Europhys. Lett. **28**, 219 (1994).

42. S. J. Duclos, K. Brister, R. C. Haddon, A. R. Kortan, and F. A. Thiel, Nature **351**, 380 (1991).

43. M. Núñez Regueiro, P. Monceau, A. Rassat, P. Bernier, and A. Zahab, Nature **354**, 289 (1991).

44. C. S. Yoo and W. J. Nellis, Science **254**, 1489 (1991).

45. F. Moshary, N. C. Chen, I. F. Silvera, C. A. Brown, H. C. Dorn, M. S. de Vries, and D. S. Bethune, Phys. Rev. Lett. **69**, 466 (1992).

46. W. A. Harrison, *Electronic structure and the properties of solids*, Dover Publications Inc. New York (1989).

47. L. C. Lew Yan Voon and L. R. Ram-Mohan, Phys. Rev. B **47**, 15500 (1993).

48. J. Yu, R. K. Kalia, P. Vashishta, Appl. Phys. Lett. **63**, 3152 (1993); Phys. Rev. B **49**, 5008 (1994).

49. X. H. Xu and G. E. Scuseria, Phys. Rev. Lett. **72**, 669 (1994); Phys. Rev. Lett. **74**, 374 (1995).

50. G. Galli and F. Mauri, Phys. Rev. Lett. **73**, 3471 (1994); J. Kim, G. Galli, J. W. Wilkins, A. Canning, J. Chem. Phys. **108** 2631 (1998).

51. S. Saito, S. Okada, S. Sawada, and N. Hamada, Phys. Rev. Lett. **75**, 685 (1995).
52. L. Chico, V. H. Crespi, L. X. Benedict, S. Louie, and M. L. Cohen, Phys. Rev. Lett. **76**, 971 (1996).

COMPUTATIONS OF HIGHER FULLERENES

Zdeněk SLANINA, Xiang ZHAO, and Eiji ŌSAWA
Department of Knowledge-Based Information Engineering
Toyohashi University of Technology, Toyohashi, Aichi 441-8122, Japan

Computational support of experiments and observations in fullerene research represents an indispensable tool for rational and efficient understanding of the entirely new class of agents for materials science. The fullerene family is expanded along two lines - derivatization of the C_{60} and C_{70} cages, and their inflation towards higher fullerenes. During recent years our knowledge of higher fullerenes has developed considerably. Several experimental groups can now isolate and characterize higher fullerenes C_n with n around and even over 90. Their ^{13}C NMR analysis is however extremely complicated, and even allows for alternative interpretations. Nevertheless over twenty stable fullerenes C_n have presently been characterized (n from 60 to 94). Numbers of topologically possible isomeric cages are high for n around 90 and thus, experiment alone cannot answer all structural questions. In fact, the total numbers of all possible cages, even if built only from pentagons and hexagons are enormous. The numbers of isomers can be scaled down dramatically by means of so-called isolated pentagon rule (IPR).

Quantum-chemical computations of higher fullerenes can be performed well at semiempirical level and to some extent also at ab initio level within the IPR concept. The computations always end with several isomers of relatively low energy. Several such mixtures of fullerene isomers have extensively been studied - C_{78}, C_{80}, C_{82}, C_{84}, C_{86}, C_{88}, and C_{90}. The computations demonstrated that the higher isomers cannot be really understood without considering temperature effects, i.e., entropy contributions. This requirement comes directly from the fact that fullerenes are synthesized at high temperatures. In some cases other than ground-state structure in fact dominates in high-temperature mixtures.

The topologically possible structures are first generated by a combinatorial treatment. The geometry optimizations can start at a molecular-mechanical level. Then the structures must be reoptimized with semiempirical methods like AM1, PM3 or SAM1. The geometry optimizations are followed by semiempirical harmonic vibrational analysis. The inter-isomeric energetics should further be checked with affordable ab initio approximations, either Hartree-Fock (HF) SCF treatments like HF/4-31G, or even a correlated procedure like B3LYP/6-31G*. Relative concentrations (mole fractions) of the fullerene isomers can be computed using their partition functions (the rigid-rotor and harmonic-oscillator approximation). The thermodynamic treatment has to include chirality contribution (in some cases it can be a decisive term). A special care has to be paid to the symmetries of the optimized structures as they come from a numerical not geometrical generation.

The article surveys the combined quantum-chemical and statistical-mechanical computations of several groups of isomers of higher fullerenes and relates them to the available experiment. In particular, it deals with the IPR isomers of C_{76} - C_{90}. Among them, surprising experimental findings for C_{80} or ample isomerism observed for C_{90} are rationalized. The computations reveal considerable thermal effects on the relative concentrations in the IPR isomeric sets of higher fullerenes. The effects result from a complex interplay between rotational, vibrational, potential-energy terms, chirality factors, and can help in elucidation of synthetic conditions for particular isomers. Such effects can never be computed if only energetics is considered while entropy terms would be neglected. On the other hand, the treatment deals with thermodynamic equilibrium (though in fact only with respect to the inter-isomeric equilibrium). It is difficult to guess a degree to which this presumption is satisfied in the experiment. However, the thermodynamic-equilibrium treatment already produced a reasonable agreement with experiment in eight isomeric systems (C_{76}, C_{78}, C_{80}, C_{82}, C_{84}, C_{86}, C_{88}, and C_{90}), with no serious failure. This relatively large set strengthens the believe in its still wider applicability and thus, in its research potential for various fullerene systems relevant to materials science.

1   Introduction

Fullerenes have represented object of a considerable interest[1] of materials science. History of the carbon clusters goes back at least to Hahn and his coworkers[2] though their upper limit was only C_{15} back in the forties. Observations[3-5] of the carbon vapor in the fifties and sixties could already reach C_{33}. However, inert gases and proper temperature/pressure regimes were not used and a presence of fullerenes was rather unlikely in the experiments. The small aggregates were also computed by Pitzer and Clementi[6], and later by Hoffmann[7] and others[8,9]. The highest level of theory then applied was represented by the MINDO/2 method, nevertheless, the computed relative energies were already essentially correct[8] (Fig. 1). Still, the golden age of the Hückel-type studies of aromatic systems missed a suggestion[10] of three-dimensional carbonaceous objects and especially the instructive visualization[11] of C_{60}. Hence, the first Hückel calculations of C_{60} were delayed[12,13] and apparently did not influence experimental studies though there was already a quite relevant suggestion by Davidson[13]: *'Should such structures or their higher homologs ever be rationally synthesized or obtained by pyrolytic routes from carbon polymers, they would be the first manifestations of authentic, discrete, three-dimensional aromaticity'.*

On the other hand, computational chemistry could apply a wide range of reliable methodology immediately after the C_{60} discovery by Kroto, Smalley and others[14] and thus, supply useful support to further advances. In particular, Huffman and Krätschmer[15] made a full use of the computationally predicted

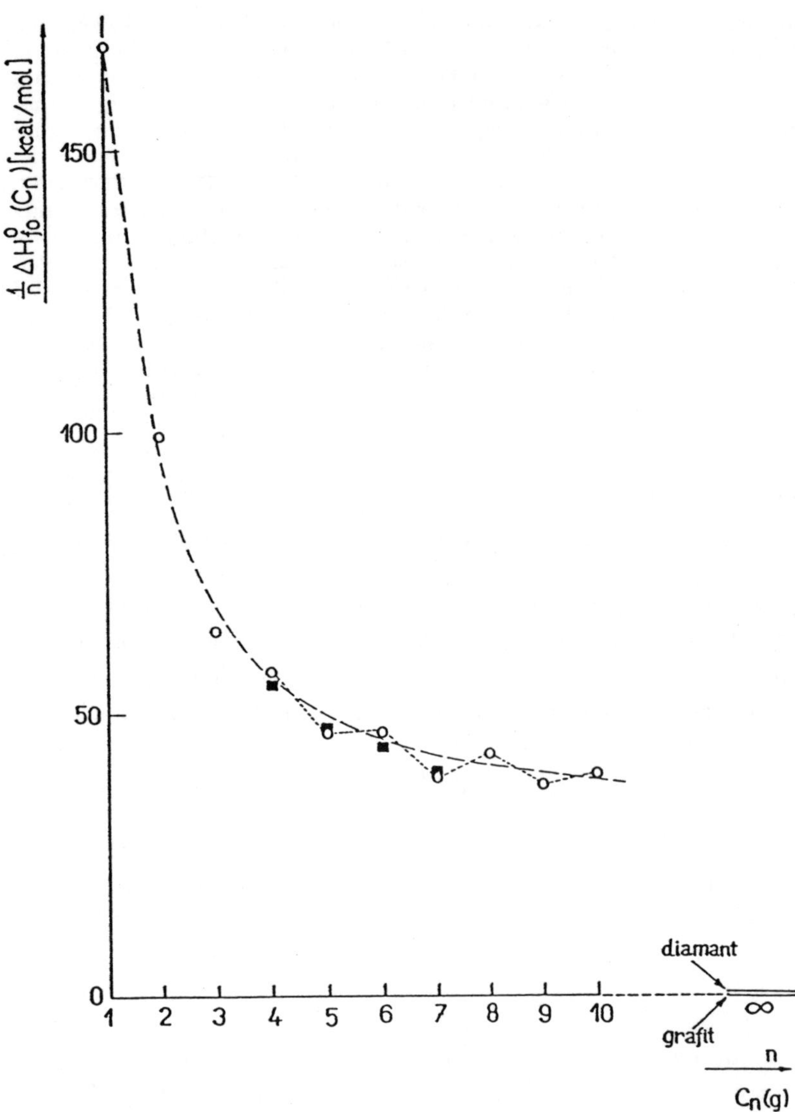

Fig. 1. The MINDO/2 computed[8] relative heats of formation at room temperature for a few first gas-phase carbon clusters (linear, cyclic, polyhedral); the Czech words grafit and diamant stand for graphite and diamond, respectively.

four IR bands in their historical carbon-arc C_{60} synthesis[16] (six quantum-chemical papers[17−22] are explicitly quoted in their recollections[15]). Vibrational analysis using semiempirical molecular orbital theory for C_{60} and especially for C_{70} was certainly close to the practical upper limit of computational chemistry in 1985 (in fact, only one vibrational analysis of C_{70} was reported[22] before the fullerene synthesis). However, in the nineties the computational-chemistry tools have been further upgraded so that they have been considerably contributing to the fullerene research avalanche.[23,24] The computational contribution is well documented by the landmark semiempirical calculations of Bakowies and Thiel[25] and *ab initio* correlated treatments by Häser *et al.*[26]

Computational chemistry has indeed been playing the role of a real partner in fullerene research so that theory and experiment are mutually complementing tools. There has been an exponential growth of the field[23,24] and there are already numerous review articles[27−64] available. Nevertheless, computations of higher fullerenes have not been really reviewed yet in spite of their application potential to materials science and other fields. This chapter tries to fill the gap, reviewing the computations and their relationships to observations.

At present over twenty stable fullerenes C_n have been identified[65−67] with n varying from 60 to 94. Elucidation of their structures has entirely been based on the isolated-pentagon-rule (IPR) conjecture[68], using the topologically generated patterns.[69−72] The lowest IPR stoichiometry which allows for a structural multiplicity is $n=76$, however, the case is controlled[73] by a Jahn-Teller effect. Anyhow, beyond $n=76$ isomerism of the IPR structures should generally play a role in observations. In fact, several such mixtures of fullerene isomers have been computed and an agreement with experiments found: C_{78} (e.g.[74−77]), C_{80} (e.g.[78−81], though experiment[81] has discovered only one C_{80} species so far), C_{82} (e.g.[82−87], at present at least two C_2 isomers confirmed experimentally[65]), C_{84} (e.g.[88−92]), C_{86} (e.g.[93−96]), C_{88} (e.g.[94−96]), and C_{90} (e.g.[97,98]). The combined quantum-chemical and statistical-mechanical computations have clearly shown that temperature effects are critically important in understanding higher fullerenes. Although the inter-isomeric separation energies are important as a starting point, they alone cannot predict the relative stabilities of the IPR isomers at elevated temperatures. As the temperatures reached in fullerene syntheses are high, entropy contributions can even over-compensate the enthalpy terms. This energy-entropy interplay and other important aspects of computational chemistry of higher fullerenes are reviewed in this chapter.

2   Combinatorial Enumerations

Enumeration and exhaustive generation of fullerene structures represent an essential prerequisite for studies of potential hypersurfaces. A particularly distinct contribution to the problem comes from Fowler *et al.*[69,70,99−117], though other groups have also participated[71,72,118−149] in the structure elucidation of various fullerene systems. The numbers and structural patterns of the IPR fullerene cages are especially important in our context, i.e., the cages in which each pentagon is surrounded he hexagons only. It is generally supposed that the IPR structures should exhibit particularly low energies. It has always been obvious that for a selected carbon content several isomeric cages are possible. If we limit ourselves to the cages built from pentagons and hexagons only, there are still several possibilities how to arrange those five- and six-membered rings. The cage isomerism first appears for C_{28} and therefrom is always present (though the IPR structures start only with C_{60}).

Enumerations of fullerene cages have developed together with other algebraic and combinatorial approaches to fullerenes. Let us mention the concept of Goldberg polyhedra, actually developed earlier in mathematics, and recently applied[101] to a systematic generation of icosahedral cages. Goldberg polyhedra are built from $20 \times (b^2 + bc + c^2)$ vertices, where b and c are non-negative integers. If $b = c$ or $bc = 0$, the point group of symmetry is I_h, otherwise the symmetry is reduced to the rotational subgroup I (though it is common to call both cases icosahedral symmetry). In particular, C_{60} comes for the choice $b = c = 1$. Other I_h symmetry cases are then, for example, $n = 20$, 80, 180, 240, 320, 500, 540, etc. The algebraic treatment can help further as it can predict which of the Goldberg structures has a closed or open shell electronic configuration at the HMO level. The governing parameter is $(b - c)$: if it is divisible by 3, the cluster has a multiple of 60 atoms and it is a closed shell; otherwise the cluster has $(60 \times n + 20)$ atoms and it is an open shell (so that Jahn-Teller distortion is predicted, resulting in a symmetry reduction and a lower stability). Hence, the treatment predicts closed shell configurations among Goldberg polyhedra for $n = 60$, 180, 240, 420, 540, etc.

Another useful concept with interesting combinatorial relationships is the Stone-Wales transformation.[119] It is a general process for rearrangements of the rings in fullerene cages. Such a pairwise interchange of two pentagons and two hexagons is also called the pyracylene transformation,[111] and it can be further generalized.[148,149] It can be visualized as a movement of two atoms in which two bonds are broken. It is not necessarily a convenient, feasible kinetic process - it is thermally (but not photochemically) forbidden.

There is still a way how to activate the kinetic process - its feasibility can originate[150] in a catalysis or even autocatalysis by free atoms present in the reaction mixture at very high temperatures. Hence, catalysis seems to be of crucial importance not only for production of nanotubes but also of fullerenes themselves. Mechanism of fullerene synthesis is virtually unknown. There are tremendous experimental difficulties in elucidation of the mechanism, and computational support and insight is thus essential. There are some general supposed steps like production of small linear, cyclic, and polycyclic species, and their combination into cages. Finally, multiple isomerizations should transform general cages into a few most stable IPR structures. Although kinetics in electronically excited states could bring some reduction of the barriers, a catalytic action is most likely to be the decisive step. The catalytic action should happen through an influence of either elemental atoms or their small clusters, in particular carbon, nitrogen, oxygen, hydrogen atoms. Graphite always contains small amounts of many chemical elements, and N_2, O_2 also come as an impurity in the inert gas used as the medium for the fullerene synthesis. During the process of graphite evaporation in the electric arc all the components are atomized. The free atoms can create intermediate complexes with the cages, even carbon atoms themselves could act this way. Preliminary computations in particular indicate[150] a catalytic activity for the complexes based on oxygen and nitrogen atoms. Again, we have to consider not only ground state but also excited potential hypersurfaces. Moreover, in addition to neutral species also positively and/or negatively charged intermediate complexes should be computed. This should elucidate the particularly potent catalytic agents, first for the fullerene synthesis and later on for a control on nanotube production. Understanding the catalytic action represents a key step in mastering fullerene synthesis. Computations are actually more suitable to solve the problem, though its complexity requires a full-scale application of sophisticated tools of computational chemistry. On the other hand, a further partial development of the computational methodology is still necessary in some respects.

Although the kinetic aspects of the fullerene synthesis are of paramount importance, we can get a useful tool even if we completely disregard the potential-energy barriers. Namely, the Stone-Wales transformation allows for a stepwise generation of various isomeric cages and can be used as a formal combinatorial structure generator.[148,149] This application represents another exhaustive or near-exhaustive generation of topologically possible cages and was recently applied to the cage-isomerism problem of C_{32}. In this way, altogether 199 C_{32} were generated[149] (it was a generalized search as four- and seven-membered rings were considered, too). However, the Stone-Wales transformations can give some insight even without considering kinetic barriers. They were for example applied to an analysis[111] of C_{84} IPR structures. The analysis demonstrated that the transformations would decompose into two disjoint families.

The leapfrog transformation[101] represents another useful concept for generation and classification of fullerene cages. It works with a topological tool named dual. Two polyhedra are defined as duals if the vertices in one polyhedron correspond to the face-centers of the other body (e.g., cube and octahedron are duals). The duals of fullerene cages have to have triangular faces and 5- and 6-coordinated vertices (for a related discussion of topological links between fullerenes and boron hydrides, see the study of Lipscomb and Massa[121]). The leapfrog transformation, producing another fullerene, consists of two steps: (i) cap each face in a fullerene by a central atom (becoming a pentagonal or hexagonal pyramid), and (ii) take the dual of the triangular polyhedron created in (i). The result is again a fullerene cage but with three times more atoms. Both starting and final fullerene cages belong to the same symmetry group. This transformation suggests[101] a $60 + 6 \times k$ stability rule (k is zero or any integer greater than 1). Hence, C_{78}, C_{84}, or C_{90} should and are important higher fullerenes.

Enumeration of all cages and of the IPR fullerene cages represents a particularly important application of the algebraic approaches. Manolopoulos et $al.$[123] introduced a concept of two-dimensional representation of fullerenes by their ring spiral, and the concept has been developed into a key tool in the enumerations. The authors noticed that fullerenes, at least below some dimension threshold, could be peeled like an orange - each face, after the first, borders its immediate predecessor, so that the the rings come off the cage in a single continuous spiral. It was not obvious if this is true for any cage regardless its dimension. It is now known that it is not - a counter-example is already known[117] - a fullerene with 380 atoms. It is the smallest tetrahedral fullerene without a spiral (T symmetry). It is not clear, however, whether this is the smallest possible unspirable fullerene of any symmetry. The C_{380} structure looks more like a giant tetrahedron than a sphere, and this shape is apparently important for its unspiralability. Therefore, we have to consider the enumeration algorithm of Fowler and Manolopoulos[69,70] derived from the ring-spiral concept as a rather practical, heuristic tool - not exact in a mathematical sense but still considerably powerful. It fails only for the n values which are not relevant to our present considerations.

The leading idea behind the ring-spiral algorithm is that the bonding topology can be reconstructed from the sequence of rings in the spiral. Thus, all possible C_n fullerene graphs can be generated by considering the ways in which 12 pentagons and $\left(\frac{n}{2} - 10\right)$ hexagons can be combined into a spiral. The combinatorial term reads:

$$S_n = \frac{\left(\frac{n}{2} + 2\right)!}{12! \times \left(\frac{n}{2} - 10\right)!},$$

(1)

though the number of the spirals, S_n, is higher than the resulting number of related fullerenes. The correspondence between ring spirals and distinct fullerenes is not 1:1, and it is instead subjected to a reduction mechanism. A spiral can be started in a number of possible ways, but this problem can be eliminated through a uniqueness test. The uniqueness test is based on the eigenvalues of the adjacency matrix. Hence, we in fact deal with spectrally-distinct fullerene isomers. Numerical imprecisions could in principle be a factor in such tests though not for smaller fullerenes. The problem of (quasi) isospectral structures should deserve interest, too. Finally, not every spiral generated can be closed.

The total number of spectrally distinct C_n fullerene isomers increases rapidly with the carbon content n. In particular, it has been established that the number of spectrally distinct C_{60} isomers is 1812. First, Manolopoulos *et al.*[123] reported 1760 isomers. Liu *et al.*[124,125] concluded 1790 isomers using an independent algorithm (they related[126] the discrepancy to the fact that the spiral algorithm produced a lower bound of the exact value). Finally, both groups arrived[127,128] to the common value of 1812. For example, for $n=$ 50, 40, 30, and 20, the computed[69,70] numbers of all isomers are 271, 40, 3, and 1, respectively. Let us stress that these counts consider only pentagon/hexagon rings, with no other cycles allowed, and a chiral pair is considered essentially as one isomer. We should also realize that, from the quantum-chemical point of view, there is not necessarily a one-to-one correspondence between topologically generated structures and local energy minima on a potential hypersurface. The structure actually optimized can be a transition state, the particular stationary point may not exist at all, or there may be more than one conformer with the required connectivity.

Odd-numbered cages are not usually considered in the enumerations. They are also known[151,152] but they cannot simply be built from three-coordinated atoms only. For example, if we consider just one two-coordinated carbon in the odd-numbered cages, and pentagons/hexagons only, the number of pentagons is reduced[129] to 10. If we allow for two tetra-coordinated and one two-coordinated carbons, the number of pentagons is 14. An extension of the IPR concept and generalized enumerations could further be developed. Decorated fullerenes represent still more derivatized cages.[130] There exist some experimental basis for the decorated species as cages built from water molecules and held together by hydrogen bonds.[131] Formally, from any fullerene we can derive a related hydrogen-bonded cage. For n water molecules, we have $\frac{3n}{2}$ bonds forming the cage, and $\frac{n}{2}$ H atoms must be directed outward (or possibly inside the cage) because they cannot be involved in hydrogen bonding. The structures can play a role in atmospheric process.

Among all possible isomers, the IPR structures are particularly important as high stability candidates. The requirement of isolated pentagons can be readily implemented into the spiral algorithm. First, of all possible spirals we can immediately eliminate those with at least one pentagon/pentagon junction. This is of course not sufficient because we still have to eliminate spirals with the secondarily generated pentagon/pentagon junctions. As already mentioned there cannot be an IPR structure for $n < 60$. This fact is obvious as we always have 12 pentagons, and 12 isolated pentagons represent $12 \times 5 = 60$ carbon atoms. Indeed, the smallest possible IPR fullerene is buckminsterfullerene C_{60}. The second smallest IPR fullerene comes for $n = 70$, and it is still a unique structure like those for $n = 72$ and 74. After this threshold the picture becomes more diversified as, for example, for $n = 76, 78, 80, 82,$ and 84, there are 2, 5, 7, 9, and 24 IPR structures, respectively. It is not necessarily a monotonous series - for $n = 86$ the number drops to 19, and then again increases: 35 and 46 IPR isomers for $n = 88$ and 90, respectively.

Other enumeration algorithms have been developed and offer complementary tools, for example the net-drawing method.[71,72,147] Another technique suggested by Liu *et al.*[124,125] generates the cages ring by ring in all possible ways starting from a seed, *e.g.*, a single pentagon. In each step, the program searches for unsaturated vertices of degree two and then adds a segment to the two unsaturated vertices to create a ring. Again, in order to identify distinct cages it is necessary to eliminate redundant occurrences using a graph-isomorphism testing program. A related generating algorithm by Dias[135] employs successive circumscribing and final capping.

Let us mention for the completeness that from the purely mathematical point of view, enumeration problems were solved by Pólya's theorem[153,154] though it is not necessarily the most practical and versatile tool. Balasubramanian[136,139,145,146] applied Pólya's theorem to various substituted fullerenes. For example, for the derivatives[136] $C_{60}X_n$ the numbers of isomers are 37, 577, and 1971076398255692 for n equal to 2, 3, and 30, respectively. Optical isomers can be distinguished or treated as one species. For example, there are 14 d,l pairs among the $C_{60}X_2$ isomers.

3 Quantum-Chemical and Statistical-Mechanical Calculations

Once the topological generation produces bonding patterns of possible structural types, quantum-chemical part can start, being later on followed by statistical-mechanical procedures. It helps to use some molecular-mechanical procedure and perform a preliminary geometry optimization,[155,156] however, it can be misleading in Jahn-Teller situations. Semiempirical quantum-chemical geometry optimizations[157,158] are then most typical step though *ab initio* Hartree-Fock Self-Consistent-Field (HF SCF) can also be performed[159] at present. There is a standard triad[160−162] of the semiempirical methods: MNDO, AM1, PM3. Recently, another method has been added[163] - SAM1 (Semi-Ab-Initio Model 1). The SAM1 method is available in the AMPAC program package,[164] other semiempirical methods also in the MOPAC package.[165] Both semiempirical and *ab initio* procedures are extensively implemented in the Gaussian series,[166] and in a simplified form, for example, in the Spartan package.[167]

The geometry optimizations are routinely carried out with the analytical energy gradient. The geometry optimization is performed in the Cartesian coordinates, i.e. without any symmetry constraints. The symmetry constraints can be imposed if internal coordinates are used, however, the option is less common at present. The computed energetics can further be checked at still more sophisticated levels of theory though with fixed geometries from a simpler procedure. Such higher treatments can either be done at *ab initio* Hartree-Fock SCF levels (like HF/3-21G, HF/4-31G, HF/6-31G*) or even with density functional theory (e.g. with Becke's three parameter functional with the non-local Lee-Yang-Parr correlation functional in the standard 6-31G* basis set, B3LYP/6-31G*). It frequently happens that the separation energetics for fullerene isomers agree quite well (Fig. 2). The optimization procedure produces stationary points, mostly local energy minima, however it can also lead to a saddle point. Hence, nature

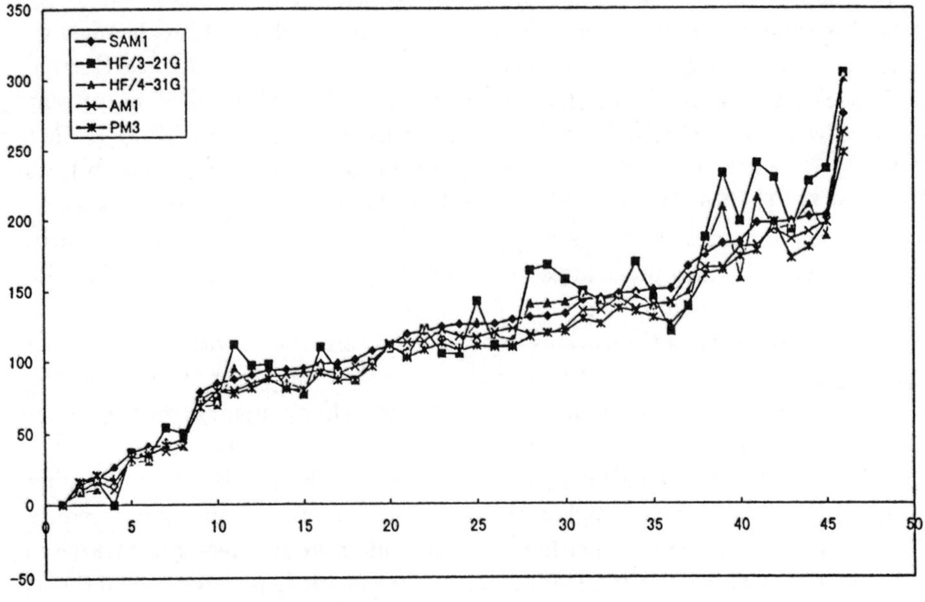

Fig. 2. The relative energetics[97] (kJ/mol) of C_{90} by various methods.

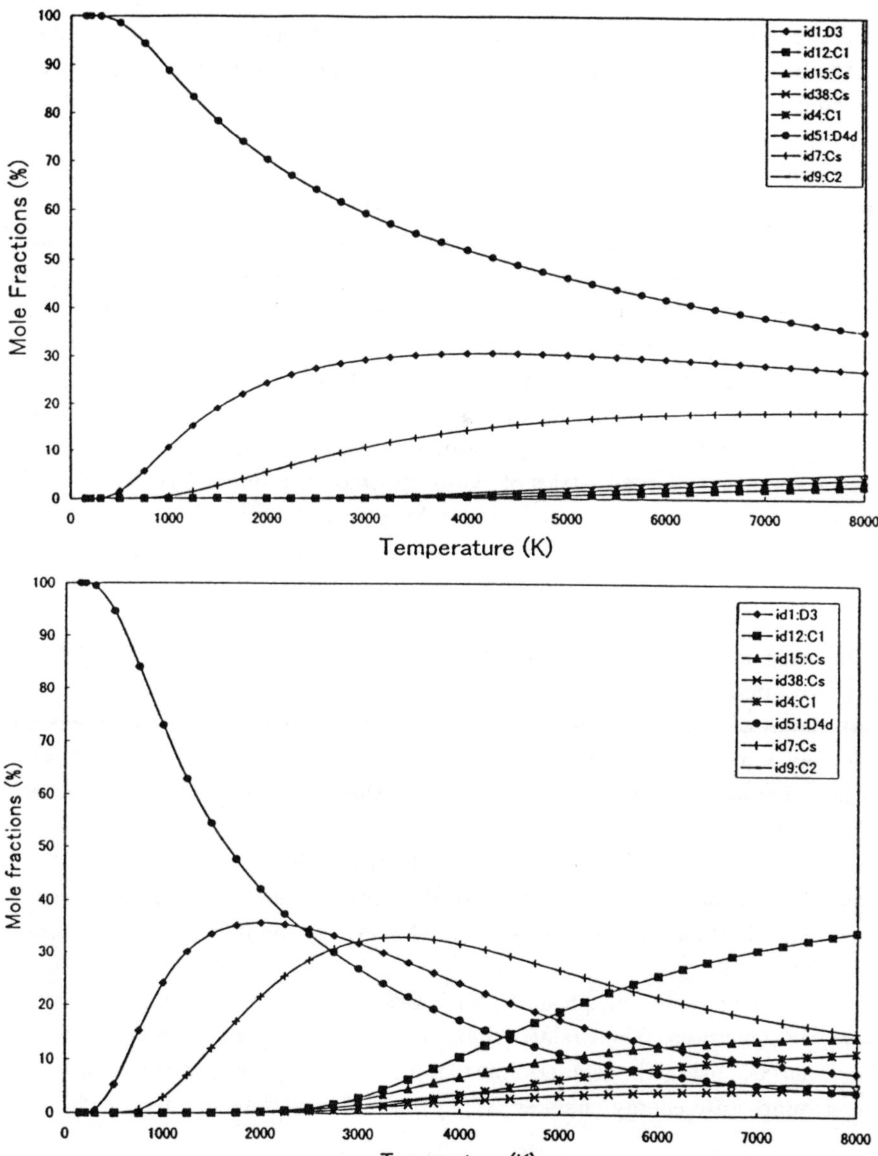

Fig. 3. The computed[149] C_{32} relative concentrations: Top - simple Boltzmann factors, Eq. (3); bottom - the proper w_i terms from Eq. 2.

of the located stationary points must be checked properly. This step is based on the harmonic vibrational analysis, mostly carried out with numerically constructed force constants (numerical differentiation of the analytical gradient). After elimination of the six modes representing overall translations and rotations, one has to check if there is any imaginary vibrational frequency - an attribute of saddle points. If all the computed vibrational frequencies are real, we deal with a local energy minimum and the structure is significant in stability reasoning. The computed vibrational frequencies can also be used in simulation of the vibrational spectra and in construction of the vibrational partition functions. For the spectra simulation the vibrational frequencies are to be combined with the computed IR intensities or Raman activities.

Although the usage of the Cartesian coordinates simplifies preparation of the input data for the geometry optimizations, it also introduces a difficulty with recognition of symmetries of the optimized structures. Symmetry of the optimized structures can be determined by a procedure[80] which considers precision of the computed coordinates as a variable parameter, ϵ. The origin of the coordinate system is placed in the center of charge - the point is only candidate for possible center of symmetry. Candidates for C_2 axes are either lines connecting any two nuclei or perpendicular bisectors of the distance between any two nuclei of the same kind. Then C_n axes with $n > 2$, S_n and S_{2n} axes, and planes of symmetry are investigated. The symmetry operations identified this way have to create one of the known symmetry groups. For each symmetry operation considered, coordinates of the interrelated atoms before and after symmetry operation are checked with respect to ϵ. If the coordinates are identical within the accuracy ϵ (i.e., the largest difference is smaller than ϵ) a symmetry element has been found at the accuracy level ϵ. This approach is more reliable than the standard symmetry search without the flexible accuracy measure.

The above steps will produce, for a fixed carbon content, a set of, say m, isomeric structures. Their relative concentrations can be expressed as the mole fractions, w_i, using the isomeric partition functions q_i. In the terms of q_i and the ground-state energy changes $\Delta H^o_{0,i}$ the mole fractions are given:[153,168,169]

$$w_i = \frac{q_i exp[-\Delta H^o_{0,i}/(RT)]}{\sum_{j=1}^m q_j exp[-\Delta H^o_{0,j}/(RT)]}, \tag{2}$$

where R is the gas constant and T the absolute temperature. The partition functions are to be constructed within the rigid-rotor and harmonic-oscillator

(RRHO) approximation. Otherwise, there is only one presumption behind Eq. 2 - the presumption of non-interacting particles (or, more specifically, the ideal gas behavior), and the condition of the inter-isomeric thermodynamic equilibrium.

Let us mention that the semiempirical quantum-chemical methods are parametrized for room temperature, i.e., they produce the conventional heats of formation at room temperature $\Delta H^o_{f,298}$ (or the related separation or relative terms $\Delta H^o_{f,298,r}$). Thus, we have to convert them to the heats of formation at the absolute zero temperature $\Delta H^o_{f,0}$, i.e., the terms which appear in Eq. 2. Finally, the vibrational zero-point energy can be extracted and the relative potential energies ΔE_r result from the treatment (they exactly correspond to the energy terms from *ab initio* computations). For methodological reasons we can also consider simple Boltzmann factors:

$$w_i^, = \frac{exp[-\Delta E^o_{r,i}/(RT)]}{\sum_{j=1}^m exp[-\Delta E^o_{r,j}/(RT)]}, \tag{3}$$

entirely based only on the potential-energy terms. It is important to realize that the simple Boltzmann factors can never cross and thus, they cannot represent well a complicated system - for an illustration, see Fig. 3.

There is still another interesting factor which should be discussed. Chirality contributions are to be considered in Eq. 2. There is no asymmetric carbon atom in conventional sense in the fullerene cages (with three coordinated carbon atoms), nevertheless, some of the structures are still chiral, i.e. they are not superimposable upon their mirror image. This structural dissymmetry can readily be recognized from the point group of symmetry as presence of no reflection symmetry, i.e. absence of rotation-reflection axes S_n: only the C_n, D_n, T, O, and I groups obey the requirement. For an enantiomeric pair its partition function q_i in Eq. 2 has to be doubled (if we assume the presence of both optical isomers, which seems natural under fullerene synthesis conditions).

4 The C_{76} IPR Isomers

Since its isolation by Ettl *et al.*[170] C_{76} has received a constant attention.[171-179] C_{76} is the smallest fullerene which allows for isomerism of the IPR structures though there are just two different IPR structures. Their topological symmetries are D_2 and T_d. However, the latter structure exhibits degenerate, partially filled frontier orbitals and thus, it has to undergo a Jahn-Teller lowering of symmetry and energy. It turns out in our AM1 computations[73,180] that this relaxation process ends in a D_{2d} symmetry. The D_{2d} isomer is located about 108 kJ/mol above the ground state in the AM1 potential energy. The separation energy derived from the heat of formation at room temperature $\Delta H^{o}_{f,298}$ is 103 kJ/mol. The geometrical distortion is quite small as can be seen from the almost identical rotational constants of the D_{2d} structure: 0.00174, 0.00173, and 0.00173 cm^{-1}. For the other species, D_2, the rotational constants are distinctly different: 0.00194, 0.00171, and 0.00156 cm^{-1}. All computed vibrational frequencies of both structures are real; hence we really deal with local energy minima after the symmetry relaxation.

The D_2 structure is in a convenient situation as it is lower in energy, it is enhanced by its chirality factor, and its low vibrational frequencies are lower than those of the D_{2d} isomer. Owing to this coincidence it must be prevailing even at elevated temperatures. Even at 4000 K the ground-state structure represents more than 95 % of the two-isomer equilibrium mixture. Hence, the D_{2d} C_{76} species can hardly be observed.

5   The C_{78} IPR Isomers

In contrast to the C_{76} case, isomerism of C_{78} has been well known[74–77,108,181] not only from computations but also from experiments. First, two isomers (D_3 and C_{2v} symmetry, ratio 1:5, the latter species is also labeled by $C_{2v}(I)$) were observed by Diederich et al.[74] Later on, Kikuchi et al.[75] reported three isomers of C_{78} with the symmetries D_3, $C_{2v}(I)$, and $C_{2v}(II)$, ratio 2:2:5. The observations were soon completed with computations[77] of the relative stabilities. In fact, it was only the third case of such computations in the fullerene field after two model studies.[182,183]

The computations of C_{78} were performed withe the MNDO, AM1, and PM3 methods. There are five IPR isomers of C_{78} and each of the topologically possible structures produces a local energy minimum. Their symmetries after the MNDO optimizations are $D_{3h}(I)$, $D_{3h}(II)$, D_3, $C_{2v}(I)$, and $C_{2v}(II)$. In all three semiempirical methods the $C_{2v}(II)$ structure represents the lowest energy minimum (in agreement with MM3 force-field results).

The global minimum represents more than 50% of the equilibrium mixture even at a temperature of 4000 K. On the contrary, the relative population of the $D_{3h}(I)$ structure is always negligible. The remaining three structures exhibit similar relative stabilities, and at high temperatures this three-membered group represents nearly 50% of the equilibrium isomeric mixture. The picture is basically the same for the three energetics considered. According to Fowler *et al.*[108] the structures $D_{3h}(\text{I})$, $D_{3h}(\text{II})$, D_3, $C_{2v}(\text{I})$, and $C_{2v}(\text{II})$ should exhibit 8, 8, 13, 21, and 22 NMR lines, respectively. Diederich *et al.*[74] isolated two C_{78} isomers which showed 13 and 21 NMR lines, respectively, and were therefore identified as the D_3 and $C_{2v}(\text{I})$ structures. The presence of only two isomers (rather than five) was rationalized by two product channels, the minor one leading to the D_3 species and the major one to the $C_{2v}(\text{I})$ species.

Kikuchi *et al.*[75] reported the observation of the D_3, $C_{2v}(\text{I})$, and $C_{2v}(\text{II})$ isomers of C_{78} in a 2:2:5 ratio which is qualitatively more consistent with the computations than the 1:5:0 ratio from Ref. 74. However, in order to get a quantitative agreement we have to reach rather high temperatures. Still, it should be realized that a relatively small change in the observed ratio can reduce the temperatures considerably. There is a room for further computational activities as also additional experimental results were reported recently.[184-186] In particular, data reported by Saunders *et al.*[87] could be consistent even with five species.

6   The C_{80} IPR Isomers

The C_{80} system allows[78-81] for seven different IPR structures. The isomers are conventionally coded by **A-G** (Fig. 4). Of the seven IPR C_{80} structures, the species **B** has topological symmetry I_h, i.e., the same as in the icosahedral C_{60}. However, according to the concept of the Goldberg polyhedra based on topological duals it has to be an open shell and it has to undergo a Jahn-Teller distortion towards lower energy and lower symmetry. Again, we can check on the rotational constants that the distortions are quite small: 0.00157, 0.00156, and 0.00155 cm^{-1}, i.e. a near-spherical top. The symmetry of the **B** structure is lowered to D_2, but still the species possesses the highest energy in the studied IPR set. The computations (both at room and at absolute zero temperature) point out the **C** isomer (D_{5d} symmetry) as the system ground state, being followed by the **A** species of a D_2 symmetry.

Fig. 5 presents the temperature evolution of the HF/4-31G computed[80] relative concentrations, w_i, for the seven-membered mixture under the thermodynamic-equilibrium conditions (Fig. 5 refers to the topological symmetries of the isomers). At very low temperatures the ground-state structure, **C** - D_{5d} symmetry, has to be prevailing, and the relative concentrations of other structures have to obey the $\Delta H^o_{f,0,rel}$ order. But at higher temperatures the pre-exponential factors in Eq. 2 become important (while importance of the exponential part

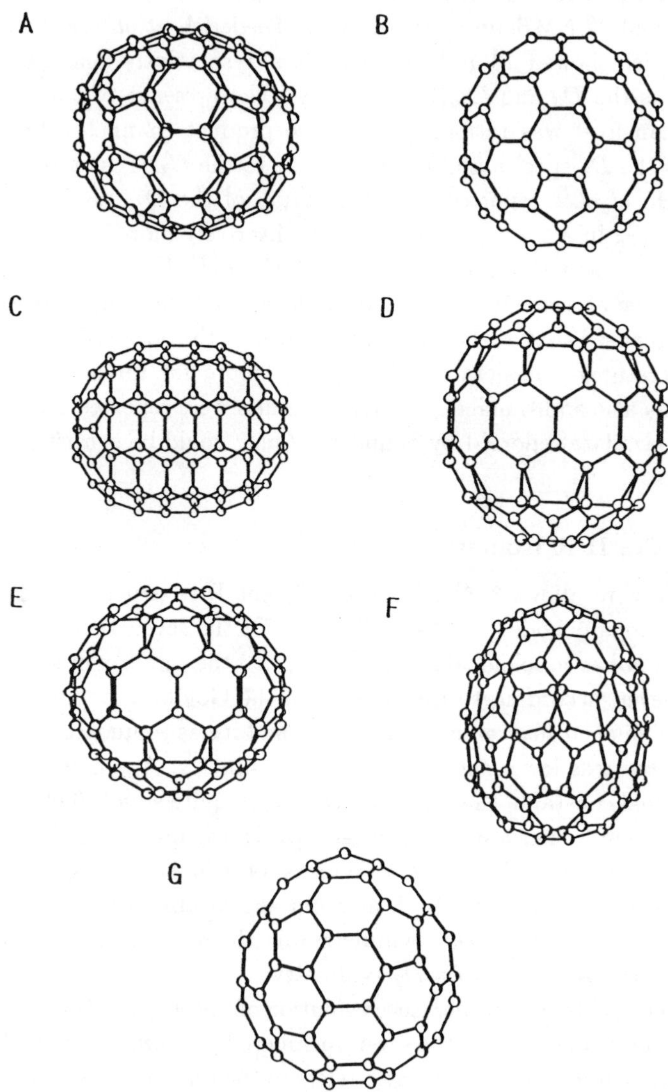

Fig. 4. The IPR structures[80] of C_{80}.

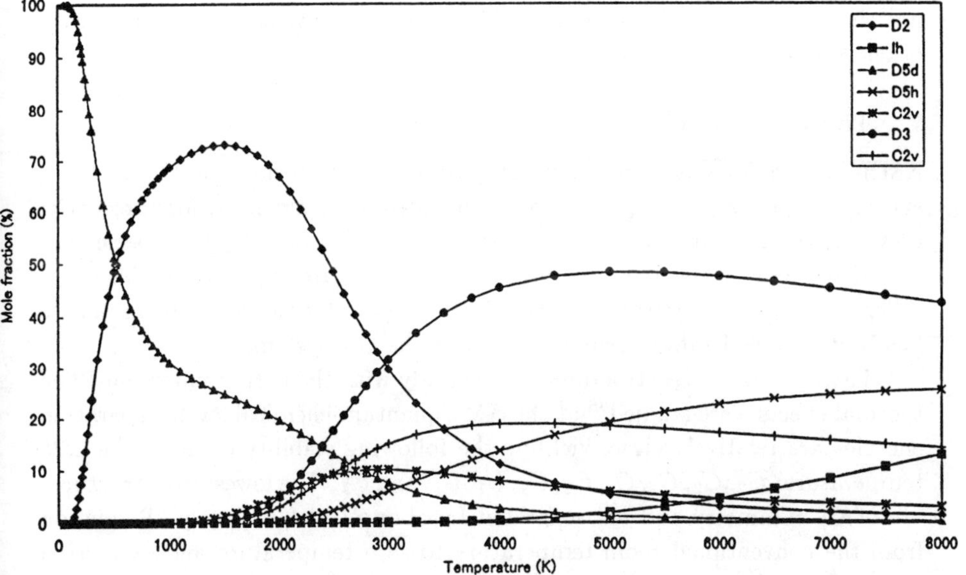

Fig. 5. The computed[96] (HF/4-31G energetics) relative concentrations of C_{80} (topological symmetries indicated).

is gradually decreased). At a relatively low temperature the **A** (D_2 symmetry) species reaches equimolarity with the **C** species, and beyond that point is always more populated. Nevertheless, at considerably higher temperatures another species becomes the most populated.

The separation energies were computed[80] with the SAM1, AM1, HF/STO-3G, HF/3-21G, and HF/4-31G approaches. The results reasonably well agree and the agreement between various types of computations supports reliability of the predicted relative concentrations. Recently, Hennrich *et al.*[81] reported NMR spectra of C_{80} consistent with the **A** (D_2 symmetry) species. According to the HF/4-31G computations[96,187] the equimolarity point between the critical **A** and **C** structures is reached already at 517 K. At a temperature of 1500 K (which probably represents a reasonable temperature for the fullerene synthesis) the **A** structure of D_2 symmetry forms about 73% of the equilibrium mixture. In fact, the **A** isomer reaches its relative concentration maximum at a temperature of 1497 K in the HF/4-31G energetics. Thus, the computations and observations point out as the most populated species a structure which is not the ground state of the system. This interesting event is known also with other chemical systems.[188]

7   The C_{82} IPR Isomers

NMR spectra for C_{82} were originally reported by Kikuchi *et al.*[75] and three symmetry species (C_2, C_{2v}, C_{3v}) were concluded in a ratio 8:1:1. More recently, however, the interpretation has been modified[65–67] - only C_2 C_{82} species are considered now while the two minor components are supposed to be other fullerenes. In fact, a remark in Ref. 65 allows even for two C_2 isomers. There has been a considerable research interest in the C_{82} system.[82–87]

There are nine C_{82} structures that comply with the IPR requirement. The thermal effects were studied[86] at the AM1 quantum-chemical level. Separation energies are relatively close, yielding the following stability order at the zero temperature: $C_2, C_2, C_s, C_s, C_2, C_s, C_1, C_s$, and C_s. The lowest two structures are of C_2 symmetry and are separated by about 17 kJ/mol only. Reduction from the conventional room temperature to zero temperature and extraction of the zero-point energy brings changes smaller than 1 kcal/mol, and changes nothing in the stability order.

At very low temperatures the ground-state structure (C_2 symmetry) is of course dominant. At moderate temperatures the other low-energy C_2 isomer represents the second most stable species. However, its population in fact exhibits a temperature maximum of about 20% and then decreases. At very high temperatures other structures come into consideration, including the third C_2 isomer. In order to correlate computations and observations, some information on the temperature history of the particular sample is needed, i.e., a representative reaction temperature. Anyhow, the group of C_2 species always represents more than 50% of the mixture (e.g., 84% at 1500 K). This result well agrees with the newer interpretation[65-67] of the experiment regardless if there is one or two observed C_2 structures. Incidentally, one could, with some difficulty, find a ground for the original interpretation (C_2, C_{2v}, C_{3v} in a ratio 8:1:1).

8 The C_{84} IPR Isomers

C_{84} fullerenes are also among the very first species observed[88] immediately after the C_{60}/C_{70} synthesis. Hence, their observations have immediately been complemented with theoretical studies[25,69,91,92,107,111,189-194] though we are facing already 24 IPR structures. In their NMR observation Kikuchi *et al.*[75] concluded two major isomers of D_2 and D_{2d} symmetry. Saunders *et al.*[87] could however recognize up to nine C_{84} isomers. The two major isomers have been separated very recently by Dennis *et al.*[195]

Bakowies *et al.*[192] supplied a complete MNDO characterization of the C_{84} IPR isomers and their study was soon completed[92] with temperature dependencies of the relative stabilities in the isomeric set. From our point of view, C_{84} is another interesting isomeric set where the lowest-energy cage is not the most populated species at the conditions of the fullerene synthesis. The two critical species are conventionally labeled[192] **23** and **22** and exhibit D_{2d} and D_2 symmetry, respectively. In the MNDO computations the lowest isomer **23** is located only about 2 kJ/mol below structure **22**. The structure highest in the energy, having also a D_2 symmetry, is located about 240 kJ/mol above the lowest minimum.

The lowest minimum **23** is in fact the most populated species only for low temperatures below room temperature - the **22/23** equimolarity point is located at a temperature of 276 K. After this relative stability interchange structure **22** becomes the most populated isomer in a wide temperature interval. Moreover, the **22/23** ratio is roughly 2:1 anywhere above 500 K - for example the ratio of the two isomers is 58.2:41.7, 41.4:20.6, and 19.0:8.84 at temperature 500, 2000, and 4000 K, respectively. In a simple model, this 2:1 ratio can in fact be identified[113] with the chirality partion function in Eq. 2. Clearly enough, there are other species exhibiting a significant mole fraction at very high temperatures - it is pretty consistent with the observations by Saunders *et al.*[87] For example, if we select a threshold w_i value of 10%, the computations point out three additional isomers of this relative stability. Competition between some species creates an interesting temperature dependency with a maximum (already seen in Fig. 5). In fact, even in very complex fullerene mixtures we can in some specific cases suggest an optimal region of temperatures in which production should be maximized. Although temperature history of the sample used in the observations of Kikuchi *et al.*[75] is virtually unknown, the computed data can match their observed ratio in a very wide temperature interval and hence, the C_{84} system represents another instructive example of an encouraging theory-experiment agreement.

9   The C_{86} IPR Isomers

As soon as the isomeric fullerene systems C_{76}, C_{78}, C_{80}, C_{82}, and C_{84} were computed, it had become rather clear that the inter-isomeric thermodynamic equilibrium presumption was actually working (though it was somewhat surprising). Hence, still higher systems should be submitted for computations as long as some observations are available - at present it extends[65-67] till at least C_{94}. C_{86} was the next system in the row.[93-96] There are nineteen topologically different C_{86} structures (Fig. 6) which obey the IPR requirement. They were computed primarily with the newest tool of semiempirical quantum chemistry, the SAM1 method.[163] Nevertheless, also the AM1, HF/STO-3G, and STO/3-21G treatments consistently produce a C_2 species (numbered **17** in Fig. 6) as the ground state of the IPR C_{86} set.

At very low temperatures the ground-state structure is of course prevailing, however, its mole fraction decreases relatively fast. At first, a C_s species reaches close to 20% (it has its maximum at a temperature of 1503 K). There is a third significant species (C_3 symmetry). Although it is higher in energy its entropy contribution allows for equimolarity (about 29% each) with the ground state at a temperature of 1739 K. Let us mention however that at this temperature two additional species exhibit relative concentrations of about 10%. Achiba *et al.*[65] concluded from ^{13}C NMR spectra presence of two C_2 isomers in their C_{86} sample (the abundance of the isomers was about 4:1). Our computations for the C_2 and C_s structures around 1500 K would agree with the observation; at higher temperatures we would predict coexistence of more than two isomers. It is however true that the computed symmetry is C_s, not C_2. This minor difference has not been clarified yet.

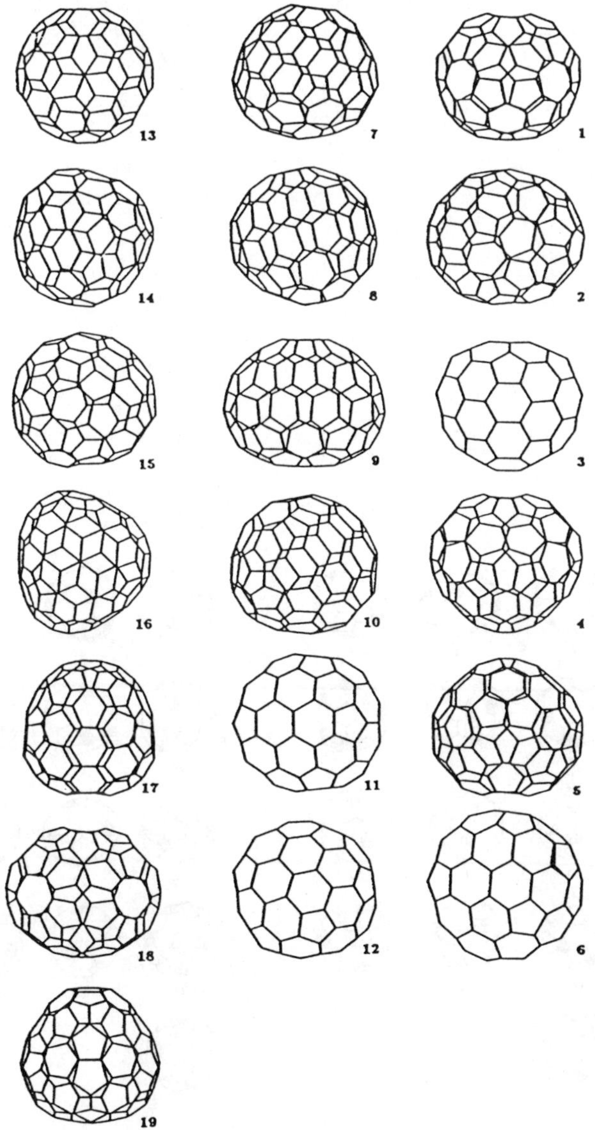

Fig. 6. The IPR structures[93] of C_{86} (numbering according to Ref. 72).

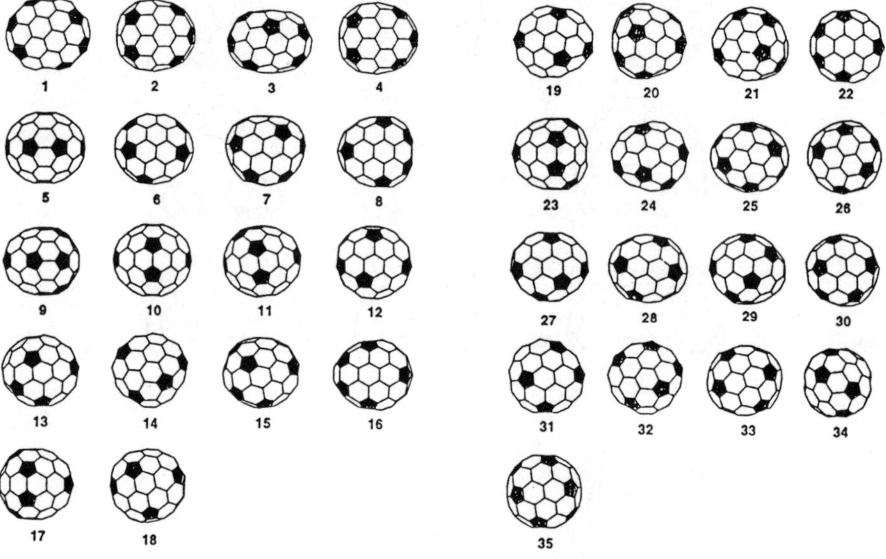

Fig. 7. The IPR structures[196] of C_{88} (numbering according to Ref. 70).

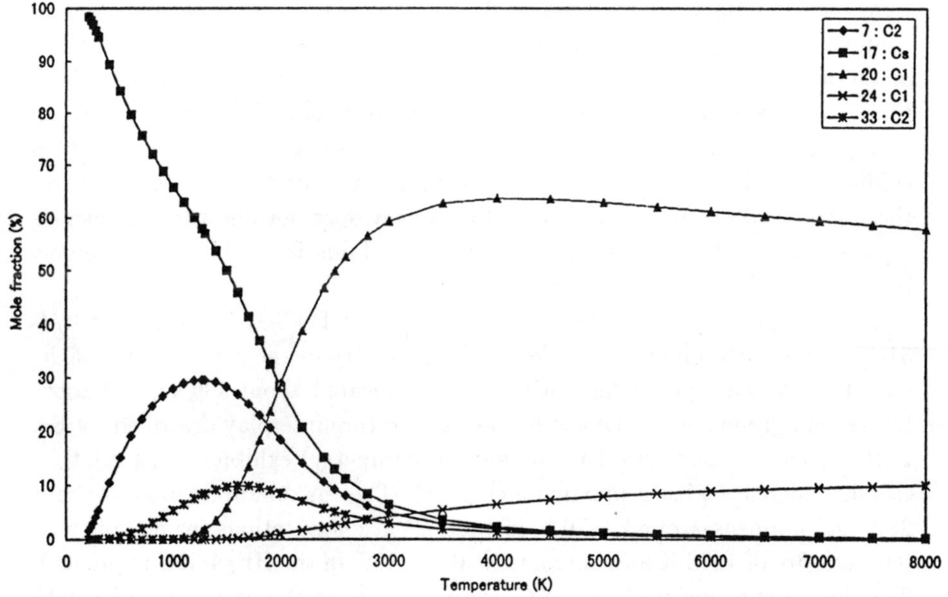

Fig. 8. The computed[196] (HF/4-31G energetics) relative concentrations of C$_{88}$.

10   The C_{88} IPR Isomers

Although there is a drop in the IPR structures when going from C_{84} to C_{86}, the C_{88} IPR system is already quite extended. There are thirty five topologically different C_{88} IPR structures (Fig. 7), nevertheless, also this system has reasonably been clarified,[94-96,196] again primarily at the SAM1 semiempirical level.

As already indicated (e.g., with the icosahedral structure for C_{80}) the original topological symmetries represent a kind of upper bound - the true symmetries after a quantum-chemical geometry optimization can be the same or lower. In particular, during quantum-chemical calculations the topological symmetry can be lowered owing to Jahn-Teller and pseudo Jahn-Teller effects, or simply owing to general energy reasons. Hence, quantum-chemical computations are then an essential tool in structure calculations as molecular mechanics cannot handle the Jahn-Teller effect. Such symmetry reductions are not rare in the fullerene field, and can frequently happen even outside a Jahn-Teller situation. For example, the icosahedral B_{32} (a topological boron analogy of C_{60}) exhibits[197] imaginary frequencies when computed *ab initio*. Similarly, hexaanion of C_{60} seems to be icosahedral (as the degenerate LUMO is just filled), but mother Nature wants it otherwise and relaxes its high symmetry with an energy gain[198] of some 60-150 kJ/mol. In the C_{88} system the SAM1 computed symmetry in five cases falls below the topological expectations. In four cases the topological symmetry itself is so low that it does not allow for degenerate representations. Hence, we meet only one pure Jahn-Teller distortion (namely from T to D_2).

Four types of computations[196] (SAM1, HF/STO-3G, HF/3-21G, HF/4-31G) consistently predict a species of C_s symmetry as the ground state of the C_{88} IPR set. The species highest in energy is located about 300 kJ/mol above the system ground state. Owing to the temperature interplay described by Eq. 2, the species second lowest in energy (C_2 symmetry) exhibits a fast relative-stability increase with a maximum (Fig. 8). The maximum fraction is about 23% and happens around 1470 K in the SAM1 computations (or already at a temperature of 1270 K and amounts to about 30% in the HF/4-31G approach). The third lowest species (again a C_2 symmetry) has a temperature profile with a maximum, too, though rather modest. There is however a structure with a steady increase which eventually becomes dominant - a C_1 symmetry, down from a C_2 topological expectation. The isomer is relatively rich in potential energy - over 100 kJ/mol above the ground state. It has an equimolarity point with the ground-state structure which comes in the HF/4-31G computations at about 1990 K with 29% for each species - see Fig. 8 (in the SAM1 energetics the equimolarity is reached at about 2240 K with about 26% for either of the two isomers). At a temperature of 3000 K the C_1 relative concentration amounts to 59 and 47% in the HF/4-31G and SAM1 energetics, respectively.

^{13}C NMR spectrum of C_{88} in solution has been interpreted by Achiba *et al.*[65] as consistent with an isomer of C_2 symmetry as a major structure. Although our computations point out even two C_2 structures, their temperature maxima appear at rather low temperatures when the ground state structure still has a strong population and should be seen in the experiment. However, if we consider temperatures above 2240 and 1990 K according to the SAM1 and HF/4-31G predictions, respectively, we shall deal with the C_1 structure as the most populated and eventually the dominant one. The structure exhibits a C_2 topological symmetry which is reduced to C_1 in the quantum-chemical optimization. Such symmetry reductions can be connected with rather small coordinate distortions. For example, the topologically icosahedral C_{80} cage exhibits[80] the Jahn-Teller geometry distortions of the order of 10^{-3} or 10^{-2} Å. Hence, one can speak on a near-icosahedral structure in this case. The C_1 C_{88} structure exhibits in our search a C_2 symmetry for $\epsilon > 4 \times 10^{-2}$ Å, too. On the other hand, the vibrational amplitudes of C_{60} from electron diffraction are still larger, and the same is true, e.g., for the vibrational amplitudes of kekulene at room temperature.[196] The data give us some ground to think on the C_1 structure as having a near C_2 symmetry. This argument would produce an agreement between theory and experiment, and it would also suggest that the sample originated at temperatures somewhat over 2000 K.

11 The C_{90} IPR Isomers

With the continuing confidence in a reasonable applicability of the inter-isomeric thermodynamic equilibrium presumption, we have treated[97,98] the C_{90} system. There are already forty six topologically different C_{90} IPR structures. The computations point out structure with a C_2 symmetry (labeled **38** and **45** according to Ref. 72 and 70, respectively - see Fig. 9) as the system ground state. There is an exception in the HF/3-21G treatment as the D_{5h} structure (only one such high-symmetry species in the set) is practically isoenergetic with the **38/45** structure (see Fig. 2).

In observations of Achiba *et al.*[65] altogether five C_{90} species were identified from ^{13}C NMR spectra (distributed in three HPLC fractions): one C_{2v}, three

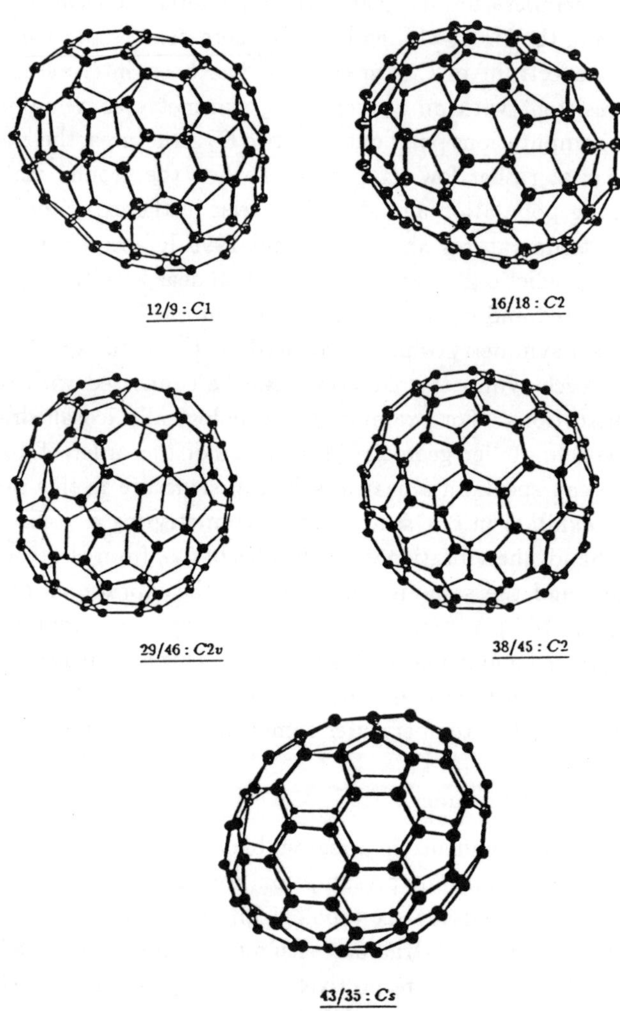

Fig. 9. The five IPR structures[98] of C_{90} significant at higher temperatures (numbering: Ref. 72/Ref. 70).

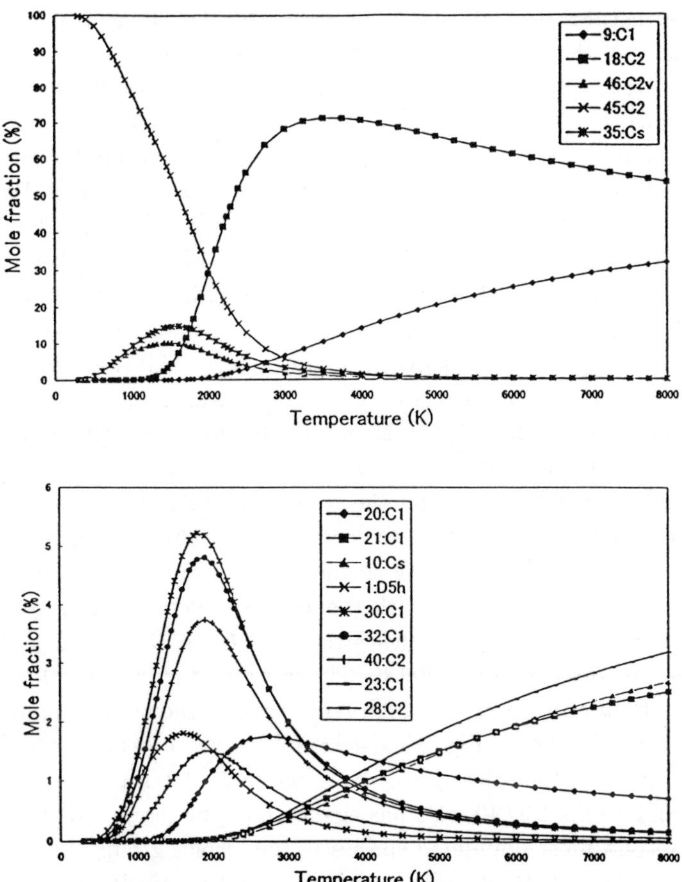

Fig. 10. The SAM1 computed[98] relative concentrations of C_{90}: Top - structures from Fig. 9; bottom - some less-populated species (numbering according to Ref. 70).

C_2, and one C_1 (recent surveys[66,67] do not mention the C_1 structure and give two C_2 structures instead, though no reason for the change is mentioned). The symmetries of the five SAM1 lowest-energy structures are (cf. Fig. 9, though only the structures really relevant at higher temperatures are shown there): C_2 (**38/45**), C_{2v} (**29/46**), C_s (**43/35**), D_{5h} (**19/1**), C_1 (**22/30**), i.e., at least two structures do not appear in the experiment. The computation-observation agreement however becomes even worse if we look into the NMR pattern of the computed C_{2v} structure - 24 lines, 3 of them weaker. The C_{2v} species considered in the experiment exhibits 25 lines, 5 of them weaker!

In the view of the above partial agreement with the experiment we really have to investigate possible temperature effects on the relative stabilities. It turns out that there are just five structures which exhibit a significant population in a high-temperature region (Fig. 10). In addition to the three structures lowest in energy (C_2 **38/45**, C_{2v} **29/46**, C_s **43/35**), two high-energy species are relevant: C_2 **16/18** ($\Delta E_r = 130$ kJ/mol) and C_1 **12/9** ($\Delta E_r = 202$ kJ/mol). Again, at very low temperatures the ground-state structure has of course to be the dominant species, however, its decrease of stability with increasing temperature is considerable. According to the SAM1 computations the ground-state isomer exhibits equimolarity with the **16/18** structure of C_2 symmetry at a temperature of 2012 K. Two other structures, C_s **43/35** and C_{2v} **29/46**, show moderate maxima close to 1500 K. The last structure of Fig. 9, C_1 **12/9**, becomes quite important at very high temperatures though its population is still too low around 1000 K. Fig. 10 also samples several less-populated structures - it is of interest that the species with the highest temperature maximum in this second group, **22/30**, also has again a C_1 symmetry, and the same is true for the second most populated species in the second group, **32/32**. Results in the the SAM1 and HF/4-31G energetics are not particularly different.

We do not know what was a temperature interval in which the C_{90} sample used in the observations was actually synthesized, we may guess the temperatures somewhere beyond 800 K. Our computations predict that at elevated temperatures we primarily deal with the five structures of the symmetries: two times C_2, C_s, C_{2v}, C_1. This SAM1 high-temperature set compares now better with the conclusion[65] from the ^{13}C NMR spectra: C_{2v}, three times C_2, C_1. However, we should remember that in the computations and experiment we in fact deal with the C_{2v} species of different NMR pattern. The experimental conclusion is based on an NMR spectrum consisting[65] of exactly 70 lines, five of them weaker. Any IPR C_{90} isomer of C_2 symmetry exhibits always 45 lines (none weaker). There are C_{2v} C_{90} structures with exactly 25 lines (5 of them weaker). This is the essence of the experimental interpretation. However, in the computations, we deal with a different C_{2v} structure - only 24 lines (3 of them weaker). Morover, we have still another species not mentioned in the experiment interpretation, a C_s structure with 46 lines (2 of them weaker). If we now combine together the NMR patterns of the two structures predicted by theory, C_{2v} and C_s, we also get 70 lines (5 of them weaker). With this alternative interpretation of the NMR spectrum[65] we reach a complete agreement of the computations with the observations. Let us note finally that the C_s species is somewhat more populated than C_{2v} which actually helps to equalize intensities of their NMR lines.

12   Concluding Remarks

As this article shows there have been productive computation-observation interactions in the fullerene research. For higher fullerenes in particular, the interactions are essential owing to the complexity. The reported considerable thermal effects on the relative concentrations in IPR sets result from an interplay between rotational, vibrational, potential-energy terms, and chirality factors. Such effects can never be seen if only energetics is considered and entropy neglected (i.e., the simple factors from Eq. 3 instead of Eq. 2). Our treatment is however based on the inter-isomeric thermodynamic equilibrium. We really do not know a degree to which this presumption is satisfied in the experiment. However, the thermodynamic-equilibrium treatment already produced a reasonable theory-experiment agreement in eight isomeric systems (C_{76} to C_{90}) with no serious failure. This relatively large set strengthens our believe in still wider applicability of the equilibrium treatment, and its further testing on C_{92} and C_{94} is in progress. This growing family of higher fullerenes will soon or later attract interest of materials science, either as a unique group of species or as a bridge towards nanostructures.

In this survey we almost exclusively treated the IPR structures only. Future research could however bring to our attention some non-IPR structures, too. After all, there is a report on observations of possible isomers of C_{60} and C_{70} by Anacleto *et al.*[199] It this is true, we still do not know their structure. It could be some non-IPR cages as computations on related silicon structures support[200] the option. However, Gotts *et al.*[201] in their famous ion chromatography experiments can see charged cyclic C_{60} structures. Hence, non-IPR solid state materials should not be excluded.

Acknowledgments

The reported research has in part been supported by the Ministry of Education, Science and Culture in Japan. X.Z. thanks the Toyota Tsusho International Scholarship Foundation for granting him a Ph.D. fellowship. The authors also wish to thank the following organizations for kindly permitting the reprinting of copyrighted material: Elsevier Scientific Publishing Company; Marcel Dekker, Inc.; The Electrochemical Society, Inc.

References

1. M. S. Dresselhaus, G. Dresselhaus and P. C. Eklund, *Science of Fullerenes and Carbon Nanotubes* (Academic Press, San Diego, 1996).
2. J. Mattauch, H. Ewald, O. Hahn and F. Strassmann, *Z. Phys.* **20**, 598 (1943).
3. R. E. Honig, *J. Chem. Phys.* **22**, 126 (1954).
4. J. Drowart, R. P. Burns, G. DeMaria and M. G. Inghram, *J. Chem. Phys.* **31**, 1131 (1959).
5. H. Hintenberger, J. Franzen and K. D. Schuy, *Z. Naturforsch.*, **18a**, 1236 (1963).
6. K. S. Pitzer and E. Clementi, *J. Am. Chem. Soc.* **81**, 4477 (1959).
7. R. Hoffmann, *Tetrahedron* **22**, 521 (1966).
8. Z. Slanina, *Ph.D. Thesis* (Czech. Acad. Sci., Prague, 1974, p. 105).
9. Z. Slanina, *Radiochem. Radioanal. Lett.* **22**, 291 (1975).
10. D. E. H. Jones, *New Scientist* **32**, 245 (1966).
11. E. Ōsawa, *Kagaku* **25**, 854 (1970); *Chem. Abstr.* **74**, 75698v (1971).
12. D. A. Bochvar and E. G. Gal'pern, *Dokl. Akad. Nauk SSSR* **209**, 610 (1973).
13. R. A. Davidson, *Theor. Chim. Acta* **58**, 193 (1981).
14. H. W. Kroto, J. R. Heath, S. C. O'Brien, R. F. Curl and R. E. Smalley, *Nature* **318**, 162 (1985).
15. D. R. Huffman and W. Krätschmer, *Materials Res. Soc. Proc.* **206**, 601 (1991).
16. W. Krätschmer, L. D. Lamb, K. Fostiropoulos and D. R. Huffman, *Nature* **347**, 354 (1990).
17. Z. C. Wu, D. A. Jelski and T. F. George, *Chem. Phys. Lett.* **137**, 291 (1987).

18. V. Elser and R. C. Haddon, *Nature* **325**, 792 (1987).

19. R. E. Stanton and M. D. Newton, *J. Phys. Chem.* **92**, 2141 (1988).

20. D. E. Weeks and W. G. Harter, *Chem. Phys. Lett.* **144**, 366 (1988).

21. D. E. Weeks and W. G. Harter, *J. Chem. Phys.* **90**, 4744 (1989).

22. J. M. Rudziński, Z. Slanina, E. Ōsawa and T. Iizuka, *8th IUPAC Conf. Phys. Org. Chem., Tokyo* 262 (1986); Z. Slanina, J. M. Rudziński, M. Togasi and E. Ōsawa, *J. Mol. Struct. (THEOCHEM)* **61**, 169 (1989).

23. T. Braun, *Angew. Chem., Int. Ed. Eng.* **31**, 588 (1992).

24. T. Braun, A. Schubert, H. Maczelka and L. Vasvári, *Fullerene Research 1985-1993* (World Scientific, Singapore, 1995).

25. D. Bakowies and W. Thiel, *J. Am. Chem. Soc.* **113**, 3704 (1991).

26. M. Häser, J. Almlöf and G. E. Scuseria, *Chem. Phys. Lett.* **181**, 497 (1991).

27. W. Weltner Jr. and R. J. van Zee, *Chem. Rev.* **89**, 1713 (1989).

28. R. F. Curl and R. E. Smalley, *Science* **242**, 1017 (1988).

29. H. W. Kroto, *Science* **242**, 1139 (1988).

30. R. C. Haddon, *Accounts Chem. Res.* **21**, 243 (1988).

31. D. A. Jelski and T. F. George, *J. Chem. Educ.* **65**, 879 (1988).

32. D. J. Klein and T. G. Schmalz, in *Quasicrystals, Networks, and Molecules of Fivefold Symmetry*, ed. I. Hargittai (VCH Publishers, New York, 1990).

33. E. Ōsawa, *Kagaku* **45**, 552 (1990).

34. D. R. Huffman, *Phys. Today* **44**, 22 (1991).

35. H. W. Kroto, A. W. Allaf and S. P. Balm, *Chem. Rev.* **91**, 1213 (1991).

36. H. W. Kroto, *Angew. Chem., Int. Ed. Engl.* **31**, 111 (1992).

37. J. R. Heath, in *ACS Symposium Series No 481 - Fullerenes. Synthesis, Properties, and Chemistry of Large Carbon Clusters*, eds. G. S. Hammond and V. J. Kuck (ACS, Washington, 1992).

38. J. E. Fischer, P. A. Heiney, D. E. Luzzi and D. E. Cox, in *ACS Symposium Series No 481 - Fullerenes. Synthesis, Properties, and Chemistry of Large Carbon Clusters*, eds. G. S. Hammond and V. J. Kuck (ACS, Washington, 1992).

39. Z. Slanina, *Chem. Listy* **86**, 327 (1992).

40. R. E. Smalley, *Accounts Chem. Res.* **25**, 98 (1992).

41. J. P. Hare and H. W. Kroto, *Accounts Chem. Res.* **25**, 106 (1992).

42. J. E. Fischer, P. A. Heiney and A. B. Smith III, *Accounts Chem. Res.* **25**, 112 (1992).

43. F. Diederich and R. L. Whetten, *Accounts Chem. Res.* **25**, 119 (1992).

44. R. C. Haddon, *Accounts Chem. Res.* **25**, 127 (1992).

45. P. J. Fagan, J. C. Calabrese and B. Malone, *Accounts Chem. Res.* **25**, 134 (1992).

46. J. H. Weaver, *Accounts Chem. Res.* **25**, 143 (1992).

47. J. M. Hawkins, *Accounts Chem. Res.* **25**, 150 (1992).

48. F. Wudl, *Accounts Chem. Res.* **25**, 157 (1992).

49. S. W. McElvany, M. M. Ross and J. H. Callahan, *Accounts Chem. Res.* **25**, 162 (1992).

50. R. D. Johnson, D. S. Bethune and C. S. Yannoni, *Accounts Chem. Res.* **25**, 169 (1992).

51. D. Koruga, S. Hameroff, J. Withers, R. Loutfy and M. Sundareshan, *Fullerene C_{60} - History, Physics, Nanobiology, Nanotechnology* (Elsevier, Amsterdam, 1993).

52. T. G. Schmalz and D. J. Klein, in *Buckminsterfullerenes*, eds. W. E. Billups and M. A. Ciufolini (VCH Publishers, New York, 1993).

53. G. E. Scuseria, in *Buckminsterfullerenes*, eds. W. E. Billups and M. A. Ciufolini (VCH Publishers, New York, 1993).

54. C. T. White, J. W. Mintmire, R. C. Mowrey, D. W. Brenner, D. H. Robertson, J. A. Harrison and B. I. Dunlap, in *Buckminsterfullerenes*, eds. W. E. Billups and M. A. Ciufolini (VCH Publishers, New York, 1993).

55. R. C. Haddon and K. Raghavachari, in *Buckminsterfullerenes*, eds. W. E. Billups and M. A. Ciufolini (VCH Publishers, New York, 1993).

56. M. L. Cohen and V. H. Crespi, in *Buckminsterfullerenes*, eds. W. E. Billups and M. A. Ciufolini (VCH Publishers, New York, 1993).

57. S. C. Erwin, in *Buckminsterfullerenes*, eds. W. E. Billups and M. A. Ciufolini (VCH Publishers, New York, 1993).

58. M. S. Dresselhaus, G. Dresselhaus and P. C. Eklund, *J. Mater. Res.* **8**, 2054 (1993).

59. J. Cioslowski, *Rev. Comput. Chem.* **4**, 1 (1993).

60. C. Z. Wang, B. L. Zhang, K. M. Ho and X. Q. Wang, *Int. J. Mod. Phys.* B **7**, 4305 (1993).

61. W. Andreoni, in *Electronic Properties of Fullerenes*, eds. H. Kuzmany, J. Fink, M. Mehring and S. Roth (Springer-Verlag, Berlin, 1993).

62. J. González, F. F. Guinea and M. A. H. Vozmediano, *Int. J. Mod. Phys.* B **7**, 4331 (1993).

63. J. Cioslowski, *Electronic Structure Calculations on Fullerenes and Their Derivatives* (Oxford University Press, Oxford, 1995).

64. Z. Slanina, S.-L. Lee and C.-H. Yu, *Rev. Comput. Chem.* **8**, 1 (1996).

65. Y. Achiba, K. Kikuchi, Y. Aihara, T. Wakabayashi, Y. Miyake and M. Kainosho, in *Science and Technology of Fullerene Materials*, eds. P. Bernier, D. S. Bethune, L. Y. Chiang, T. W. Ebbesen, R. M. Metzger and J. W. Mintmire (Materials Research Society, Pittsburgh, 1995).

66. Y. Achiba, K. Kikuchi, Y. Aihara, T. Wakabayashi, Y. Miyake and M. Kainosho, in *The Chemical Physics of Fullerenes 10 (and 5) Years Later*, ed. W. Andreoni (Kluwer Academic Publishers, Dordrecht, 1996).

67. Y. Achiba, *Kagaku* **52** (5), 15 (1997).

68. H. W. Kroto, *Nature* **329**, 529 (1987).

69. D. E. Manolopoulos and P. W. Fowler, *J. Chem. Phys.* **96**, 7603 (1992).

70. P. W. Fowler and D. E. Manolopoulos, *An Atlas of Fullerenes* (Clarendon Press, Oxford, 1995).

71. M. Yoshida and E. Ōsawa, *Bull. Chem. Soc. Jpn.* **68**, 2073 (1995).

72. M. Yoshida, *Ph.D. Thesis* (Toyohashi University of Technology, Toyohashi, 1996).

73. Z. Slanina, M.-L. Sun, S.-L. Lee and L. Adamowicz, in *Recent Advances in the Chemistry and Physics of Fullerenes and Related Materials, Vol. 2*, eds. K. M. Kadish and R. S. Ruoff (The Electrochemical Society, Pennington, 1995).

74. F. Diederich, R. L. Whetten, C. Thilgen, R. Ettl, I. Chao and M. M. Alvarez, *Science* **254**, 1768 (1991).

75. K. Kikuchi, N. Nakahara, T. Wakabayashi, S. Suzuki, H. Shiromaru, Y. Miyake, K. Saito, I. Ikemoto, M. Kainosho and Y. Achiba, *Nature* **357**, 142 (1992).

76. D. Bakowies, A. Geleßus and W. Thiel, *Chem. Phys. Lett.* **197**, 324 (1992).

77. Z. Slanina, J.-P. François, D. Bakowies and W. Thiel, *J. Mol. Struct. (THEOCHEM)* **279**, 213 (1993).

78. S. J. Woo, E. Kim and Y. H. Lee, *Phys. Rev. B* **47**, 6721 (1993).

79. K. Nakao, N. Kurita and M. Fujita, *Phys. Rev. B* **49**, 11415 (1994).

80. M.-L. Sun, Z. Slanina, S.-L. Lee, F. Uhlík and L. Adamowicz, *Chem. Phys. Lett.* **246**, 66 (1995).

81. F. H. Hennrich, R. H. Michel, A. Fischer, S. Richard-Schneider, S. Gilb, M. M. Kappes, D. Fuchs, M. Bürk, K. Kobayashi and S. Nagase, *Angew. Chem., Int. Ed. Engl.* **35**, 1732 (1996).

82. K. Kikuchi, N. Nakahara, T. Wakabayashi, M. Honda, H. Matsumiya, T.

Moriwaki, S. Suzuki, H. Shiromaru, K. Saito, K. Yamauchi, I. Ikemoto and Y. Achiba, *Chem. Phys. Lett.* **188**, 177 (1992).

83. S. Nagase, K. Kobayashi, T. Kato and Y. Achiba, *Chem. Phys. Lett.* **201**, 475 (1993).

84. G. Orlandi, F. Zerbetto and P. W. Fowler, *J. Phys. Chem.* **97**, 13575 (1993).

85. S. Nagase and K. Kobayashi, *Chem. Phys. Lett.* **214**, 57 (1993).

86. Z. Slanina, S.-L. Lee, K. Kobayashi and S. Nagase, *J. Mol. Struct. (THEOCHEM)* **339**, 89 (1995).

87. M. Saunders, H. A. Jiménez-Vázquez, R. J. Cross, W. E. Billups, C. Gesenberg, A. Gonzalez, W. Luo, R. C. Haddon, F. Diederich and A. Herrmann, *J. Am. Chem. Soc.* **117**, 9305 (1995).

88. F. Diederich, R. Ettl, Y. Rubin, R. L. Whetten, R. Beck, M. Alvarez, S. Anz, D. Sensharma, F. Wudl, K. C. Khemani and A. Koch, *Science* **252**, 548 (1991).

89. R. D. Beck, P. S. John, M. M. Alvarez, F. Diederich and R. L. Whetten, *J. Phys. Chem.* **95**, 8402 (1991).

90. K. Kikuchi, N. Nakahara, M. Honda, S. Suzuki, K. Saito, H. Shiromaru, K. Yamauchi, I. Ikemoto, T. Kuramochi, S. Hino and Y. Achiba, *Chem. Lett.* 1607 (1991).

91. K. Raghavachari and C. M. Rohlfing, *J. Phys. Chem.* **95**, 5768 (1991).

92. Z. Slanina, J.-P. François, M. Kolb, D. Bakowies and W. Thiel, *Fullerene Sci. Technol.* **1**, 221 (1993).

93. Z. Slanina, S.-L. Lee, M. Yoshida and E. Ōsawa, *Chem. Phys.* **209**, 13 (1996).

94. Z. Slanina, S.-L. Lee, M. Yoshida and E. Ōsawa, in *Physics and Chemistry of Fullerenes and Their Derivatives*, eds. H. Kuzmany, J. Fink, M. Mehring and S. Roth (World Sci. Publ., Singapore, 1996).

95. Z. Slanina, S.-L. Lee, M. Yoshida and E. Ōsawa, in *Recent Advances in the Chemistry and Physics of Fullerenes and Related Materials, Vol. 3*, eds. K. M. Kadish and R. S. Ruoff (The Electrochemical Society, Pennington, 1996).

96. Z. Slanina, S.-L. Lee and L. Adamowicz, *Int. J. Quantum Chem.* **63**, 529 (1997).

97. Z. Slanina, X. Zhao, S.-L. Lee and E. Ōsawa, in *Recent Advances in the Chemistry and Physics of Fullerenes and Related Materials, Vol. 4*, eds. K. M. Kadish and R. S. Ruoff (The Electrochemical Society, Pennington,

1997).

98. Z. Slanina, X. Zhao, S.-L. Lee and E. Ōsawa, *Chem. Phys.* **219**, 193 (1997).

99. P. W. Fowler and J. Woolrich, *Chem. Phys. Lett.* **127**, 78 (1986).

100. P. W. Fowler, *Chem. Phys. Lett.* **131**, 444 (1986).

101. P. W. Fowler and J. I. Steer, *J. Chem. Soc., Chem. Commun.* 1403 (1987).

102. P. W. Fowler, J. E. Cremona and J. I. Steer, *Theor. Chim. Acta* **73**, 1 (1988).

103. A. Ceulemans and P. W. Fowler, *Phys. Rev. A* **39**, 481 (1989).

104. A. Ceulemans and P. W. Fowler, *J. Chem. Phys.* **93**, 1221 (1990).

105. P. W. Fowler, *J. Chem. Soc., Faraday Trans.* **86**, 2073 (1990).

106. P. Fowler, *Nature* **350**, 20 (1991).

107. P. W. Fowler, *J. Chem. Soc., Faraday Trans.* **87**, 1945 (1991).

108. P. W. Fowler, R. C. Batten and D. E. Manolopoulos, *J. Chem. Soc., Faraday Trans.* **87**, 3103 (1991).

109. A. Ceulemans and P. W. Fowler, *Nature* **353**, 52 (1991).

110. P. W. Fowler and D. E. Manolopoulos, *Nature* **355**, 428 (1992).

111. P. W. Fowler, D. E. Manolopoulos and R. P. Ryan, *J. Chem. Soc., Chem. Commun.* 408 (1992).

112. P. W. Fowler, *J. Chem. Soc., Perkin Trans. II* 145 (1992).

113. P. W. Fowler, D. E. Manolopoulos and R. P. Ryan, *Carbon* **30**, 1235 (1992).

114. D. E. Manolopoulos, D. R. Woodall and P. W. Fowler, *J. Chem. Soc., Faraday Trans.* **88**, 2427 (1992).

115. P. W. Fowler and V. Morvan, *J. Chem. Soc., Faraday Trans.* **88**, 2631 (1992).

116. P. W. Fowler, D. E. Manolopoulos, D. B. Redmond and R. P. Ryan, *Chem. Phys. Lett.* **202**, 371 (1993).

117. D. E. Manolopoulos and P. W. Fowler, *Chem. Phys. Lett.* **204**, 1 (1993).

118. S.-L. Lee, *Theor. Chim. Acta* **81**, 185 (1992).

119. A. J. Stone and D. J. Wales, *Chem. Phys. Lett.* **128**, 501 (1986).

120. C. Coulombeau and A. Rassat, *J. Chim. Phys.* **88**, 173 (1991).

121. W. N. Lipscomb and L. Massa, *Inorg. Chem.* **31**, 2297 (1992).

122. A. C. Tang, Q. S. Li, C. W. Liu and J. Li, *Chem. Phys. Lett.* **201**, 465 (1993).

123. D. E. Manolopoulos, J. C. May and S. E. Down, *Chem. Phys. Lett.* **181**,

105 (1991).

124. X. Liu, D. J. Klein, T. G. Schmalz and W. A. Seitz, *J. Comput. Chem.* **12**, 1252 (1991).

125. X. Liu, D. J. Klein, W. A. Seitz and T. G. Schmalz, *J. Comput. Chem.* **12**, 1265 (1991).

126. X. Liu, T. G. Schmalz and D. J. Klein, *Chem. Phys. Lett.* **188**, 550 (1992).

127. D. E. Manolopoulos, *Chem. Phys. Lett.* **192**, 330 (1992).

128. X. Liu, T. G. Schmalz and D. J. Klein, *Chem. Phys. Lett.* **192**, 331 (1992).

129. M.-L. Sun, *M.A. Thesis* (National Chung-Cheng University, Chia-Yi, 1995).

130. P.W. Fowler, C. M. Quinn and D. B. Redmond, *J. Chem. Phys.* **95**, 7678 (1991).

131. S. Wei, Z. Shi and A. W. Castleman Jr., *J. Chem. Phys.* **94**, 3268 (1991).

132. D. J. Klein, W. A. Seitz and T. G. Schmalz, *J. Phys. Chem.* **97**, 1231 (1993).

133. D. J. Klein, *Chem. Phys. Lett.* **217**, 261 (1994).

134. A. T. Balaban, D. J. Klein and C. A. Folden, *Chem. Phys. Lett.* **217**, 266 (1994).

135. J. R. Dias, *Chem. Phys. Lett.* **209**, 439 (1993).

136. K. Balasubramanian, *Chem. Phys. Lett.* **182**, 257 (1991).

137. K. Balasubramanian, *Chem. Phys. Lett.* **183**, 292 (1991).

138. K. Balasubramanian, *J. Chem. Inf. Comput. Sci.* **32**, 47 (1992).

139. K. Balasubramanian, *Chem. Phys. Lett.* **198**, 577 (1992).

140. K. Balasubramanian, *Chem. Phys. Lett.* **197**, 55 (1992).

141. K. Balasubramanian, *J. Mol. Spectr.* **157**, 254 (1992).

142. K. Balasubramanian, *Chem. Phys. Lett.* **201**, 306 (1993).

143. K. Balasubramanian, *Chem. Phys. Lett.* **206**, 210 (1993).

144. K. Balasubramanian, *Chem. Phys. Lett.* **202**, 399 (1993).

145. K. Balasubramanian, *J. Phys. Chem.* **97**, 6990 (1993).

146. K. Balasubramanian, *J. Phys. Chem.* **97**, 4647 (1993).

147. M. Yoshida and E. Ōsawa, *Bull. Chem. Soc. Jpn.* **68**, 2083 (1995).

148. H. Ueno, *M.A. Thesis* (Toyohashi University of Technology, Toyohashi, 1997).

149. X. Zhao, H. Ueno, Z. Slanina and E. Ōsawa, in *Recent Advances in the Chemistry and Physics of Fullerenes and Related Materials, Vol. 5*, eds.

K. M. Kadish and R. S. Ruoff (The Electrochemical Society, Pennington, 1997).

150. E. Ōsawa, Z. Slanina, K. Honda and X. Zhao, in *Recent Advances in the Chemistry and Physics of Fullerenes and Related Materials, Vol. 5*, eds. K. M. Kadish and R. S. Ruoff (The Electrochemical Society, Pennington, 1997).

151. S. W. McElvany, J. H. Callahan, M. M. Ross, L. D. Lamb and D. R. Huffman, *Science* **260**, 1632 (1993).

152. J.-P. Deng, D.-D. Ju, G.-R. Her, C.-Y. Mou, C.-J. Chen, Y.-Y. Lin and C.-C. Han, *J. Phys. Chem.* **97**, 11575 (1993).

153. Z. Slanina, *Contemporary Theory of Chemical Isomerism* (Kluwer, Dordrecht, 1986).

154. S. Fujita, *Symmetry and Combinatorial Enumeration in Chemistry* (Springer-Verlag, Berlin, 1991).

155. T. Clark, *A Handbook of Computational Chemistry. A Practical Guide to Chemical Structure and Energy Calculations* (Wiley, New York, 1985).

156. D. B. Boyd, *Rev. Comput. Chem.* **1**, 321 (1990).

157. J. J. P. Stewart, *Rev. Comput. Chem.* **1**, 45 (1990).

158. M. C. Zerner, *Rev. Comput. Chem.* **2**, 313 (1991).

159. W. J. Hehre, L. Radom, P. R. von Schleyer and J. A. Pople, *Ab Initio Molecular Orbital Theory* (J. Wiley, New York, 1986).

160. M. J. S. Dewar and W. Thiel, *J. Am. Chem. Soc.* **99**, 4899 (1977).

161. M. J. S. Dewar, E. G. Zoebisch, E. F. Healy and J. J. P. Stewart, *J. Am. Chem. Soc.* **107**, 3902 (1985).

162. J. J. P. Stewart, *J. Comput. Chem.* **10**, 209 (1989).

163. M. J. S. Dewar, C. Jie and J. Yu, *Tetrahedron* **49**, 5003 (1993).

164. *AMPAC 5.0* (Semichem, Shavnee, KS, 1995).

165. J. J. P. Stewart, *MOPAC 5.0* (QCPE 455, Indiana University, 1990).

166. M. J. Frisch, G. W. Trucks, H. B. Schlegel, P. M. W. Gill, B. G. Johnson, M. A. Robb, J. R. Cheeseman, T. Keith, G. A. Petersson, J. A. Montgomery, K. Raghavachari, M. A. Al-Laham, V. G. Zakrzewski, J. V. Ortiz, J. B. Foresman, J. Cioslowski, B. B. Stefanov, A. Nanayakkara, M. Challacombe, C. Y. Peng, P. Y. Ayala, W. Chen, M. W. Wong, J. L. Andres, E. S. Replogle, R. Gomperts, R. L. Martin, D. J. Fox, J. S. Binkley, D. J. Defrees, J. Baker, J. P. Stewart, M. Head-Gordon, C. Gonzalez and J. A. Pople, *Gaussian 94, Revision E.2* (Gaussian, Inc., Pittsburgh PA, 1995).

167. W. J. Hehre, L. D. Burke and A. J. Schusterman, *Spartan* (Wavefunction,

Inc., Irvine, 1993).

168. Z. Slanina, *Int. Rev. Phys. Chem.* **6**, 251 (1987).

169. Z. Slanina, *Theor. Chim. Acta* **83**, 257 (1992).

170. R. Ettl, I. Chao, F. N. Diederich and R. L. Whetten, *Nature* **353**, 149 (1991).

171. D. E. Manolopoulos, *J. Chem. Soc., Faraday Trans.* **87**, 2861 (1991).

172. H. P. Cheng and R. L. Whetten, *Chem. Phys. Lett.* **197**, 44 (1992).

173. Q. Li, F. Wudl, C. Thilgen, R. L. Whetten and F. Diederich, *J. Am. Chem. Soc.* **114**, 3994 (1992).

174. S. Hino, K. Matsumoto, S. Hasegawa, H. Inokuchi, T. Morikawa, T. Takahashi, K. Seki, K. Kikuchi, S. Suzuki, I. Ikemoto and Y. Achiba, *Chem. Phys. Lett.* **197**, 38 (1992).

175. G. Orlandi, F. Zerbetto, P. W. Fowler and D. E. Manolopoulos, *Chem. Phys. Lett.* **208**, 441 (1993).

176. J. M. Hawkins and A. Meyer, *Science* **260**, 1918 (1993).

177. S. J. Austin, P. W. Fowler, G. Orlandi, D. E. Manolopoulos and F. Zerbetto, *Chem. Phys. Lett.* **226**, 219 (1994).

178. R. H. Michel, H. Schreiber, R. Gierden, F. Hennrich, J. Rockenberger, R. D. Beck, M. M. Kappes, C. Lehner, P. Adelmann and J. F. Armbruster, *Ber. Bunsenges. Phys. Chem.* **98**, 975 (1994).

179. R. H. Michel, M. M. Kappes, P. Adelmann and G. Roth, *Angew. Chem., Int. Ed. Engl.* **33**, 1651 (1994).

180. Z. Slanina and S.-L. Lee, *In HPC-ASIA '95, Electronic Proceedings* (National Center for High-Performance Computing, Hsinchu, Taiwan, 1995).

181. D. Bakowies and W. Thiel, *Chem. Phys.* **151**, 309 (1991).

182. Z. Slanina, L. Adamowicz, D. Bakowies and W. Thiel, *Thermochim. Acta* **202**, 249 (1992).

183. Z. Slanina and L. Adamowicz, *Fullerene Sci. Technol.* **1**, 1 (1993).

184. R. Taylor, G. J. Langley, A. G. Avent, T. J. S. Dennis, H. W. Kroto and D. R. M. Walton, *J. Chem. Soc., Perkin Trans. 2* 1029 (1993).

185. T. Wakabayashi, K. Kikuchi, S. Suzuki, H. Shiromaru and Y. Achiba, *J. Phys. Chem.* **98**, 3090 (1994).

186. M. Benz, M. Fanti, P. W. Fowler, D. Fuchs, M. M. Kappes, C. Lehner, R. H. Michel, G. Orlandi and F. Zerbetto, *J. Phys. Chem.* **100**, 13399 (1996).

187. S.-L. Lee, M.-L. Sun and Z. Slanina, *Int. J. Quantum Chem., Quantum Chem. Symp.* **30**, 355 (1996).

188. Z. Slanina, *J. Mol. Struct. (THEOCHEM)* **65**, 143 (1990).

189. F. Negri, G. Orlandi and F. Zerbetto, *Chem. Phys. Lett.* **189**, 495 (1992).

190. K. Raghavachari, *Chem. Phys. Lett.* **190**, 397 (1992).

191. X.-Q. Wang, C. Z. Wang, B. L. Zhang and K. M. Ho, *Phys. Rev. Lett.* **69**, 69 (1992).

192. D. Bakowies, M. Kolb, W. Thiel, S. Richard, R. Ahlrichs and M. M. Kappes, *Chem. Phys. Lett.* **200**, 411 (1992).

193. Z. Slanina and S.-L. Lee, *J. Mol. Struct. (THEOCHEM)* **333**, 153 (1995).

194. Z. Slanina and S.-L. Lee, *NanoStruct. Mater.* **4**, 39 (1994).

195. T. J. S. Dennis, T. Kai, T. Tomiyama and H. Shinohara, *to be published*.

196. Z. Slanina, F. Uhlík, M. Yoshida and E. Ōsawa, *Fullerene Sci. Technol.* (in press).

197. Z. Slanina, M.-L. Sun and S.-L. Lee, *NanoStruct. Mater.* **8**, 623 (1997).

198. Z. Slanina, S.-L. Lee, F. Uhlík and L. Adamowicz, in *Physics and Chemistry of Fullerenes and Their Derivatives*, eds. H. Kuzmany, J. Fink, M. Mehring and S. Roth (World Sci. Publ., Singapore, 1996).

199. J. F. Anacleto, H. Perreault, R. K. Boyd, S. Pleasance, M. A. Quilliam, P. G. Sim, J. B. Howard, Y. Makarovsky and A. L. Lafleur, *Rapid Commun. Mass. Spectr.* **6**, 214 (1992).

200. Z. Slanina, S.-L. Lee, K. Kobayashi and S. Nagase, *J. Mol. Struct. (THEOCHEM)* **312**, 175 (1994).

201. N. G. Gotts, G. von Helden and M. T. Bowers, *Int. J. Mass Spectr. Ion Process.* **149/150**, 217 (1995).

RELAXATIONS OF CHARGE TRANSFER AND PHOTOEXCITATION IN C_{60} AND POLYMERS

X. Sun

Institute for Molecular Science, Okazaki 444, Japan
and
*Department of Physics, Fudan University, Shanghai 200433, China**
E-mail:xinsun@fudan.ac.cn

G. P. Zhang

Max-Planck-Institut für Mikrostrukturphysik, Weinberg 2, D-06120 Halle, Germany
E-mail, zhang@mpi-halle.mpg.de

Thomas F. George

Departments of Chemistry and Physics, University of Wisconsin-Stevens Point,
Steven Point, Wisconsin 54481-3897, USA
E-mail: tgeorge@uwsp.edu

R. T. Fu

*Department of Physics, Fudan University, Shanghai 200433, China**
and
National Laboratory of Infrared Physics, Academia Sinica, Shanghai 200083, China
E-mail:rtfu@fudan.ac.cn

After an electron is transferred to C_{60} or C_{60} is photoexcited, due to the Jahn-Teller instability, the bond structure of C_{60} is distorted while its symmetry is reduced from I_h to D_{5d}. Meanwhile, there emerge two localized electronic states between LUMO and HOMO. The dynamical processes of both charge transfer and photoexcitation in C_{60} are studied numerically. It is found that the relaxation of photoexcitation is three times faster than that of charge transfer. Similar relaxation processes take place in polymers. Since a polymer is quasi-one dimensional, it is found that the lattice fluctuation has a substantial impact on the relaxation time.

1   Introduction

The discovery of C_{60}, known as buckyball and belonging to the class of molecules called fullerenes, has opened a new interdisciplinary field in physics, chemistry and material science [1,2]. Behind its peculiar phenomena in superconductivity [3-6], nonlinear optics [7-10], magnetism [11-12], photoconductivity [3], etc., charge transfer (CT) and photoexcitation (PE) in C_{60} play an essential role. Therefore, in order to understand the mechanism of the physical and chemical properties of fullerene, it is needed to study thoroughly these two fundamental processes: CT and PE in C_{60}, especially in the dynamical aspect.

The experiments of photoinduced charge transfer in C_{60}-DMA [14] and C_{60}-MEH-PPV [15-19] show that the relaxation time for photoinduced charge transfer is on a sub-picosecond scale. Theoretical studies by means of dynamical simulation further reveal the details of the relaxation processes in CT [20-23] and PE [24-26] in C_{60}. It is found that although both CT and PE are ultrafast processes, the relaxation of PE is much faster than that of CT.

This article comprehensively surveys the relaxation processes in C_{60} and a polymer. In Section 2, the bond distortion and localized electronic states of C_{60} after CT or PE are described. The relaxation processes of CT and PE in C_{60} are demonstrated in Section 3. In Section 4 the photoinduced ultrafast relaxation in a polymer is presented. The conclusion is given in Section 5.

2   Bond distortion and localized states of charged or excited C_{60}

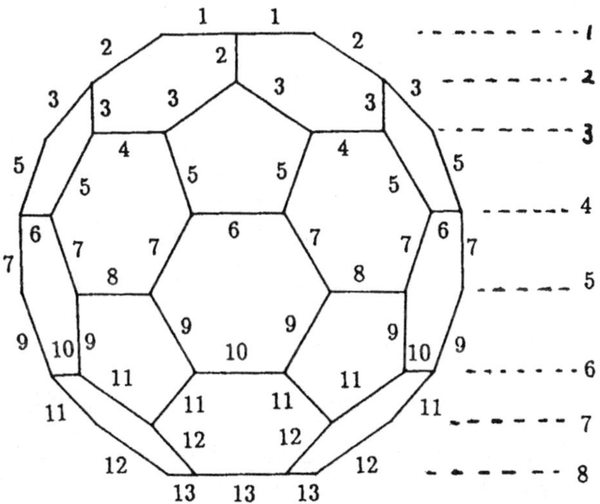

FIG. 1. Laminar structure of C_{60}. (The numbers are explained in the text in the paragraph following Eq. (7).)

Neutral C_{60} has the structure of a truncated icosahedron with the symmetry I_h, which consists of 12 pentagons separated by 20 hexagons, as shown in Fig. 1. There are 60 vertices in this structure, where each vertex is occupied by one carbon atom, and all these 60 carbon atoms are equivalent to each other. This feature has been verified by the NMR spectrum, where there is only one line at 142.68 ppm [27]. There exists bond alternation in the bond structure of C_{60}: the border between a pentagon and hexagon is a single bond with bond length of 1.433 Å, while the border between two adjacent hexagons is a double bond with bond length of 1.389 Å [28]. Electronically, the HOMO level is five-fold degenerate and LUMO three-fold. The separation between LUMO and HOMO is 1.9 eV [29].

The bond structure and electronic states of C_{60} can be modeled by a tight-binding Hamiltonian [30-34],

$$\hat{H} = -\sum_{i,j,s} t_{ij}(\hat{C}^{\dagger}_{i,s}\hat{C}_{j,s} + h.c.) + \frac{K}{2}\sum_{i,j}(|\,\mathbf{r}_i - \mathbf{r}_j\,| - \mathrm{D}_0)^2 \,, \tag{1}$$

where t_{ij} is the hopping integral between the nearest-neighbor atoms sitting at $\mathbf{r}_i$ and $\mathbf{r}_j$, and $\hat{C}^{\dagger}_{i,s}$ and $\hat{C}_{i,s}$ are the creation and annihilation operators of electron in atom i with spin s, respectively. The second term in (1) is the elastic energy, with the spring constant K.

The hopping integral t_{ij} depends on the distance $D_{ij} = |\mathbf{r}_i - \mathbf{r}_j|$ between atoms i and j. Since the atom displacement $\mathbf{Q}_i = \mathbf{r}_i - \mathbf{r}_i^0$, where $\mathbf{r}_i^0$ is the equilibrium position of atom i in neutral C_{60}, is much smaller than the bond length, the change of the distance between atoms i and j, which results from the atom displacement, can be expressed as,

$$\Delta D_{ij} = |\mathbf{r}_i - \mathbf{r}_j| - |\mathbf{r}_i^0 - \mathbf{r}_j^0| = \mathbf{n}_{ij} \cdot (\mathbf{Q}_i - \mathbf{Q}_j), \tag{2}$$

where $\mathbf{n}_{ij}$ is the unit vector in the direction from site i to site j, $\mathbf{n}_{ij} = (\mathbf{r}_i^0 - \mathbf{r}_j^0)/|\mathbf{r}_i^0 - \mathbf{r}_j^0|$.

Then the hopping integral can be expanded as,

$$t_{ij}(D_{ij}) = t_{ij}(|\mathbf{r}_i^0 - \mathbf{r}_j^0| + \Delta D_{ij}) = t_0 + \alpha \mathbf{n}_{ij} \cdot (\mathbf{Q}_i - \mathbf{Q}_j), \tag{3}$$

where $t_0 = t_{ij}(|\mathbf{r}_i^0 - \mathbf{r}_j^0|)$, $\alpha = dt_{ij}/dD_{ij}$. Substituting Eq. (3) into Eq. (1), the Hamiltonian reads

$$\hat{H} = -\sum_{i,j,s} t_0(\hat{C}^{\dagger}_{i,s}\hat{C}_{j,s} + h.c.) - \sum_{i,j,s} \alpha \mathbf{n}_{ij} \cdot (\mathbf{Q}_i - \mathbf{Q}_j)(\hat{C}^{\dagger}_{i,s}\hat{C}_{j,s} + h.c.)$$

$$+ \frac{K}{2}\sum_{i,j}(|\,\mathbf{r}_i - \mathbf{r}_j\,| - \mathrm{D}_0)^2 \,, \tag{4}$$

where the second term is the electron-phonon interaction, with α as the coupling constant. We have determined the parameters by fitting the bond lengths and the difference between HOMO and LUMO levels. The parameters take the following values: $t_0 = 1.8$ eV, $\alpha = 3.5$ eV/Å, $K = 30.0$ eV/Å^2, $D_0 = 1.54$ Å.

In any atom configuration $\{\mathbf{r}_i\}$, the electronic energy spectrum and eigenfunctions can be obtained by numerically solving the equation

$$H\Psi_\nu = \mathcal{E}_\nu(\{\mathbf{r}_i\})\Psi_\nu \tag{5}$$

where the energy spectrum $\mathcal{E}_\nu(\{\mathbf{r}_i\})$ is dependent on the configuration $\{\mathbf{r}_i\}$. The total energy $E(\{\mathbf{r}_i\})$ is

$$E(\{\mathbf{r}_i\}) = \sum_{occ} \mathcal{E}_\nu(\{\mathbf{r}_i\}) + \frac{K}{2}\sum_{ij}(|\mathbf{r}_i - \mathbf{r}_j| - D_0)^2 . \tag{6}$$

The equilibrium positions of sixty carbon atoms $\{\mathbf{r}_i\}$ (i=1, 2, 3, ..., 60) can be determined by minimizing the total energy $E(\mathbf{r}_i)$

$$\frac{\delta E(\{\mathbf{r}_i\})}{\delta \mathbf{r}_i} = 0. \tag{7}$$

In the case of neutral C_{60}, this gives a single-double bond structure with symmetry I_h, where the lengths of the single and double bonds are 1.43 and 1.39 Å, respectively, and the energy difference between LUMO and HOMO is 1.83 eV. These results fit the experimental results quite well.

After charge transfer or photoexcitation, due to Jahn-Teller instability, the bond structure is distorted. It should be mentioned that the bond distortion possesses a laminar structure [30]. In order to demonstrate the layer structure more distinctly, it is suitable to orient the buckyball in such a way that its top and bottom are pentagons, which is shown in Fig. 1. In this picture, the buckyball consists of eight layers, where the layer numbers are indicated in the right column of Fig. 1. Meanwhile, the bonds are divided into 13 layers, where the number attached to the bonds indicates the ordinates of the bond layers. Such distortion reduces the symmetry from I_h to D_{5d}, which has been observed in experiments [35, 36]. The bond distortion has the following remarkable features: (a) in each layer, all the bonds have same change in their bond lengths, where the layer dependence of the bond distortion is shown in Fig. 2; (b) the changes of bond lengths are symmetrical with respect to the equator; (c) the length changes in the 6th-8th layers are much larger than those in other layer, i.e. the bond distortion is mainly concentrated in the equatorial area.

Electronically the bond distortion partly breaks the degeneracy and shifts the levels as shown in Fig. 3. The original five-fold degenerate HOMO level H_u is split into three levels E_{1u}, E_{2u} and A_{1u}. Both E_{1u} and E_{2u} are two-fold degenerate and close to the original H_u level. But A_{1u}, whose wave function

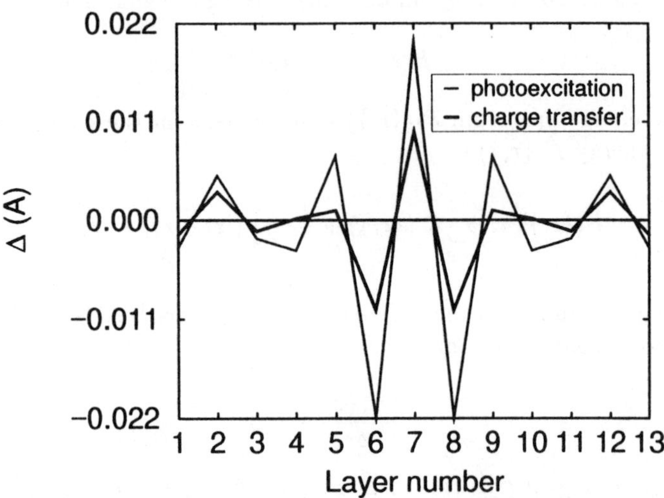

FIG. 2. Layer dependence of changes of bond lengths. The thin line refers to photoexcitation, and the thick line to charge transfer.

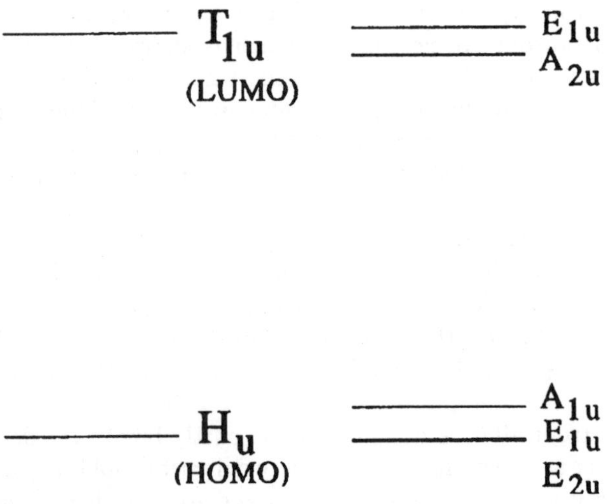

$$T_{1u} \quad (\text{LUMO}) \qquad \frac{E_{1u}}{A_{2u}}$$

$$H_u \quad (\text{HOMO}) \qquad \frac{A_{1u}}{\frac{E_{1u}}{E_{2u}}}$$

FIG. 3. Splits of HOMO and LUMO.

is denoted as Ψ_d, is apparently raised up from HOMO level. The separation between A_{1u} and H_u is 0.06 eV for charge transfer and 0.12 eV for photoexcitation, respectively. The original three-fold degenerate LUMO level T_{1u} is split into two levels yE_{1u} and A_{2u}. E_{1u} is two-fold degenerate and very close to the original T_{1u} level. But A_{2u}, whose wave function is denoted as Ψ_u, is apparently pulled down from LUMO level. The separation between A_{2u} and T_{1u} is 0.05 eV for charge transfer and 0.11 eV for photoexcitation, respectively.

It is noticeable that the states Ψ_d and Ψ_u are different from other electronic states in the sense that they are localized in the equatorial area. Actually, Ψ_d and Ψ_u are self-trapping electronic bound states associated with the bond distortion induced by charge transfer or photoexcitation.

3   Relaxations of charge transfer and photoexcitation in C_{60}

3.1   Dynamical equation

Now we turn to consider the dynamical processes of CT and PE. When one extra electron is transferred to neutral C_{60}, or C_{60} is excited, the original equilibrium structure with symmetry I_h can not be held any longer, and all atoms will adjust their positions by altering bond lengths. The force $\mathbf{f}_i$ which exerts on atom i can be calculated from

$$\mathbf{f}_i = -\frac{\delta E(\{\mathbf{r}_i\})}{\delta \mathbf{r}_i} \ . \tag{8}$$

Since the carbon atom is much heavier that the electron, it is reasonable to neglect the quantum effect of the atom vibrations; thus the classical dynamical equations can be applied to the motions of the carbon atoms [37,38]. In C_{60} there are ten Jahn-Teller active modes: $A_g(1)$, $A_g(2)$, and $H_g(1)$, $H_g(2)$, $H_g(3)$, $H_g(4)$, $H_g(5)$, $H_g(6)$, $H_g(7)$, $H_g(8)$. Since the electron-phonon coupling for pure radial modes such as $A_g(1)$ (496 cm^{-1}) is very small [39], the radial relaxation is much smaller than the tangential ones, so that eventually the carbon atoms move mainly on the surface. Hence two polar coordinates θ_i and ϕ_i are sufficient to describe the position of the i-th atom, and the dynamical equations for each atom read as

$$\ddot{\theta}_i - \dot{\phi}_i^2 \sin\theta_i \cos\theta_i + \lambda\dot{\theta}_i = -\frac{1}{m}\frac{1}{r_0^2}F(\theta)$$

$$\ddot{\phi}_i + 2\dot{\theta}_i\dot{\phi}_i \cot\theta_i + \lambda\sin\theta_i\dot{\phi}_i = -\frac{1}{m}\frac{1}{r_0^2\sin^2\theta_i}F(\phi) \ , \tag{9}$$

where the dot denotes the derivative with respect to time, λ is the damping and m the mass of the carbon atom, and

$$F(\theta) = \frac{\delta E(\{\theta_j, \phi_j\})}{\delta \theta_i},$$

$$F(\phi) = \frac{\delta E(\{\theta_j, \phi_j\})}{\delta \phi_i}. \tag{10}$$

Subject to the above forces, the atoms will move away from the original positions $\{\theta_i^0, \phi_i^0\}$ of the neutral C_{60}. The initial conditions are

$$\theta_i|_{t=0} = \theta_i^0 \ , \ \phi_i|_{t=0} = \phi_i^0,$$
$$\dot{\theta}_i|_{t=0} = 0 \ , \ \dot{\phi}_i|_{t=0} = 0. \tag{11}$$

The combined dynamical equations (9) and (10) can be tackled numerically step by step. Each step should last a very short interval τ so that the changes of the forces in this interval can be ignored. The positions $\{\theta_i, \phi_i\}$ and velocities $\{\dot{\theta}_i, \dot{\phi}_i\}$ at the end of each step serve as the initial conditions of the next step. Consequently, the positions of all atoms at any time can be figured out, and the whole picture of the time-dependent lattice relaxation is obtained.

Since the vibration period τ_0 of the atoms in C_{60} is about 10^{-14} sec, the interval τ and damping λ should satisfy,

$$\tau << \tau_0 << \frac{1}{\lambda} \tag{12}$$

In our calculation, we take $\tau = 2.11$ fs, $\lambda = 100/$fs.

3.2   Relaxation of charge transfer

In CT, one electron enters LUMO, then the atoms start to move while the bond lengths begin to change. The numerical results of the evolution of the bond distortion in charge transfer are shown in Fig. 4. The series of pictures Figs. 4(a)-(e) represent the configurations at times $t = 40, 80, 120, 160$ and $350\ \tau$, respectively. It can be seen that, in the early stages, the changes of the bond lengths are rather diverse, and even the bonds in the same layer get different modifications. The bond lengths in the 7th layer vibrate around 0.01 Å while those in the 6th and 8th layers vibrate around -0.01 Å. The bonds in the 4th and 5th layers, or equivalently the 10th and 9th layers, behave differently. The bonds in these two layers vibrate with opposite phases.

By tracing the main axis along which all atoms oscillate, we find that it may be shifted to other equivalent axes (there are six equivalent axes all together).

This resembles the conclusion by Wang *et al.* [40], who claimed that the axis of the reduced symmetry D_{5d} could be changed to other equivalent ones through tunnelling with the aid of lattice fluctuation. Our observation reveals that the axis changing is damping-dependent. With light damping this change is

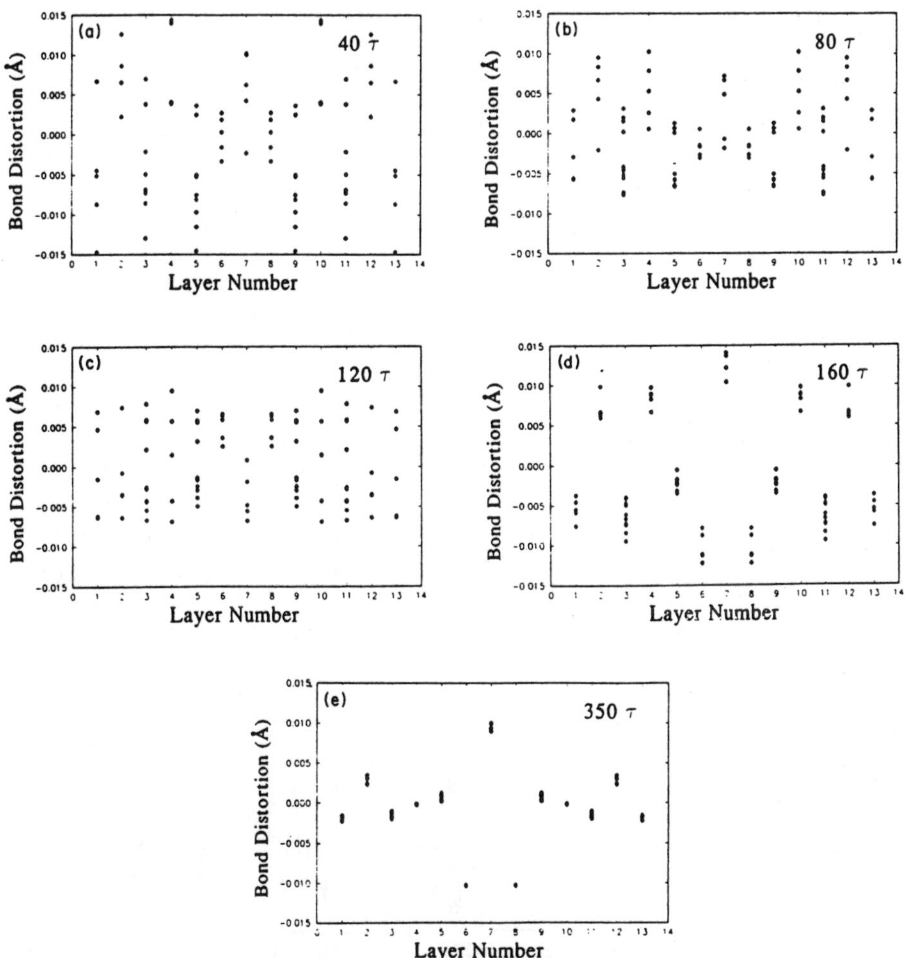

FIG. 4. Time dependence of the changes of the bonds. (a)-(e) display the bond alternations at times 40, 80, 120, 160 and 350 τ, respectively. The layer number refers to the bond layer (see Fig. 1).

frequent, while the heavy damping largely suppresses this change.

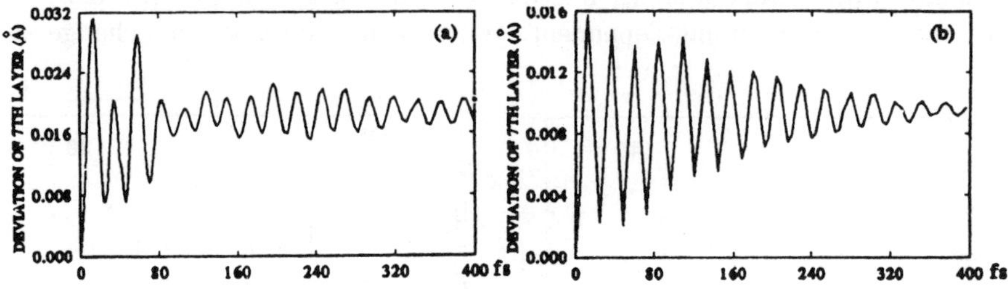

FIG. 5. Time dependence of the average change of the bond lengths in the seventh layer. (a) PE, (b) CT.

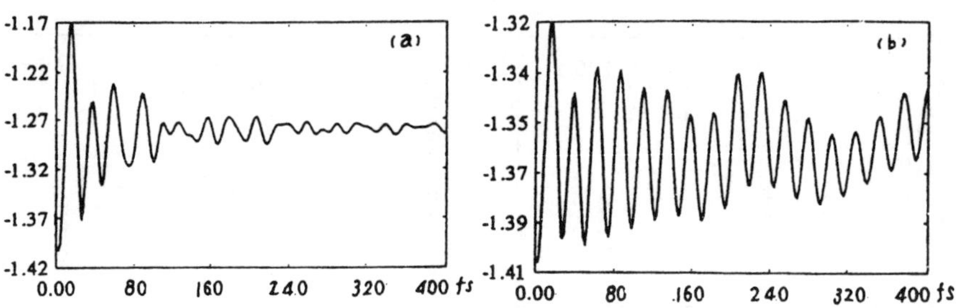

FIG. 6. Time evolution of the level A_{1u}. (a) PE. (b) CT.

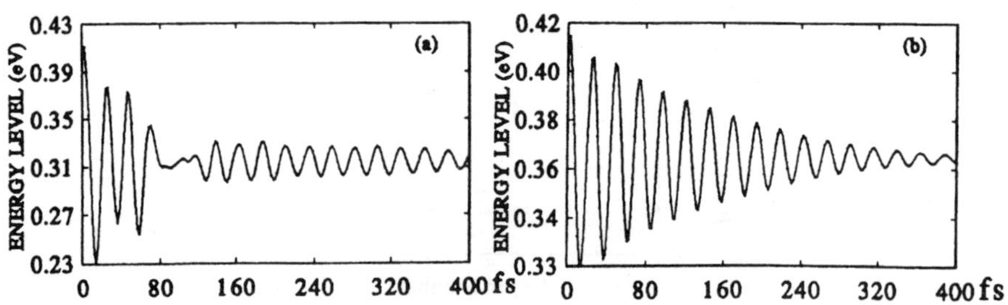

FIG. 7. Time evolution of the level A_{2u}. (a) PE. (b) CT.

Figure 4 shows that, after about 160 τ, the outline of the laminar structure roughly appears, where the modifications of different bonds in the same layer are getting close to each other (see Fig. 4(d)). After that, the bond distortions gradually reach the equilibrium structure of C_{60}^{-}, which is shown in Fig. 2. By inspecting the evolution process, one can estimate the relaxation time T_{CT} to be 300 fs.

To make the simulation more convincing, we have chosen several different time intervals and dampings to check our results. We have found that the relaxation process does not depend on the selection of the step interval τ as long as τ is much smaller than the vibration period τ_0, and the same for λ as long as $\frac{1}{\lambda} >> \tau_0$. These features conclude our simulation is reliable.

As shown in Fig. 2, the bond distortion appears in the equatorial area near the seventh layer, which provides a good opportunity to visualize the time-dependent change of the bond length in that layer. Since during the relaxation the different bonds in the 7th layer may have different changes in their bond length, one needs to take the average over all the bonds in the 7th layer. The time-dependence of the average change of the bond length in the 7th layer is plotted in Fig. 5(b), which again indicates the relaxation time T_{CT} to be 300 fs.

As mentioned in Sec. 2, there are two prominent features for CT or PE. One is the bond distortion, the other is the emergence of two localized electronic states A_{1u} and A_{2u} shown in Fig. 3. Figures 4 and 5(a) demonstrate the time evolution of the bond distortion for CT. The formations of A_{1u} and A_{2u} for CT with time are shown in Figs. 6(b) and 7(b), respectively. These two figures also show that the relaxation time of CT is about 300 fs.

3.3   Relaxation of photoexcitation

Upon photoexcitation, an electron goes from HOMO to LUMO. The initial condition is that one electron resides in LUMO and one hole is left in HOMO. Analogously, we solve the dynamical equations (9) and (10) to obtain the time evolution of the bond distortion and the levels A_{1u} and A_{2u}.

Figure 8 shows the time evolution of the changes of the bond lengths for PE. At earlier stages [e.g., $t = 20\,\tau$, Fig. 8(b)], the changes of the bonds are rather diverse, and even the bonds in the same layer experience very different modifications, which is similar to the relaxation process of CT. After about 50 τ, the typical laminar structure appears roughly, and the changes of different bonds in the same layer get close to each other [see Fig. 8(d)]. From this evolution, one can estimate that the relaxation time T_{PE} is about 90 fs, and at $t = 120\,\tau$, the bond distortion [see Fig. 8(e)] is very close to the equilibrium configuration of excited C_{60} [see Fig. 2]. In comparison to the relaxation time T_{CT} of CT, which is estimated as 0.3 ps in Sec. 3.1, T_{PE} is much shorter.

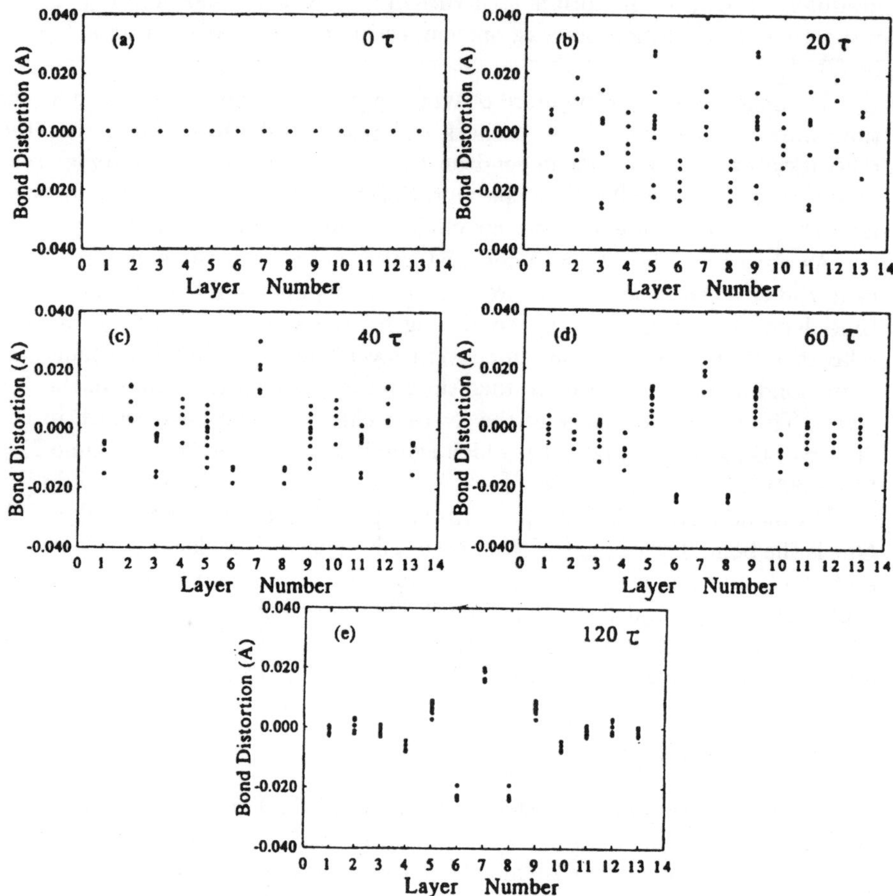

FIG. 8. For photoexcitation, the changes of bond length at the following times: 0, 20, 40, 60, 120 τ.

In analogy to CT, the bond distortion of PE is also localized around the 7th layer. The time dependence of the average change of bond lengths in the 7th layer is plotted in Fig. 5(a), which shows more apparently that the relaxation time of PE is about 90 fs. As for the formation of the localized electronic states

A_{1u} and A_{2u} in PE, their time evolutions are demonstrated in Figs. 6(a) and 7(a), respectively. Both of them confirm that the relaxation time of PE is 90 fs. Both the bond distortion and electronic states show that the relaxation in PE is much quicker than that in CT. In Figs. 5-7, (a) and (b) refer to PE and CT, respectively. These figures further justify the difference between CT and PE. However this large difference between PE and CT is not a surprising result. Actually, the same phenomenon has been observed in polymers [37]. In trans-polyacetylene, there are also two kinds of relaxations: (a) the relaxation of charge transfer, i.e., one electron (or hole) is transferred into the conduction band (or valence band), where the electron (hole) interacts with the lattice to form a negative (or positive) polaron, and its relaxation time is about 0.2 ps; and (b) the relaxation of photoexcitation, where an electron-hole pair decays into a soliton and antisoliton pair. The relaxation time is about 40 fs, which is 5 times faster that CT.

Such a prominent difference must have some physical reasons. Although a complete understanding is not available, one plausible reason is that in the case of photoexcitation, an electron resides in the LUMO state while the hole is in HOMO, such that both the electron and the hole exert the forces on the atoms. Under these forces, the atoms move quickly to a new equilibrium configuration. However, in the formation of a polaron, only an electron is added in the LUMO, such that the force is relatively weaker than that of the former. Therefore, the relaxation of PE is faster than that of CT. This can also be used to explain why the distortion of the relaxation is almost twice as large as that of charge transfer (see Fig. 2).

3.4 Normal mode analysis

The clear difference between the relaxations of CT and PE motivates us to perform a more detailed study. Moreover, from the experiments it is known that for five H_g modes, the electron-phonon coupling is different, while almost all the theoretical studies show that all the couplings are basically same. To reconcile this apparent discrepancy, we resort to a normal mode analysis. Such an analysis has been proved very powerful in the short-time dynamical behavior of liquids. However, a traditional normal mode analysis expands the temporary potential energy ("temporary" in the sense that the atomic positions are a function of time) in the basis of normal modes, where imaginary frequencies would appear, which diminishes its efficiency. We overcome this difficulty by expanding the kinetic energy. The main procedure is the following.

We build the normal mode basis by directly diagonalizing the force matrix.

The kinetic energy $E_{kinetic}$ is expanded as

$$E_{kinetic} = \sum_\nu E_\nu^{kinetic},\tag{13}$$

where $E_\nu^{kinetic}$ is the kinetic energy for the normal mode ν. Hence we can monitor the change of $E_\nu^{kinetic}$ to see whether a specific mode ν can be excited or not during the relaxation.

From the simulation we found that the active modes are either A_g or H_g symmetric, which is consistent with the theoretical prediction, but the main characters of these two modes are different. The A_g modes exchange the kinetic energy (KE) with the system almost harmonically though the behavior is slightly different for charge transfer and photoexcitation, while a clear decay is observed for the H_g modes. The H_g modes account for the main characteristic of the relaxation. The reason for the different behaviors between the A_g and H_g modes is that they have very different eigenvectors. For the H_g mode a laminar structure can be noted if the bonds are distorted along the eigenvectors (see Fig. 9(a)). For the A_g mode the symmetry of bond structure retains even if the bond is deformed along to the eigenvector of that mode (see Fig. 9(b)).

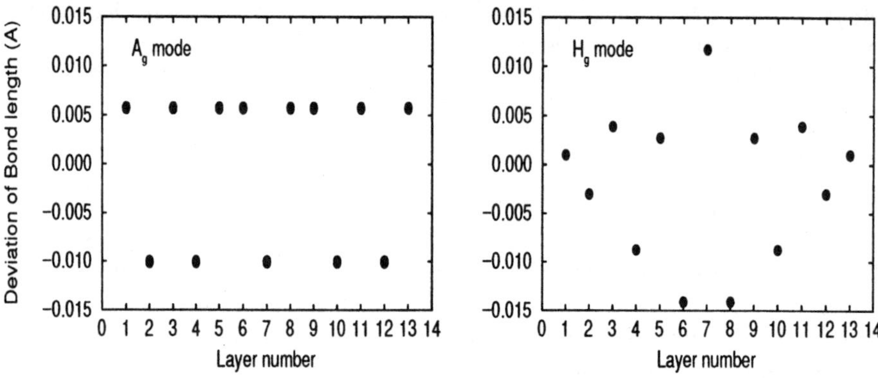

FIG. 9. For the A_g and H_g modes, the eigenvectors are different. A laminar structure can be noted for H_g modes, which is absent for the A_g mode.

The whole relaxation process is controlled by a few dominant modes. We plot two dominant normal modes for both CT and PE. In Fig. 10 the kinetic energies for $H_g(1)$ and $H_g(2)$ for CT and PE are shown, respectively. For PE, from the right panel of Fig. 10, one can see that after 50 fs, the KE of the $H_g(2)$

mode decreases while the KE of the $H_g(1)$ mode increases sharply. Then both $H_g(2)$ and $H_g(1)$ control the relaxation process. After 100 fs, a clear decay for $H_g(1)$ is observed. From the above, we know that at this time the relaxation process is accomplished. The relaxation process is very different from that of CT. The left panel of Fig. 10 displays the KE of $H_g(2)$ versus time for CT. In the beginning (less than 50 fs), this mode shares the maximum KE and thus controls the early behavior of the relaxation process. All the atoms vibrate chiefly along the framework of the $H_g(2)$ mode. After 400 fs, the KE of the $H_g(2)$ mode decreases while the KE of the $H_g(1)$ mode increases. Now both $H_g(1)$ and $H_g(2)$ become the dominant modes and almost equally important. All the atoms, after 400 fs, vibrate along the $H_g(1)$ and $H_g(2)$ modes. This

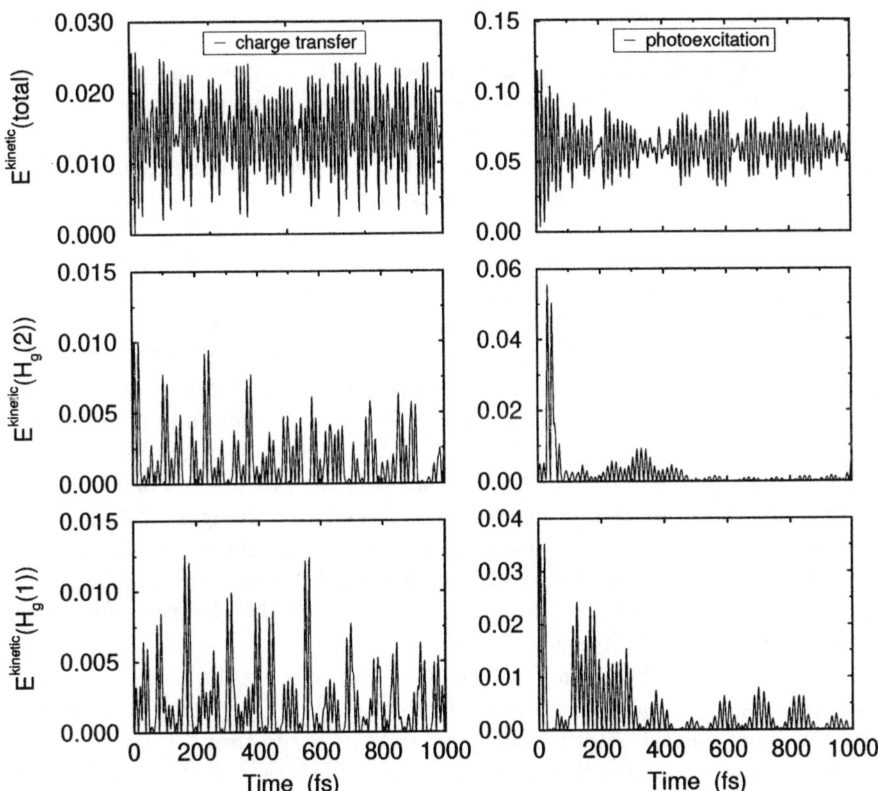

FIG. 10. The first row shows the total kinetic energy for CT and PE. The second and third rows refer to the kinetic energies for $H_g(2)$ and $H_g(1)$, respectively.

change actually signifies that the relaxation has come to an end, as one can see from the above (see Sec. 3.3).

Carefully analyzing those active H_g modes, we find that the different relaxation times for CT and PE actually result from their different dependences on the H_g modes. The relaxation becomes faster if the dominant modes lose the energy more quickly. Additionally, due to the fact that we have observed only one or two dominant H_g modes are strongly invoked, it is suggested that only one or two H_g modes could have larger electron-phonon coupling, while for the rest of them the coupling is relatively small, which is consistent with the experimental results.

3.5   Photoinduced charge transfer

Photoinduced charge transfer (PCT) in C_{60}, due to its potential applications, has been extensively studied. However, our understanding is still very limited, especially regarding what mainly affects the charge transfer process. Sariciftci [16] employed an interesting mechanism: PCT depends on the ionization potential I_D^* of the excited donor, the electron affinity A_A of the acceptor, the Coulombic energy U_c of the separate ions (including polarization effects) and structural relaxation of the polymer background. PCT can not occur unless these quantities satisfy the following inequality: $I_D^* - A_A - U_c < 0$. This inequality explains PCT well if the photon energy $\hbar\omega$ is larger than 2.0 eV. However, for the region with photon energy between 1.3 and 2.0 eV, PCT is difficult to understand from the above inequality as the donor now is not excited. Here, based on our calculation, we suggest another important factor, namely the affinity of the excited acceptor C_{60}. Naturally a complete understanding of PCT is very difficult, but we simplify the process and only consider whether the electron affinity (EA) of the excited C_{60} changes. The increase of electron affinity favors the above inequality, so that the excited C_{60} could easily get electrons from the polymers. We employ the single configuration interaction (SCI) to calculate how EA changes with the electron interaction and electron-phonon coupling.

Our results are shown in Fig. 11. One can observe the electron affinity really increases, compared with that of the unexcited C_{60}, which is consistent with the experimental observations. Experiments verify that the photoexcited C_{60} is a stronger acceptor than that in the ground state. EA strongly depends on the electron-phonon coupling but weakly relies on the electron-electron interaction.

Hence one may conclude that the above inequality should be generalized as $I_D^* - A_A^* - U_c < 0$. Although superficially such generalization is slight, the main physics is different. With this new inequality, we can explain the whole spectrum convincingly, especially when the donor is unexcited. The increased EA plays a critical role in PCT. At 1.3 eV, with the increase of the concentration of C_{60} from 1% to 50%, the photoconductivity increases many orders of magnitude. This can be understood as follows: the higher the concentration of C_{60}, the more fullerenes can get electrons, with the photoconductivity eventually increasing sharply at 1.3 eV. This effect is noticeable when the concentration of C_{60} is 50%, where the conductivity due to the increase of EA of the excited acceptor C_{60} has enhanced so much that the improvement of conductivity, which results from the increase of the donor ionization in the region of $\hbar\omega$ larger than 2.0 eV, becomes a small portion of the whole spectrum.

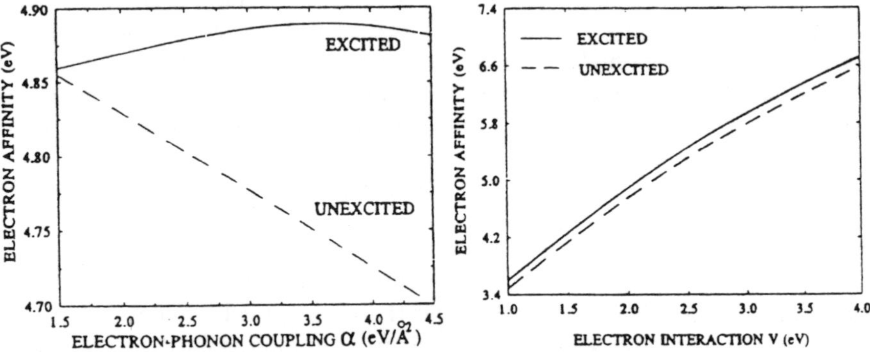

FIG. 11. The increase of both the electron-phonon coupling and electron-electron interaction enhances EA. The solid line is for the affinity of the excited C_{60}, while the dashed line is for that of the unexcited C_{60}.

4   Relaxation of photoexcitation in polymers

In this section, the effect of lattice fluctuation on photoinduced relaxation processes in polymers [41] are studied. Photoexcitation in a polymer causes lattice distortion u_n (self-trapping), which produces a soliton-antisoliton pair (with degenerate ground state) or bipolaron (with nondegenerate ground state). These relaxation processes can be simulated by using the SSH model [42]. Here, only the case with a degenerate ground state is considered.

When the phonon frequency is much smaller than the electronic energy gap, the scattering of phonons with electrons on the Fermi surface is so small that the phonon can propagate freely besides thermal scattering. The perturbation theory is then equivalent to an electron system with a random static potential, which is described by a random displacement u_n, with $< u_n > = u_0$. It is easy to show that, for dispersionless phonons, the potential has Gaussian (white noise) correlation at temperature $T = 0$,

$$< u_n u_{n'} > = u_0^2 + (\delta u)^2 \delta_{n,n'} . \tag{14}$$

Here, bond correlation represents bond disorder, or alternatively off-diagonal disorder. δu is the amplitude of zero-point motion or zero-point energy of a phonon. For a mono-mode and dispersionless phonon, $\delta u = (\hbar/2M\omega_{2k_F})^{1/2}$. In trans-polyacetylene, $\delta u = 0.03$ Å, which is comparable to u_0.

The relaxation of the photoexcitation with zero-point motion can be simulated by the dynamical equations described in Sec. 3. First, set an initial lattice configuration according to Eq. (14), and take δu to be an adjustable parameter for describing the strength of lattice fluctuation. Then, when the polymer is photoexcited, there appears an electron-hole pair. Due to the electron-phonon interaction, the original bond structure is unstable and the atoms are subjected to the forces

$$F_n = -\delta E_{tot}(\{u_n\})/\delta u_n, \tag{15}$$

where $E_{tot} = \sum_{occ} \mathcal{E}_\alpha(\{u_n\}) + \frac{K}{2} \sum_n (u_n - u_{n+1})^2$.

The equation of motion can be integrated within the adiabatic approximation since the Peierls gap is much larger than the optical phonon energy. The relaxation process from electron-hole pair to soliton-antisoliton pair is demonstrated by two evolutions: (1) the evolution of the top state of the valence band, which splits from the band edge and evolves to a mid-gap state associated with soliton creation; and (2) the evolution of the lattice configuration $\psi_n = (-1)^n u_n$.

For the case without lattice fluctuation ($\delta u = 0$), the evolutions of the electron state and lattice configuration are shown in Fig. 12. In the early stage, about 100 fs, the electron-hole pair remains free and no apparent lattice distortion emerges. After that, an irreversible process from free electron-hole pair to soliton-antisoliton pair occurs. However, when the lattice fluctuation is taken into account, the stage with the free electron-hole pair no longer exists. This remarkable change is shown in Fig. 13, where the strength of the lattice fluctuation is $\delta = 0.001$ Å. From Fig. 13, it can be seen, in the presence of the lattice fluctuation, once the electron-hole pair is created, the self-trapping process starts immediately, and the soliton-antisoliton configuration and mid-gap state is well established at about 60 fs; without the lattice fluctuation, it needs more than 160 fs to reach such a configuration. In Fig. 13, the lattice fluctuation $\delta = 0.001$ Å is quite small, but it already shows its significant effect on the relaxation process. This main feature remains even if we vary δu from 0.0001 Å to 0.04 Å. These results conclude that the lattice fluctuation plays a substantial role in this ultrafast process. It suppresses the oscillation of the free electron-hole pair and invokes the creation of the soliton-antisoliton pair.

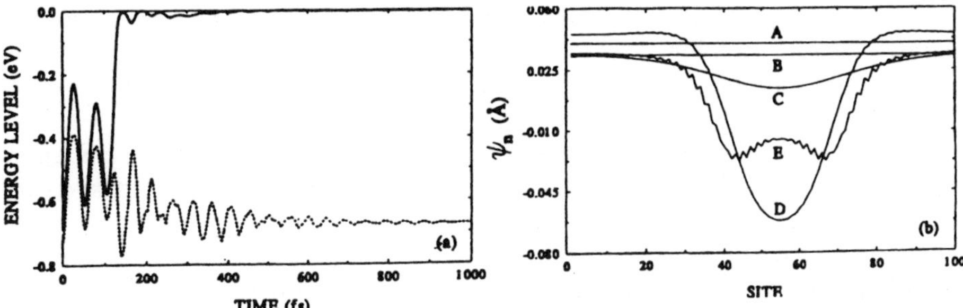

FIG. 12. The relaxation process for $\delta u = 0$. (a) Time dependence of the energy levels of the state at the top of the valence band (solid curve) and the next state in the valence band (dashed curve); (b) lattice configuration, where curves A, B, C, D and E correspond to $t = 0$, 100, 120, 140 and 160 fs, respectively.

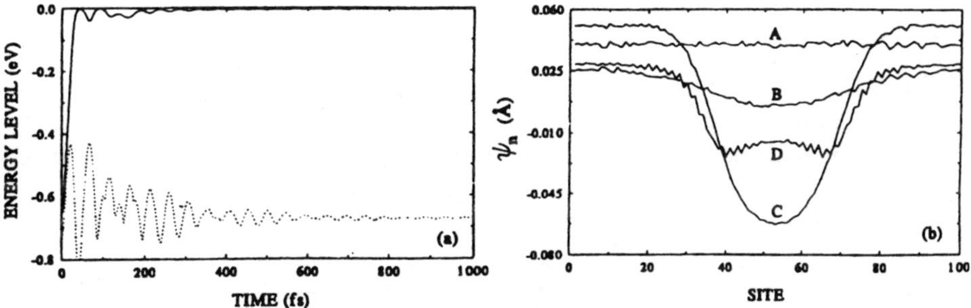

FIG. 13. The relaxation process for $\delta u = 0.001$ Å. (a) Time dependence of the energy levels of the state at the top of the valence band (solid curve) and the next state in the valence band (dashed curve); (b) lattice configuration, where curves A, B, C and D correspond to $t = 0$, 20, 40 and 60 fs, respectively.

Ishida [43] attributed the observed long relaxation time [44] to the free exciton oscillation. The above simulation shows that this attribution is valid only in the absence of the lattice fluctuation. However, for real low-dimensional materials, the fluctuation can not be ignored, so that this premise is not true. Actually, the role of the lattice fluctuation in the relaxation process is similar to that of the condensation kernel when a gas condenses to a liquid. If the condensation kernel is absent, the gas can not condense even below the condensation temperature. Similarly, without the lattice fluctuation, more time is needed to find the "kernel" for breaking the symmetry. In a real material, the lattice fluctuation provides such "kernels", and the oscillation of free excitons can not be found experimentally. Since there is no barrier between the free state and self-trapping state in a one-dimensional system, the relaxation time for an electron-hole pair is comparable with the period of lattice vibration (40 fs).

Experiments show that the relaxation time for exciton trapping is about 100-150 fs [44]. This can be explained as follows: The photoexcited free excitons are coupled with the optical phonon, and at the beginning, the trapped excitons have not relaxed to the bottom of the potential curve, so that the binding energy of trapped exciton remains as the kinetic energy of the lattice oscillation. Later they relax to the bottom of the potential by emitting phonons. The time of phonon emission is determined by the energy redistribution rate from the strongly coupled phonon to the other low-frequency phonon modes. The observed relaxation time is caused by both trapping and phonon emission.

5   Conclusion

From the simulations of the relaxation processes in both C_{60} and polymers, two prominent features are found. First the relaxation of photoexcitation is three times faster than that of charge transfer. This distinct difference in C_{60} is more clearly demonstrated in Fig. 14, where the ordinate is the level of the localized state A_{1u} and the abscissa the level of A_{2u}. Each dot represents the

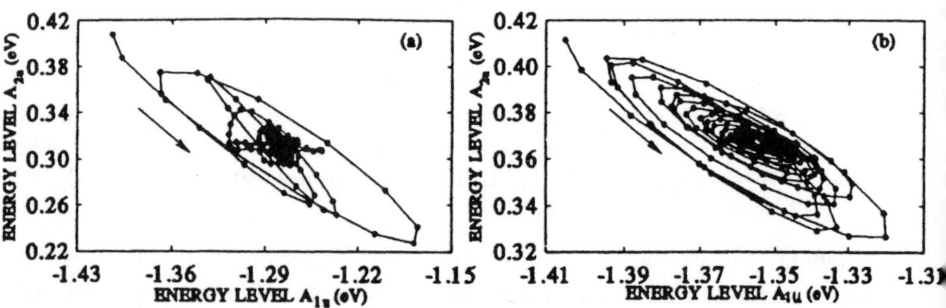

FIG. 14. Two gap states A_{1u} and A_{2u} evolve with time. (a) PE; (b) CT. The arrow denotes the time direction.

values of the levels A_{1u} and A_{2u} at different times; for instance, the dot in the upper left corner represents the levels at $t = 0$. The arrow denotes the direction of advancing time, and the interval between two successive dots is 1 τ. The centers of the dark area in Figs. 14(a) and (b) are the final positions of the gap states A_{1u} and A_{2u}, respectively. From Fig. 14(a), one can observe that for PE it takes only about three turns to reach the dark area. On the other hand, for CT (see Fig. 14(b)), it takes more than ten turns to get the dark area. Secondly the lattice fluctuation has a substantial impact on the relaxation process as it suppresses the free exciton oscillation and promotes the self-trapping.

Acknowledgments

We would like to express our sincere thanks to Dr. R. L. Fu and Mr. X. S. Rao for providing relevant data. This work is supported by the Chinese National Science Foundation and Project 863.

* Permanent address.

References

1. H. W. Kroto, J. R. Heath, S. C. O'Brien, R. F. Curl and R. E. Smalley, Nature **318**, 162 (1985)
2. W. Kräschmer, L. D. Lamb, K. Fostiropoulos and D. R. Hoffman, Nature **347**, 354 (1990)
3. R. C. Haddon *et al.*, Nature **350**, 320 (1991)
4. A. F. Hebard *et al.*, Nature **350**, 600 (1991)
5. M. J. Russeinsky *et al.*, Phys. Rev. Lett. **66**, 2830 (1991)
6. K. Holczer *et al.*, Science **252**, 1154 (1991)
7. H. Hoshi *et al.*, Jpn. J. Appl. Phys. **30**, L1397 (1991)
8. G. B. Talapatra *et al.*, J. Phys. Chem. **96**, 5206 (1992)
9. Z. H. Kataki, Chem. Phys. Lett. **188**, 892 (1992)
10. J. S. Metch *et al.*, Chem. Phys. Lett. **197**, 26 (1992)
11. P.-M. Allemand *et al.*, Science **253**, 301 (1991)
12. F. Wudl and J. D. Thompson, J. Phys. Chem. Sol. **53**, 1449 (1992)
13. Y. Wang, Nature **356**, 585 (1992)
14. R. J. Sension, A. Z. Szarka, G. R. Smith and R. M Hochstrasser, Chem. Phys. Lett. **185**, 179 (1991)
15. N. S. Sariciftci, L. Smilowitz, A. J. Heeger and F. Wudl, Science **258**, 1474 (1992)
16. N. S. Sariciftci and A. J. Heeger, Int. J. Mod. Phys. B **8**, 237 (1994)

17. A. Watanabe and U.Zto, J. Phys. Chem. **98**, 7736 (1994)
18. D. M. Guldi, P. Neta and K.-D. Asmus, J. Phys. Chem. **98**, 4617 (1994)
19. K. N. Grzeskowiak and A. J. Heeger, J. Phys. Chem. **98**, 5661 (1994)
20. G. P. Zhang, R. T. Fu, X. Sun, D. L. Lin and Thomas F. George, Phys. Rev. B **50**, 11976 (1994)
21. R. T. Fu, Z. G. Yu, Q. F. Huang, X. Sun and R. L. Fu, Chin. Phys. Lett. **11**, 133 (1994)
22. G. P. Zhang, R. T. Fu, X. Sun, X. F. Zong, K. H. Lee and T. Y. Park, J. Phys. Chem. **99**, 12301 (1995)
23. G. P. Zhang, X. Sun, X. F. Zong, L. Pandey and T. F. George, Phys. Lett. A **220**, 275 (1996)
24. X. Sun, G. P. Zhang, Y. S. Ma, K. H. Lee, T. Y. Park, T. F. George and L. Pandey Phys. Rev. B **53**, 15481 (1996)
25. G. P. Zhang, Y. S. Ma, X. Sun and Y. Takahash, Chin. Phys. Lett. **12**, 665 (1995)
26. G. P. Zhang, X. Sun, T. F. George and L. N. Pandey, J. Chem. Phys. **106**, 6398 (1997)
27. R. Taylor, J. P. Hare, Ala'a K. Abdul-Sada and H. W. Kroto, J. Chem. Soc. Chem Commun. 1423 (1990). H. Ajie *et al.*, J. Phys. Chem. **94**, 8631 (1990)
28. J. M. Hawkins, A. Meyer, T. A. Lewis, S. Loren and F. J. Hollander, Science **252**, 312 (1991)
29. J. H. Weaver, Phys. Rev. Lett. **66**, 1741 (1991)
30. R. T. Fu, Z. Chen, R. L. Fu and X. Sun, Chin. Phys. Lett. **9**, 541 (1992)
31. V. de Coulon, J. M. Martins and F. Reuse, Phys. Rev. B **45**, 1367 (1992)
32. B. Friedman, Phys. Rev. B **45**, 1454 (1992)
33. H. Harigaya, Phys. Rev. B **45**, 13676 (1992)
34. X. Sun *et al.*, Synth. Met. **55**, 2979 (1993)
35. X. Wei and Z. V. Vardeny, Phys. Rev. B **52**, R2317 (1995)
36. B. Gotschy, M. Keil, H. Klos and J. Rystau, Solid State Commun. **92**, 935 (1994)
37. W. P. Su and J. R. Schrieffer, Proc. Natl. Acad. Sci. (USA) **77**, 5626 (1980)
38. J. T. Gammel, Phys. Rev. B **45**, 6408 (1992)
39. M. Schlüter, M. Lanno, M. Needels, G. A. Baraff and D. Tománek, Phys. Rev. Lett. **68**, 526 (1992)
40. C. L. Wang *et al.*, unpublished
41. X. S. Rao, R. T. Fu, X. Sun and K. Nasu, Phys. Lett. A **226**, 383 (1997)
42. A. J. Heeger, S. Kivelson, J. R. Schriefer and W.-P. Su, Rev. Mod. Phys. **60**, 781 (1988)

43. K. Ishida, Solid State Commun. **90**, 89 (1994)

44. M. Yoshizuwa, A. Yasuda and T. Kubayashi, Appl. Phys. B **53**, 296 (1991)

IONIC CHARGE TRANSPORT IN MOLECULAR MATERIALS: POLYMER ELECTROLYTES

MARK A. RATNER

Department of Chemistry

Northwestern University

Evanston, Illinois 60208-3113

Ionic conduction in polymeric materials provides a striking challenge to the use of theoretical modeling to understand chemical processes. The materials are glass forming, dynamically disordered, hybrid organic/inorganic structures. They correspond to the concentrated regime of classical electrolytes, and are complicated by the polymer, which acts as an immobile solvent. These materials are of substantial interest in many electrochemical applications, including particularly batteries and fuel cells. Therefore, not surprisingly, there is extensive experimental investigation of the nature of ion transport in these structures. Theoretical modeling, as described here, is substantially more primitive. Nevertheless, several important advances have been made, including electronic structure studies of local ion solvation, molecular dynamics simulation of solvation structures and elementary diffusion mechanisms, dynamic percolation pictures for relating relaxation and diffusion behavior, and hopping models to study the effects of coulomb interaction and trapping. These advances are briefly described. Some of their important implications, particularly insights about mechanism and suggestions for design of new materials, are discussed.

I. Introduction

Traditional synthetic organic chemistry has perfected the preparation of molecules by design. Theory now plays a substantial role here, largely in conformational analysis, in comparison of experimentally measured and computed properties, and in prediction of stable and unstable structures.[1] The preparation and optimization of materials is a much more complex issue, since structure can exist on length scales from angstroms to millimeters, and since it is often non-covalent interactions that determine the appropriate structural motifs. The role of theory in optimizing and understanding materials structures and properties, accordingly, is more complex than the simple straightforward molecular calculations that are so useful in small molecule problems.[2]

These issues in understanding the structure of materials become even greater challenges when one is interested in materials "by design" - that is, in being able to predict the structure and/or properties of new materials. The arsenal of theoretical techniques (electronic structure calculations, MonteCarlo studies, molecular dynamics simulations, ab-initio molecular dynamics, hopping models, molecular modeling, etc.) can all be of use in understanding problems as complicated as, say, signal transduction in biological systems, artificial photosynthesis, photorefractive materials, functional block polymers or synthetic electrolytes.

Solid materials can, to some extent, be categorized as molecular or non-molecular (ionic, metallic, atomic). Examples of the former category would include proteins, plastics, wax, ice or soaps. The latter includes steel, salt, diamonds and rubies. This categorization is not a strict one, and such materials as bones, teeth, frozen sea water or reinforced concrete are hybrids between the two categories. A number of synthetic materials are also hybrids, and understanding and optimization of this category is one of the great challenges in understanding properties of materials.

We consider here an example category of materials, polymer-based solid electrolytes.[3-8] These materials are of interest in the context of mobile power - that is, of batteries and fuel cells. While traditional batteries use liquid electrolytes, solid electrolytes have been known from ancient times. Their first careful study was provided (as was nearly everything else in the area of electrolytic substances) by Faraday,[9] who observed solid electrolyte behavior in a number of materials, and characterized the nature of the conduction and its thermal variation. Synthetic materials exhibiting solid electrolyte character began with work at Ford Motor Company on oxide systems.[10] Polymer-based solid electrolyte materials first developed roughly 20 years ago,[4] and their study has expanded very substantially in the intervening time period.

Polymer-based, solvent free polymer electrolytes have many advantages for advanced batteries: they are light weight, inexpensive, safe, shape compliant, environmentally benign, and exhibit quite high energy densities. Therefore, their optimization and understanding is an important challenge to materials chemistry.

There are also a number of vexing problems with polymer electrolytes, especially for lithium batteries. The two most important ones are reduced power density (the battery can discharge only slowly, essentially because of conduction limitations within the electrolyte and across the electrode/electrolyte interface) and safety issues following re-charge (finally divided metallic lithium is an extremely aggressive substance, and will react, often violently, with any basic materials, like an oxide or a sulfide).[11]

Theory and modeling can contribute substantially to the understanding and optimization of polymer electrolyte materials. In this chapter, we will overview some of the progress that has been made in this area, discuss some suggestions for new materials that arise directly from theoretical study, and suggest several avenues of substantial challenge in the area of theory and modeling of polymer electrolyte materials.

Because both structure and properties are important, nearly the entire arsenal of theoretical methods can be brought to bear on the problem of polymer electrolytes. One of the things that makes this study difficult is indeed the substantial scales of both distance and time that are of interest in these materials: the primary coordination of the mobile ions is of the order of angstroms, but polymer assembly structures can be characterized on length scales up to microns, and transport in batteries must occur over macroscopic distances of at least 50 microns. Polymers exhibit characteristic relaxation times ranging over at least 12 orders of magnitude, from localized vibrational relaxation to large scale segmental motions and reptations. While complete understanding of the material over these time and length scales is an impractical goal, sufficient understanding to describe transport mechanistically and suggest a means for its improvement is an area in which some progress has been made, and more can be anticipated.

In Section II, some necessary experimental background on polymer electrolytes is presented. Section III deals with structural issues, and with applications of electronic structure methods and modeling techniques. Section IV discusses charge transport in polymer electrolytes, and its modeling using methods of varying levels of sophistication and exactness. Section V discusses some suggestions for modification and improvement of polymer electrolyte materials, based on theoretical study. Finally, some overall remarks are presented in Section VI.

2. Experimental Background: Polymer/Salt Complex Electrolytes

Polar polymers can solvate either cations (by Lewis base sites on the polymer), or anions (by Lewis acid sites). Polyethylene oxide (PEO) is a commercial, highly polar polymer. It is characterized by low glass transition temperature, high stability, and strong Lewis basicity at the etheric oxygen. PEO forms complexes with many salts. In the 1970's, it was shown that uni/univalent salts with relatively low cohesive energies would dissolve in PEO, forming polymer/salt complexes.[4,12,13] Upon this dissolution, ionic conductivity in the temperature range above the melting point at roughly 65 degrees C increases by several orders of magnitude, attaining values near 10^{-5}S/cm. These polymer/salt complex electrolytes are of primary interest as electrolytes in lithium or sodium advanced batteries.

Several issues arise in the understanding and mechanistic modeling of these polymer/salt complex electrolytes. The most important are the formation of the complex, structure of the complex, and the transport processes within the complex.

2.1 Formation of polymer/salt complexes

The formation of a complex between the salt and the polymer can be schematically represented as in Eq. 1.

$$(PEO)_n + MX \rightarrow PEO_n \cdot MX \tag{1}$$

The fact that chemical reaction has indeed occurred is seen by the increase in the glass transition temperature of the product (by roughly 40°C), change in density, change in vibrational spectroscopy and change in transport properties.[3-8,11-13] Thermodynamically, the driving force for the completion of Eq. 1 is the complexation of the alkali cation M^+ by the etheric oxygen base sites. This complexation is largely independent of the anion, and salts with very high lattice energies (such as sodium chloride or lithium fluoride) will not form polymer salt complexes in this way.[12,13] Fig. 1 shows the thermodynamic balances appropriate for polymer/salt complex formation; indeed, we observe that very high lattice energy salts do not dissolve to form complexes.

FIGURE 1: Salts That Form Complex Polymeric Electrolytes with PEO*

	Li^+	Na^+	K^+	Rb^+	Cs^+
F^-	no 1036	no 923	no 821	no 785	no 740
Cl^-	yes 853	no 786	no 715	no 689	no 659
Br^-	yes 807	yes 747	no 682	no 660	no 631
I^-	yes 757	yes 704	yes 644	yes 630	yes 604
SCN^-	yes 807	yes 682	yes 619	yes 616	yes 568
$CF_3SO_3^-$	yes 725	yes 650	yes 605	yes 585	yes 550

*The numbers reported are the lattice energies of the salts (in kJ/mol). "Yes" indicates polymer-salt complex formation and "no" indicates the lack of complex formation. The stair-step line indicates the division between complex formation and separate phases.

FIG. 1. Thermodynamics of complex formation. The ability of a particular salt to form a complex is compared with its lattice energy. Since complex formation is driven by Lewis acid/base formation with the etheric oxygens, salts with high lattice energies will not form complexes. From Ref. 12.

2.2 Structure

Polymer/salt complexes are examples of hybrid molecular/ionic materials, in which structure needs to be defined on several different length scales. Locally, x-ray studies,[14,15] supplemented by vibrational spectroscopy,[3,16,17] by nmr,[18] by exafs[19] and other structural probes suggest that cations are coordinated both by solvent oxygens and by anions. The alkali coordination shell is always extensive; generally the alkali is either five or six coordinate.

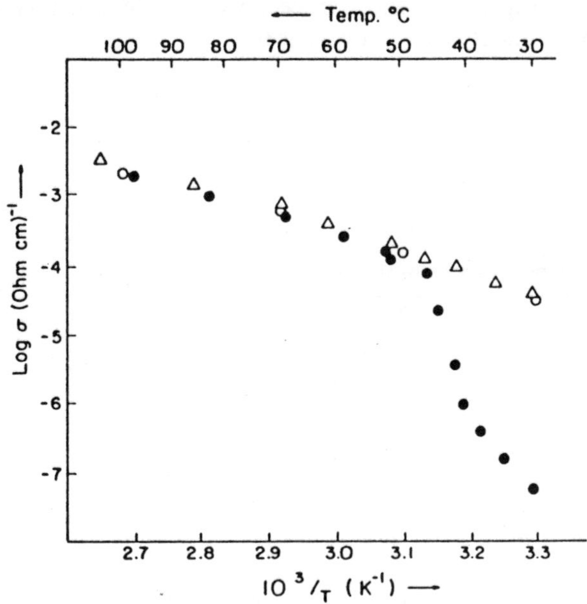

FIG. 2. Demonstration that only the amorphous region exhibits facile conduction. The conductivity of a sodium salt complex electrolyte as a function of inverse temperature. The super cooled amorphous phase, prepared by rapid quenching from high temperature, conducts better than the equilibrium phase of the same temperature. From Ref. 12.

On slightly longer length scales, it has been observed[20,21] (Fig. 2) that conduction occurs only in the amorphous, continuum phase: the phase diagrams[22,23] (Fig. 3) show that crystalline phases in the PEO/salt complex materials exist, but the conductivity in these crystalline phases is very much smaller than in the elastomeric continuum amorphous phase.

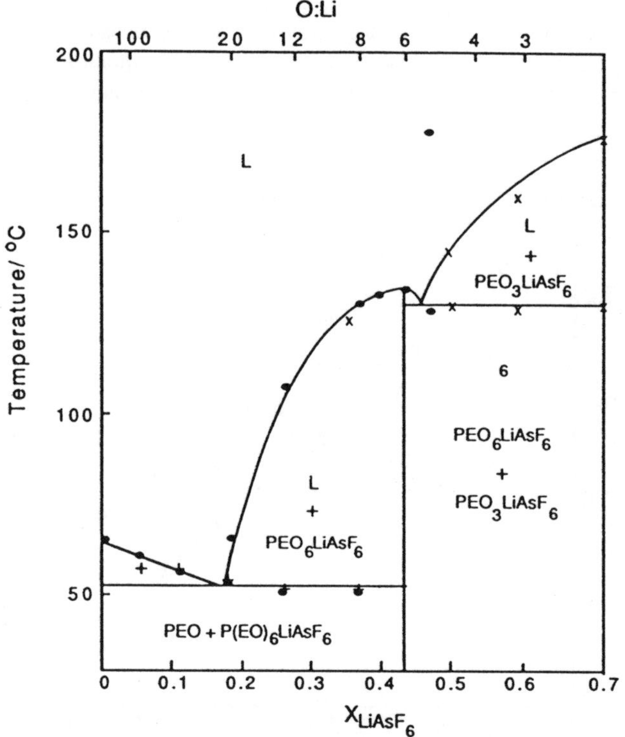

FIG. 3. A characteristic phase diagram for a polymer/salt complexes. From Ref. 3. Note the complex phase structure, including crystalline and amorphous phases.

Many different polymer/salt complexes have been prepared, with different salts and different polymers.[24] It is generally found that conduction occurs only in the amorphous phases, and that structures always show tight coordination spheres around the mobile cations.

2.3 Transport

The most striking aspect of polymer/salt complex electrolytes involves their transport properties. Fig. 4 shows a characteristic thermal dependence of the conduction.[3-8,11] Note the strongly bent curve in Arrhenius coordinates. Usually the conductivity σ is fit by the so-called VTF equation[25,26]

$$\sigma = A \exp\{-B/k_B(T-T_0)\} \qquad (2)$$

Here k_B is Boltzmann's constant and T_0 is the so-called equilibrium glass transition temperature, an empirical parameter which is ordinarily about 50 degrees below the kinetically-determined glass transition temperature. B is pseudo activation energy, and A a prefactor.   As Fig. 4 indicates, the VTF equation actually describes conductivity very well in many polymer/salt complex materials.

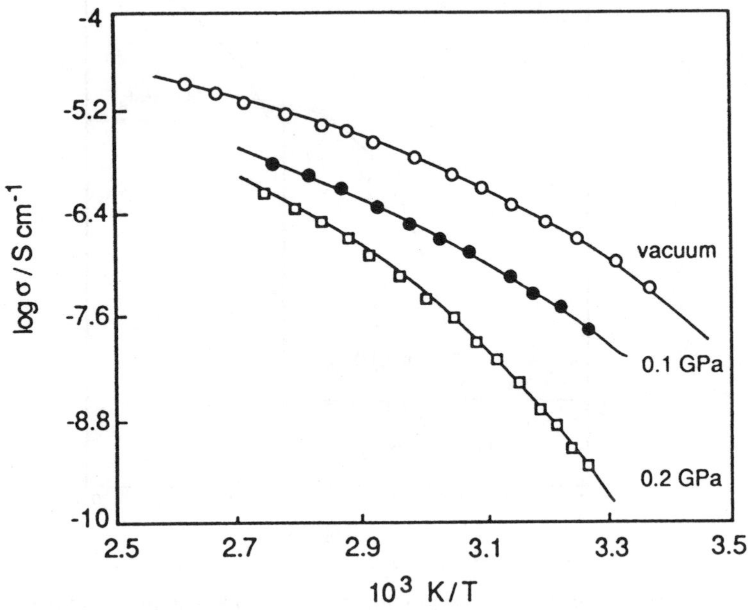

FIG. 4. Demonstration of the VTF equation. The conductivity for a lithium triflate complex is plotted against inverse temperature in Arrhenius coordinates. Note the curved plot on Arrhenius coordinates. From Ref. 3.

Thus conduction occurs only above the glass transition temperature. It occurs only in the amorphous phase, and it depends on the glass transition temperature of the complex. It also depends on the concentration - Fig. 5 shows the results of early work on a phosphazene electrolyte.[27]   Note that, with increasing salt concentration, the glass transition temperature increases monotonically. This would suggest, from Eq. 2, that eventually the conductivity should decrease with increasing salt concentration, as the material becomes stiffer.   This is indeed observed.

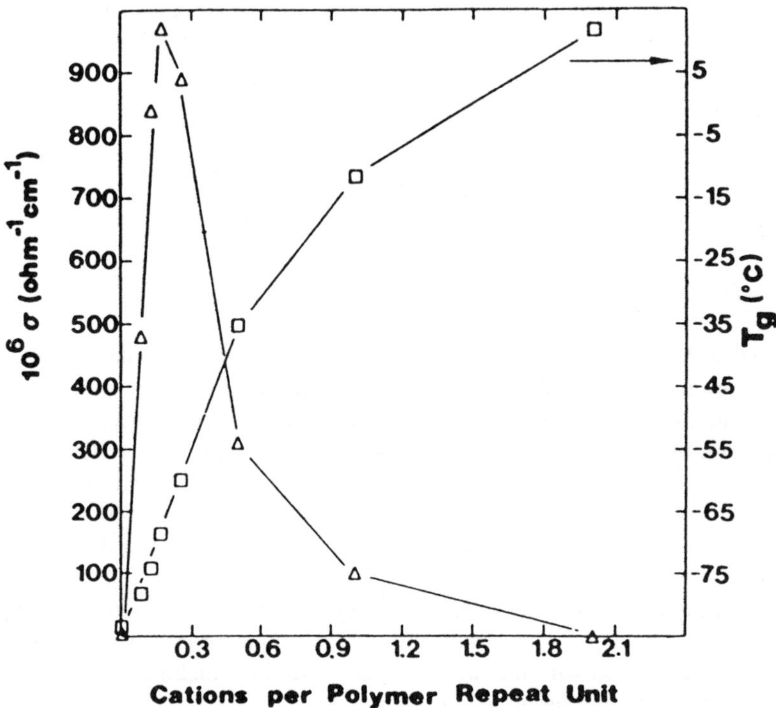

FIG. 5. The joint dependence of glass transition temperature of the complex and the conductivity, for silver salt complexes in MEEP. Note that as the concentration of salt increases, the conductivity first increases (more carriers appear) and finally decreases (as the glass transition temperature increases, so that polymer dynamics limit conduction). From Ref. 12.

Measurement of the transference number t_+ (the percentage of the current carried by cations) is difficult; indeed, in these concentrated electrolytes with immobile solvent it is not clear how to interpret either the measurement or the value of the transference number.[13,28] Nevertheless, it seems quite clear that the mobility of the cations is generally smaller than the anions. This is understood locally in terms of tighter complexation - the cations are held more strongly by the Lewis base sites than the anions are by the polar backbone.

There exists a very extensive phenomenology of polymer electrolyte materials.[3,5-8,11] We have presented here just enough data to serve as the background for the theoretical questions and the extensive modeling that have been completed in an attempt to understand structure and transport of polymer electrolyte materials. Theoretical efforts have focused on understanding and optimization of two critical transport properties, conductivity and cation transference number.

3. Theoretical Analysis: Structure

Since structure in polymer electrolytes exists on several distance scales, structural modeling is done using several methods that are appropriate for such different distances.

3.1 Ab-initio Calculations: Local Coordination

Ab-initio electronic structure methods, so useful in small molecule studies, are still fairly rare in modeling of extended materials. In the case of polymer electrolytes, electronic structure methods are important for understanding ion pair energetics,[29,30] complexation energetics,[30] and possible clustering structures. Curtiss[30,31] has presented ab-initio electronic structure calculations on the local coordination of cations in model ether compounds. He finds that chelation effects are important, and that the coordination environment of the cation exhibits strong shell-like effects, such that breaking a coordination bond between the alkali and the oxygen is energetically quite unfavorable (bond energy of roughly 15 Kcal/mol). This is in agreement with much earlier results of ion cyclotron resonance spectroscopy on the bond energies,[32] and with the models[33] put forward for mobility, which argue that at most one bond can be broken at a time.

Ab-initio calculations have also been used to study ion pairing energetics.[29,31] In the absence of solvent effects, ion pairing energies are strongly negative; this suggests that all ions remain paired. Inclusion of solvent effects is more complicated, and has not been extensively reported in the ab-initio literature.

Fig. 6 shows some results of Curtiss' calculations. The preference for cation full coordination is clear, and the predicted structures are of interest. Experimentally, Fig. 7 shows the results of a fiber x-ray study of a highly concentrated salt solution, corresponding to three etheric oxygens per formula unit of salt.[14] This material is crystalline, and does not exhibit high ionic conduction. Nevertheless, the structure is interesting: note that the cation is roughly five fold coordinated, but that two of these five coordinate sites correspond to iodide anions. This combination of ion pairing and etheric coordination is expected on thermodynamic grounds; ion pairing has been observed[14-17] in polymer electrolytes and the extent of ion pairing and its relation to the conduction mechanism is an important issue (discussed more extensively in Section IV).

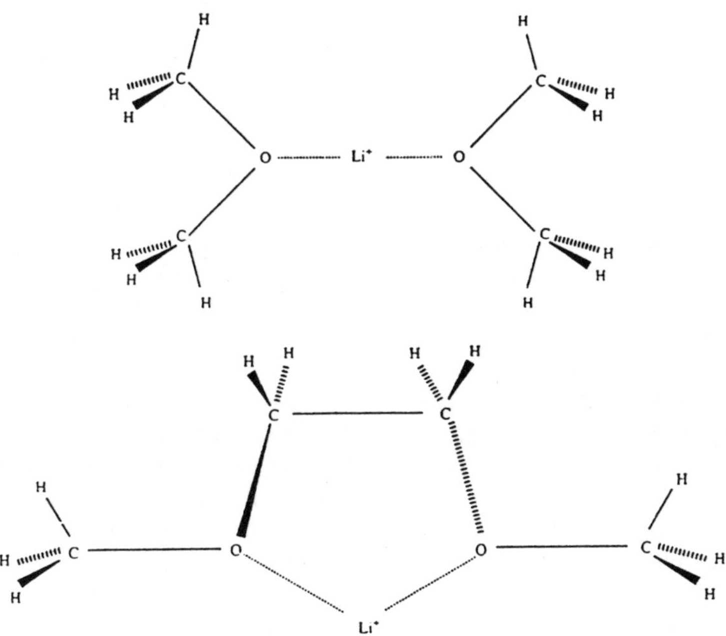

FIG. 6. Some complexation structures calculated using ab-initio methods. The alkali ion interaction with the oxygen of Lewis acid/base type. From Ref. 30.

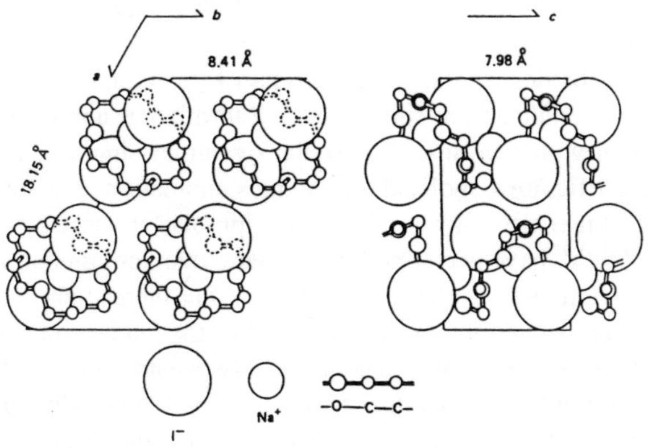

FIG. 7. The structure of 3:1 sodium iodide/PEO. This crystalline phase is non-conductive, but the structure demonstrates the joint complexation of the cation by anion and etheric oxygen. From Ref. 14.

3.2 *Fundamental Steps in Ion Diffusion: Molecular Dynamics Simulations*

There are several difficulties in applying molecular dynamics[34,35] to polymer-electrolyte-based batteries. The first involves study of interfacial effects,[36] to account properly for the screening, double layers, and dynamical polarization that occur at the interface.[37] Even at the simple level of conductivity calculation, the difficulty is that polymer electrolytes exhibit relatively low conductivities, compared to aqueous solution. With a characteristic diffusion coefficient of, say $10^{-7} cm^2/sec$, diffusion of an ion over 10 Å requires $\sim x^2/t \approx 10^{-7} sec$ seconds. For a simulation with a characteristic time scale of $10^{-15} sec$, this requires monitoring the particle for 10^8 time steps. Since it requires many such events to describe a conductivity or diffusivity, actual calculations of conductivity or diffusivities using molecular dynamics are still not reliable.

A more serious difficulty has to do with structure. Since polymer electrolytes in their conductive regions are elastomeric amorphous continua rather than crystalline, structure is only well defined at very short lengths, and becomes statistical beyond the first coordination environment. This means that a proper simulation must evaluate dynamically the actual structure in which transport occurs. Such a simulation places huge demands both on the potential energy surface assumed and on the attainment of appropriate equilibrium structures. Relaxation phenomena in glass formers characteristically exhibit stretched rather than Debye relaxation behavior, so that the KWW form

$$\phi(t) = \exp\{ -(t/\tau_o)^\beta \} \tag{3}$$

becomes appropriate as the relaxation function.[38,39,40,41] Here the parameter τ_0 is the mean relaxation time, and the β describes the distribution ($\beta = 1$ for Debye behavior). This behavior, characteristic of glass formers, means that there will be relaxation times extending from subpicosecond to minutes and beyond; permitting all such relaxation within ordinary molecular dynamics simulation is impossible.

Despite these difficulties, molecular dynamics simulations remain very useful in understanding polymer electrolytes. Groups in Uppsala,[43] Evanston,[44] Minneapolis,[45] London[46] and Zurich[47] have performed molecular dynamics simulations on polymer electrolytes at various levels of sophistication. One interesting result from these simulations is demonstrated in Figures 8A and 8B: the fundamental motion process of a single cation indeed involves breakage of only one coordination bond at a time. This coordination bond could be either to an anion or to a coordinating oxygen, but the energetic cost of breaking more than one coordination bond is simply prohibitive. Therefore the essential motion of the ion in these materials does not[48] involve hopping in the usual sense. In a hopping situation, the mobile ions spend most time vibrating around a particular site and much less time jumping between the sites. Here, the motion is more continuum like or liquid like, with the dynamical motions of the solvent environment intimately coupled[39] with the diffusive motions of the ion. This concept is particularly important, since it is responsible for the dynamic coupling between the relaxation motions of the host and the conductive or diffusive motions of the ions. This coupling, in turn, is responsible for the reduced conductivity compared to crystalline solid electrolytes of β alumina type,[49] for the characteristic WLF behavior of Fig. 4, and for the power density bottleneck in polymer electrolyte batteries. It is, indeed, fundamental to ion diffusion in homogeneous dense materials with strong Lewis acid/base interactions. Overcoming it is one of the major challenges in designing improved polymer electrolytes.

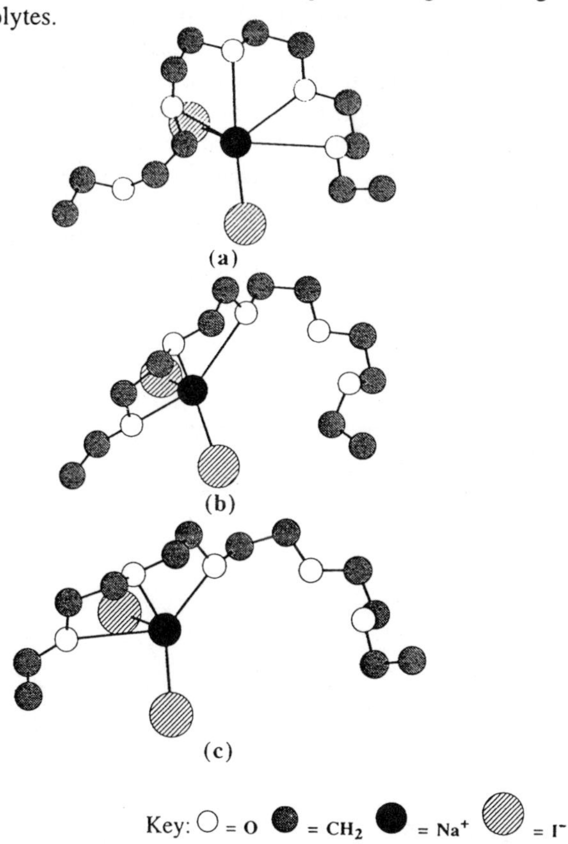

FIG. 8. Cation jump between neighboring sites on the PEO chain. "Bonds" to the cation indicate those oxygen and iodine ligands that are found within 3.8Å of the sodium. The same atoms are displayed and the viewing angles are identical. The sequence is taken from the local environment of Na in the simulation of the $PEO_{20}NaI$ system at 500 K at (a) 325 ps, (b) 326 ps, and (c) 327 ps. From the molecular dynamics study in Ref. 43.

3.3 *Ion Clustering and Temperature Dependence*

Given the enthalpy advantage of forming ion pairs or ion clusters, and given the high ion concentration of most polymer electrolyte materials (with mean separations between ions less than 10Å), one suspects that ion pairing and clustering would be important.   Indeed, the fiber x-ray structure of Fig. 7 shows that the first coordination shell for the cation contains two anions. Experimental work using Raman spectroscopy[17,50,51] shows quite definitively (Fig. 9) that ion paring and clustering occur, and that the number of contact ion pairs increases with temperature.   Molecular dynamics simulations using simple solvent models (dimethyl ether, diglyme) or even Stockmayer particles, show the fundamental nature of this transition.[44] Fig. 10 shows some results of such molecular dynamics simulations. Calculated here is the potential of mean force acting between sodium cation and iodide anions (the potentials assumed are very simple, and do not include polarization). Note that the potential of mean force is deeper at high than at low temperatures; this means (and the molecular dynamics simulations indeed show) that at high temperatures the ion pairing is greater than at low temperatures. This is contrary to the result of a simple argument involving activation energies,[52] that suggested a decrease of ion pairing with increased temperature.

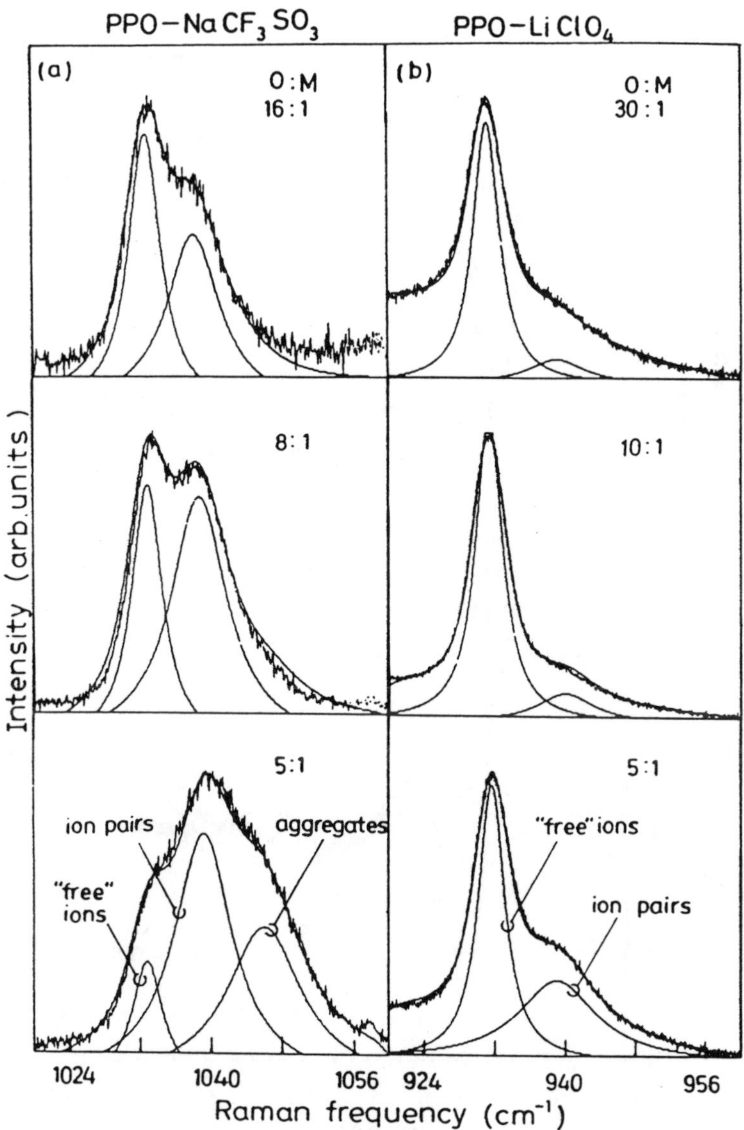

FIG. 9. Raman spectroscopic investigation of ion pairing and clustering. The two peaks are assigned to isolated triflate ions and paired triflate ions. From Ref. 17.

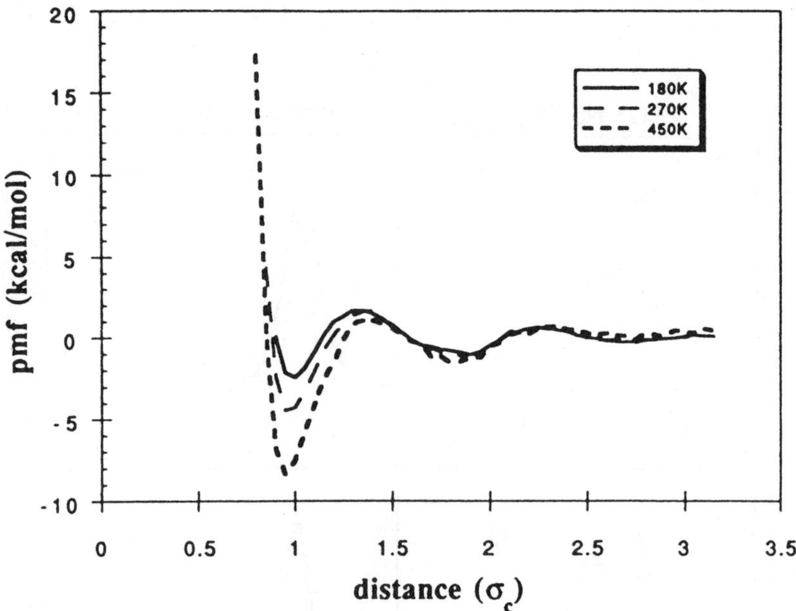

FIG. 10. Potential of mean force calculated using molecular dynamics simulations, for a dimethylether sodium iodide structure. The potential of mean force becomes deeper (more trapping or ion pairing) as temperature increases. From Ref. 44.

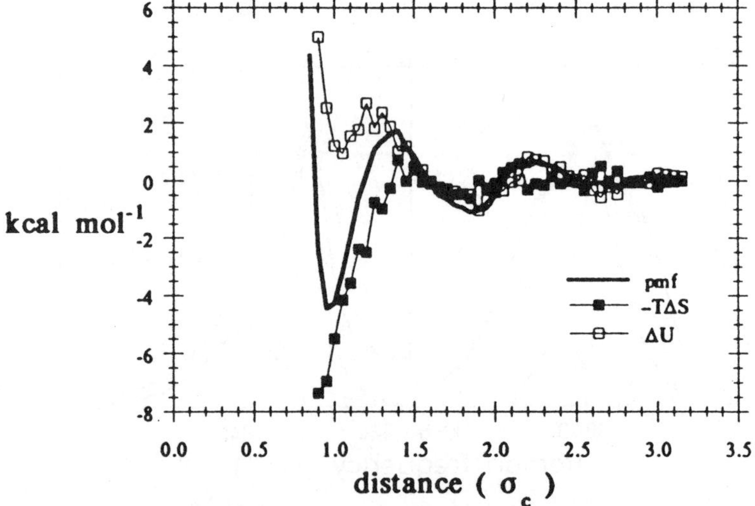

FIG. 11. Break-up of the potential of mean force into entropic and enthalpic components. Note that it is the entropy that, largely, drives the temperature dependence of the ion pairing. From Ref. 44.

The simplest way to understand this thermal dependence of the clustering phenomenon is to break the potential of mean force into its entropy and enthalpy components. Fig. 11 shows such an analysis: the driving force for ion clustering is clearly entropic, as well as enthalpic. As temperature increases, entropic effects are increased. Therefore the entropy advantage of forming ion clusters (essentially more free configuration entropy is available to the solvent if it is not ordered by the presence of the ionic solute) is responsible for the segregation. Macroscopically, this manifests itself as the "salting out" phenomenon;[53] that is, as temperatures of polymer/salt complexes are increased, one observes the formation of microcrystallites of salt. This corresponds to formation of macroscopic clusters, essentially for the same entropic reason as suggested by the potential of mean force analysis.

Applications of molecular dynamics arguments have helped to clarify local structure and dynamics. They also suggest a strong solvent/ion dynamical coupling. The difficult problem of actually simulating the behavior of a real battery, or even of a continuum electrolyte, remains difficult because of extended relaxation time scales, long range coulomb interactions, and dynamical coupling. Characterizing longer range structure either experimentally or computationally is a challenging issue in amorphous materials, and this remains one of the challenges for the future.

4. Ion Transport: Dynamical Simulations

One important aspect of achieving materials by design is sufficient mechanistic understanding of the response property that structural or compositional modifications can be suggested to improve the material. In polymer electrolytes, the crucial performance parameters include conductivity, cation transport number, interfacial stability and structural integrity. Without sufficient ionic conductivity, however, the other performance measures are really not relevant: an electrolyte should move ions. Accordingly, most of the mechanistic analysis of polymer electrolytes has been devoted to the conductivity. The different model studies can be differentiated on the basis of the methodologies adopted.

4.1 Molecular Dynamics: Insights and Difficulties

We have already discussed in Section III above the difficulties involved in using molecular dynamics to study conduction in polymer electrolytes: the conductivity is relatively low, so that the associated simulation time scales are quite long. The structure is poorly defined, and relaxation processes can be slow, so that approaching equilibrium structures, and doing computer transport within them, is problematic.

Halley[45] and Neyertz[43] have taken interesting approaches to this problem. Neyertz uses oligomers, rather than polymers, and uses MonteCarlo averaging to prepare what appear to be equilibrium structures. Halley prepares polymers by forming chemical bonds among monomers whose equilibrium properties are computed more easily. In both these cases, the prepared polymeric structure is then permitted to relax for many time steps in molecular dynamics, and then analysis of the transport is begun.

Complete calculation of conductivity or transport number has not yet been reported, largely because of the time scale difficulty already discussed. The most striking results of the Neyertz analysis are discussed in Section III: the coordination sphere around the cation changes only gradually, so that ion hopping is not really a correct picture for ion transport. Halley finds that, again, hopping is not really an appropriate picture. He suggests that large sections of the structure must rearrange for ion transport to occur; this is consistent with the strong coordination of the cation by the Lewis base sites on the polymer backbone.

Payne et al have discussed the actual conduction mechanism in small molecule ether electrolytes.[44] They point out that at high concentrations, the Kohlrausch form

$$\sigma = \sum_i n_i \mu_i q_i \tag{4}$$

is poorly defined; here σ, n_i, μ_i and q_i are respectively the conductivity, the concentration of species of type i, the mobility of that species and the charge of that species. The Kohlrausch form is correct in the infinite dilution limit, but when the concentrations become high, ions will move within clusters or among clusters or as clusters; in any of these cases, the identification of mobility for a particular charged species, which is required for the Kohlrausch relationship to hold, fails. Payne et al present analyses of the conduction in terms of individual ion motions, cluster motions and intercluster motions, and suggest that in these concentrated electrolyte solutions, the actual charge transport is not necessarily fruitfully treated in terms of isolated cation or anion motions.

These molecular dynamics insights are important and significant. The slightly unsatisfactory situation that neither transport number nor conductivity can be directly calculated reflects the difficulties that are involved in a concentrated electrolyte solution in the presence of an immobile solvent with extended relaxation processes. This is a challenging problem.

4.2 Hopping Models, Coupling and Dynamic Bond Percolation

Coarse-grained models are often useful in chemical dynamics. Hopping models, master equation analysis and state models represent approximations to the correct dynamics, approximations in which a kind of coarse graining is applied to the state structure of the system.[54] In framework solid electrolytes, in which hopping processes clearly occur and one can distinguish between intersite hopping times and site localization times, hopping models have been extensively and usefully employed.[55] In polymer electrolytes, the coupled motion of ionic diffusion and local site relaxation suggests that hopping models are not entirely valid. On the other hand, one can always impose a lattice structure upon the polymeric continuum, and define populations of ions within lattice cells. Motion among those lattice cells can then be written in a master equation form; one must then establish rules for motion among cells, for its dependence upon the potential, and for the appropriate boundary conditions. This sort of analysis will work if the situation is sufficiently strongly damped (high viscosity) that momentum correlations are not important. For dense fluids, like polymer/salt complex electrolytes, this situation does occur, and therefore one might suspect that a modified hopping model that takes into account the coupling between solvent relaxation and ion motion would in fact be valid.

The dynamic bond percolation model and its generalization to the dynamic disorder hopping, do precisely such an analysis.[56-60] The treatment assumes that the master equation holds, in the form

$$\frac{dP_i}{dt} = \sum_j \{ W_{ij}P_i - W_{ji}P_j \} \tag{5}$$

Here P_i is the probability of finding the mobile ion at the ith site; W_{ij} is the probability per unit time of a jump from site i to site j. The difference between dynamic disordered hopping and ordinary hopping is that the hopping rates W_{ij} are themselves time dependent, and evolve on the time scale of relaxation of the polymer host.

The formal model of Eq. 5 becomes specific to a percolation situation when it is assumed that the W_{ij} have only two values: each W_{ij} can be either 0 (probability of 1-f) or a constant value, w (probability f). The assignment of any given link (called bond in this model) between cells i and j as available (value w) are unavailable (value 0) itself evolves in time; the simplest assumption is that the values of the bonds as available or unavailable is reassigned randomly after a renewal time, τ_{ren}. This simple model, called the dynamic bond percolation model, is then characterized by the two time scales τ_{ren} and the hopping time, 1/w. The renewal time, in turn, is determined by the local microviscosity of the solvent:[58] the rate of renewal is the inverse local relaxation time, that determines bond breakage near the mobile ion.

Extensive applications of dynamic bond percolation models[57-66] have resulted in understanding of the frequency dependence of ion motion, the intrinsic coupling between host microviscosity and polymer electrolyte conduction, relationships between relaxation and viscosity, modeling of the stoichiometry dependence of the conduction and so forth. As examples of some of the physical implications of the dynamic bond percolation model, one can cite the following:

1. In the dynamic percolation picture, the static percolation threshold disappears: mean square displacement is always proportional to time, and the diffusion coefficient is related to the inverse average renewal time. Analytic results are obtainable in one dimension, for which one obtains the simple form

$$D = \frac{1}{2} \langle r^2 \rangle_0 / \bar{\tau}_{ren} \tag{6}$$

Here $\bar{\tau}_{ren}$ is the average renewal time, while the expectation value in the numerator is the mean square distance that ion diffuses within a renewal epic - that is, without lattice renewal occurring.

2. An analytic continuation formula can be rigorously obtained, that relates the diffusion coefficient in the presence of renewal to that of the absence of renewal, in the form

$$D(\omega) = D_0(\omega - i/\bar{\tau}_{ren}) \tag{7}$$

$D_0(\omega)$ is the diffusion coefficient as the function of frequency in the absence of renewal, while $D(\omega)$ is the diffusion coefficient as a function of frequency in the renewing lattice.

3. Although the original dynamic percolation picture did not include interionic interaction, adding other potential terms to the hopping model is straightforward: one then simply uses the usual energy weighting, such that

$$W_{ij} = W_{ij}^0 \exp\{\Delta\varepsilon_{ij}/k_B T\} \tag{8}$$

Here W_{ij}^0 is the jumping rate between sites i and j in the absence of energetic effects, while $\Delta\varepsilon_{ij}$ is the energy difference between the states before and after an ion jumps from site i to site j. This Metropolis-type weighting guarantees the approach to thermal equilibrium. One very important result[67-69] of such simulations is shown in Fig. 12. This shows the diffusion coefficient as a function of the renewal time, for different values of the dielectric constant (that is, for different screening of the coulomb potential between ions). The two slopes show that, indeed, diffusion coefficient is proportional to inverse renewal time. They also demonstrate an increased screening (increased dielectric constant) will reduce ion trapping effects, and increase the diffusion coefficient.

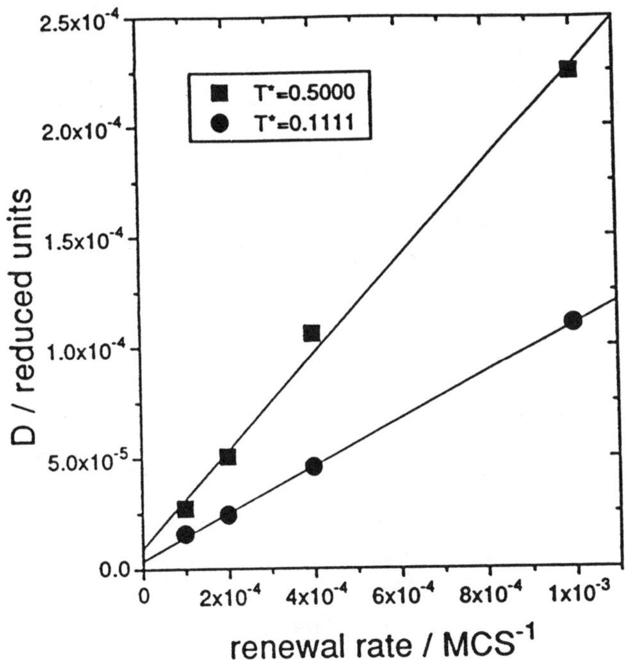

FIG. 12. Dependence of the diffusion coefficient for a lattice model with dynamic bond percolation and ion pairing, as a function of the renewal or relaxation time of the polymer host. The straight lines demonstrate renewal limited behavior (diffusion) versus renewal frequency, suggested by Eq. 9. The two different slopes demonstrate the strong ion trapping effects on conduction. From Ref. 69.

5. A general result of dynamic percolation models, that is not limited to polymer electrolytes, concerns the importance of renewal in determining mean square displacement. Fig. 13 shows that the difference between concave upwards and concave downwards behavior for the mean square displacement as a function of the renewal within each renewal interval, is really all that distinguishes ballistic type motion (appropriate for wide band conduction in metals) and over-damped processes (appropriate for ion motion in dense fluids).[65] In both cases, the overall diffusion will be proportional to the inverse mean renewal, according to Eq. 6 (in one dimension).

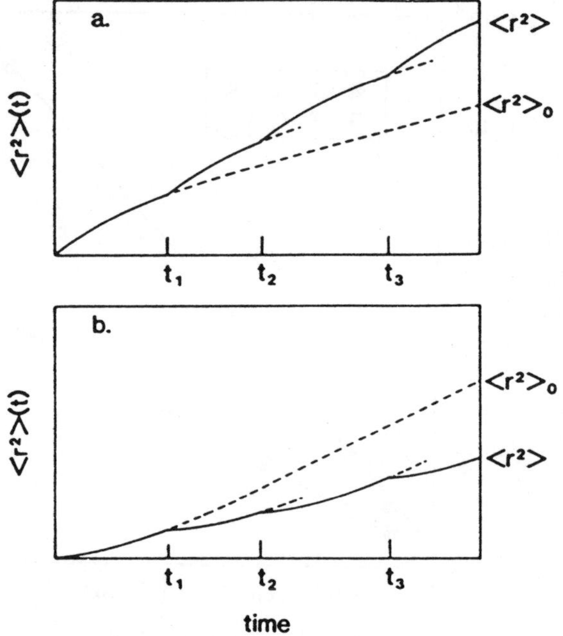

FIG. 13. Generalized dynamic hopping models, demonstrating the effect of renewal on the transport. The concave upward lines are for situations in which the motion within a renewal time is ballistic; those that are concave downwards are for viscosity dominated diffusive behavior within each interval. In either case, the overall slope of the average diffusion is proportional to time. From Ref. 65.

The dynamic percolation model is a very oversimplified picture, but describes the ionic transport process in terms of a lattice picture, and three characteristic parameters: the percentage of available bonds f, the hopping rate w, and the mean renewal rate time, $\bar{\tau}_{ren}$. These parameters can, in turn, be deduced from microscopic simulation over shorter time scales. This characteristic analysis, in which correct microscopic behavior on short time scales determines effective parameters for characterizing the motion of longer time scales, occurs in classical statistical mechanics in the evolution from newtonian behavior through Langevin equation to the smaller Smoluchowski (over-damped) limit to the master equation, hopping model.[54,70] It is fair to anticipate that similar understandings for other materials systems of interest, ranging from binding through self-assembly structures to transport processes and structural reorganizations, can again be approached using a hierarchy of progressively cruder and rougher models.

5. Experimental Implications: Improved Polymer Electrolytes

Based on the various theoretical and modeling studies that have been completed, it is possible to make some significant statements on the design of improved polymer electrolyte materials. Some of these suggestions have been acted upon historically, and have produced substantially better electrolytes. Others remain as design ideas, whose experimental investigation is very much incomplete.

The coarsest grained model that we have discussed, the dynamic bond percolation picture, makes several strong suggestions about optimization of polymer electrolyte materials. If we combine the very simple dynamic renewal ideas with the Nernst-Einstein relationship (that essentially ignores ion clustering effects) the approximate relation

$$\sigma = C \; n_{EFF} \, q^2 <\tau^2>_0 / 6 k_B T \overline{\tau}_{ren} \tag{9}$$

for the conductivity, σ, is obtained. Here the symbols have the same meaning as in Eq. 6, with n_{EFF} the effective number of carriers. Using this relationship, and the results of other modeling studies, several important experimental design suggestions follow directly.

1. Decreased glass transition temperature for the polymer host will increase the conductivity. This follows directly from the WLF relationship of Eq. 2, and is perhaps the most important and significant suggestion to arise from the modeling. Ambient temperature electrolytes that substantially improve conductivity, compared to PEO, have been prepared using the notion of reduced glass transition temperature. These materials include comb polymers based on siloxanes and on phosphazenes,[27,72] as well as aperiodic PEO[73] (PEO with the occasional methylene inserted to break the crystallinity and reduce the glass transition temperature). These techniques have yielded significantly increased (by roughly two orders of magnitude) ambient temperature polymer electrolyte materials, using the same salts as the parent PEO structures. Above the PEO melting point, the advantages of these different backbone structures are reduced, as the denominators in Eq. 2 increase.

The highest conductivity ambient temperature polymer electrolytes are indeed those based on the comb polymer motif, with low glass transition temperature structures.

2. Addition of plasticizers or low molecular weight compounds to form gels will lower the glass transition temperature, and therefore increase the conductivity.[3,74,75] Indeed, these gel electrolytes exhibit very high conductivities (near 10^{-3}S/cm at room temperature). Plasticizers and gel formers include ethylene carbonate and propylene carbonate, and these gel materials are of very substantial interest both in

lithium metal batteries and in lithium ion batteries. Their very substantially lowered T_g increases the conductivity for the same reason that the choice of a softer host polymer does: this follows directly from the WLF form of Eq. 2.

3. Reducing pairing or clustering of the effective ionic carriers will increase conduction. This follows from the n_{EFF} term in Eq. 9: increasing clustering will tie down the effective carriers, reducing diffusivity and conductivity. Several schemes can be suggested for reducing the amount of paring or clustering.

a.   Increase the dielectric constant of the host polymer. This approach is problematic, since generally high dielectric constant materials also are stiff polymers, with high glass transition temperatures; the glass transition temperature effect, being exponential (Eq. 2) dominates the linear effect in n_{EFF} of Eq. 9. Accordingly, for example, carbonate polymers have shown very poor room temperature conductivities, despite having higher dielectric constants than the polyethers.[76]

b.   Select anions that form weak ion pairs, because of low effective basicity. This has been an extremely effective approach. Such anions include delocalized aromatic anions such as tetraphenylborate, or anions whose basicity is substantially reduced by inductive effects. The imide and carbide salts (structure B, C) are probably the most important in this regard.[77] Indeed, use of the imide salt increases the conductivity of lithium PEO electrolytes above the PEO melting point by more than an order of magnitude, and accordingly the imide salts are being actively pursued for technological applications in automotive traction batteries.

c.   Optimize ionic concentrations to minimize pairing or clustering. Clearly, if no salt is present there can be no clustering, but there can also be no conduction. Fig. 5 showed that increasing salt concentration will first lead to an increase in conduction, but eventually the conductivity will decrease as the material becomes stiffer (including raised glass transition) and as ion pairing or clustering increases. Simulations have shown that at high concentrations, the single ion motion between electrodes is irrelevant, and that most conduction arises from intercluster and intra-cluster ion diffusions.[44] More clustering generally increases intercluster transport distance, and therefore decreases the conductivity.

d.   By introducing cation encrypting agents, ion clustering can be reduced and conductivity increased. This is indeed so; both crown ethers[78] and cryptands[79] (structures D, E) have been used to increase the conductivity of both polymer/salt complex electrolytes and polyelectrolytes. Fig. 14 shows the results from Shriver's laboratory using polyelectrolyte materials.[80] Note that addition of the cryptand molecule decreases effective ion pairing and increases conduction; notice also that this increase is smaller in the aluminosilicate structures (in which ion pairing was originally lower, and therefore the conduction was originally higher) than it is in the aluminates. This difference is significant, and indicates that increasing the number of ionic carriers can increase the conductivity to a relatively limited extent, compared to modification of the intrinsic mobility that arises from the glass transition temperature effect above.

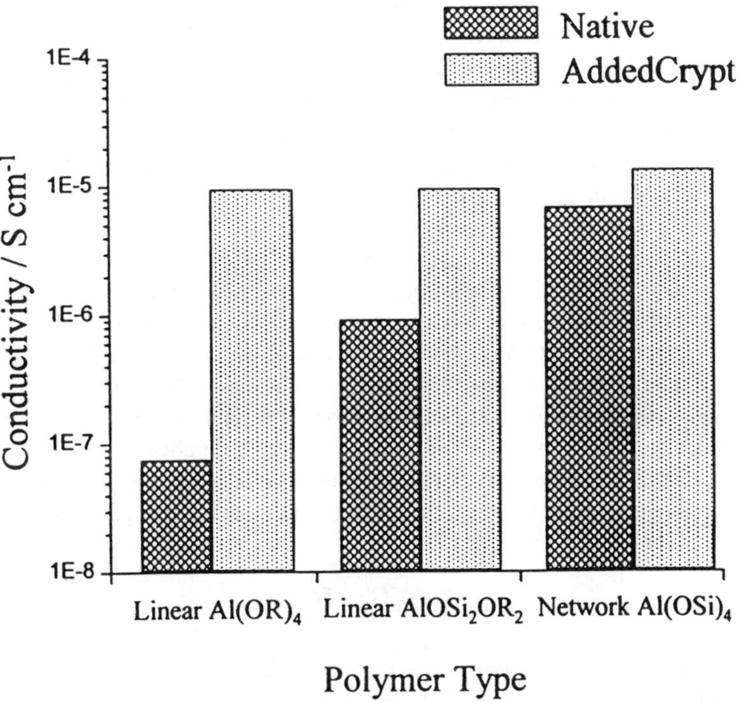

FIG. 14. Reduction of ion pairing in polyelectrolytes causes increased conduction. Measured conductivities for a series of aluminate and alumino silicate structures, with the addition of a cryptand species, that surrounds the sodium cation and reduces ion pairing. Note that with decreased Lewis site basicity (decreased intrinsic ion pairing) the effect of the added crypt is reduced. (D.F. Shriver, unpublished)

4. Decreased basicity of localized anion structures can increase conduction in polyelectrolytes. This statement is essentially the polyelectrolyte generalization of the statement 3b above. Accordingly, Fig. 14 indeed shows that using aluminosilicates, in which the local electron density and basicity are lower than in the parent aluminates, decreases ion clustering and increases conduction.[80] There has been less attention paid to the polyelectrolyte area than to the polymer/salt electrolyte area, both experimentally and theoretically, but one might expect this to change in the future (polyelectrolytes will exhibit unit transference number for the cation, without self polarization effects, over-voltages, and irreversible precipitation reactions). Thus far, their conductivities have been lower than the polymer/salt complexes. This difference can be understood theoretically using hopping models,[68] but improved polyelectrolytes would be more technologically useful than polymer/salt complex electrolytes themselves.

5.   Homogeneous polymer/salt complex electrolytes and polyelectrolytes are expected to show maximal conductivities limited by the coupling[39,81] between ion motion and polymer structural relaxation, as suggested by Eqs. 2 and 6. This in turn suggests that increased conductivity in materials might well be approached by preparing more complex structures, structures that are not homogenous.

a. Composite materials based on suspensions of non-electrolytes. Extensive experimental work[82-89] on clay/polymer composites, on silica/polymer composites, on alumina/polymer composites, and on composites with other dispersions show two attractive properties: first, the conductivity reduction compared to the pure polymer electrolyte may be quite small - see Fig. 15. Addition of the hard inorganic filler substantially improves the yield and flow properties, and can make the electrolyte stiff enough to resist structural deformation in actual application. The simplest Maxwell model (essentially based on the idea of mixing conductor with non-conductor) suggests reduction[90] of the conductivity in an amount proportional to the volume fraction of dispersant:

$$\sigma = \sigma_p(1 - 1.5\phi_2) \tag{10}$$

Here ϕ_2 is the volume fraction of the non-conductive dispersant, while σ_p is the conductivity of the pure polymer electrolyte.

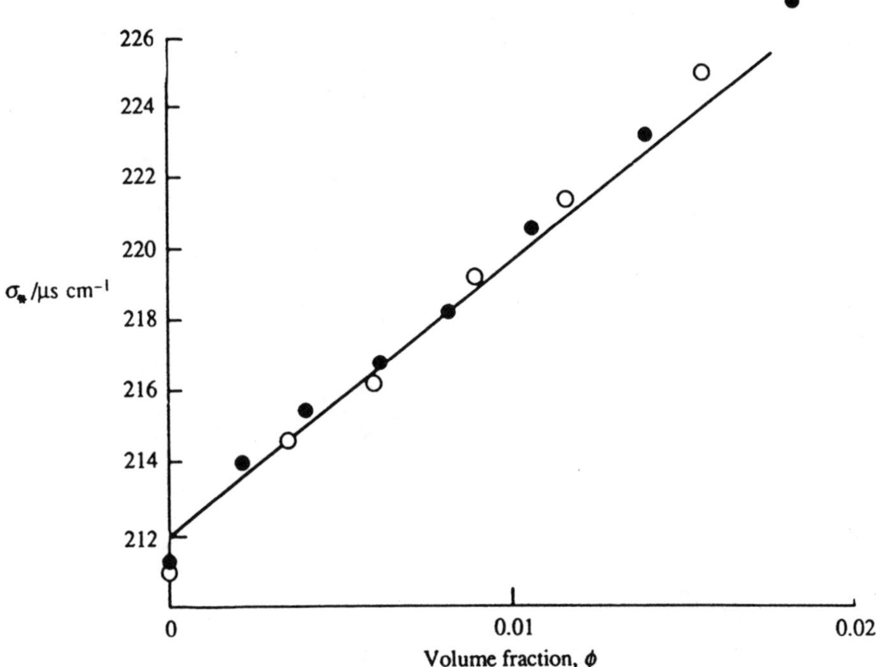

FIG. 15. Conductivity as a function of the volume fraction of filler ions, for a liquid electrolyte consisting of hydrochloric acid filled with latex particles. Note that the conductivity <u>increases</u> as the mole fraction of conducting species (the HCℓ solution) decreases. From Ref. 90.

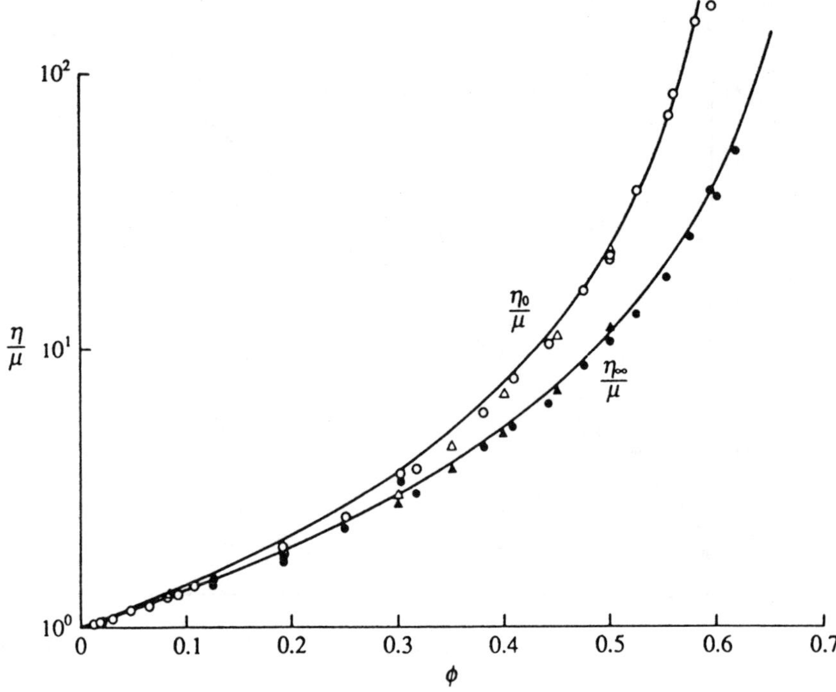

FIG. 16. Dependence of the shear viscosity as a function of volume fraction of filler, for addition of silica fillers to fluid solutions. The hard particles eventually dominate the small wave vector shear viscosity. From Ref. 90.

One might expect to do better. In the analogous area of liquid electrolytes, the study of electrokinetic data suggests that under certain conditions addition of non-conductive filler particles actually increases the conductivity, essentially due to polarization effects around the particles.[90,91] In the area of solid electrolytes, the so-called Liang effect, extensively observed in heavy metal salt conductors, is manifested when addition of filler particles such as alumina or silica increase the conductivity of salts such as lithium iodide.[92-94] This increased conductivity arises essentially from double layer/space charge effects at the interface.

At the interface between the dispersant and the polymer, an inhomogeneity clearly occurs. It is possible that in this inhomogeneous region both the electrostatic effects leading to ion clustering and the viscous drag effects leading to coupling of polymer relaxation dynamics and ion transport can be relaxed; under these conditions, an electrokinetic increment of the conductivity[90] might be expected. While this has been observed[91] in liquid electrolytes (Fig. 16), it is still unseen in polymer electrolytes. Finding conditions for such an increment in the conduction would be a substantial step forward in electrolyte design.

b. Conductive rubbers, based on molten salts, should exhibit high conductivities. Angell pointed out[95] that if one continues a diagram like that of Fig. 5 far enough, then if the complexes form between the polymer and a low melting salt, the conductivity should eventually begin to increase again, as the salt concentration increases and heads towards the highly conductive molten salt phase. One could then prepare a solid electrolyte, based essentially on polymer in salt rather than salt in polymer, that exhibits the high conductivity of molten salt and the dimensional stability of the polymer complex. These materials have been prepared in several laboratories; Fig. 17 shows the original Angell argument and some of the earlier results.[95] The idea is a very attractive one, since the conductivity increments can be quite high, and the cost might be quite low if appropriate salts were chosen. Moreover, such an electrolyte would be much more like a molten salt, and therefore might avoid some of the polarization and instability problems of polymer electrolytes.

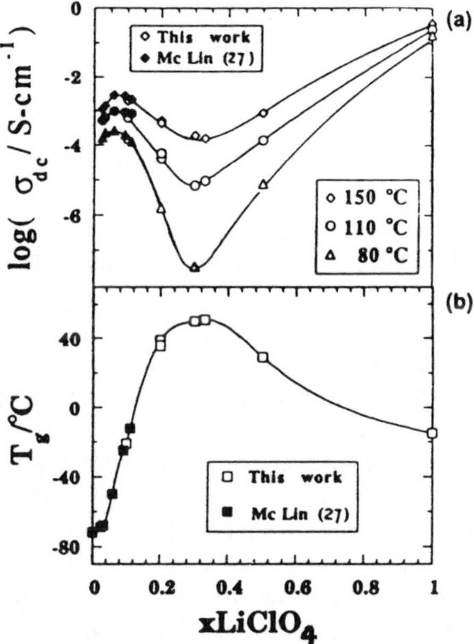

FIG. 17. Demonstration of the ionic rubber region of the phase diagram. For relatively small mole fractions of added salt, the salt in polymer material exhibits increasing glass transition temperature, and decreasing conduction, as salt is added. On the salt rich end (polymer in salt), the material behaves essentially like the molten salt, exhibiting higher conduction. The material is lithium perchlorate in poly (propylene oxide). From Ref. 95.

There has not been extensive pursuit of this rubber electrolyte idea - most of the salts originally chosen are themselves thermodynamically unstable (perchlorates and chlorates), but recent work from Shriver's laboratory[96] has used the imide salts (B above). These initial results are quite promising, and the rubbery materials will be of substantial interest in the future.

6. Comments: Materials Design Using Theoretical Methodologies

It is perhaps instructive, in approaching the design of improved materials, to examine the situation in certain biological systems. The development of monoclonal antibodies is based on the idea of a transition state mimic - that is, if one can produce a molecular species whose structure in some sense mimics the expected transition state for a desired biochemical transformation, then one can train an enzyme to recognize that transition state, thereby lowering the barrier and catalyzing the biochemical reaction. Similarly, mechanism based inhibitors in protein systems[97] are based on the idea that, by understanding the different mechanistic steps in the function of an enzyme, one can introduce chemical species that will bind irreversibly to one of the intermediates. The important point about both of these approaches, in our current context, is that they are based on mechanism: understanding the mechanism of a particular process permits its control.

This mechanism based approach can be used for understanding and improvement of materials response properties such as linear or nonlinear optical properties, mechanical properties, conduction properties, structural relaxation properties, light scattering properties, elasticity or wetting properties, biochemical binding properties, etc. Once again, however, the notions of understanding materials behavior and improvement starting with mechanism can be important. Table 1 sketches a small number of areas of current interest in materials design in which an understanding of the mechanism of materials response has led to, and will continue to lead to, preparation of improved functional materials.

Understanding the response mechanism of materials requires many theoretical tools. These range in sophistication and elegance from so-called ab-initio molecular dynamics or full quantum dynamics to advanced electronic structure methodologies (at the ab-initio level, and including correlation effects) through applications of semiempirical electronic structure theory to molecular dynamics and Monte Carlo methods, to hopping models, Langevin and Brownian descriptions to static and dynamic Ising models and other highly coarse grained pictures.

Many of these methods are either ill-defined or inappropriate for studies of single molecule behavior. They are needed for understanding materials, however, because both the structure and the dynamics of materials systems are substantially more complex than those associated with individual molecular responses. As discussed in Section I, both structure and dynamical response in characteristic molecular materials extend over many orders of magnitude in space and time dimensions respectively, so that a particular theoretical technique that may be very useful on particular scales (say for quantum dynamics for small systems at subpico- second times) is essentially useless for understanding a different property of precisely the same material (such as the creep behavior in a solid polymer).

As the picture of any material becomes more complex, so the palette of theoretical techniques must become more extensive. Nevertheless, several of the ideas for mechanistic understanding and optimization of small molecules, ideas that arose originally in the fields of physical organic chemistry and of quantum chemistry, remain crucial for the understanding and optimization of materials systems. The study and understanding of mechanism and its modification by changes in chemical structure is a very powerful concept, one that extends from molecular electronic structure methods and quantum dynamics of individual molecules through quantitative structure activity relations and the various correlative schemes of physical organic chemistry, physical inorganic chemistry and physical biochemistry. The very successful application of these mechanism-based ideas in materials is a relatively new theme, but one that will be critical in the chemical, biological and materials sciences into the 21st century.

Acknowledgements

In the work on polymer electrolytes, I have been privileged to work with several outstanding scientists. I'm particularly grateful to Du Shriver, Abe Nitzan, Mark Lonergan, Maria Forsyth, Steve Druger, Simon deLeeuw, John Perram, Caroline Harris, Kate Doan, Brian Papke, Ryan Dupon, Marvin Silverberg, Vilia Payne, and Charlie Hardy. I am grateful to the ARO and to the DOE/LBL advanced batteries program, as well as the Northwestern Materials Research Center funded by the NSF/MRL office, for support of this research.

TABLE I

EXAMPLES OF MECHANISM-BASED
MODIFICATION OF MATERIALS

Property	Mechanism	Alteration
Second-order nonlinear optics	Intramolecular β; strong charge-transfer transitions	Pi-type donor, acceptor conjugated chromophore
Electronic resistivity	Peierls localization	Attach S, Se for dimensional cross conductivity
Picosecond optical limiting	Reverse saturable absorption	Facilitate intersystems crossing by heavy-atom addition
Block polymer domain segregation	Thermodynamic stabilization	Appropriate functional substitution
Ionic localization by radioisotope stabilization	Very slow diffusion in glasses	Low polarizability, high T_g oxide glasses
Glassy electrolytes	Polarizability reduction of diffusion barriers	Chalcogenide glasses
Polymer electrolytes	Increase σ and t_+	Section V of this paper

References

1. W.J. Hehre, L. Radom, P.R. Schleyer and J.A. Pople, Ab-Initio Molecular Orbital Theory, (Wiley, New York, 1986).
2. A.P. Alivisatos, P.F. Barbara, A.W. Castleman, J. Chang, D.A. Dixon, M.L. Klein, G.M. McLendon, J.S. Miller, M.A. Ratner, P.J. Rossky, S.I. Stupp and M.E. Thompson, *J. Phys. Chem.*, submitted.
3. F.M. Gray, *Solid Polymer Electrolytes* (VCH, New York, 1991).
4. M.B. Armand, J.M. Chabagno and M.J. Duclot, in *Fast Ion Transport in Solids* (Edited by P. Vashishta, J.N. Mundy and G.K. Shenoy), p. 131. North-Holland, New York (1979); P.V. Wright, in *Polymer Electrolyte Reviews-2* (Edited by J.R. MacCallum and C.A. Vincent), p. 61. Elsevier, London (1987); D.E. Fenton, J.M. Parker and P.V. Wright, *Polymer* **14**, 589 (1973).
5. J.R. MacCallum and C.A. Vincent, *Polymer Electrolyte Reviews 1* (Elsevier, London, 1987); J.R. MacCallum, and C.A. Vincent, *Polymer Electrolyte Reviews 2* (Elsevier, London, 1989).
6. J.S. Tonge and D.F. Shriver, in *Polymers for Electronic Applications*, edited by J.H. Lai (CRC, Boca Raton, FL, 1989), Vol. 157; C.A. Vincent, *Electrochim. Acta* **40**, 2035 (1995).
7. M. Ratner, in *Polymer Electrolyte Reviews 1* (Edited by J.R. MacCallum and C.A. Vincent) Vol. 1, pp. 173. Elsevier, New York, (1987).
8. J.O. Thomas, Ed. Fifth International Conf. on Polymer Electrolytes (Elsevier, in press).
9. M.A. Ratner and A. Nitzan, Discuss. Faraday Soc. **88**, 19 (1989).
10. Y.F.Y. Yao and J.T. Kummer, *J. Inorg. Nuc. Chem.* **29**, 2453 (1967).
11. S. Greenbaum, Ed. *Electrochim. Acta* **40**, #13/14 (1995).
12. M.A. Ratner and D.F. Shriver, *Chem. Revs.* **88**, 109 (1988).
13. M. Armand, *Faraday Disc.* **88**, 65 (1989).
14. Y. Chatani and S. Okamura, *Polymer* **28**, 1815 (1987).
15. P. Lightfoot, J.L. Nowinski and P.G. Bruce, *J. Am. Chem. Soc.* **116**, 7469 (1994); P.G. Bruce, *Electrochim. Acta* **40**, 2077 (1995).
16. R. Dupon, B.L. Papke, M.A. Ratner, D.H. Whitmore and D.F. Shriver, *J. Am. Chem. Soc.* **104**, 6247 (1982).
17. L.M. Torell and S. Schantz, in *Polymer Electrolyte Reviews*; (Edited by J.R. MacCallum and C.A. Vincent) Vol. 2, p. 1. Elsevier Applied Sciences: London (1989); S. Schantz, L.M. Torell and J.R. Stevens, *J. Chem. Phys.* **94**, 6862 (1991).
18. S.G. Greenbaum, J.J. Wilson, M.C. Wintersgill, and J.J. Fontanella in *Second International Symposium on Polymer Electrolytes*, edited by B. Scrosati (Elsevier, University of Siena, 1989), pp. 35-48.
19. R.J. Latham, R.G. Linford, R.A. Pynenburg, W.S. Schlindwein, *J. Electrochem.*

Soc. **240**, 1056 (1993).

20. C. Berthier, W. Gorecki, M. Minier, M.B. Armand, J.M. Chabagno and P. Rigaud, *Sol. St. Ionics* **11**, 91 (1983).

21. L.C. Hardy and D.F. Shriver, *Macromolecules* **17**, 975 (1984).

22. C.D. Robitaille and D. Fauteux, *J. Electrochem. Soc.* **133**, 315 (1986).

23. P.R. Sorensen and T. Jacobsen, *Polym. Bull.* **9**, 47 (1981); ref. 3, chap. 4.

24. R. Koksbang, I.I. Olsen and D. Shackle, *Sol. St. Ionics* **69**, 320 (1984).

25. P.G. Bruce and C.A. Vincent, *J. Chem. Soc. Faraday Trans.* **89**, 3187 (1993).

26. H. Vogel, *Phys. Z.* **22**, 645 (1921); V.G. Tammann and W. Hesse, *Anorg. Allg. Chem.* **156**, 245 (1926); G.S. Fulcher, *J. Am. Ceram. Soc.* **8**, 339 (1925).

27. P.M. Blonsky and D.F. Shriver, *J. Am. Chem. Soc.* **106**, 6854 (1984).

28. M. Doyle and J. Newman, *Electrochimica Acta*, **40**, pp. 219-2196 (1995); J. Newman, *Electrochemical Systems*, Prentice-Hall, Englewood Cliffs, NJ (1991).

29. P. Teeters, S.L. Stewart and L. Svoboda, *Sol. St. Ionics* **28/30**, 1054 (1988); J. Francisco and M. Williams, *J. Phys. Chem.* **94**, 8522 (1990).

30. P.T. Boinske, L. Curtiss, J.W. Halley, B. Lin and A. Sutjianto, *J. Comp.-Aided Design*, **3**, 3851 (1996).

31. L. Curtiss, personal communcation.

32. P. Kebarle, *Ann. Revs. Phys. Chem.* **28**, 445 (1977).

33. B.L. Papke, M.A. Ratner and D.F. Shriver, *J. Electrochem. Soc.* **129**, 1694 (1982).

34. M.P. Allen and D.J. Tildesley, Computer Simulation of Liquids (Oxford University Press, New York, 1987).

35. D. Frenkel and B. Smit, Understanding Molecular Simulation (Academic, 1996).

36. L. Christie, P. Los and P.G. Bruce, *Electrochimica Acta*, **40**, 2159-2164 (1995); E. Peled, D. Golodnitsky, G. Ardel and V. Eshkenazy, *Electrochimica Acta*, **40**, 2197-2204 (1995).

37. L. deJonghe, personal communication.

38. A.L. Tipton, M.C. Lonergan, M.A. Ratner, D.F. Shriver, T. Wong and K. Han, *J. Phys. Chem.* **98**, 4148 (1994).

39. R. Bergman, A. Brodin, D. Engberg, Q. Lu, C.A. Angell and L.M. Torell, *Electrochimica Acta*, **40**, 2049-1055 (1995); L.M. Torell and C.A. Angell, *British Polymer Journal*, **20**, 173 (1988); C.A. Angell, *Solid State Ionics*, **9/10** 3 (1983); C.A. Angell, *Solid State Ionics*, **18/19**, 72 (1986).

40. M.T. Cicerone, M.D. Ediger, *J. Chem. Phys.* **104**, 7210 (1996).

41. K.L. Ngai, *Phys. Rev.* **B48**, 13481 (1983).

42. R. Bohmer, K.L. Ngai, C.A. Angell and D.J. Plazek, *J. Chem. Phys.* **99**, 4201 (1993).

43. S. Neyertz, D. Brown and J.O. Thomas, *Electrochimica Acta*, **40**, 2063-2069 (1995); S. Neyertz and D. Brown, *J. Chem. Phys.* **104**, 3797 (1996).

44. V.A. Payne, J-H Xu, M. Forsyth, M.A. Ratner, D.F. Shriver and S.W. deLeeuw, *Electrochimica Acta*, **40**, 2987-2091 (1995); V.A. Payne, M.C. Lonergan, M. Forsyth, M.A. Ratner, D.F. Shriver, S.W. deLeeuw, and J.W. Perram, *Solid State Ionics*, **81**, 171-181 (1995); M. Forsyth, V.A. Payne, M.A. Ratner, S.W. deLeeuw and D.F. Shriver, *Mat. Res. Soc. Symp. Proc.* **210**, 203 (1991); M. Forsyth, V.A. Payne, M.A. Ratner, S.W. deLeeuw, and D.F. Shriver, *Solid State Ionics*, **53-56**, 1011 (1992); V.A. Payne, M. Forsyth, M.A. Ratner, D.F. Shriver and S.W. deLeeuw, *J. Chem. Phys.*, **100**, 5201 (1994).

45. J.W. Halley, personal communication.

46. C.R.A. Catlow and G.E. Mills, *Electrochimica Acta*, **40**, 2057-2062 (1995).

47. F. Müller-Plathe, *Acta Polymer*, **45**, 259 (1994).

48. J.W. Halley, to be published.

49. cf. e.g. G.A. Nazri, J.-M. Tarascon and M. Schreiber, eds. *Solid State Ionics* **IV** (MRS, 1995, V. 369).

50. W. Huang, R. Frech, P. Johansson and J. Lindgren, *Electrochimica Acta*, **40**, 2147-2151 (1995).

51. R. Frech and W. Huang, *Solid State Ionics*, **72**, 103 (1994); W. Huang and R. Frech, *Polymer*, **35**, 235 (1994).

52. H. Cheradame, in H. Benoit & P. Rempp, IUPAC *Macromolecules*, Pergamon, London, 1992.

53. C.C. Wintersgill, J.J. Fontanella, S.G. Greenbaum and K.J. Adamic, *Br. Polym. J.* **20**, 195 (1988).

54. Stochastic Processes in Chemical Physics, Ed. I. Oppenheim, K.E. Shuler and G. H. Weiss (MIT Press, 1977).

55. W.Dietrich, B. Fulde and I. Peschel, *Adv. Phys.* **29**, 527 (1980).

56. S.D. Druger, A. Nitzan, and M.A. Ratner, *J. Chem. Phys.* **79**, 3133 (1983).

57. S.D. Druger, *J. Chem. Phys.* **95**, 2169 (1991); S.D Druger, *J. Chem. Phys.* **100**, 3979 (1994).

58. S.D. Druger, M.A. Ratner, and A. Nitzan, *Solid State Ionics*, **9/10**, 1115 (1983).

59. A number of related treatments occur in discussion of transport in disordered media; for example, A.K. Harrison and R. Zwanzig, *Phys. Rev. A* **32**, 1972 (1985); R. Hilfer and R. Orbach, *Chem. Phys.* **128**, 275 (1988); R. Granek and A. Nitzan, *J. Chem. Phys.* **97**, 3823 (1992); R.F. Loring, *ibid.* **94**, (1991).

60. S.D. Druger, M.A. Ratner, and A. Nitzan, *Phys. Rev. B* **31**, 3939 (1985).

61. S.D. Druger, M.A. Ratner, and A. Nitzan, *Solid State Ionics* **18/19**, 106 (1986).

62. S.M. Ansari et al., *Solid State Ionics* **17**, 101 (1985).

63. M.C. Lonergan, A. Nitzan, M.A. Ratner and D.F. Shriver, *J. Chem. Phys.* **103**, 3253-3261, (1995).

64. M.A. Ratner and A. Nitzan, *J. Phys. Chem.* **98**, 1765 (1994).

65. S.D. Druger and M.A. Ratner, *Chem. Phys. Lett.* **151**, 434 (1988).

66. M.C. Lonergan, Ph.D. Thesis, Northwestern University, 1994.

67. M.C. Lonergan, J.W. Perram, M.A. Ratner, D.F. Shriver, *J. Chem. Phys.* **98**, 4937 (1993).

68. M.C. Lonergan and M.A. Ratner, to be published.

69. M.C. Lonergan, D.F. Shriver, and M.A. Ratner, *Electrochemica Acta*, **40** 2041-2048 (1995).

70. N. Wax, ed., Selected papers on Noise and Stochastic Processes (Dover, New York, 1954).

71. K. Nagaoka, H. Naruse, I. Shinohara and M. Watanabe, *J. Poly. Sci. Poly. Lett. Ed.* **27**, 659 (1984); D. Fish, I.M. Khan, J. Smid, *Makromol. Chem. Rapid Comm.* **7**, 115 (1986); R.S. Spindler & D.F. Shriver, *J. Am. Chem. Soc.* **110**, 3036 (1988).

72. H.R. Allcock, P.E. Austin, T.X. Neenan, J.T. Sisco, P.M. Blonsky and D.F. Shriver, *Macromols.* **19**, 1508 (1986).

73. C.V. Nicholas, D.J. Wilson, C. Booth and J.R.M. Giles, *Brit. Polym. J.* **20**, 2890 (1985).

74. K.M. Abraham and M. Alamgir, *J. Electrochem. Soc.* **137**, 1657 (1990); P.E. Stallworth, S.G. Greenbaum, F. Croces, S. Slant and M. Salomon, *Electrochimica Acta*, **40**, 2137-2141 (1995); K.M. Abraham, *Electrochemica Acta* **38**, 1681 (1993); K.M. Abraham and M. Alamgir, *Journal of Power Sources* **43-44**, 195 (1993).

75. D.R. MacFarlane, J. Sun, P. Meakin, P. Fasoulopoulos, J. Hey and M. Forsyth, *Electrochimica Acta*, **40**, 2131-2136 (1995).

76. M. Forsyth, A.L. Tipton, D.F. Shriver, M.A. Ratner and D.R. MacFarlane, *Sol. St. Ionics* **99**, 257 (1997).

77. J.F. LeNest, S. Callens, A. Ghandini and M. Armand, *Electrochim. Acta.* **37**, 1585 (1992).

78. M.L. Kaplan, E.A. Reitman, R.J. Cava, L.K. Holtz, E.A. Chandross, *Sol. St. Ionics* **25**, 37 (1987).

79. K. Chen, S. Ganathiappan and D.F. Shriver, *Chem. Mater.* **1**, 483 (1989); G. Zhou, I.M. Khan and J. Smid, *Macromols.* **26**, 2202 (1993).

80. J.C. Hutchinson & D.F. Shriver, to be published.

81. M.G. McLin and C.A. Angell, *Sol. St. Ionics* **53/56**, 1027 (1992).

82. J.E. Weston and B.C.H. Steele, *Sol. St. Ionics*, **7**, 25 (1982).

83. J. Plocharski and W. Wieczorek, *Sol. St. Ionics*, **25/30**, 977 (1988).

84. F. Capuano, F. Croce and B. Scrosati, *J. Electrochem. Soc.* **138**, 1918 (1994)

85. D. Golodnitzky, G. Ardel and E. Peled, *Sol. St. Ionics* **85**, 231 (1996).

86. Y. Matsuo and J. Kuwano, *Sol. St. Ionics* **79**, 295 (1995).

87. G.K.R. Senadeera, M.A. Careem, S. Skaarup and K. West, *Sol. St. Ionics* **85**, 37 (1996).

88. P. Aranda, B. Casal, J.C. Galvan and E. Ruiz-Hitzky, in P. Bernier, ed., Chemical Physics of Intercalation II, 397 (Plenum, New York, 1993).

89. D. Fauteux, A. Massucco, A.Van Buren and M. Shi, *Electrochem. Acta* **40**, 2185 (1995).

90. W.B. Russel, D.A. Saville and W.R. Schowalter, Colloidal Dispersions (Cambridge, 1991).

91. J. Fan and P.S. Fedkiw, *J. Electrochem. Soc.* **144**, 399 (1997).

92. R.A. Vaia, S. Vasudevan, W. Krawiec, L.G. Scanlon and S.P. Giannelis, *Adv. Mat.* **7**, 154 (1995).

93. W. Wieczorek, J.R. Stephens and Z. Florjanczyk, *Sol. St. Ionics* **85**, 67 (1996).

94. J. Maier, *Sol. St. Ionics* **86/88**, 55 (1996).

95. Q. Lu, E. Sanchez and C.A. Angell, *Electrochimica Acta* **40**, 2239-2244 (1995); C.A. Angell, J. Fan, C. Liu, Q. Lu, E. Sanchez and K. Xu, *Sol. St. Ionics* **69**, 343 (1994).

96. D.F. Shriver and R. Dillon, to be published.

97. R.B. Silverman, *Methods Enzymol.* **249**, 240 (1995).

COMPUTATIONAL APPROACHES IN OPTICS OF FRACTAL CLUSTERS

Vadim A. MARKEL and Vladimir M. SHALAEV

Department of Physics, New Mexico State University,
Las Cruces, NM 88003, USA
Email: vmarkel@nmsu.edu

Theoretical and computational approaches to investigating optical properties of fractal clusters built from many small identical particles are reviewed and illustrated by numerical examples.

1 Introduction

The physical and geometrical properties of fractal clusters attracted a persistently growing attention of researchers in the past two decades.[1-3] The reason for this is two-fold. First, it turns out that, in many cases, natural processes of aggregation of small particles lead to formation of fractal clusters rather than regular structures.[4,5] Examples include aggregation of colloidal particles in solutions,[6-10] formation of fractal soot from little carbon spherules in the process of incomplete combustion of carbohydrates,[11-16] and growth of cold-deposited self-affine films.[17,18] The second reason is that the properties of fractal clusters are very rich in physics and different from those of either bulk material, or isolated particles (monomers).[19] In this chapter we focus on theoretical and computational approaches to calculating optical characteristics of fractal clusters, such as the optical cross sections, and review some of the results.

The most simple and extensively used model for fractal clusters is a collection of identical spherically symmetrical particles (monomers) that form a self-supporting geometrical structure. It is convenient to think of the monomers as of identical rigid spheres that form a bond on contact. A cluster is self-supporting if each monomer is attached to the rest of the cluster by one or more bonds. Fractal clusters are classified as geometrical (built as a result of a deterministic iteration process) or random. Most clusters in nature are random.

The fractal dimension of a cluster, D, can be found from the relation between the number of particles in a cluster, N, and its gyration radius, R_g:

$$N = (R_g/R_0)^D \, , \tag{1}$$

where R_0 is a constant of the order of the minimum separation between monomers. Another definition of the fractal dimension utilizes the pair density-density correlation function $\langle \rho(\mathbf{r})\rho(\mathbf{r} + \mathbf{R})\rangle$:

$$\langle \rho(\mathbf{r})\rho(\mathbf{r} + \mathbf{R})\rangle \propto R^{D-3} \ \text{ if } R_0 \ll r \ll R_g \ . \tag{2}$$

Typically, both definitions give very close values for the fractal dimension. However, a difference can be seen in very large clusters, when multiscaling can be present.[20,21]

The major models for computer simulation of random fractal clusters are the cluster-cluster aggregation (Meakin model),[22,23] the diffusion limited aggregation (Witten-Sander model),[24] and the percolation model.[25] The first two models involve aggregation processes and, therefore, are dynamical. The third model is static in nature. The majority of natural clusters can be accurately described by one of these models with some variations. More details about the aggregation algorithms as well as further references can be found in the reviews by Jullien and Botet.[4,5]

Random fractal clusters are very complex systems built from simple elementary blocks. It is important to emphasize that the rich and complicated properties of fractal clusters are determined by their geometrical structure rather than by the structure of each elementary block (monomer). In the formulation of a typical problem in optics of fractal clusters, the properties of monomers and the law of their interaction with the incident field and with each other are known, while the properties of a cluster as a whole must be found.

The early advances in the study of nonresonant scattering of light and X-rays by fractal clusters were due to Bale and Schmidt,[26] Berry and Percival[27] and Martin and Hurd.[28] Since that, the theory of nonresonant scattering by fractals was developed in great detail.[12,13,15,21,29–36]

The nontrivial fluctuative nature of fractals is fully manifested when one considers the resonant optical properties of fractal clusters. The foundations of the theory of resonant interaction of light with fractals were built by Butenko, Shalaev and Stockman,[37,38] Markel, Muratov, Stockman and George,[39] Shalaev, Botet and Jullien[40] and Stockman, George and Shalaev.[41] Localization of light and strong enhancement of local fields in fractals were intensively studied[37–39,42–51] As was originally predicted,[38,42,44] strong fluctuations of local fields in fractals result in giant enhancement of nonlinear susceptibilities.[19,46,52] A 10^6 enhancement of the degenerate four-wave mixing in fractal silver clusters was observed experimentally.[53,54] Wavelength- and polarization-selective spectral holes in the absorption spectra of fractal metal clusters in colloid solutions were induced by intense laser pulses.[6,9,55] Other strongly enhanced nonlinearities in absorption and refraction by fractal clusters were also observed.[10]

Practically, all the existing analytical solutions in optics of random fractal clusters are approximate, and the limits of their applicability are usually not obvious *a priori*. In many instances, such solutions don't exist at all. This makes numerical methods to be of great importance. While many theoretical methods, such as various scaling theories,[39–41, 44, 47, 49, 50, 56, 57] explicitly make use of the fractal geometry of clusters, most computational methods are designed for clusters of arbitrary geometry. However, when one is interested in performance characteristics of a numerical method, such as convergence, accuracy or accumulation of the round off errors, the fractal geometry can become important.

2   Basic Theoretical Approaches

In this Section we consider a cluster of N monomers located in the points $\mathbf{r}_i$, $i = 1, ..., N$ and interacting with an incident plane monochromatic wave of the form

$$\mathbf{E}_{inc}(\mathbf{r}, t) = \mathbf{E}_0 \exp(i\omega t - i\mathbf{k} \cdot \mathbf{r}) \; . \tag{3}$$

The factor $\exp(i\omega t)$ is common for all time varying fields and will be omitted below.

2.1   Coupled Dipoles

In this model, each monomer in a cluster is considered to be a point dipole with polarizability α, located at the point $\mathbf{r}_i$ (at the center of the respective spherical monomer). The dipole moment of the ith monomer, $\mathbf{d}_i$, is proportional to the local field at the point $\mathbf{r}_i$ which is a superposition of the incident field and all the secondary fields scattered by other dipoles. Therefore, the dipole moments of the monomers are coupled to the incident field and to each other as described by the coupled dipole equation (CDE):

$$\mathbf{d}_i = \alpha \left[\mathbf{E}_{inc}(\mathbf{r}_i) + \sum_{j \neq i}^{N} \hat{G}(\mathbf{r}_i - \mathbf{r}_j)\mathbf{d}_j \right] \; . \tag{4}$$

Here the term $\hat{G}(\mathbf{r}_i - \mathbf{r}_j)\mathbf{d}_j$ gives the dipole radiation field created by the dipole $\mathbf{d}_j$ at the point $\mathbf{r}_i$, and $\hat{G}(\mathbf{r})$ is the regular part of the free space dyadic Green's function:

$$G_{\alpha\beta}(\mathbf{r}) = k^3 \left[A(kr)\delta_{\alpha\beta} + B(kr)r_\alpha r_\beta / r^2 \right] \; , \tag{5}$$

$$A(x) = [x^{-1} + ix^{-2} - x^{-3}] \exp(ix) \,, \tag{6}$$

$$B(x) = [-x^{-1} - 3ix^{-2} + 3x^{-3}] \exp(ix) \,, \tag{7}$$

where $\hat{G}\mathbf{d} = G_{\alpha\beta}d_\beta$, the Greek indices stand for the Cartesian components of vectors and summation over repeated indices is implied. The operator $\hat{G}$ is completely symmetrical: $\hat{G}(\mathbf{r}) = \hat{G}(-\mathbf{r})$, $G_{\alpha\beta}(\mathbf{r}) = G_{\beta\alpha}(\mathbf{r})$. While the monomer size is assumed to be small compared to the wavelength, the overall cluster size is, in general, arbitrary. That is why the near-, intermediate- and far-zone terms are included in formulas (6),(7).

The CDE is a system of $3N$ linear equations that can be solved to find the dipole moments $\mathbf{d}_i$. The scattering amplitude is expressed through the dipole moments as

$$\mathbf{f}(\mathbf{k}') = k^2 \sum_{i=1}^{N} \left[\mathbf{d}_i - (\mathbf{d}_i \cdot \mathbf{k}')\mathbf{k}'/k^2\right] \exp(-i\mathbf{k}' \cdot \mathbf{r}_i) \,, \tag{8}$$

where $\mathbf{k}'$ is the scattered wave vector (which gives the direction of scattering) and $|\mathbf{k}'| = |\mathbf{k}|$. The cross sections of extinction, scattering and absorption can be found from the optical theorem:

$$\sigma_e = \frac{4\pi}{k} \frac{\mathrm{Im}[\mathbf{f}(\mathbf{k}) \cdot \mathbf{E}_0^\star]}{|\mathbf{E}_0|^2} = \frac{4\pi k}{|\mathbf{E}_0|^2} \mathrm{Im} \sum_{i=1}^{N} \mathbf{d}_i \cdot \mathbf{E}_{inc}^*(\mathbf{r}_i) \,, \tag{9}$$

$$\frac{d\sigma_s}{d\Omega} = |\mathbf{f}(\mathbf{k})|^2 \;;\quad \sigma_s = \int |\mathbf{f}(\mathbf{k})|^2 d\Omega \,, \tag{10}$$

$$\sigma_a = \sigma_e - \sigma_s \,. \tag{11}$$

It can be shown[58] by direct integration according to (10) over the spatial angles Ω, and with the use of equations (4)-(8), that the integral scattering and absorption cross sections can be expressed through the dipole moments as

$$\sigma_s = \frac{4\pi k}{|\mathbf{E}_0|^2} \sum_{i=1}^{N} \left\{ \mathrm{Im}\left[\mathbf{d}_i \cdot \mathbf{E}_{inc}^*(\mathbf{r}_i)\right] - y_a |\mathbf{d}_i|^2 \right\} \,, \tag{12}$$

$$\sigma_a = \frac{4\pi k}{|\mathbf{E}_0|^2} y_a \sum_{i=1}^{N} |\mathbf{d}_i|^2 \,, \tag{13}$$

$$y_a = -\mathrm{Im}\left(\frac{1}{\alpha}\right) - \frac{2k^3}{3} \geq 0 \ . \tag{14}$$

Note that the constant y_a is non-negatively defined[59] for any physically reasonable α. The ratio $3y_a/2k^3$ characterizes the relative strength of absorption by a single isolated monomer. However, it was shown[59] that the resonance absorption of a cluster can be large even when $3y_a/2k^3 \ll 1$.

2.1.1 *Polarizability of a Monomer*

In order to solve the CDE (4), one needs to specify not only the location of monomers, but also the polarizability α, which plays the role of the coupling constant. In this Subsection we discuss the methods for defining α.

The CDE was originally proposed by Purcell and Pennypacker[60] for numerical analysis of scattering and absorption of light by nonspherical dielectric particles. In the formulation of Purcell and Pennypacker, a dielectric particle is represented by an array of point dipoles placed on a cubic lattice and restricted by the surface of the particle. The polarizability of an elementary dipole, α, is defined by the Clausius-Mossotti relation. Equivalently, it is equal to the polarizability of a little sphere of such a radius that the total volume of all the spheres is equal to the total volume of the particle under investigation. To satisfy this equality, the following relation between the lattice period, a, and the radius of a sphere, R_m, must hold: $a^3 = 4\pi R_m^3/3$. Note that two spheres of the radius R_m placed in the neighboring sites of a lattice with the period a would geometrically intersect because $a/R_m = (4\pi/3)^{1/3} \approx 1.612 < 2$. The polarizability of an elementary dipole in the Purcell-Pennypacker model is given by the Lorenz-Lorentz formula[60] with the correction for radiative reaction introduced later by Draine:[61]

$$\alpha = \frac{\alpha_{LL}}{1 - i(2k^3/3)\alpha_{LL}} \ , \tag{15}$$

$$\alpha_{LL} = R_m^3 \frac{\epsilon - 1}{\epsilon + 2} \ , \tag{16}$$

where ϵ is the dielectric constant of the material and α_{LL} is the Lorenz-Lorentz polarizability without the radiation correction. Note that the polarizability written in the form (15) provides the positive value of y_a (14). More sophisticated formulas for α, containing several first terms in the expansion of the polarizability with respect to the parameter kR_m, can be found in papers by Lakhtakia,[62,63] Draine and Goodman[64] and in the references therein.

In the case of fractal clusters the situation is different from the described above. Instead of the imaginary spheres, we have real touching spherical monomers that form a cluster. Of course, these monomers cannot intersect geometrically. Let the radius of the real monomers that can be seen experimentally be R_{exp}, and the distance between two neighboring monomers be a_{exp}; for touching spheres, $a_{exp}/R_{exp} = 2$. However, it was shown both theoretically[65,66] and experimentally[67] that the CDE (4) with $a_{exp}/R_{exp} = 2$, α given by (15) and (16), $a = a_{exp}$ and $R_m = R_{exp}$ yields incorrect results. The reason for this is that the dipole field generated by one monomer inside the other one is not homogeneous; it is much stronger near the point where the monomers touch than in the centers of the neighboring monomer. (Strictly speaking, when the external field is not homogeneous inside a dielectric sphere, one cannot restrict consideration to only dipole moments; the coupled multipole approach is discussed below). Effectively, by replacing two touching spheres by two point dipoles located in their centers, we underestimate the strength of their interaction.

To compensate the above fact, a model for a fractal cluster was introduced[68] in which neighboring spheres were allowed to intersect geometrically. The radius of these spheres, as well as the distance between two neighboring monomers, are chosen to be different from the experimental ones: $R_m \neq R_{exp}$, $a \neq a_{exp}$, but it is required that the ratio a/R_m is equal to $(4\pi/3)^{1/3} \approx 1.612$, the same as in the Purcell and Pennypacker model.[60] Note that a close value for the above ratio, $a/R_m \approx 1.688$, was obtained from the condition that an infinite linear chain of polarizable spherules has the correct depolarization coefficient.[69] The second equation for R_m and a can be obtained from the optically important condition that the model cluster has the same fractal dimension, radius of gyration and total volume as the experimental one. The two equations can be satisfied simultaneously for nontrivial fractal clusters (with $D < 3$) and lead to

$$R_m = R_{exp}(\pi/6)^{D/[3(3-D)]} \; ; \;\; N = N_{exp}(6/\pi)^{D/(3-D)} \, , \qquad (17)$$

where N_{exp} and N are the number of monomers in the original and in the model cluster, respectively. The invariance of the radius of gyration, R_g, follows from (1) and (17).

The above model was shown to be in a good agreement with experimental spectra of fractal cluster.[68] However, for the practical application of this model, one must take into account that the dielectric constant ϵ used in formulas (15),(16) is usually tabulated for bulk samples. When very small monomers are considered, a correction to ϵ connected with the finite-size effects can be significant.[68]

2.1.2   Eigenmode Expansion

The CDE (4) takes an elegant form when written in matrix notations. We introduce a linear complex vector space C^{3N} and $3N$-dimensional vectors of the dipole moments and incident fields according to[a] $|d\rangle = (\mathbf{d}_1, \mathbf{d}_2, ..., \mathbf{d}_N)$ and $|E\rangle = (\mathbf{E}_{inc}(\mathbf{r}_1), \mathbf{E}_{inc}(\mathbf{r}_2), ..., \mathbf{E}_{inc}(\mathbf{r}_N))$. We also define an orthonormal basis $|i\alpha\rangle$ in C^{3N}, such that the Cartesian components of the dipole moments are expressed in this basis as $d_{i\alpha} = \langle i\alpha|d\rangle$. The linear operator W acts on the vector of dipole moments $|d\rangle$ according to the rule: $\langle i\alpha|W|d\rangle = \sum_{\beta j} G_{\alpha\beta}(\mathbf{r}_i - \mathbf{r}_j)d_{j\beta}$. Then Eq. (4) can be written as:

$$|d\rangle = \alpha\left(|E\rangle + W|d\rangle\right) \tag{18}$$

and the expressions for the optical cross sections acquire the form:

$$\sigma_e = \frac{4\pi k}{|\mathbf{E}_0|^2}\mathrm{Im}\langle E|d\rangle \ , \tag{19}$$

$$\sigma_a = \frac{4\pi k y_a}{|\mathbf{E}_0|^2}\langle d|d\rangle \ . \tag{20}$$

The interaction matrix W is complex and symmetrical, and, consequently, non-Hermitian. Therefore, its eigenvectors are not orthogonal, and, moreover, may not form a complete basis in C^{3N}. However, it can be shown[58] that the eigenvectors are linearly independent (and, therefore, form a basis) if W is not degenerate, or if the nature of its degeneracy is "geometrical" (due to a certain symmetry of the cluster under consideration) rather than "random". In most practical cases, we can define a complete basis of eigenvectors, $|n\rangle$, corresponding to the eigenvectors w_n:

$$W|n\rangle = w_n|n\rangle \ . \tag{21}$$

For complex symmetrical matrices, the orthogonality rule ($\langle m|n\rangle = \delta_{mn}$) is replaced by[58]

$$\langle \bar{m}|n\rangle = 0 \ \ \text{if} \ m \neq n \ , \tag{22}$$

where bar denotes complex conjugation of all elements of a vector. Thus, $|n\rangle$ denotes a column vector, $\langle n|$ denotes a row vector with the complex conjugated elements, and $\langle \bar{n}|$ denotes a row vector with exactly the same elements as $|n\rangle$.

---

[a]From this point, we will use Roman letters to denote $3N$-dimensional vectors and $3N \times 3N$-dimensional matrices, to distinguish them from scalars that are typeset in italics and usual 3-dimensional vectors typeset in bold letters.

We adopt the usual normalization of the eigenvectors, $\langle n|n\rangle = 1$, but $\langle\bar{n}|n\rangle$ is not equal to unity and can be, in general, complex.

The representation of the unity operator in the basis $|n\rangle$ is

$$I = \sum_{n=1}^{3N} \frac{|n\rangle\langle\bar{n}|}{\langle\bar{n}|n\rangle} \ . \tag{23}$$

The eigenvector expansion for the solution to the CDE, $|d\rangle$, can be easily obtained with the use of the above equality:

$$|d\rangle = \sum_{n=1}^{3N} \frac{|n\rangle\langle\bar{n}|E\rangle}{\langle\bar{n}|n\rangle(1/\alpha - w_n)} \ . \tag{24}$$

Equations (19),(20), (24) give the general form of the dependence of the solutions to the CDE and the optical cross sections on the polarizability α. While the direct numerical application of the eigenvector expansion may be not practical (except for the quasistatic limit, see Section 2.1.3 below), it is useful for analysis of different approximations.

The following two exact properties[58] of the eigenvalues of the CDE can be useful for assessment of accuracy of different numerical methods for diagonalization of W:

$$\sum_{n=1}^{3N} w_n = 0 \ ; \quad -2k^3/3 \leq \mathrm{Im}\, w_n \leq (3N - 1)2k^3/3 \ . \tag{25}$$

2.1.3 *Quasistatic Approximation*

If not only the monomers, but the whole cluster is small compared to the incident wavelength $\lambda = 2\pi c/\omega$, one can neglect the intermediate- and far-zone terms in the interaction operator (5)-(7). Therefore, the matrix W becomes real and Hermitian:

$$\langle i\alpha|\mathrm{W}^{(0)}|j\beta\rangle = \frac{-\delta_{\alpha\beta}|\mathbf{r}_i - \mathbf{r}_j|^2 + 3(r_{i\alpha} - r_{j\alpha})(r_{i\beta} - r_{j\beta})}{|\mathbf{r}_i - \mathbf{r}_j|^5} \ , \tag{26}$$

where $\mathrm{W}^{(0)}$ is the interaction matrix in the quasistatic limit. Analogously, the exponential factor $\exp(i\mathbf{k}\cdot\mathbf{r}_i)$ can be set to unity in the expression for the incident wave (3). The quasistatic free term $|E^{(0)}\rangle$ is defined by $\langle i\alpha|E^{(0)}\rangle = E_{0\alpha}$.

The quasistatic eigenvectors, which we denote $|n^{(0)}\rangle$, become orthogonal and real. Moreover, as can be easily seen from equation (5), the eigenvectors and the corresponding real eigenvalues, $w_n^{(0)}$, do not depend on ω.

Now the eigenvector expansion for the solution of the CDE takes the form

$$|d\rangle = \sum_{n=1}^{3N} \frac{|n^{(0)}\rangle\langle n^{(0)}|E^{(0)}\rangle}{1/\alpha - w_n^{(0)}} \tag{27}$$

and can be effectively utilized in numerical calculations. Indeed, the only source of the dependence on the optical frequency ω in (27) is $\alpha = \alpha(\omega)$. The eigenvectors and the eigenvalues depend only on the geometry of a particular cluster. This means that it is sufficient to diagonalize the interaction matrix $W^{(0)}$ just once, and then the solution for any frequency can be obtained by a simple summation according to (27).

The quasistatic approximation can be formulated either in the strong or in the weak form. The strong form is obtained by formally setting $k = 0$ in formulas (5)-(7). In this approximation, extinction and absorption cross sections are equal, as follows from formulas (14), (19),(20) and the exact relation $\text{Im}\langle E^{(0)}|d\rangle = -\text{Im}(1/\alpha)\langle d|d\rangle$ that can be found[39] with the use of (27):

$$\sigma_e = \sigma_a = \frac{4\pi k}{|E_0|^2}\text{Im}\sum_{n=1}^{3N} \frac{|\langle E^{(0)}|n^{(0)}\rangle|^2}{1/\alpha - w_n^{(0)}} . \tag{28}$$

Consequently, the scattering cross section is zero in the strong form of the approximation. Physically, the strong form is valid when the absorption parameter $3y_a/2k^3$ is large, which means that absorption is much larger than scattering.

The strong form of the quasistatic approximation is inadequate for description of scattering. The intuitive idea that the scattering cross section is proportional to the square modulus of the total dipole moment of a cluster (which can be calculated in the assumption that $k = 0$) was shown to be incorrect.[58] Also, the strong form of the quasistatic approximation breaks dramatically even for the absorption cross section when the parameter $3y_a/2k^3$ is not large enough. Although it may seem that the above parameter is always large for small monomers (proportional to $(\lambda/R_m)^3$), analysis of formulas (14)-(16) shows that this might be not the case when $\text{Im}\epsilon/|\epsilon - 1|^2 \ll 1$. The last inequality can hold either for highly transparent materials ($\text{Im}\epsilon \approx 0$), or for materials with large $|\epsilon - 1|$, such as metals.

In the weak form of the quasistatic approximation, one can still use the eigenvectors $|n^{(0)}\rangle$, but it is necessary to keep a few first terms of the expansion of w_n and $|E\rangle$ in terms of the powers of k. The expansion of the eigenvalues can be done with the use of the standard perturbation theory[58] using the orthonormal set of $|n^{(0)}\rangle$ as an unperturbed basis. The results (in the first order of the perturbation theory) are summarized below:

$$w_n = w_n^{(0)} + i\frac{2k^3}{3}\left(|\mathbf{D}_n|^2 - 1\right) , \qquad (29)$$

where $\mathbf{D}_n$ is the total dipole moment of the nth eigenmode defined by

$$D_{n\alpha} = \sum_{i=1}^{N}\langle i\alpha|n^{(0)}\rangle . \qquad (30)$$

The expansion of the free term is trivial:

$$\langle i\alpha|E\rangle = E_{0\alpha}\left[1 + i(\mathbf{k}\cdot\mathbf{r}_i)\right] . \qquad (31)$$

It is important to keep the second term in expansion (31) in the case of "antisymmetrical states",[58] when $\langle E^{(0)}|n^{(0)}\rangle = 0$.

Finally, the expressions for the optical cross sections are

$$\sigma_a = \frac{4\pi k}{|\mathbf{E}_0|^2}\sum_{n=1}^{3N}|\langle E|n^{(0)}\rangle|^2\frac{y_a}{|1/\alpha - w_n|^2} , \qquad (32)$$

$$\sigma_s = \frac{4\pi k}{|\mathbf{E}_0|^2}\sum_{n=1}^{3N}|\langle E|n^{(0)}\rangle|^2\frac{2k^3/3 + \mathrm{Im}w_n}{|1/\alpha - w_n|^2} = \frac{8\pi k^4}{3|\mathbf{E}_0|^2}\sum_{n=1}^{3N}|\langle E|n^{(0)}\rangle|^2\frac{|\mathbf{D}_n|^2}{|1/\alpha - w_n|^2} . \qquad (33)$$

As follows from the analysis of the scattering cross section (33), a cluster cannot be replaced by a single particle with an effective total dipole moment $\mathbf{D}^{tot} = \sum_i \mathbf{d}_i$, even when the cluster size is much smaller than λ. Instead, different eigenmodes scatter independently, without mutual interference. The classical relation $\sigma_s = 8\pi k^4|\mathbf{D}^{tot}|^2/3|\mathbf{E}_0|^2$ holds if only one eigenmode can be effectively excited, as in the case of a homogeneous distribution of a large number of monomers inside a spherical volume.

2.1.4 *Born Approximation and Mean Field Theories*

The Born expansion for the solution $|d\rangle$ can be obtained by iterating equation (18):[70,71]

$$|d\rangle = \alpha \sum_{k=0}^{\infty} (\alpha W)^k |E\rangle \ . \tag{34}$$

It can be easily verified that expansion (34) converges to the exact solution (24) if $|\alpha w_n| < 1 \ \forall n$.[71]

A rough estimate of the boundaries of the eigenvalues of W can be made using the Gershgorin theorem which states that each eigenvalue of a matrix $A = \{a_{ij}\}$ lies in at least one of the Gershgorin disks $|z - a_{ii}| < \sum_{j \neq i} |a_{ij}|$ in the complex z-plane. It should be noted that the interaction matrix W has zero diagonal elements and, therefore, its Gershgorin disks are all concentric. An estimate of the Gershgorin boundaries can be done with the use of a fractal density distribution function, replacing the above summation by integration. The convergence of the Born expansion can strongly depend on the fractal dimension and the cluster size. Especially, this is important when the complete interaction operator (5)-(7) is used. On the other hand, in the quasistatic limit the bounds for eigenvalues are typically independent of the cluster size if $D < 3$. This is explained by the fact that the static dipole field is integrable with any reasonable fractal density distribution function at large r. For example, quasistatic calculations[68] for 3-dimensional lattice cluster-cluster aggregates with the lattice period a showed that $\max(|a^3 w_n|) \approx 6$.

The first Born approximation can be used when $|\alpha w_n| \ll 1 \ \forall n$. Then $|d\rangle = \alpha|E\rangle$. Using $\langle E|E\rangle = N|\mathbf{E}_0|^2$, we find that $\sigma_e = 4\pi k N \mathrm{Im}\alpha$ and $\sigma_a = 4\pi k y_a N |\alpha|^2$, as one would expect for a collection of non-interacting monomers. Although the integral scattering cross section is given by a formula for N independent scatterers, $\sigma_s = (8\pi k^4/3)N|\alpha|^2$, the differential cross section carries information about the geometrical structure of a cluster. Indeed, as follows from (8) and (10), the differential scattering cross section is proportional to the the function $I(\mathbf{q})$, known as the structure factor:[13, 28]

$$\frac{d\sigma_s}{d\Omega} \propto I(\mathbf{q}) = \left| \sum_{i=1}^{N} \exp\left(i\mathbf{q} \cdot \mathbf{r}_i\right) \right|^2 \ ; \quad \mathbf{q} = \mathbf{k} - \mathbf{k}' \ . \tag{35}$$

For clusters that are spherically symmetrical on average, the structure factor can be averaged over spatial rotations:

$$\overline{I(\mathbf{q})} = N + N(N-1)\phi_2(q) \ ; \quad \phi_2(q) = \int_0^{\infty} p_2(r) \frac{\sin qr}{qr} dr \ , \tag{36}$$

where the horizontal line denotes ensemble averaging and $p_2(r)$ is the probability density to find two different monomers in a cluster separated by the distance

r. As follows from (2), $p_2(r)$ in fractal clusters has a scaling behavior in the intermediate asymptote region $R_0 \ll r \ll R_g$: $p_2(r) \propto r^{D-1}$. If $D < 2$ and $q \gg 1/R_g$, the integral in (36) converges at the upper limit while r is still much smaller than R_g to produce the well-known result $\overline{I(\mathbf{q})} = N^2(1/N + c/q^D)$, where c is a constant depending on D. It was also shown that the dispersion of the average scattered intensity, $\langle I^2 \rangle - \langle I \rangle^2$, can be expressed similarly to (36) with the use of the fourth-order density correlation function.[21, 36]

The mean-field approximation proposed for optical scattering by fractal clusters by Berry and Percival[27] is physically distinct from the Born expansion, but is very similar mathematically to the first Born approximation. In the mean-field theory, interaction between monomers can be arbitrarily strong and infinite order of multiple scattering is allowed. The main assumption of the mean-field approximation is that the ensemble average of the term $\mathrm{W}|\mathrm{d}\rangle$ in equation (18) can be performed as $\overline{\mathrm{W}|\mathrm{d}\rangle} = \overline{\mathrm{W}} \ \overline{|\mathrm{d}\rangle}$. In other words, W and $|\mathrm{d}\rangle$ are assumed to be statistically independent.

For an ensemble of clusters that are spherically symmetrical on average, $\overline{\mathrm{W}} = Q$, where Q is a scalar, because there is no selected direction in space. Performing an ensemble averaging of equation (18), we obtain:

$$\overline{|\mathrm{d}\rangle} = \frac{\alpha|\mathrm{E}\rangle}{1 - Q\alpha} \, . \tag{37}$$

The above expression is similar to the first term in the Born expansion for $|\mathrm{d}\rangle$ (34), except for the constant factor $(1 - Q\alpha)^{-1}$. The integral optical cross sections are different in the mean-field approximation from those in the first Born approximation, but the dependence of the scattered intensity on $q = |\mathbf{k} - \mathbf{k}'|$ is exactly the same. Comparison of formula (37) with the exact solution (24) reveals that the mean-field approximation is equivalent to the assumption that all the eigenvalues of W are equal: $w_n = Q$, which, in turn, is possible only if W is a scalar.

When $|Q\alpha|$ is not small, multiple scattering is important in a cluster. Berry and Percival showed by averaging W with a fractal distribution function[27] that the character of multiple scattering is distinctly different when $D < 2$ and $D > 2$. In the first case, multiple scattering is negligibly small if kR_m is much smaller than unity for any cluster size. Physical interpretation of this fact is that a cluster with $D < 2$ is always optically transparent. When $D > 2$, multiple scattering becomes important if $N \sim 1/(kR_m)^{D/(D-2)}$, or, equivalently, $R_g \sim R_m/(kR_m)^{D-2}$.

2.2   Coupled Multipoles vs Coupled Dipoles

The fact that higher multipolar modes must be taken into account when the interacting monomers are close to each other was noted by Gerardy and Ausloos,[65] Sansonetti and Furdyna[67] and Claro.[66,72,73]   As was mentioned in Section 2.1.1, the point dipole approximation is inadequate for description of touching spheres. The reason for this is that even if the wavelength of the incident radiation is infinite, the spheres themselves induce highly inhomogeneous fields inside each other which, in turn, excite all the higher multipole moments. It is also incorrect to assume that the fields that act on the point dipoles in the coupled dipoles model are the same as in the centers of the respective spherules; the dipole approximation underestimates the interaction strength between two neighboring monomers.

To overcome the limitations of the coupled dipole approximation, an approach for rigorous numerical solution of the electromagnetic problem of an agglomerate of dielectric spheres interacting with an incident wave and each other has been developed.[65,66,72–79] We will refer to this approach as to the coupled multipole method. The essence of this method is to expand the EM field inside each sphere, and the field scattered by each sphere, in vector spherical harmonics, and to match the boundary condition on all surfaces of discontinuity. Below, we summarize briefly the mathematical formalism of coupled multipoles.

The fields induced inside the ith sphere, $\mathbf{E}_i^{(int)}(\mathbf{r})$, and scattered by the ith sphere, $\mathbf{E}_i^{(scatt)}(\mathbf{r})$, can be always expanded in the series involving the vector spherical harmonics, $\mathbf{M}_{mn}^{(k)}(\mathbf{r})$ and $\mathbf{N}_{mn}^{(k)}(\mathbf{r})$, where n takes the integer values from 1 to infinity, m takes the integer values from $-n$ to n, and the superscript k denotes the type of Bessel function which governs the radial dependence of the corresponding harmonic. In accordance with the commonly accepted convention, $k = 1$ corresponds to the spherical Bessel function of the first kind j_n, $k = 2$ - to the spherical Bessel function of the second kind y_n, and $k = 3, 4$ - to the spherical Hankel functions of the first and the second kind, $h_n^{(1)}$ and $h_n^{(2)}$, respectively. A comprehensive review of the vector spherical harmonics and their properties can be found in the book by Bohren and Huffman.[80] The expansion for the scattered and internal fields has the following form:

$$\mathbf{E}_i^{(scatt)}(\mathbf{r}) = \sum_{n=1}^{\infty} \sum_{m=-n}^{n} \left[a_{i,mn}\mathbf{N}_{mn}^{(3)}(\mathbf{r} - \mathbf{r}_i) + b_{i,mn}\mathbf{M}_{mn}^{(3)}(\mathbf{r} - \mathbf{r}_i) \right] , \qquad (38)$$

$$\mathbf{E}_i^{(int)}(\mathbf{r}) = \sum_{n=1}^{\infty} \sum_{m=-n}^{n} \left[c_{i,mn} \mathbf{N}_{mn}^{(1)}(\mathbf{r} - \mathbf{r}_i) + d_{i,mn} \mathbf{M}_{mn}^{(1)}(\mathbf{r} - \mathbf{r}_i) \right] , \qquad (39)$$

where $a_{i,mn}, b_{i,mn}, c_{i,mn}$ and $d_{i,mn}$ are the unknown coefficients, and the arguments $\mathbf{r} - \mathbf{r}_i$ in the right-hand sides of (38) and (39) reflect the fact that the corresponding spherical harmonics must be calculated in the system of coordinates with the origin at $\mathbf{r}_i$; at the same time the functions $\mathbf{M}_{mn}^{(k)}(\mathbf{r})$ and $\mathbf{N}_{mn}^{(k)}(\mathbf{r})$ are universal and independent of the origin. Note that the internal field expansion contains only the spherical Bessel functions of the first kind, j_n, $(k = 1)$ that are finite at $r = 0$, while the scattered field expansion contains only the spherical Hankel functions of the first kind, $h_n^{(1)}$ $(k = 3)$ that have the form of outgoing waves. Expressions for the scattered and internal magnetic field can be obtained by taking a curl of (38) and (39) according to $\mathbf{H} = ik^{-1}\mathrm{curl}\mathbf{E}$.

The incident wave is usually supposed to be linearly polarized along the x-axis and propagating in the z direction. Then the expansion for $\mathbf{E}_{inc}(\mathbf{r}) = \mathbf{E}_0 \exp(i\mathbf{k} \cdot \mathbf{r})$ takes the form:[80]

$$\mathbf{E}_{inc}(\mathbf{r}) = \exp(i\mathbf{k} \cdot \mathbf{r}_i) \sum_{n=1}^{\infty} \frac{-i^{n+1}|\mathbf{E}_0|(2n+1)}{2n(n+1)} \left[\mathbf{M}_{1n}^{(1)}(\mathbf{r} - \mathbf{r}_i) - \right.$$
$$\left. \mathbf{M}_{-1n}^{(1)}(\mathbf{r} - \mathbf{r}_i) + \mathbf{N}_{1n}^{(1)}(\mathbf{r} - \mathbf{r}_i) + \mathbf{N}_{-1n}^{(1)}(\mathbf{r} - \mathbf{r}_i) \right] . \qquad (40)$$

The expression in the right-hand side of (40) does not actually depend on $\mathbf{r}_i$, but provides an expansion in terms of harmonics centered in the point $\mathbf{r}_i$. Note that not all possible spherical harmonics are present in expansion (40). It contains only harmonics with $m = \pm 1$, which is a general property of monochromatic waves. For the scattering problem involving one sphere, only the types of harmonics present in (40) can be excited. However, for the problem involving several spheres, this is not the case, because the secondary scattered waves contain the spherical harmonics of all possible kinds. Again, an expansion for the magnetic field can be obtained by taking a curl of equation (40)

The most challenging task in the coupled multipole method is to expand the spherical harmonics centered around the ith monomer in terms the spherical harmonics centered around the jth monomer. The above procedure is necessary for the boundary conditions consideration, because the field in each spherical region must be represented only in terms of spherical harmonics centered in the origin of this region. In general, it is possible to write[76]

$$\mathbf{M}_{mn}^{(3)}(\mathbf{r} - \mathbf{r}_i) = \sum_{n'=1}^{\infty} \sum_{m'=-n'}^{m'=n'} \left[A_{mn}^{m'n'}(\mathbf{r}_i - \mathbf{r}_j)\mathbf{M}_{m'n'}^{(1)}(\mathbf{r} - \mathbf{r}_j) + \right.$$
$$\left. B_{mn}^{m'n'}(\mathbf{r}_i - \mathbf{r}_j)\mathbf{N}_{m'n'}^{(1)}(\mathbf{r} - \mathbf{r}_j) \right] . \qquad (41)$$

An analogous expansion can be written for the outgoing harmonic $\mathbf{M}_{mn}^{(3)}(\mathbf{r} - \mathbf{r}_i)$; it has exactly the same form as (41), but coefficients $A_{mn}^{m'n'}$ and $B_{mn}^{m'n'}$ change places. Although the above procedure is simple and physically transparent, the coefficients $A_{mn}^{m'n'}$ and $B_{mn}^{m'n'}$ are rather complicated functions of the Clebsch-Gordan coefficients and involve integrals of the Legendre polynomials that cannot be calculated analytically in the general case. We will not adduce them here due to their complexity; the readers interested in numerical implementation of the coupled multipole method can find the necessary coefficients in the original papers by Fuller[76,77] and Xu.[78]

For each surface of discontinuity, and for each mode defined by the pair of indices (mn), there are four unknown coefficients, $a_{i,mn}, b_{i,mn}, c_{i,mn}$ and $d_{i,mn}$. Correspondingly, there are four boundary condition equations for the tangential components of the electric and magnetic fields. Note that the internal field in each monomer is given only by expansion (39), while the external field is given by a sum of the incident wave expansion (40) and the sum of all secondary scattered waves (39), where the outgoing spherical harmonics are expanded according to (41). Application of the boundary conditions leads to the following system of linear equations[76] for the scattering coefficients $a_{i,mn}$ and $b_{i,mn}$:

$$a_{i,mn} = a_{i,mn}^{(0)} \left[p_{i,mn} + \sum_{j \neq i, m'n'} A_{mn}^{m'n'}(\mathbf{r}_i - \mathbf{r}_j)a_{j,m'n'} + B_{mn}^{m'n'}(\mathbf{r}_i - \mathbf{r}_j)b_{j,m'n'} \right] ,$$
$$(42)$$

$$b_{i,mn} = b_{i,mn}^{(0)} \left[q_{i,mn} + \sum_{j \neq i, m'n'} B_{mn}^{m'n'}(\mathbf{r}_i - \mathbf{r}_j)a_{j,m'n'} + A_{mn}^{m'n'}(\mathbf{r}_i - \mathbf{r}_j)b_{j,m'n'} \right] ,$$
$$(43)$$

where $p_{i,mn}$ and $q_{i,mn}$ are both proportional to the exponential factor $\exp(i\mathbf{k} \cdot \mathbf{r}_i))$.

The coupled multipole equation (42),(43) is infinite-dimensional, as the order n can be arbitrary large. Physically, it is usually possible to introduce a reasonable cut-off for n. It was shown on the example of a rigorous solution for two touching spheres that such a cut-off can be related to nonlocality effects in the dielectric function, such as a finite wavelength of an electron in a medium. This provides that the harmonics with characteristic features smaller than the electron wavelength cannot be effectively excited. If we denote the cut-off for n as n_{max}, the total number of the coupled multipole equations becomes $Nn_{max}(4n_{max} + 5)(n_{max} + 1)/6 \sim 2Nn_{max}^3$.

The above fact makes practical application of the coupled multipole method for any system with a reasonably large number of monomers, when the fractal geometry is well manifested, an extremely difficult task. Taking into account that the number of operations necessary to solve a system of L linear equations scales as L^3 for direct methods and as L^2 for most iterative methods, the dependence of the computer time on the maximum order n_{max} is given by n_{max}^6 in the most favorable case. Apart from this fact, calculation of the coefficients in equations (42),(43) is a separate computational problem and must be carried out for each wavelength.

When comparing the coupled dipoles and the coupled multipoles methods, we must take into account the following factors. The coupled dipoles formalism treats inaccurately only the interactions between those monomers that are closer to each other than approximately two respective diameters. Thus, most pair interactions in a large cluster are described by the CDE satisfactorily. The model cluster with overlapping neighboring spherules introduced in Section 2.1.1 makes a reasonable correction to the dipole interaction while preserving the geometrical structure, size and total volume of a cluster. The coupled dipole model is very simple from both the theoretical and the computational points of view, with the number of equations scaling as N. In the coupled multipoles model, the number of equations scales as Nn_{max}^3, and it is usually not clear *a priori*, how large the cut-off value, n_{max} should be. It was shown[75] that infinite orders of n are required for the case of two touching spheres, unless nonlocality of the dielectric function is taken into account. However, most calculations based on the coupled multipoles method stop at some relatively small orders of n, or involve a relatively small number of spheres. Nevertheless, the coupled multipoles method is very important as an independent rigorous verification of the coupled dipole method and yields some effects that cannot be described in the dipole approximation.

2.3   Iterative and renormalization methods

A number of methods based on the real space renormalization group[81,82] has been proposed for calculating the optical responses of fractal structures[83,84] and conductivity of percolation two-component composites.[85]

In the most pure form, the renormalization methods can be applied to geometrical clusters that are mathematically self-similar (unlike the random clusters which are self-similar only in the statistical sense). The geometrical clusters are usually built with the use of a certain "generator" which is replicated at each step to form a larger cluster, which, in turn, is used as a generator for the next step, and so on.[83] The main idea of the method is to assume that, at each stage, the structure used as a unit for the next stage can be replaced by an equivalent sphere with some polarizability which depends on the unit polarizability at the previous stage and on the geometry. This leads to a recursion relation for the polarizability, or any other optical characteristic, of the generated cluster.

The method of renormalization was successfully applied not only to geometrical clusters, but also to disordered (percolation) systems.[84]

3   Numerical methods

In this Section we discuss only the numerical methods that can be applied within the coupled dipoles approximation. The numerical implementations of the coupled multipoles and other methods are too vast and diverse and cannot be reviewed here due to the limited volume of this chapter. On the other hand, the coupled dipoles method is most practical when one is interested in the influence of the large scale geometrical structure of a cluster on its optical properties, because the dipole approximation becomes accurate for monomers separated by a distance larger than their size.

3.1   The quasistatic approximation

3.1.1   Jacobi Diagonalization

In the quasistatic approximation, one of the most powerful numerical approaches is diagonalization of the Hermitian matrix $W^{(0)}$. This can be done by the standard Jacoby method which proved to be very accurate and robust for clusters with the number of monomers up to $N = 1,000$. The diagonalization needs to be done only once for each cluster.

In the strong form of the quasistatic approximation, one can be interested in the spectral dependence of the extinction (or, equivalently, the absorption) cross section. The latter can be calculated with the use of (27) for each λ (or for each value of α) by simple summation, once the diagonalization of $W^{(0)}$ is done. Note that for calculation of the cross sections, it is sufficient to store in memory or on disk only the scalar products $\langle E^{(0)}|n^{(0)}\rangle$ and the corresponding eigenvalues instead of the complete set of the eigenvectors. However, in this case the cross sections can be found only for a specific polarization of the incident light, defined by $\langle i\alpha|E^{(0)}\rangle = E_{0\alpha}$. A more general approach is to define three vectors $|E_x^{(0)}\rangle$, $|E_y^{(0)}\rangle$ and $|E_z^{(0)}\rangle$, for the x, y or z polarization of the incident light, respectively, and to store the scalar products of these vectors and $|n^{(0)}\rangle$ for each n. Then the extinction cross section for the cluster under consideration can be easily calculated for any polarization of the incident light (including elliptical polarization) and any value of α by summation according to (27). We emphasize that the set of values $w_n^{(0)}$ and $\langle E_\alpha^{(0)}|n^{(0)}\rangle$, for $\alpha = x, y, z$ and $n = 1, ..., 3N$ can be easily stored on disk for consequent calculations.

In the weak form of the quasistatic approximation, we must use formulas (32) and (33) for the optical cross sections. Therefore, in addition to the above values, one needs to store the value of $|\mathbf{D}_n|^2$ for each eigenmode. As follows from (30), $\mathbf{D}_n$ is also α-independent. Further, in most random systems there are no antisymmetrical eigenstates, such that $|\langle E^{(0)}|n^{(0)}\rangle| \ll |\mathbf{E}_0|$. Then it is possible to replace $|E\rangle$ in (32),(33) by $|E^{(0)}\rangle$. If an antisymmetrical eigenstate exists, one must take into account the second term in expansion (31) which is, actually, λ-dependent. However, this dependence is trivial ($\sim 1/\lambda$) and the scalar product $\langle E|n^{(0)}\rangle$ can be easily calculated for any λ if it is known for a certain value λ_0 (note that the propagation direction of the incident wave must be specified in this case).

If, however, one is interested in the local fields or dipole moments rather than in the integral cross sections (this is important when nonlinear optical responses must be calculated), the complete set of eigenvectors of $W^{(0)}$ must be stored. For example, the set of eigenvectors for $N = 1,000$ would require 36Mb of storage. If the disk space is limited, the physical value in question can be calculated by the same program, while the eigenvectors are still in the computer memory. The disadvantage of this approach is that after the program is terminated, an important information contained in the eigenvectors is lost and should a need to find another independent physical value arise, the time extensive part of the calculations would have to be repeated.

3.1.2 *Lanczos Algorithm*

When the dimensionality of the system, $L = 3N$, is very large, the complete diagonalization of $W^{(0)}$ carries excessive information and is not necessary. Indeed, as was discussed in Section 2.1.4, the Gershgorin bounds for the eigenvalues do not depend in the quasistatic limit on the number of monomers, N. This means that the number of resonance eigenvalues that lie inside the homogeneous width $\Gamma = |\mathrm{Im}(1/\alpha)|$, such that $|\mathrm{Re}(1/\alpha) - w_n^{(0)}| < \Gamma$ (see equation (27)), grows approximately proportionally to N. Therefore, if one is interested only in integral cross sections rather than in the distribution of local fields, it is sufficient to sample only such number of eigenstates that the density of eigenvalues in the interval $(w_{min}^{(0)}, w_{max}^{(0)})$ becomes larger than $1/\Gamma$. A method based on this observation was proposed.[68] For simplicity, we discuss below the strong form of the quasistatic approximation.

As follows from (18),(28), the extinction and absorption cross sections in the strong form of the quasistatic approximation are given by an average of the form $\langle \mathrm{E}^{(0)} | [1/\alpha - \mathrm{W}^{(0)}]^{-1} | \mathrm{E}^{(0)} \rangle$. (The matrix $[1/\alpha - \mathrm{W}^{(0)}]^{-1}$ is given by the expansion $\sum_n |n^{(0)}\rangle\langle n^{(0)}| [1/\alpha - w_n^{(0)}]^{-1}$.) It was shown[86] that the above average can be written as a continued fraction that formally terminates after L levels:

$$\langle \mathrm{E}^{(0)} | [1/\alpha - \mathrm{W}^{(0)}]^{-1} | \mathrm{E}^0 \rangle = \cfrac{N|\mathbf{E}_0|^2}{1/\alpha - \eta_0 - \cfrac{\beta_1^2}{1/\alpha - \eta_1 - \cfrac{\beta_2^2}{1/\alpha - \eta_2 \cdots}}} . \tag{44}$$

The η's and β's are determined by the basic Lanczos recursion relation:[87]

$$\mathrm{W}|\mathrm{u}_i\rangle = \beta_i|\mathrm{u}_{i-1}\rangle + \eta_i|\mathrm{u}_i\rangle + \beta_{i+1}|\mathrm{u}_{i+1}\rangle , \tag{45}$$

where $\beta_0 = 0$, $|\mathrm{u}_{-1}\rangle = |\mathrm{u}_{L+1}\rangle = 0$, $|\mathrm{u}_0\rangle = |\mathrm{E}^{(0)}\rangle$ and β's are calculated from the condition that the vectors $|\mathrm{u}_i\rangle$ are orthonormal. For approximate calculations, the fraction is terminated after n steps, where $n \ll L$, and the last fraction in the sequence has the form

$$\cdots \frac{\beta_{n-1}^2}{1/\alpha - \eta_{n-1} - \beta_n^2 g(\alpha)} , \tag{46}$$

where the terminator $g(\alpha)$ is taken as the Green's function for a constant chain[86] of vectors $|\mathrm{u}_i\rangle$:

$$g(\alpha) = \frac{1/\alpha - a - \sqrt{(1/\alpha - a)^2 - 4b^2}}{2b^2} ; \tag{47}$$

$$a = (w_{max}^{(0)} + w_{min}^{(0)})/2 ; \quad a = (w_{max}^{(0)} - w_{min}^{(0)})/4 \tag{48}$$

and the minimum and maximum eigenvalues, $w_{min}^{(0)}$ and $w_{max}^{(0)}$, can be easily determined by diagonalizing the $n \times n$ symmetric tridiagonal matrix made up of the η's on the diagonal and the β's on the first off diagonal.

The computationally intensive part of this calculation is the matrix vector multiplication, performed at each level. Therefore the total number of operations required is $\sim L^2 n$ where n is the number of levels in the continuous fraction (44). The number of iterations required for convergence is proportional to $1/\Gamma$, which seems to be the common property of many iterative methods applied to the CDE.

3.2 Beyond the quasistatic approximation

Beyond the quasistatic limit, the diagonalization of the interaction matrix is computationally inefficient because its eigenvalues and eigenvectors depend explicitly of λ. Therefore, the computationally intensive part must be repeated for any new wavelength. On the other hand, inversion of matrix is a somewhat simpler task than resolving the eigenproblem. Below, we review a direct and an iterative method for inversion of the complex symmetrical matrix that proved to be robust and efficient for the solution of the CDE.

3.2.1 Method of the Square Root

An excellent direct method for inversion of a complex symmetrical matrix is the LU expansion, known also as the method of the square root. We summarize it briefly below.

Consider an $L \times L$ symmetrical matrix $\mathrm{A} = \{a_{ij}\}$ with complex elements[b]. It is always possible to represent this matrix as a product of a lower and upper triangular matrices (hence, the term LU expansion) as $\mathrm{A} = \mathrm{T}'\mathrm{T}$, where $t_{ij} = 0$ if $j > i$, and prime denotes transposition. An iterative relation for the elements t_{ij}, $j \leq i$ can be easily derived from the equation

$$a_{ij} = \sum_{k=1}^{\max(i,j)} t_{ik} t_{jk} \ . \tag{49}$$

Starting with

$$t_{11} = \sqrt{a_{11}} \ ; \quad t_{i1} = a_{i1} t_{11}^{-1} \ \text{ for } i > 1 \ , \tag{50}$$

and using the recursion

$$t_{jj} = \sqrt{a_{jj} - \sum_{k=1}^{j-1} t_{jk}^2} \ ; \quad t_{ij} = \left[a_{ij} - \sum_{k=1}^{j-1} t_{ik} t_{jk} \right] t_{jj}^{-1} \ \text{ for } i > j \ , \tag{51}$$

---

[b]In our case, $\mathrm{A} = 1/\alpha - \mathrm{W}$ and $L = 3N$.

one can find all the elements t_{ij} in approximately $L^3/6$ floating-point operations. There is also L calculations of a square root involved, but this number is insignificant. Note also that all the elements t_{jj} are not equal to zero if the determinant of A is not equal to zero. The first step in the iterative process (50),(51) is also always possible because $a_{11} = 1/\alpha$ and, consequently, $t_{11} \neq 0$.

Once the elements t_{ij} are found, the equation $T'T|x\rangle = |b\rangle$ can be easily solved in two steps. First, we denote $|y\rangle \equiv T|x\rangle$ and solve $T'|y\rangle = |b\rangle$ for $|y\rangle$ in L^2 steps (this can be easily done because the matrix T' is triangular with all the elements above the main diagonal equal to zero). Second, the equation $T|x\rangle = |y\rangle$ is solved for $|x\rangle$ in exactly the same way.

Because the method of the square root utilizes effectively the symmetry of A, it is approximately 6 times faster than the standard Gaussian method or any other direct method with $\approx L^3$ operations. It is also important that the matrices A and T can be stored in the same array, which allows to reduce memory equirements.

3.2.2 *Conjugate Gradient Method*

The conjugate gradient method was applied by Draine[61] in the Purcell-Pennypacker model for interstellar graphite grains and later by Draine and Flatau.[88] The algorithm of Draine is a special case of a more general conjugate gradient algorithm described by Petravic and Kuo-Petravic.[89] The implementation of Draine of the conjugate gradient method proved to be very robust for the particular problem of the coupled dipoles equation, and a number of improvements that can increase the computation speed by 2 or 3 times has been reported.[90,91]

The conjugate gradient method is based on minimizing the functional $F(|x\rangle) = \||A|x\rangle - |b\rangle\|$, where $|x\rangle$ is the unknown vector and $|b\rangle$ is the free term. At each step, the components of $|x\rangle$ are changed along the gradient of F in the complex L-dimensional space. Unlike some simpler methods based on the fastest descent in the L-dimensional space, the conjugate gradient method always formally converges to the exact solution after L steps; however, it can accurately approximate the solution much earlier. Similar to most other iterative methods, the convergence of the conjugate gradient method depends on the spectral range of the interaction matrix, $w_{max} - w_{min}$, and the quality of resonance of an isolated particle, $1/\Gamma$. The general rule is that the number of iterations must be not smaller than[c] $(w_{max} - w_{min})/\Gamma$. While the value of $\Gamma = |\mathrm{Im}(1/\alpha)|$ depends only on the material properties, $(w_{max} - w_{min})/\Gamma$ is

---

[c]The convergence can be much faster in the cases with special symmetry, when many eigenvectors of A are orthogonal to $|b\rangle$. If $|b\rangle$ coincides with one of the eigenvectors, the convergence takes place in just one iteration.

defined by the geometry. Typically, the latter value is larger in more dense structures with higher fractal dimension.

The basic conjugate-gradient algorithm was described in detail by Draine.[61] To find a solution to the equation $A|x\rangle = |b\rangle$, we start with an initial guess $|x_0\rangle$ and make the following initial settings:

$$|z\rangle = A^\dagger|b\rangle \,, \tag{52}$$

$$|p_0\rangle = |g_0\rangle = |z\rangle - A^\dagger A|x_0\rangle \,, \tag{53}$$

$$|w_0\rangle = A|x_0\rangle \,, \tag{54}$$

$$|v_0\rangle = A|p_0\rangle \,, \tag{55}$$

where $\dagger$ denotes Hermitian conjugation. Then, starting with $i = 0$, the consequitive orders of approximation are calculated according to

$$\alpha_i = \langle g_i|g_i\rangle / \langle v_i|v_i\rangle \,, \tag{56}$$

$$|x_{i+1}\rangle = |x_i\rangle + \alpha_i|p_i\rangle \,, \tag{57}$$

$$|w_{i+1}\rangle = |w_i\rangle + \alpha_i|v_i\rangle \,, \tag{58}$$

$$|g_{i+1}\rangle = |z\rangle - A^\dagger|w_{i+1}\rangle \,, \tag{59}$$

$$\beta_i = \langle g_{i+1}|g_{i+1}\rangle / \langle g_i|g_i\rangle \,, \tag{60}$$

$$|p_{i+1}\rangle = |g_{i+1}\rangle + \beta_i|p_i\rangle \,, \tag{61}$$

$$|v_{i+1}\rangle = A|g_{i+1}\rangle + \beta_i|v_i\rangle \,. \tag{62}$$

The choice of the initial guess is important for the convergence. One possibility is to use the first Born approximation as the initial guess. If the calculation is repeated for a sequence of λ's, the result of the previous calculation can be used as the initial guess for the next one. This technique can effectively increase the number of points in λ calculated in a given time. To further improve the convergence, a correction to the initial guess can be made to account for a change in λ. For example, a solution $|d_1\rangle$ corresponding to $\lambda = \lambda_1$ is used as an initial guess for the calculation with $\lambda = \lambda_2$. The corrected components of the initial guess can be found according to $\langle i\alpha|d_2^{initial}\rangle = \langle i\alpha|d_1\rangle \exp[i(\lambda_1/\lambda_2 - 1)\mathbf{k}_1 \cdot \mathbf{r}_i]$.

In some cases the convergence of integral characteristics, such as the optical cross sections, is an insufficient indication of reaching a good approximation to the solution. The discrepancy of the equation must be controlled independently. The functional $F(|d\rangle)/\langle E|E\rangle$ can serve as a convenient measure of the discrepancy and should be compared to unity. We'll see below on the example of a calculation which uses the solution for a slightly different λ as the initial guess that the convergence of the cross sections can be reached significantly earlier than the discrepancy becomes much smaller than unity.

4   Linear Optical Properties

Below we illustrate the theoretical and computational approaches discussed in the previous sections by some numerical examples.

4.1   Calculations in the Quasistatic Limit

We start with the coupled dipoles calculations of linear optical characteristics in the quasistatic limit. First, we introduce some useful variables. The *spectral variable* is defined as $-\mathrm{Re}(1/\alpha)$. In the vicinity of an isolated resonance the polarizability α can be written as

$$\alpha = \frac{R_m^3 \omega_m}{(\omega - \omega_0) - i\gamma} \, , \tag{63}$$

where ω_0 is the resonance frequency and γ is the spectral width of the resonance. The quantity Γ introduced above is related to γ by $\Gamma = (\gamma/\omega_m)R_m^{-3}$. As follows from (63), the spectral variable is expressed as $-\mathrm{Re}(1/\alpha) = (\omega - \omega_0)/\omega_m R_m^3$ and is proportional to the detuning from the resonance. The introduced spectral variable provides a convenient way to describe the system in the most general terms, without specifying the material properties.

All the clusters used in our calculations were built on a simple cubic lattice with a unit step. This means that the physical quantities of the dimensionality of length, such as R_m, are measured in lattice units. In particular, this concerns the polarizability α and the eigenvalues $w_n^{(0)}$. In fact, we will use below the dimensionless values a^3/α and $a^3 w_n^{(0)}$, where the lattice unit a is set to unity.

We also introduce the average polarizability of a cluster as a whole, χ, according to $\chi = \langle \mathbf{E}^{(0)} | [1/\alpha - \mathbf{W}^{(0)}]^{-1} | \mathbf{E}^{(0)} \rangle$. The extinction cross section is given by $\sigma_e = (4\pi k/|\mathbf{E}_0|^2)\mathrm{Im}\chi$ according to (18) and (19).

In Fig. 1 we show the results of the quasistatic calculations for the value $\mathrm{Im}\chi$ as a function of the spectral variable $-\mathrm{Re}(1/\alpha)$. The calculation was performed for several ensembles of 3-dimensional cluster-cluster aggregates with $D \approx 1.78$. Calculations for the first two ensembles that contained random clusters with $N = 500$ and $N = 1,000$, respectively, were done using the complete Jacoby diagonalization of $\mathbf{W}^{(0)}$ and subsequent use of formula (27). The homogeneous width Γ was set to 0.1 in these calculations. As one can see from the figure, the value $(w_{max}^{(0)} - w_{min}^{(0)})/\Gamma$ was approximately equal to 120. The

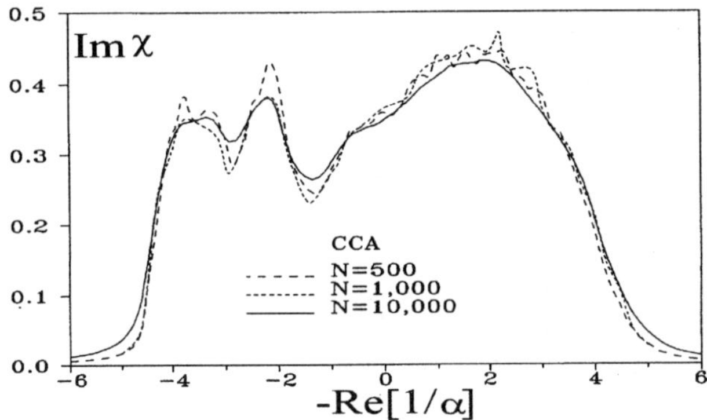

Figure 1: The spectral dependence of the imaginary part of the average cluster polarizability, χ, as a function of the spectral variable $-\mathrm{Re}(1/\alpha)$ for random cluster-cluster aggregates (CCA's) with different number of monomers, N. The curves for $N = 500$ and $N = 1,000$ are calculated using the Jacobi diagonalization and the curve for $N = 10,000$ - using the iterative algorithm based on the Lanczos recursion (Section 3.1.2).

total number of eigenstates in the case of $N = 1,000$ was equal to $3,000$ and far exceeded the above value.

The third curve in Fig. 1 was calculated using the iterative method based on the Lanczos recursion (Section 3.1.2) for an ensemble of clusters with $N = 10,000$. The value of Γ was set in this calculation to 0.2, and the number of iterations was equal to ≈ 100. We can estimate that $(w_{max}^{(0)} - w_{min}^{(0)})/\Gamma \approx 60$ in this case, which is not much less than the number of iterations, but a very good agreement between the exact and approximate calculations is achieved. Moreover, the Lanczos calculations are free of the random fluctuations seen in the first two curves, which might appear due to insufficient statistical averaging. We can also conclude that in the quasistatic approximation the spectrum does not depend on the number of particles in a cluster, as long as it is large enough. This can be not so if the complete interaction operator (5)-(7) is used instead of the quasistatic one.

In Fig. 2 we compare the spectra of the imaginary part of the average polarizability, Imχ, and the density of eigenvalues, ν, calculated as the number of eigenvalues in a unit interval, for cluster-cluster aggregates and for nonfractal random close-packed clusters. The nonfractal clusters were obtained by randomly placing little hard spheres (one after another) in a spherical volume until there was no more available space. This procedure allowed to achieve a fairly high filled volume fraction of ≈ 0.44 (compare to ≈ 0.52 for close-packed

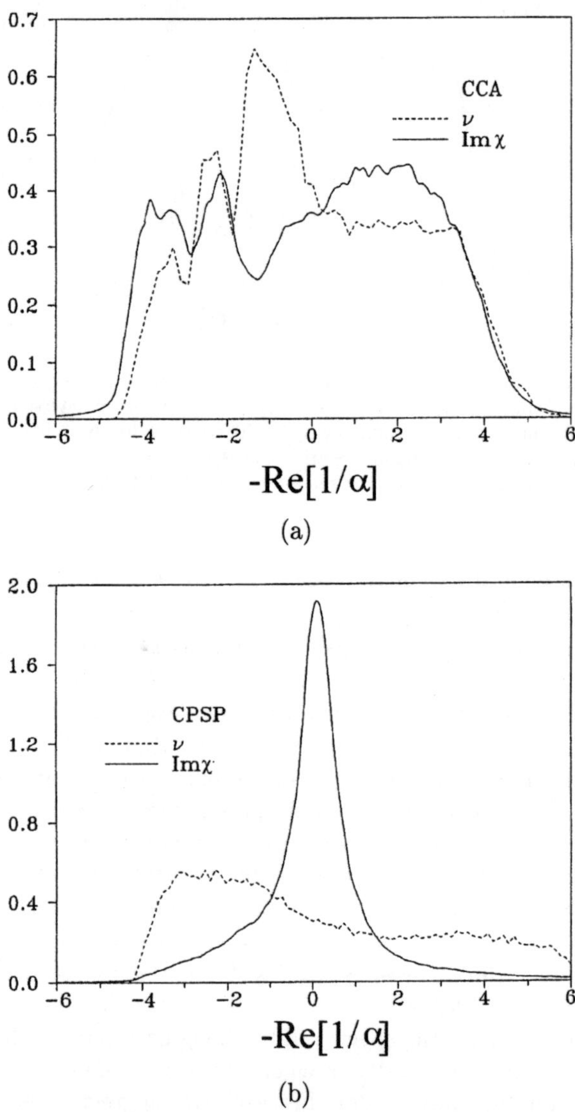

Figure 2: Spectral dependence of the imaginary part of the average cluster polarizability, χ, as a function of the spectral variable $-\mathrm{Re}(1/\alpha)$ compared to the density of eigenvalues, $\nu = \nu\left(w^{(0)} = -\mathrm{Re}(1/\alpha)\right)$. (a): random cluster-cluster aggregate (CCA's); (b): close-packed sphere of particles (CPSP). Calculations are done by the Jacoby diagonalization.

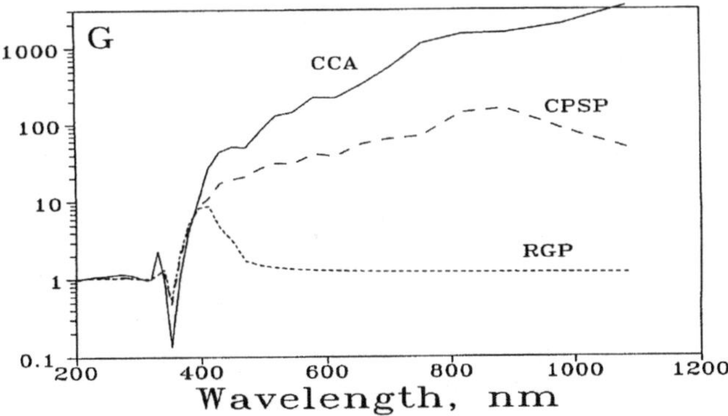

Figure 3: Local field enhancement factor, $G = \overline{|\mathbf{E}_i|^2}/|\mathbf{E}_0|^2$, as a function of the wavelength for different types of clusters built from silver monomers. CCA - cluster-cluster aggregates ($D \approx 1.78$); CPSP - close-packed sphere of particles ($D = 3$); RGP - random gas of particles ($D = 3$). Calculations are done by the Jacoby diagonalization.

spheres on a cubic lattice). The main conclusion that can be made from comparison of Figs. 2a and 2b is that in the case of fractal clusters the width of the absorption spectrum is approximately equal to the spectral width of the interaction matrix, $w_{max}^{(0)} - w_{min}^{(0)}$, while for the trivial clusters it is much smaller. Physically, this means that selection rules play a more important role in trivial clusters than in fractal ones. While in fractal clusters all eigenmodes can be effectively excited by resonance radiation, in trivial clusters the selection rules suppress excitation of all eigenmodes except for the ones with the resonance frequency close to the resonance frequency of an isolated monomer, ω_0.

In Fig. 3 we plot the local field enhancement factor, $G = \overline{|\mathbf{E}_i|^2}/|\mathbf{E}_0|^2$, as a function of the wavelength for different types of clusters built from silver monomers. Here $\mathbf{E}_i$ is the local field at the ith site in a cluster, and the horizontal line denotes statistical averaging. In this calculation, we used experimental optical constants for silver[92] to calculate the polarizability $\alpha(\lambda)$ with the use of (15),(16) as discussed in Section 2.1.1. Note that the value of R_m in (15),(16) was chosen from $a^3 = (4\pi/3)R_m^3$, and the lattice step, a, was set to 8nm. Three different types of clusters are compared. The first type is the cluster-cluster aggregates (CCA). The second type is the non-fractal close-packed sphere of particles (CPSP) described above. And the third type is also nonfractal random gas of particles (RGP) distributed randomly in a volume that would be occupied by a CCA cluster with the same number of monomers. In all cases, monomers were allowed to approach each other no closer than the lattice unit, a. Thus, the CPSP's and the RGP's differ only in the occupied volume fraction. In the CPSP model, monomers are randomly placed in a fixed spherical volume until there is some space available, which leads asymptotically to a constant occupied volume fraction, independent of the number of monomers. In the RGP model, the fraction decreases with N and is asymptotically zero.

Fig. 3 demonstrates that the fluctuative nature of fractals results in very strong fluctuations of the local fields. The fluctuations increase towards the infrared. This occurs because both the localization of fractal eigenmodes and their quality factor increase in the long-wavelength part of the spectrum.[68] This fact is especially important in nonlinear optics where the fluctuations of the local field can play the prevailing role.[19,93]

4.2   *Conjugate Gradient Calculations*

Now, we turn to the calculations beyond the quasistatic limit. Below, we illustrate the conjugate gradient method described in Section 3.2.2. The calculations are performed for a quasi 2-dimensional cluster-cluster aggregate built form $10,000$ monomers. The original 3-dimensional CCA was dropped on a plain surface and collapsed so that there is no empty spaces beneath the monomers. This model is relevant to the near-field scanning optical microscopy of fractals.[48] For convenience, we consider a reference frame in which the above surface coincides with the xy-plane, and the z-axis is perpendicular to it. The resultant cluster is not spherically symmetrical, so that one anticipates that its spectrum can depend not only on the polarization of the incident light, but also on its direction of propagation. Note that the last effect is not present in the quasistatic approximation. The cluster is built from silver monomers of the radius $R_m = 5$nm and the lattice step is $a = 8$nm (which provides that $a/R_m \approx (4\pi/3)^{1/3}$). Again, the experimental optical constants of silver[92] were used to calculate $\alpha(\lambda)$.

In Fig. 4 we illustrate the convergence of the conjugate gradient method for different values of λ. In this particular calculation, the incident light propagated in the z-direction and was polarized in the x-direction. For $\lambda = 420, 520, 620$nm and 720nm, the first Born approximation was used as an initial guess. For $\lambda = 800$nm, the initial guess was calculated on an earlier stage using $\lambda = 780$nm. For each λ, the iterations were terminated when the discrepancy of the equation became smaller than $2 \cdot 10^{-3}$. It can be easily seen from Fig. 4 that the convergence of the method becomes slower for larger λ's. This is explained by the fact that the resonance quality factor of silver nanoparticles increases towards the infrared. When the result of a preceding calculation is

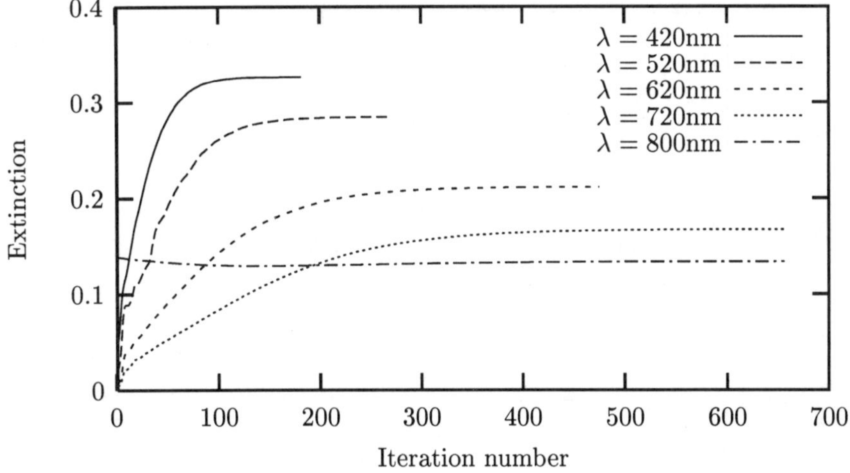

Figure 4: Convergence of the conjugate gradient method for different values of λ

used as an initial guess, the convergence is, apparently, much faster. However, an independent control of the discrepancy is always required. The calculations showed that while the convergence of the integral cross sections was already achieved, the distribution of local fields was still calculated not accurately enough, and, correspondingly, the discrepancy was large.

In Fig. 5 we plot the spectral dependence of the extinction, absorption and scattering efficiencies, defined as $Q = \sigma/\pi R_m^2 N$ for the cluster described above and for different polarizations and propagation directions of the incident wave. The most characteristic feature of these spectra is the broad long-wavelength wing. An isolated monomer has a resonance peak located near $\lambda = 400$nm with the halfwidth of approximately 50nm. As was discussed in Section 4.1, the external radiation at a certain frequency $\omega \neq \omega_0$ can resonancely excite different fractal eigenmodes, while in trivial structures such excitation is prohibited by the selection rules, unless $\omega \approx \omega_0$ (here ω_0 is the resonance frequency of an isolated monomer). It happens so that in silver clusters many fractal eigenmodes can be excited in the red and near infrared regions. This fact explains the existence of the long-wavelength tails in the spectra in Fig. 5. In the short-wavelength region ($\lambda < 300$nm), the nonresonance (bulk) absorption of silver becomes prevailing, and the fractal modes are not excited.

Another characteristic feature is the dependence of the spectra on the po-

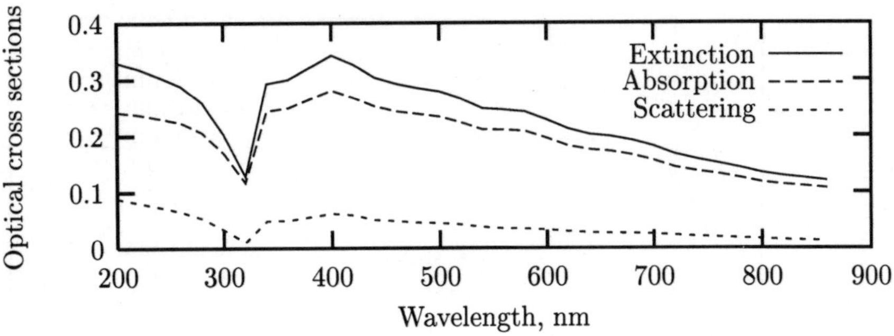

Figure 5: Extinction, absorption and scattering efficiencies as functions of the wavelength calculated by the conjugate gradient method for different polarizations and propagation directions of the incident light. (a) Propagation - x, polarization - z; (b) Propagation - z, polarization - x; (c) Propagation - y, polarization - x.

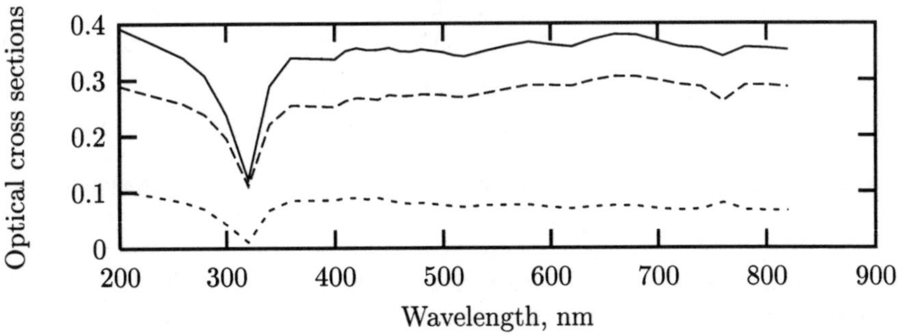

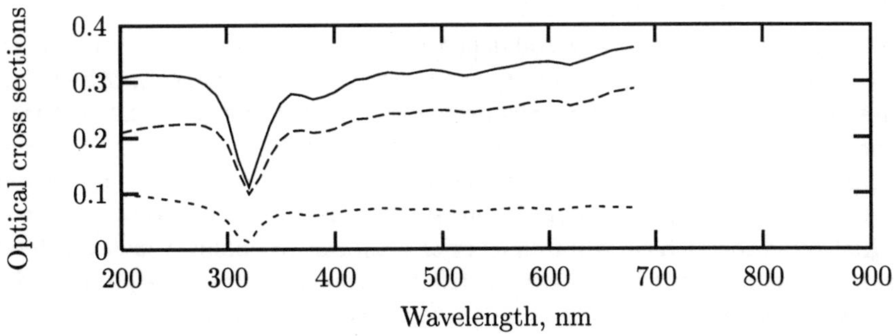

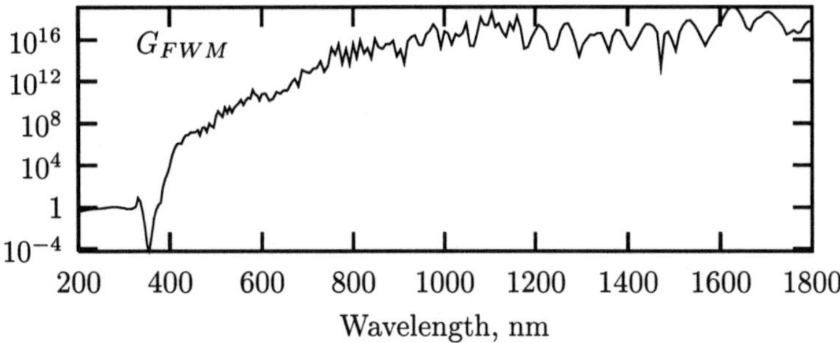

Figure 6: Enhancement factor for the degenerate four wave mixing, G_{FWM} as a function of λ for random 3-dimensional cluster cluster aggregates. The calculation was done in the quasistatic limit by Jacoby diagonalization.

larization and propagation direction of the incident light. Note that in Figs. 5b and 5c the polarization of light is the same (in the surface plane), but the propagation direction is different. The influence of the propagation direction on the optical cross sections is apparent, although the polarization dependence is stronger (compare Figs. 5a and 5b). Note that the dependence on the propagation direction cannot be obtained within the quasistatic approximation.

5　Nonlinear Optical Properties

Calculation of local fields is very important in nonlinear optics. Direct computational methods, rather than the iterative ones, are, typically, more appropriate for calculation of the local fields. Many iterative methods, such as the Lanczos-based algorithm described in Section 3.1.2, do not allow calculation of the local fields. Even if they do, as the conjugate gradient method (see Section 3.2.2), it is very difficult to control convergence of the local field at each point in space. Instead, some integral characteristics such as the discrepancy or the convergence of optical cross sections are used to check the accuracy of the obtained solutions. However, even one strong fluctuation in the local field can give a significant input to the generated nonlinear signal.

Below we show the results of quasistatic calculations of the enhancement factor for the degenerate four-wave mixing. This is a nonlinear optical effect of the third order, and the enhancement factor can be defined as[93]

$$G_{FWM} = \left| \frac{\overline{|\mathbf{E}_i|^2 \mathbf{E}_i^2}}{\mathbf{E}_0^4} \right|^2 . \tag{64}$$

The calculations were carried out for 3-dimensional cluster-cluster aggregates with $N = 1,000$ by the direct Jacoby diagonalization. As above, the clusters were considered to be built from silver monomers of the radius $R_m = 5$nm and the experimental optical constants of silver were used.

Because of the high order of the optical nonlinearity involved, the enhancement factor is extremely sensitive to fluctuations in the local field. In the infrared region, where these fluctuations are strong due to high quality of fractal resonances, the average enhancement can reach the enormous value of 10^{16} and the local enhancements can reach even much higher values.

Acknowledgments

This work was supported in part by the National Science Foundation under grants DMR-950050 and DMR-9623663, NATO under the grant CRG-950097 and the Environmental Protection Agency under the grant R822658-01-0. Also, acknowledgment is made to the donors of the Petroleum Research Fund, administered by the ACS, for partial support of this research.

References

1. B. B. Mandelbrot, *The fractal geometry of nature* (Freeman, San Francisco, 1982).
2. *Fractals in physics*, edited by L. Pietroniero and E. Tosatti (North-Holland, Amsterdam, 1986).
3. B. Sapoval, *Fractals* (Aditech, Paris, 1990).
4. R. Botet and R. Jullien, Phase Transitions **24-26**, 691 (1990).
5. R. Jullien and R. Botet, *Aggregation and fractal aggregates* (World Scientific, Singapore, 1987).
6. A. V. Karpov *et al.*, JETP Lett. **48**, 571 (1988).
7. A. I. Plekhanov, G. L. Plotnikov, and V. P. Safonov, Opt. Spectrosc. (USSR) **71**, 451 (1991).
8. Y. E. Danilova, A. I. Plekhanov, and V. P. Safonov, Physica A **185**, 61 (1992).
9. Y. E. Danilova, S. G. Rautian, and V. P. Safonov, Bulletin of the Russian Acad. Sci. - Physics **60**, 374 (1996).
10. Y. E. Danilova, V. P. Drachev, S. V. Perminov, and V. P. Safonov, Bulletin of the Russian Acad. Sci. - Physics **60**, 342 (1996).

11. C. W. Bruce, T. F. Stromberg, K. P. Gurton, and J. B. Mozer, Appl. Opt. **30**, 1537 (1991).

12. N. Lu and C. M. Sorensen, Phys. Rev. E **50**, 3109 (1994).

13. J. Cai, N. Lu, and C. M. Sorensen, J. Colloid Interface Sci. **171**, 470 (1995).

14. E. F. Mikhailov and S. S. Vlasenko, Physics-Uspekhi **165**, 253 (1995).

15. S. D. Andreev, L. S. Ivlev, E. F. Mikhailov, and A. A. Kiselev, Atmos. Oceanic Opt. **8**, 355 (1995).

16. S. D. Andreev and E. F. Mikhailov, Bulletin of the Russian Acad. Sci. **32**, 743 (1996).

17. R. Chiarello, V. Panella, and J. Krim, Phys. Rev. Lett. **67**, 3408 (1991).

18. J. Krim, I. Heyvaert, C. Van Haesendonck, and Y. Bruynseraede, Phys. Rev. Lett. **70**, 57 (1993).

19. V. M. Shalaev, Phys. Rep. **272**, 61 (1996).

20. C. Amitrano, A. Coniglio, P. Meakin, and M. Zannetti, Phys. Rev. B **44**, 4974 (1991).

21. V. A. Markel, V. M. Shalaev, E. Y. Poliakov, and T. F. George, Phys. Rev. E **55**, 7313 (1997).

22. P. Meakin, Phys. Rev. Lett. **51**, 1119 (1983).

23. M. Kolb, R. Botet, and R. Jullien, Phys. Rev. Lett. **51**, 1123 (1983).

24. T. A. Witten and L. M. Sander, Phys. Rev. Lett. **47**, 1400 (1981).

25. D. Stauffer and A. Aharony, *Introduction to percolation theory*, 2 ed. (Taylor and Francis, Philadelphia, 1991).

26. H. D. Bale and P. W. Schmidt, Phys. Rev. Lett. **53**, 596 (1984).

27. M. V. Berry and I. C. Percival, Optica Acta **33**, 577 (1986).

28. J. E. Martin and A. J. Hurd, J. Appl. Cryst. **20**, 61 (1987).

29. M. Y. Lin *et al.*, J. Colloid Interface Sci. **137**, 263 (1990).

30. M. Carpineti, M. Giglio, and V. Degiorgio, Il Nuovo Cimento **16D**, 1243 (1994).

31. M. Carpineti, M. Giglio, and V. Degiorgio, Phys. Rev. E **51**, 590 (1995).

32. D. Asnaghi, M. Carpiteti, M. Giglio, and A. Vailati, Physica A **213**, 148 (1995).

33. F. Sciortino, A. Belloni, and P. Tartaglia, Phys. Rev. E **52**, 4068 (1995).

34. N. G. Khlebtsov and A. G. Mel'nikov, Opt. Spectrosc. (USSR) **79**, 656 (1995).

35. N. G. Khlebtsov, Colloid J. **58**, 100 (1996).

36. V. A. Markel, V. M. Shalaev, E. Y. Poliakov, and T. F. George, J. Opt. Soc. Am. A **14**, 60 (1997).

37. V. M. Shalaev and M. I. Stockman, Z. Phys. D **10**, 71 (1988).

38. A. V. Butenko, V. M. Shalaev, and M. I. Stockman, Z. Phys. D **10**, 81

(1988).

39. V. A. Markel, L. S. Muratov, M. I. Stockman, and T. F. George, Phys. Rev. B **43**, 8183 (1991).

40. V. M. Shalaev, R. Botet, and R. Jullien, Phys. Rev. B **44**, 12216 (1991).

41. M. I. Stockman, T. F. George, and V. M. Shalaev, Phys. Rev. B **44**, 115 (1991).

42. V. M. Shalaev, M. I. Stockman, and R. Botet, Physica A **185**, 181 (1992).

43. V. M. Shalaev, M. Moskovits, A. A. Golubentsev, and S. John, Physica A **191**, 352 (1992).

44. M. I. Stockman *et al.*, Phys. Rev. B **46**, 2821 (1992).

45. V. M. Shalaev, R. Botet, and A. V. Butenko, Phys. Rev. B **48**, 6662 (1993).

46. V. M. Shalaev *et al.*, Physica A **207**, 197 (1994).

47. M. I. Stockman, L. N. Pandey, L. S. Muratov, and T. F. George, Phys. Rev. Lett. **72**, 2486 (1994).

48. D. P. Tsai *et al.*, Phys. Rev. Lett. **72**, 4149 (1994); V. M. Shalaev and M. Moskovits, Phys. Rev. Lett. **75**, 2451 (1995).

49. M. I. Stockman, L. N. Pandey, and T. F. George, Phys. Rev. B **51**, 185 (1995).

50. M. I. Stockman, L. N. Pandey, and T. F. George, Phys. Rev. B **53**, 2183 (1996).

51. V. M. Shalaev, R. Botet, J. Mercer, and E. B. Stechel, Phys. Rev. B **54**, 1 (1996).

52. E. Y. Poliakov, V. M. Shalaev, V. A. Markel, and R. Botet, Opt. Lett. **21**, 1628 (1996).

53. S. G. Rautian *et al.*, JETP Lett. **47**, 243 (1988).

54. A. V. Butenko *et al.*, Z. Phys. D **17**, 283 (1990).

55. Y. E. Danilova, V. A. Markel, and V. P. Safonov, Atmos. Oceanic Opt. **6**, 821 (1993).

56. S. Alexander and R. Orbach, J. Phys. Lett. (Paris) **43**, L625 (1982).

57. S. Alexander, Phys. Rev. B **40**, 7953 (1989).

58. V. A. Markel, J. Opt. Soc. Am. B **12**, 1783 (1995).

59. V. A. Markel, J. Mod. Opt. **39**, 853 (1992).

60. E. M. Purcell and C. R. Pennypacker, Astrophys. J. **186**, 705 (1973).

61. B. T. Draine, Astrophys. J. **333**, 848 (1988).

62. A. Lakhtakia, Int. J. Mod. Phys. **3**, 583 (1992).

63. A. Lakhtakia, Int. J. Infrared Millimeter Waves **13**, 869 (1992).

64. B. T. Draine and J. Goodman, Astrophys. J. **405**, 685 (1993).

65. J. M. Gerardy and M. Ausloos, Phys. Rev. B **22**, 4950 (1980).

66. F. Claro, Phys. Rev. B **25**, 7875 (1982).

67. J. E. Sansonetti and J. K. Furdyna, Phys. Rev. B **22**, 2866 (1980).

68. V. A. Markel *et al.*, Phys. Rev. B **53**, 2425 (1996).

69. V. A. Markel, J. Mod. Opt. **40**, 2281 (1993).

70. P. Chiappetta, J. Phys. A **13**, 2101 (1980).

71. S. B. Singham and C. F. Bohren, J. Opt. Soc. Am. A **11**, 1867 (1988).

72. F. Claro, Phys. Rev. B **30**, 4989 (1984).

73. F. Claro, Solid State Comm. **49**, 229 (1984).

74. R. Rojas and F. Claro, Phys. Rev. B **34**, 3730 (1986).

75. R. Fuchs and F. Claro, Phys. Rev. B **35**, 3722 (1987).

76. K. A. Fuller, J. Opt. Soc. Am. A **11**, 3251 (1994).

77. K. A. Fuller, J. Opt. Soc. Am. A **12**, 881 (1995).

78. Y.-l. Xu, Appl. Opt. **34**, 4573 (1995).

79. T. Lemaire, J. Opt. Soc. Am. A **14**, 470 (1997).

80. C. F. Bohren and D. R. Huffman, *Absorption and scattering of light by small particles* (John Wiley & Sons, New York, 1983).

81. P. J. Reynolds, W. Klein, and H. E. Stanley, J. Phys. **10**, L167 (1977).

82. A. K. Sarychev, Sov. Phys. JETP **45**, 524 (1977).

83. F. Claro and R. Fuchs, Phys. Rev. B **44**, 4109 (1991).

84. F. Brouers *et al.*, Phys. Rev. B **55**, 13234 (1997).

85. J. Bernasconi, Phys. Rev. B **18**, 2185 (1978).

86. R. Haydock, in *Solid state physics*, edited by H. Ehrenreich, F. Seitz, and D. Turnbull (Academic Press, New York, 1980), Vol. 35, Chap. The recursive solution of the Schrodinger equation, pp. 215–294.

87. J. K. Collum and R. A. Willowghby, *Lanczos algorithm for large symmetric eigenvalue computations* (Birkhauser, Boston, 1985), Vol. 1.

88. B. Draine and P. Flatau, J. Opt. Soc. Am. A **11**, 1491 (1994).

89. M. Petravic and J. Kuo-Petravic, J. Comput. Phys. **32**, 263 (1979).

90. K. Lumme and J. Rahola, Astrophys. J. **425**, 653 (1994).

91. P. J. Flatau, Opt. Lett. **22**, 1205 (1997).

92. P. B. Johnson and R. W. Christy, Phys. Rev. B **6**, 4370 (1972).

93. V. M. Shalaev, E. Y. Poliakov, and V. A. Markel, Phys. Rev. B **53**, 2437 (1996).

LOCAL FIELDS' LOCALIZATION AND CHAOS AND NONLINEAR-OPTICAL ENHANCEMENT IN COMPOSITES

MARK I. STOCKMAN

Department of Physics and Astronomy, Georgia State University, Atlanta, GA 30303
E-mail: mstockman@gsu.edu
http://www.phy-astr.gsu.edu/stockman

The paper is devoted to linear and nonlinear optical properties of disordered clusters and nanocomposites. Linear and nonlinear optical polarizabilities of large disordered clusters, fractal clusters in particular, and susceptibilities of nanocomposites are found and calculated numerically. A spectral theory with dipole interaction is used to obtain quantitative results. Major properties of systems under consideration are giant fluctuations and chaos of local fields that cause strong enhancement (by many orders of magnitude) of nonlinear optical responses. The enhancement and fluctuations properties of the local fields are intimately interrelated to the inhomogeneous localization of the systems' eigenmodes ("plasmons"). Due to these fluctuations, mean-field theory completely fails to describe optical polarizabilities.

1. Introduction

Clusters and nanocomposites belong to so-called nanostructured materials. Such materials typically are nanoparticles either bound to each other by covalent or van der Waals bonds, or dispersed in a host medium. Description of electromagnetic properties of such system is a long-standing problem going back to such names as Maxwell Garnett[1], Lorentz[2] and Bruggeman[3].

Properties of such materials may be dramatically different from those of bulk materials with identical chemical composition. A characteristic property of such systems is confinement of electrons, phonons, electric fields, etc., in small spatial regions. Such a confinement, in particular, modifies spectral properties (shifts quantum levels and changes transition probabilities), and also changes the interaction between the constituent particles. As we will be discussing in this paper, local (near-zone) electromagnetic fields are strongly fluctuating in space. Their magnitude is greatly (by orders of magnitude) enhanced with respect to the external (exciting) fields.

A phenomenon closely related to the enhancement and fluctuations of local fields is localization of elementary excitations (eigenmodes) in the composites[4-8]. The relevant excitations are polar waves that are traditionally called plasmons (this term originates from theory of metallic nanoparticles containing electron plasma, but is now often used in application to other nanocomposites). Plasmon-resonant properties leading to enhancement of local fields are especially pronounced in some metallic (especially silver, gold, or platinum) colloidal clusters, metal nanocomposites and rough surfaces. A typical example of such responses is surface-enhanced Raman scattering (see, e.g. a review of Ref. 9 and reference therein).

The most pronounced effect of the fluctuating local fields is on nonlinear optical susceptibilities. The reason for that can be understood qualitatively. Imaging two fields with the same average intensity $I_1 \propto \langle \mathbf{E}^2 \rangle$. For the sake of argument, let us say, the first field has the same constant intensity I_1 in $N \gg 1$ points, and the second is strongly localized at one point where its intensity then should be $I_2 = \langle NI_1 \rangle$. Consider a n-th order nonlinearity where the nonlinear response is proportional to $\langle \mathbf{E}^{2n} \rangle \propto \langle I^n \rangle$. The ratio of the nonlinear response for the first (constant) field is proportional to $\frac{1}{N}(NI_1^n) = I_1^n$. In contrast, the response to the second (strongly localized) field is $\frac{1}{N}(NI_1)^n = N^{n-1}I_1^n$. In such a way, the enhancement coefficient (the ratio of the nonlinear response in the second case to that in the first case is N^{n-1}. Hence the localization has a potential to bring about strongly enhanced nonlinear responses where the enhancement increases with the order of nonlinearity and the degree of localization (spatial fluctuations).

To maximize this effect, our goal is to find systems with the maximum spatial fluctuations of the local densities. We certainly expect that the density fluctuations will cause correspondingly large fluctuations of the local fields.

There exists a class of systems that stands out in this respect. These are self-similar (fractal) systems, which (on average) repeat themselves at different scales. In other words, looking at such a system and not seeing its boundaries (neither at the maximum or a minimum scales), one cannot say what fraction of the system is observed, and what is the actual size of the objects seen. For such systems, the number N of constituent particles (monomers) contained within a radius R scales as

$$N \cong \left(\frac{R}{R_0} \right)^D , \tag{1}$$

where R_0 is a typical distance between monomers, and D is the fractal (Hausdorff) dimension of the cluster. The density of monomers as given by Eq. (1) is asymptotically zero for large clusters,

$$\rho \cong R_0^{-3}\left(\frac{R}{R_0}\right)^{D-3} \to 0$$

(2)

Fig. 1 Cluster-cluster aggregate (CCA) of $N = 1000$ monomers.

However, this does not mean that the interaction between monomers can be neglected. The underlying reason for that is a strong correlation between monomers in a cluster, with the pair-pair correlation function scaling similar to Eq. (2). Thus we have a unique system whose macroscopic density is asymptotically zero, but the interaction inside the system is strong. This idea has been proposed by us in an earlier papers[4,10]

An example of a fractal cluster, obtained by cluster-cluster aggregation[11] (CCA) is shown in Fig. 1. This figure illustrates many properties of fractal clusters and composites, including the low overall density and strong correlation in the positions of monomers. We will use CCA clusters (composites) throughout the paper as a model of self-similar (fractal) systems.

To avoid possible misunderstanding, we point out that other, non-fractal systems also possess enhanced optical responses, especially those that are tailored to have optimally-chosen dielectric properties changing in space[12.] The physical origin of the enhancement in this case is the same: the local fields change in space, and consequently the higher moments of their intensity are increased. As an example of such systems we will consider a random Maxwell Garnett composite (dielectric or metallic spheres embedded in a host medium at random positions). Such a composite has earlier been considered in a mean-field approximation[13]. We will consider below a model of such composites where the inclusion spheres are positioned on a cubic lattice and call it a random lattice gas (RLG). We will use RLG as a model of random but not fractal composites.

2. Equations Governing Optical (Dipolar) Responses

We concentrate on the dipole-dipole interaction, which is a universal interaction between polarizable particles. We consider a cluster (or, a composite) whose particles (called below monomers) are positioned at points $\mathbf{r}_i$.

We assume that the system (a cluster or composite) is subjected to the electric field $\mathbf{E}$ of the incident optical wave. This field induces the dipole $d_{i\alpha}$ at an i-th monomer (here $\alpha = x, y, z$ denotes the Cartesian components of the vector, and similar notations will be used for other vectors). The dipole moments satisfy well-known system of equations

$$\alpha_0^{-1} d_{i\alpha} = E_{i\alpha}^{(0)} - \sum_{j=1}^{N} \left(\delta_{\alpha\beta} - 3\frac{\left(r_{ij}\right)_\alpha \left(r_{ij}\right)_\beta}{r_{ij}^2} \right) \frac{d_{j\beta}}{r_{ij}^3} .$$

(3)

Here $E_{i\alpha}^{(0)}$ is the wave-field amplitude at the i-th monomer, $\mathbf{r}_{ij} = \mathbf{r}_i - \mathbf{r}_j$ is the relative vector between the i-th and j-th monomers, and α_0 is the dipole polarizability of the monomer. We assume that the size of the system is much less than the wavelength of the exciting wave, and therefore the exciting field $E_\alpha^{(0)}$ is the same for all the monomers of a cluster. We note that the dipole interaction is not valid in the close vicinity of a monomer. Our choice of interaction is justified if intermediate-to-large scales predominantly contribute to the properties under consideration.

3. Scaling and Spectral Representation

The scaling theory of the optical response of fractal clusters has been developed in Refs. 5 and 6. We note that a similar spectral approach has been independently introduced R.Fuchs and collaborators.[14,15] There exists also a general spectral approach[16] that can be shown to reduce to our approach if the dipole approximation is applied.

The material properties of the system enter Eq. (3) only *via* the combination Zr_{ij}^3, where we have introduced the notation $Z \equiv \alpha_0^{-1}$. This along with the (approximate) self-similarity of the system is a prerequisite for scaling in terms of the spectral variable Z. A principal requirement for the scaling of a certain physical quantity F

is that the system eigenmodes contributing to F should have their localization radii L intermediate between the maximum scale (size of the cluster R_c) and the minimum scale, R_0. Then this quantity will not depend on any external length, leading to scaling.

Because the quantity F is not sensitive to the maximum scale R_c, it should have the functional dependence $F = F(ZR_0^3)$. On the other hand, the eigenmodes contributing to F are insensitive to a much smaller minimum scale. Therefore the dependence on R_0 can be only power (scaling), and consequently

$$F(Z) \propto \left(ZR_0^3\right)^{\gamma},$$

(4)

where γ is some scaling index.

In accord with the above arguments, it is convenient to express all results not in terms of frequency, but in terms of Z, separating the imaginary and real parts, $Z = -X - i\delta$. The choice of signs in this expression makes the dissipation parameter δ positive, while the spectral parameter X is positive when the frequency is blue-shifted from the plasmon resonance, and negative otherwise. For the sake of reference, we give here the expressions for X and δ for a metallic nanosphere in the Drude model,

$$X = \frac{1}{R_m^3}\frac{\left(\varepsilon_0 + 2\varepsilon_a\right)^{3/2}}{3\varepsilon_a\omega_p}\left(\omega - \omega_s\right); \quad \delta = \frac{1}{R_m^3}\frac{\left(\varepsilon_0 + 2\varepsilon_a\right)^{3/2}}{3\varepsilon_a\omega_p}\frac{\gamma}{2},$$

(5)

where ε_0 is the intersubband dielectric constant of the metal, ε_a is the ambient dielectric constant, ω_p is the metal plasma frequency, $\omega_s = \dfrac{\omega_p}{\sqrt{\varepsilon_0 + 2\varepsilon_a}}$ is the surface plasmon frequency, and R_m is the nanosphere's radius.

The spectral dependence of X and δ for silver is illustrated in Fig. 2. The most important feature in the figure is that in the yellow-red region of visible light, the real part of the polarizability greatly exceeds its imaginary part. Their ratio has the meaning of the quality factor of the surface-plasmon resonance Q,

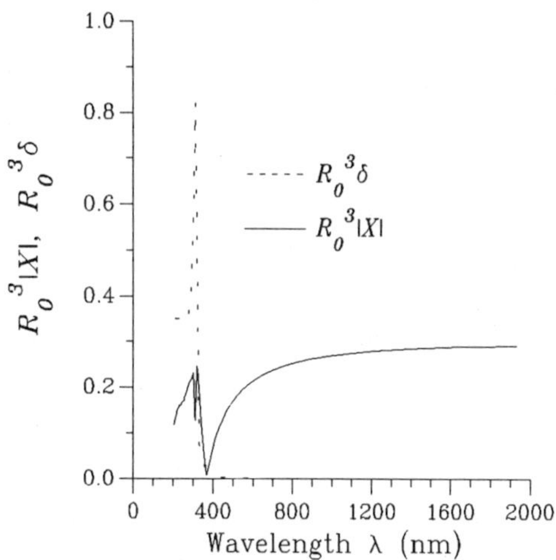

Fig. 2. Wavelength dependence of the spectral parameter X and the dissipation parameter δ.

$$Q = |X|\big/\delta \tag{6}$$

This factor shows how many times on the order of magnitude the amplitude of the local field in a vicinity of a resonant monomer exceeds that of the exciting field. The fact that for many metals Q may be large (as large as $\cong 10^2$) plays an important role in the theory since the enhancement of the optical responses is a resonant phenomenon, and strong dissipation would completely suppress it.

To introduce the spectral representation,[5,6] we will rewrite Eq.(3) as one equation in the $3N$-dimensional space. To do so, we introduce $3N$-dimensional vectors $|d)$, $\left|E^{(0)}\right)$,..., whose projections give the physical vectors

$$(i\alpha|d) = d_{i\alpha}, \quad \left(i\alpha\left|E^{(0)}\right.\right) = E_{i\alpha}^{(0)}, \ldots \tag{7}$$

Equations (3) then acquire the form

$$(Z + W)|d) = \left|E^{(0)}\right) \tag{8}$$

where W is the dipole-dipole interaction operator with the matrix elements

$$(i\alpha|W|j\beta) = \begin{cases} \left(\delta_{\alpha\beta} - 3\dfrac{\left(r_{ij}\right)_\alpha \left(r_{ij}\right)_\beta}{r_{ij}^2}\right)\dfrac{1}{r_{ij}^3}, & i \neq j \\ 0, & i = j. \end{cases} \tag{9}$$

We introduce the eigenmodes (plasmons) $|n)$ $(n = 1,\ldots,3N)$ as the eigenvectors of the W-operator,

$$W|n) = w_n |n),$$

(10)

where w_n are the corresponding eigenvalues. Practically, the eigenvalue problem (10) can be solved numerically for any given cluster. Having done so, one can calculate the Green's function

$$\mathbf{G}_{i\alpha,j\beta} = \sum_{n=1}^{3N} \frac{(i\alpha|n)(j\beta|n)}{Z + w_n},$$

(11)

which carries the maximum information on the spectrum and linear response of the system.

4. Linear Optical Responses

The polarizability of a cluster or finite volume of a composite α and its density of eigenmodes V are expressed in terms of $\mathbf{G}$ as

$$\alpha_{\alpha\beta} = \sum_{i,j} \mathbf{G}_{i\alpha,j\beta}, \quad \rho = \sum_{i,j} \mathbf{G}_{i\alpha,i\alpha},$$

(12)

where summation over repeated vector indices is implied. The dielectric constant of a cluster (composite) is given by

$$\varepsilon = \varepsilon_h \left(1 + 4\pi\alpha \frac{N}{V} \right),$$

(13)

where ε_h is the dielectric constant of a host, V is the volume occupied by the

cluster (composite) and $\alpha = \left\langle \frac{1}{3N} \sum_{i,j} \mathbf{G}_{i\beta,j\beta} \right\rangle$ is a polarizability of a monomer

in the cluster (composite). Below we will consider results of numerical computations using Eqs.(13) and (14) and will compare them to some analytical predictions.

First, let us consider scaling predictions. For this purpose we have to invoke a large magnitude of the quality factor of the optical resonance (6), $Q \gg 1$. In this case, a

dependence of type (4) becomes $F(X) \propto \left(R_0^3 |X| \right)^\gamma$. We have introduced[5,6] such a

dependence for $\alpha_{\alpha\beta}(X)$ and $\rho(X)$ and argued that the two quantities have the

same scaling,

$$\mathrm{Im}\,\alpha(X) \cong \mathrm{Im}\,\rho(X) \cong R_0^3 \left| R_0^3 X \right|^{d_o-1} , \tag{14}$$

where d_o is an index that we called the optical spectral dimension. We have also argued that the physical range of d_o is $1 \ge d_o \ge 0$.

The strong localization has been essential for the derivation of Eq.(12). It implies that all eigenmodes (at least all contributing eigenmodes) of a cluster are strongly localized The strong localization, as discussed by Alexander[17], means that for any given frequency parameter X there exists only one characteristic length L_X of these eigenmodes playing the role of simultaneously their wavelength and their localization length. Using scale invariance arguments, we have shown[5,6] that L_X should scale as

$$L_X \cong R \left| R_0^3 X \right|^{\frac{d_o-1}{3-D}} . \tag{15}$$

We have subjected the scaling predictions of Eqs. (14) and (15) to an extensive comparison with the results of large-scale computations.[7] The results were quite unexpected. One of those, a polarizability and eigenmode density for cluster-cluster aggregates (CCA), is shown in Fig. 3. The conclusion that one can draw from the figure is that neither the polarizability, nor density of eigenmodes scale. Interestingly enough, they still appear to be quite close to each other, supporting the conclusion of Refs.5 and 6 that all eigenmodes of a fractal cluster contribute (almost) equally to its optical absorption. Similar results have been obtained[7] for other types of clusters.

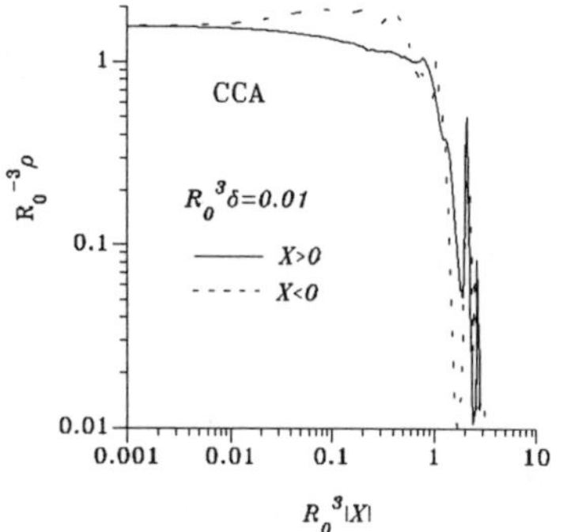

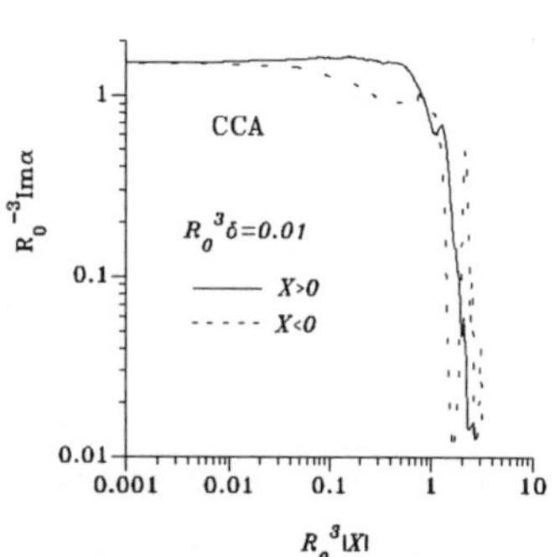

Fig. 3. Numerically obtained polarizability (upper panel) and density of eigenmodes (lower panel) for cluster-cluster aggregates.

Another relation to check is that of Eq.(15). First, one has to formulate how to calculate the localization radius We use the definition of Ref.7,

$$L_X = \frac{\sum_n \rho_n L_n}{\sum_n \rho_n}, \quad \text{where} \quad L_n = \sum_{i\beta} \mathbf{r}_i^2 \left(n|i\alpha\right)^2 - \left(\sum_{i\beta} \mathbf{r}_i \left(n|i\alpha\right)^2\right)^2 \tag{16}$$

where $\rho_n = \left[\left(X - w_n\right)^2 + \delta^2\right]^{-1}$, L_n is the localization radius of a given eigenmode, and L_X is the localization radius at a given frequency. The computed dependence of L_X is shown in Fig. 4. As one can clearly see, there is no scaling in these data too. This finding is in contradiction with the conclusion of Ref.18 (precision of our calculations is much higher than that of Ref.18).

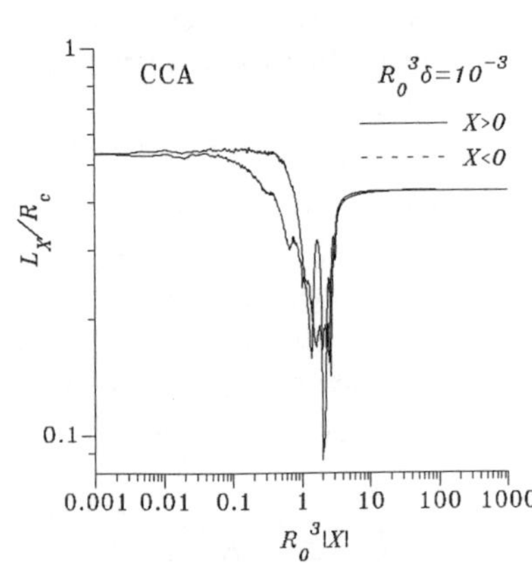

Fig. 4. Localization (coherence) radius L_X as a function of the spectral parameter X.

Evident failure of the scaling implies that at least one of the assumptions lying at its foundation is incorrect. Because we used the same model (dipole-dipole) for both the scaling theory and the numerical computations, the non-applicability of the model to system is out of question. In our consideration we consistently used high values of Q, so that the condition $Q >> 1$ is also satisfied. The only cause of the failure of scaling appears to be the strong localization assumption.

Now we will compare the results of our theory to a mean field approximation known as Maxwell Garnett formula (or an equivalent Lorentz-Lorenz formula), see, e.g., Ref. 13. This comparison is shown in. Fig. 5. As we see, the mean field theory gives a poor description of actual dielectric constant, especially in the region of the resonant absorption of the inclusions, where $\varepsilon_p < 0$.

Undoubtedly, there should be reasons for the failure of both scaling theory and mean field theory. As we understand now, two interrelated phenomena can be blamed for this failure. These are inhomogeneous localization of eigenmodes and giant

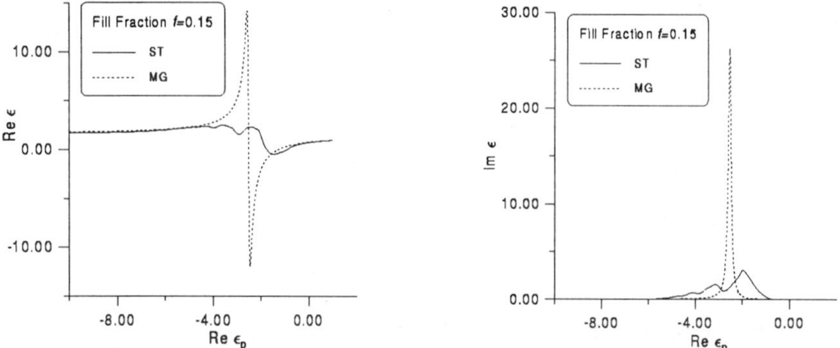

Fig. 5. Comparison of the dielectric polarizability of a composite in present spectral theory (ST) with the result of the Maxwell Garnett formula (MG). The relative dielectric constant of a composite $\mathcal{E}$ is shown as a function of the dielectric constant of the inclusion particles $\mathcal{E}_p$.

fluctuations of the local fields. In the case of the inhomogeneous localization there are eigenmodes of all localization radii from the minimum distance between monomers (inclusions) R_0 to the total size of the system R_c . Both of these two extremes render the scaling theory inapplicable. Obviously, the strong fluctuations contradict to the basic assumption of the mean field theory.

5. Inhomogeneous Localization of Eigenmodes

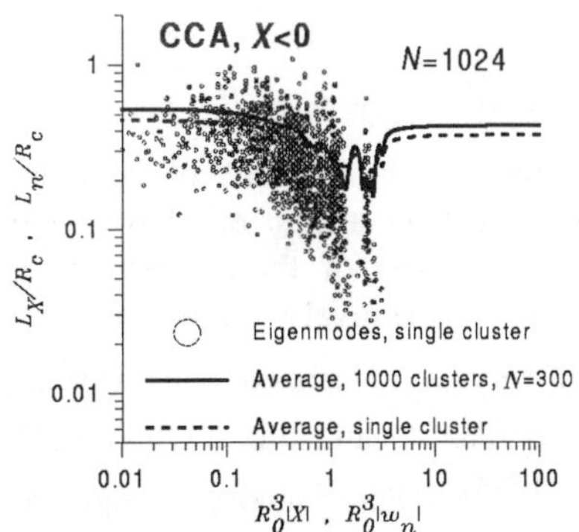

Fig. 6. All eigenmodes with negative eigenvalues for a CCA cluster of $N = 1024$ monomers.

To introduce the inhomogeneous localization[8,19,20,] we consider all eigenmodes of a single cluster. In Fig. 6 we show a special plot where each eigenmode is represented by a point in the coordinates its localization length L_n *versus* the frequency parameter X . As one can see, at any frequency (value of X) within a wide range of X , the eigenmodes have a very broad spectrum of their localization radii L_X , from the minimum scale $\simeq R_0$ of the distance between the monomers to the maximum scale $\cong R_c$ of the cluster total radius.

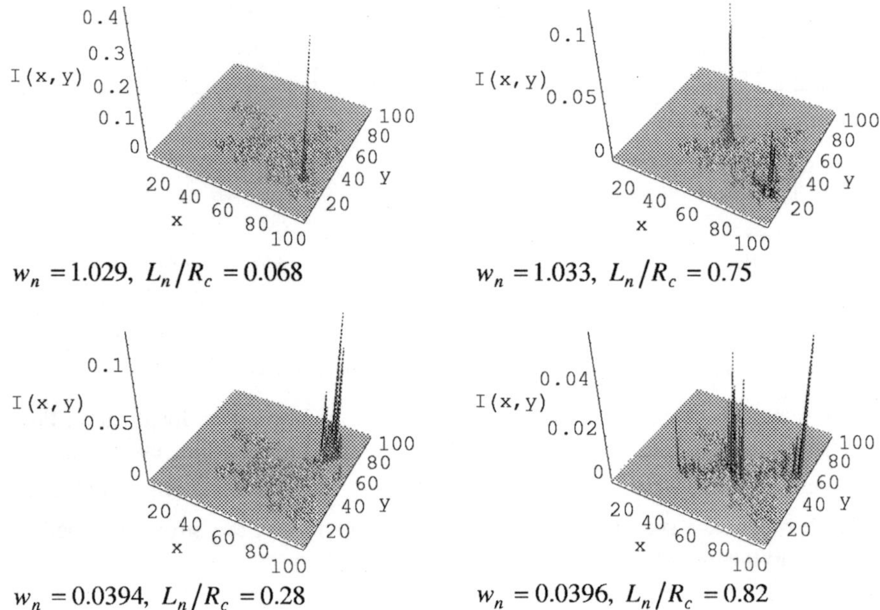

$$w_n = 1.029, \ L_n/R_c = 0.068 \qquad\qquad w_n = 1.033, \ L_n/R_c = 0.75$$

$$w_n = 0.0394, \ L_n/R_c = 0.28 \qquad\qquad w_n = 0.0396, \ L_n/R_c = 0.82$$

Fig. 7. Spatial distribution of the local-field intensities for an individual CCA cluster ($N = 1500$) shown over the two-dimensional projection of the cluster for the eigenvalues w_n (in the units of R_0^{-3}) as indicated. The coordinates are shown in units of the lattice spacing R_0 . The value of the gyration radius of the individual eigenmodes is given relative to the cluster radius R_c .

It is also useful to take a closer look at some members of the ensemble of eigenmodes of a cluster. We take two pairs of eigenmodes that have almost equal frequencies (within a pair). The spatial distribution of the intensity of these eigenmodes is presented in Fig. 7. As we can see, at a comparatively large eigenvalues ($R_0^3 w_n = 1.029$), the eigenmode is indeed a very well localized, demonstrating a very sharp peak over just a few monomers. The localization radius for this eigenmode is indeed very small, $L_n / R_c = 0.034$. Surprisingly, an eigenmode with a very close frequency ($R_0^3 w_n = 1.033$) is almost completely delocalized (for this eigenmode, $L_n / R_c = 0.75$). However, the examination of its spatial profile shows that in reality this eigenmode consists of two sharp peaks separated by a distance on the order of the total size of the cluster. As we go closer to the plasmon resonance (smaller $\left|w_n\right|$), the minimum width of the peaks increases, as one would expect in view of Eq. (15). At the same time, the internal structure of the eigenmode remains highly irregular with large spatial fluctuations. Again, the eigenmodes at the same frequency possess very different localization radii.

The behavior described above is very unusual. For the known localization patterns, eigenmodes are strongly localized for short wavelength (high frequencies) and delocalized in the long-wavelength wing. The physical reason for this is that a long wave (wavelength $\lambda \gg$ the typical size of the scattering inhomogeneities) sees almost homogeneous medium and propagates almost freely. In contrast, a short wave is strongly scattered from inhomogeneities with sizes on the order of λ (strongly here means that the scattering length itself is on the order of λ). Thus, there exists the mobility edge, i.e., a frequency above which waves are localized and below which they propagate.

This logic leading to the existence of the mobility edge is obviously inapplicable to fractals. They are self-similar systems and, therefore, do not possess any characteristic scattering length. For any eigenmode wavelength λ, there always exist inhomogeneities of the sizes comparable to λ. This may suggest that all the eigenmodes are strongly localized, as assumed in Refs.5 and 6. However, the result presented above and all numerical modeling of Refs.7, 8, 19, and 20 have shown that the strong localization is not the case. In actuality, the inhomogeneous localization[8] takes place, where the eigenmodes in a wide range of the localization radii coexist at any given frequency.

To examine statistical properties of an eigenmode distribution, we introduce the distribution function $P(L, X)$, which is the probability density that an eigenmode at a given X has the localization radius of L,

$$P(L, X) = \left\langle \sum_n \delta(L - L_n)\delta(X - w_n) \right\rangle, \tag{17}$$

We show this distribution calculated for CCA clusters in Fig. 8 and for random lattice gas (RLG) composites in Fig. 9.

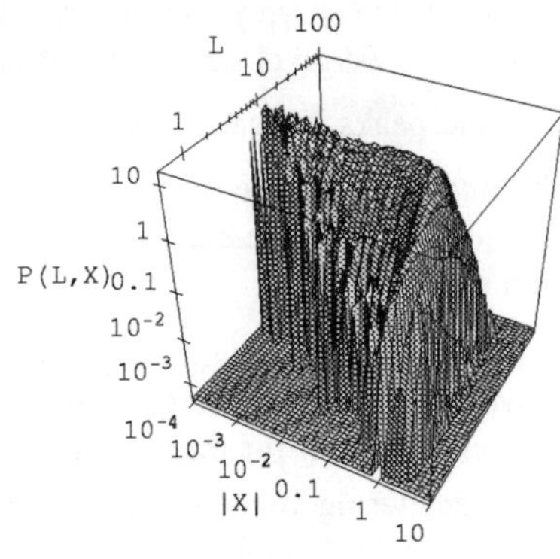

CCA

$N = 1500$

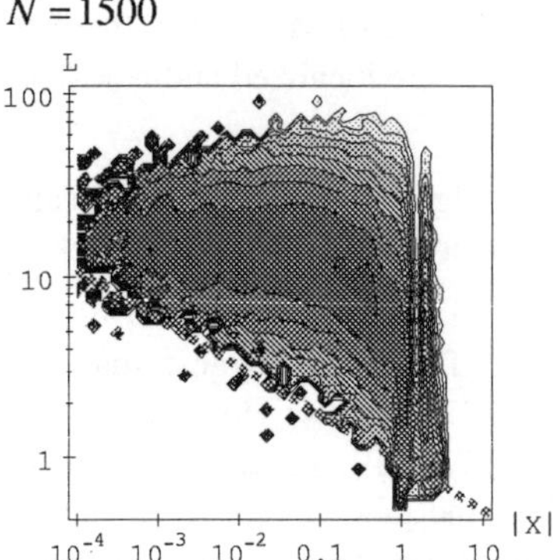

Fig. 8. Localization-length distribution $P(L, X)$ of eigenmodes for CCA clusters ($N = 1500$). The position of the lower X cutoff is qualitatively illustrated by the dashed bold line.

The most conspicuous feature of the distribution of Fig. 8 is its very large width. This width extends from almost the total size of the system R_c to some minimum cut-off size l_X that is a function of frequency $\omega(X)$. The cut-off is clearly seen in Fig. 8 where it is also indicated in the lower panel by a thick dotted line. The distribution width is so large that its characterization of the by a single dispersion relation L_X [see Eq. (16)] is absolutely insufficient. For most of the spectral region, the cut-off length l_X by magnitude is intermediate between the maximum and minimum scales, R_c and R_0. This, along with the self-similarity of the clusters, suggests that l_X scales with X, i.e., $l_X \propto |X|^\lambda$. Indeed, Fig. 8 supports the possibility of such a scaling with the corresponding index $\lambda \approx -0.25$. This illustrates general property of the inhomogeneous localization of eigenmodes for fractal (self-consistent) clusters and composites.

A different situation exists for non-fractal composites, as one can see with an example of the RLG shown in Fig. 9.

The distribution for $|X| \geq 0.1$ is similar to that of CCA (Fig. 8) characteristic of inhomogeneous localization. The major distinction from Fig. 8 is that the distribution in Fig. 9 shows the complete delocalization of the eigenmodes for $|X| \leq 0.01$ that appears in a narrow range. Such a delocalization is expected for the low-$|X|$ part of the spectrum, i.e., at frequencies close to the plasmon resonance of the individual

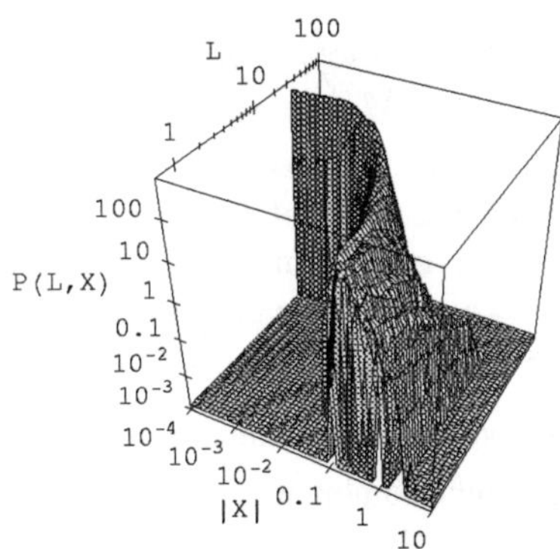

RLG

$N = 1500$

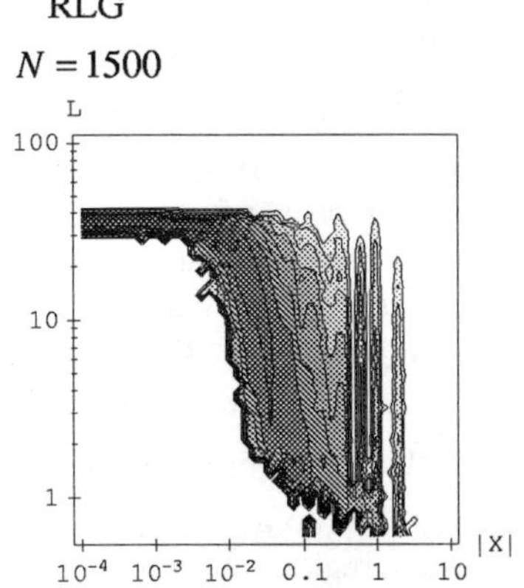

Fig. 9. Same as in Fig. 8 but for a random lattice gas (RLG) composite.

inclusions (monomers). In contrast, there is no such delocalization for fractal (CCA) clusters, as seen in Fig. 8.

6. Giant Fluctuations of Local Fields and Enhancement of Non-Radiative Photoprocesses

The picture of the intensities in any individual eigenmode (see Fig. 7) shows very large random changes of the intensity from one monomer to another, i.e., fluctuations in space. When a cluster is subjected to an external exciting radiation, its response is due to the excitations of eigenmodes. Therefore, we may expect that the eigenmode fluctuations will cause strong fluctuations of the local fields at individual monomers. We have investigated this phenomenon in Ref.21.

The local field at an ith monomer is expressed in terms of the Green's function (12),

$$E_{i\alpha} = Z \sum_{j\beta} \mathbf{G}_{i\alpha,j\beta} E_{\beta}^{(0)} \qquad (18)$$

Then the local field-intensity enhancement coefficient G_i for an ith monomer and the corresponding distribution function $P(G)$ are defined as

$$G_i = \frac{|\mathbf{E}_i|^2}{|\mathbf{E}^{(0)}|^2}, \quad P(G) = \left\langle \frac{1}{N} \sum_i \delta(G - G_i) \right\rangle \qquad (19)$$

We introduce also the nth moment of this distribution:

$$M_n = \langle G^n \rangle = \int P(G) G^n dG \qquad (20)$$

By its physical meaning, M_n is the enhancement coefficient of an nth -order non-radiative (i.e., without emissions of photons) nonlinear photoprocess. If, for instance, a molecule is attached to a monomer of a cluster, then M_n shows how many times the rate of its n -photon optical excitation exceeds such a quantity for an isolated molecule. A similar estimate is valid for the enhancement of a composite consisting of the nonlinear matrix and resonant inclusion clusters.

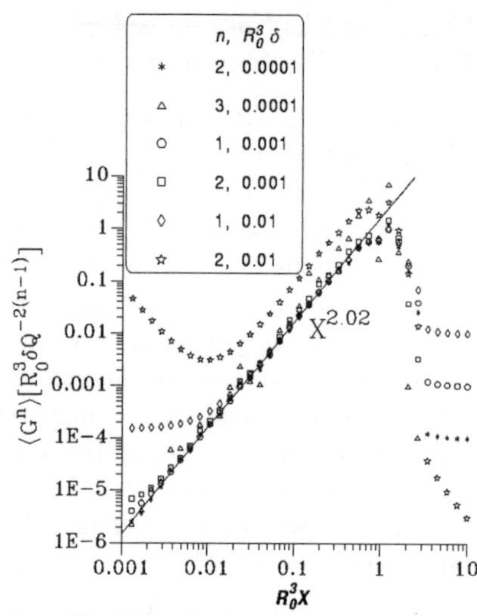

Fig. 10. Normalized enhancement factors $\left\langle G^n \right\rangle \delta Q^{-2(n-1)}$ as functions of X for CCA clusters for the values of δ and n shown.

The spectral dependencies of M_n for different combination of the degree of nonlinearity n and the dissipation parameter δ are shown in Fig. 10. The data in this figure are scaled by the factor $R_0^3 \delta Q^{-2(n-1)}$, with the resonant quality factor given by Eq. (6) The most remarkable feature of Fig. 10 is an almost perfect collapse of the data into a universal curve in an intermediate region of X for the case of very low dissipation ($R_0^3 \delta < 0.01$). Moreover, this curve is actually close to a straight line in the intermediate region indicating a scaling behavior of $M_n(X)$. We conjecture this scaling as the dependence $M_n \cong Q^{2(n-1)} M_1$.

For the first moment M_1 we have previously obtained[5,6] the *exact* relation $M_1 = \left(X^2 + \delta^2 \right) \mathrm{Im}\, \alpha / \delta$. Because the absorption $\mathrm{Im}\, \alpha$ does not scale in X (see Sec.4), the enhancement coefficient M_n should not scale either. However, the dependence $\mathrm{Im}\, \alpha(X)$ in the intermediate region of X is flat (see Fig. 3). Therefore, for $Q \gg 1$ the apparent scaling in X takes place with a trivial index of $2n$,

$$M_n \cong Q^{2n-1}|X|\mathrm{Im}\,\alpha = \frac{X^{2n}}{\delta^{2n-1}}\mathrm{Im}\,\alpha \propto X^{2n}, \tag{21}$$

in agreement with Fig. 10.

The major result of Ref. 21, given by Eq.(21) is that the excitation rate of a non-radiative nth -order nonlinear photoprocess in the vicinity of a disordered cluster is resonantly enhanced by a factor of $M_n \cong Q^{2n-1}$. This quantity can be understood qualitatively in the following way. For each of the n photons absorbed by a resonant monomer, the excitation probability (rate) is increased by a factor of $\cong Q^2$ (proportional to the local field *intensity*), therefore the total rate is increased by a factor of $\cong Q^{2n}$. However, the fraction of monomers that are resonant is small, $\cong Q^{-1}$. Consequently, the resulting enhancement factor is $M_n \cong Q^{2n-1}$, in agreement with Eq. (21).

For instance, for silver in the red spectral region $Q \cong 30$ (see, e.g., Ref. 22), so each succeeding order of the nonlinearity gives enhancement by a factor of $Q^2 \cong 1000$. We emphasize that the origin of this enhancement is the high-quality optical resonance in the monomers modified by the cluster. Among interesting effects related to the enhanced non-radiative excitation, we mention one, the selective photomodification of silver clusters.[23,24]

We have considered above the moments (averaged powers) of the local fields. Now we consider another characteristic of the fluctuations, the distribution function $P(G)$ (19) of the local-field intensity. Because the change of the minimum scale R_0 implies the change of the local fields, the distribution function is likely to scale in some intermediate range,

$$P(G) \cong G^{\varepsilon}, \tag{22}$$

Due to scale invariance, the index ε does not depend on the minimum scale R_0. Consequently, ε does not depend on frequency (the spectral parameter X) either.

A simple model that allows one to calculate the scaling index ε is the binary approximation.[5,6] In this approximation, the eigenmode is localized at only a pair of the monomers. In this case we have found[21] the enhancement factor G_i for a pair separated by a distance r , located at an angle Θ to the direction of the exciting field,

$$G_i(r,\Theta) = \frac{\delta^2 \sin^2 \Theta}{\left(X - r^{-3}\right)^2 + \delta^2} + \frac{\delta^2 \cos^2 \Theta}{\left(X + 2r^{-3}\right)^2 + \delta^2} \,. \tag{23}$$

Using this, we calculate the distribution function as

$$P(G) = \int \delta\left(G - G_i(r,\Theta)\right) C(r) d^3 r \,. \tag{24}$$

Here $C(r) \propto r^{D-1}$ is the density correlation function of the cluster. Taking into account those large values of G are of interest, we obtain from Eq. (24) that

$$P(G) \propto G^{-3/2} \tag{25}$$

Thus, in the binary approximation we obtain a *universal* scaling index of $-3/2$. Surprisingly, this index does not depend on the cluster's dimension D. Its value is determined merely by the vector nature of the fields.

It is interesting to compare both the scaling prediction and the calculated value of the index $\varepsilon = -3/2$ with the numerical results. These are shown in Fig. 11 for CCA clusters and for random walk (RW) clusters. The main feature is an unusually wide distribution. The local intensities are on the order of the exciting intensity ($G = 1$), as well as three orders of magnitude smaller or greater. This feature is referred to as giant fluctuations of the local field. The regions of high intensity are responsible for enhanced nonlinear responses. They have also been observed directly with the scanning photon-tunneling microscope.[25] The value of the index ε

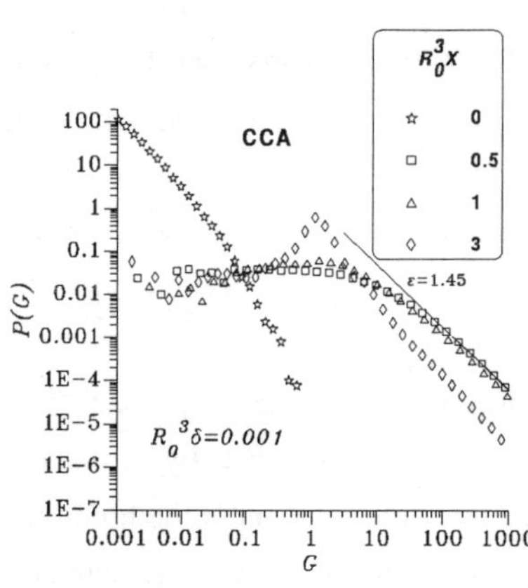

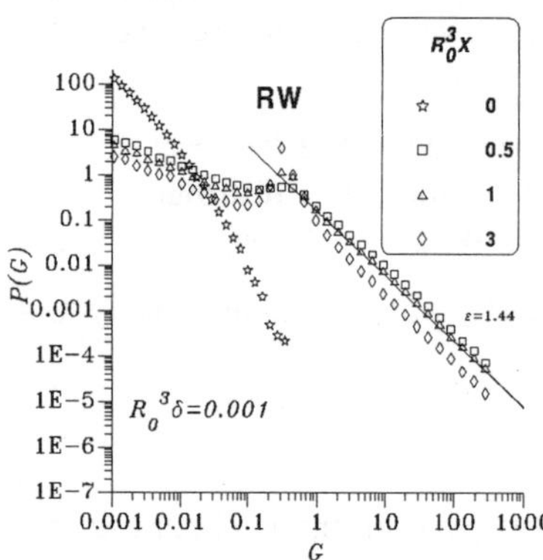

Fig. 11. Distribution function of the local field intensity $P(G)$ calculated for $R_0^3 \delta = 0.001$, for the values of X shown. The data on the upper panel are for CCA clusters and on the lower panel for random walk (RW) clusters.

is indeed almost independent from the frequency (parameter X), as expected from the scaling theory. Interestingly enough, these values ($\varepsilon = 1.45$ for CCA and $\varepsilon = 1.44$ for RW) are quite close to the prediction of the binary theory ($\varepsilon = -3/2$). This agreement is unexpected because there are no grounds to believe that the binary theory is applicable in a wide range of frequencies.

7. Chaos of Eigenmodes

The eigenmode equation (10) has the same form as the quantum-mechanical Schrödinger equation. In quantum mechanics it is not uncommon that highly-excited states or states of complex systems possess chaotic behavior (see, e.g., Refs. 26,27). Similar situation one may expect for eigenmodes of large disordered clusters and composites. The extreme sensitivity of the individual eigenmodes to a very small change of their frequency that is discussed in Sec. 5 and illustrated by Fig. 7 is a direct indication of such chaos. Even more than individual eigenmodes, statistical properties of chaotic eigenstates are of great interest. The giant fluctuations of the local fields discussed above in Sec. 6 provide one of the statistical descriptions. In this section we will consider spatial correlations of the chaotic eigenmodes.[28,29]

A principal property that distinguishes this problem from quantum-mechanical chaos is a long-range nature of the dipole-dipole interaction. A similar tight-binding problem of quantum mechanics (Anderson model) is usually formulated with only next-neighbor hopping. In studied quantum-mechanical problems, chaotic quantum states do not possess long-range spatial correlations.[27,30]

The long ranged interaction on one hand tends to induce the long-range spatial correlations. On the other hand, it may tend to establish a mean field, suppress fluctuations, and eliminate the chaos. As we demonstrate below in this section, either of those trends may dominate, depending on the system geometry and spectral region. We expect that chaos is the most pronounced in clusters and composites with fractal geometry. The rationale for it is the following. A mean field is established and spatial chaos is eliminated when the correlation range of eigenmodes exceeds a characteristic size of the density variations in the system. However, fractal (self-similar) geometry implies that the system repeat itself on all spatial scales and, consequently, there exists no such characteristic spatial scale. This is a prerequisite for the coexistence of chaos and long-range correlations.

To characterize the spatial correlations of eigenmode amplitudes, we introduce the amplitude correlation function (also called dynamic form factor),

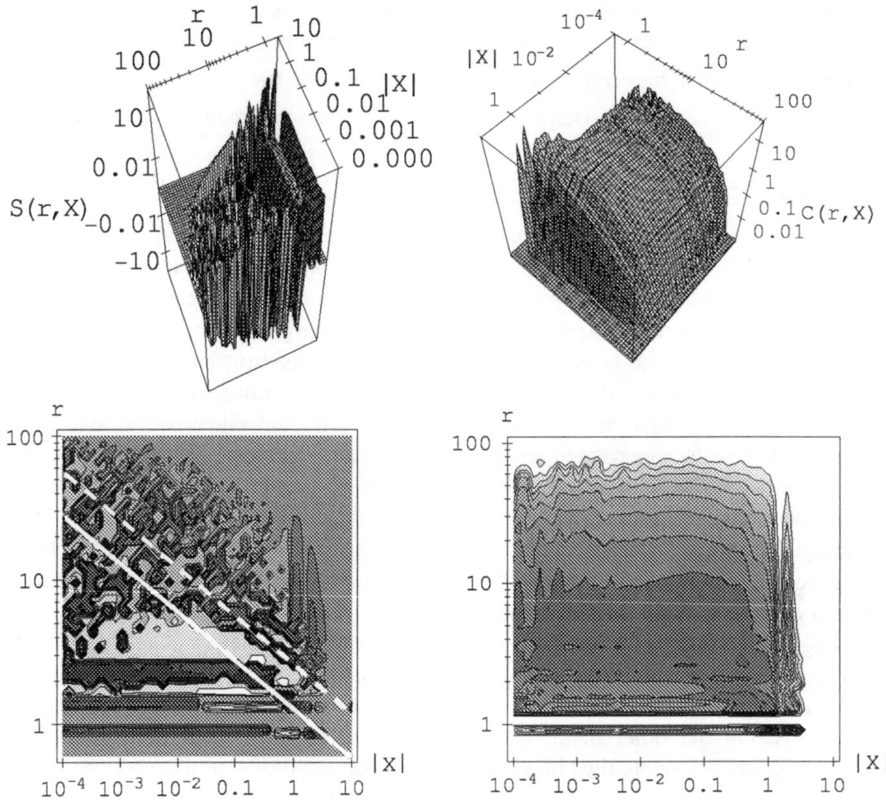

Fig. 12. Dynamic form factor $S(r, X)$ for CCA composite (the number of inclusions $N = 1500$, averaged over an ensemble of 300 composites. The upper panel shows a 3-dimensional representation of the function, while the lower panel is the corresponding contour map. The vertical scale is pseudo-logarithmic to show simultaneously positive and negative values of $S(r, X)$. To obtain it, a small region of the plot for $|S(r, X)| \leq 10^{-4}$ is removed. The function plotted is $\log[10^4 |S(r, X)|]\text{sgn}[S(r, X)]$. The horizontal scales are logarithmic.

Fig. 13. Intensity correlation function $C(r, X)$ calculated for CCA composite ($N = 1500$) averaged over ensemble of 300 systems. The upper panel is the function plotted in the triple-logarithmic scale and the lower panel is the corresponding contour map.

$$S_{\alpha\beta}(\mathbf{r}, X) = \frac{1}{\pi} \mathrm{Im} \sum_{i,j} G_{i\alpha, j\beta} \, \delta(\mathbf{r} - \mathbf{r}_{ij}) = \left\langle \sum_{n,i,j} (i\alpha|n)(j\beta|n) \, (\mathbf{r} - \mathbf{r}_{ij}) \, \delta(X - w_n) \right\rangle . \quad (26)$$

This function gives the correlation factor of amplitudes $(i\alpha|n)$ and $(j\beta|n)$ of two eigenmodes with polarizations α and β at two points separated by a spatial interval of $\mathbf{r}$ at an eigenvalue ("frequency") of X, averaged over ensemble of the systems. Similarly, we introduce a correlation function of the intensities of eigenmodes,

$$C(\mathbf{r}, X) = \left\langle \sum_{n,i,j} (i\alpha|n)^2 (j\beta|n)^2 \, \delta(\mathbf{r} - \mathbf{r}_{ij}) \, \delta(X - w_n) \right\rangle . \quad (27)$$

In a similar way, this function yields a correlation of two eigenmode intensities, $(i\alpha|n)^2$ and $(j\beta|n)^2$.

We have calculated[28,29] the above-defined correlation functions for an ensemble of CCA clusters (composites) with $N = 1500$ monomers (inclusions). The corresponding result for $S(\mathbf{r}, X) = \frac{1}{3} S_{\phi\beta}(\mathbf{r}, X)$ is shown in Fig. 12. We see a

developed pattern of irregular, chaotic correlations in almost the whole spatial-spectral region. The landscaped observed in Fig. 12 well deserves the name of a "devil's hill", where narrow regions of positive and negative correlations are interwoven, resulting in a turbulence-like pattern. This chaotic behavior is indeed a reflection of the chaos of individual eigenmodes. However, this chaos is in some sense stronger, because it is in a quantity averaged over an ensemble of statistically independent composites (consequently, it is fully reproducible). The deterministically chaotic pattern of Fig. 12 is likely to depend on a specific topology of the composite (CCA clusters). The straight lines shown in Fig. 12 are given by the binary-ternary approximation, see the discussion of Fig. 14 below.

To distinguish between fluctuations of phase and amplitude in the formation of the devil's hill in Fig. 12, we will compare it to the second-order correlation function $C(\mathbf{r}, X)$ shown in Fig. 13. We see that that this function differs dramatically from the dynamic form factor. The relief in Fig. 13 is very smooth, in contrast to Fig. 12. This implies that the devil's hill is formed due to spatial fluctuations of the phases of individual eigenmodes, while their amplitudes are smooth functions in space-frequency domain. Another important feature present in Fig. 13 is the long range of the correlation. The correlation decay in r is indeed weaker than any power.

As we discussed above in this section, self-similarity (fractality) of the CCA composite is of principal importance for the coexistence of chaos and long-range correlations of the eigenmodes. To emphasize this point, we show in Fig. 14 the results for the dynamic form factor $S(r, X)$ for RLG composites. We see that for extremely large eigenvalues, $|X| \geq 1$, the form factor $S(r, X)$ has a quasi-random structure that is reproducible under the ensemble averaging. This structure is similar to what is observed for CCA composites (cp. Fig. 12). In the middle of the spectral region, $1 \geq |X| \geq .003$, the form factor is dominated by a few branches of excitation. The strongest one is marked on the lower panel by a solid white line $X = -2r^{-3}$ that is the corresponding branch of binary approximation[5,6]. In this approximation, each eigenmode consists of two excited regions ("hot spots") and the form factor is given by

$$S(r, X) = \delta(r)\rho(X) + f(r)[\delta(X + 2r^{-3}) - \delta(X - 2r^{-3}) + 2\delta(X - r^{-3}) - 2\delta(X + r^{-3})],$$

$$(28)$$

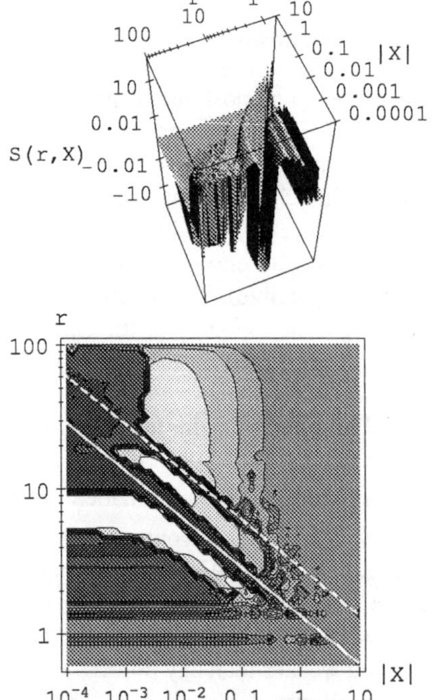

Fig. 14. The same as in Fig. 12, but for RLG composite.

where $f(r)$ is a smooth distribution of the inter-monomer distances, $f(r) = N\langle\delta(r - r_{ij})\rangle$. The weaker branch, marked by the dotted line $X = -(1 + \sqrt{57})r^{-3}$, is due to the ternary excitations, i.e., eigenvectors consisting of three hot spots.

At $X \approx 0.003$, we see a sharp transition to delocalization, where the positive-correlation region becomes uniform, spreading over most of the system. This transition occurs when the correlation length becomes comparable to the size of the system R_c, i.e. at $X \cong R_c^{-3} \propto N$. Consequently, this transition is mesoscopic, i.e., it phases out as size of the system becomes very large. This is in contrast to an Anderson transition where local density of scatters is the determining parameter. This difference from Anderson localization is due to the long-range interaction in our case. As a result, most of the eigenvectors are not propagating waves, but change from binary/ternary excitations to delocalized surface plasmons as $|X|$ decreases.

To briefly conclude this section, chaos of eigenmodes is present for both conventional and fractal geometries of the composites. However, fractal composites demonstrate this chaos in the whole spectral region, while conventional composites do only in the extremes of the spectrum. For conventional geometries (RLG composites in particular), the eigenmodes are dominated by binary/ternary excitations with a mesoscopic transition to uniform surface plasmons near plasmon resonance of the inclusion particles (monomers), i.e. at $|X| \to 0$. In the next section we will study the role of the fluctuations and chaos of eigenmodes on the nonlinear radiative photoprocesses.

8. Enhancement of Radiative Photoprocesses

Among enhanced radiative photoprocesses, one of the best studied experimentally is surface-enhanced Raman scattering (SERS).[9] The intensity of Raman scattering for molecules adsorbed at rough surfaces or colloidal-metal particles is known to be greatly enhanced, by a factor of up to 10^6. There are two limiting cases of SERS. In the first case, the Raman shift is very large, much greater than the homogeneous width of monomer's absorption spectrum. In this case, as we have shown[22] the exact expression of the SERS enhancement factor G^{RS} can be obtained as

$$G^{RS} = \frac{X^2 + \delta^2}{\delta} \operatorname{Im} \alpha \approx Q|X| \operatorname{Im} \alpha$$

(29)

This result predicts that the on the order of magnitude the enhancement is the same (by the factor of $\cong Q$) as for non-radiative photoprocesses of the second order. Qualitatively, we may interpret this result in the following way. For a large frequency shift, the outgoing photon is out of resonance and does not considerably interact with the system. Correspondingly, the entire enhancement is due to the local field of the absorbed photons, the same as for non-radiative processes [cp. the discussion after Eq.(21)].

Much more dramatic enhancement is found[22] for the second case, namely that of Raman frequency shifts smaller than monomer's linewidth. This situation is characteristic of the most typical and interesting cases of SERS. We have shown[22] that in this case there exists an approximate expression for the enhancement coefficient G^{RS} of SERS,

$$G^{RS} \approx \frac{1}{2} Q^3 R_0^3 |X| \operatorname{Im} \alpha = \frac{1}{2}\left(\frac{X}{\delta}\right)^3 R_0^3 |X| \operatorname{Im} \alpha$$

$$\tag{30}$$

In this case we have $G^{RS} \cong Q^3$, where the physical interpretation is the following.

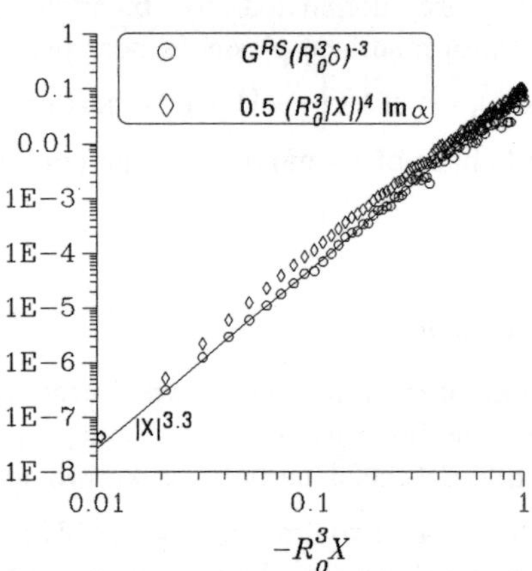

Fig. 15. Scaled enhancement coefficient of SERS from silver colloid clusters in comparison with the prediction of Eq.(30). The theoretical dependence is calculated for CCA clusters.

Enhancement by $\cong Q^2$ is due to the first power of the local-field intensity. Another power $\cong Q^2$ appears because the outgoing photon is not emitted freely, but in the resonant environment. Finally, one power of Q vanishes because only a small fraction $\cong 1/Q$ of the monomers is resonant to the exciting radiation.

We compare the functional dependence predicted by Eq. (30) with the numerical simulation in Fig. 15. As we see, this equation describes the numerically found dependencies quite well, especially the dependence on X.

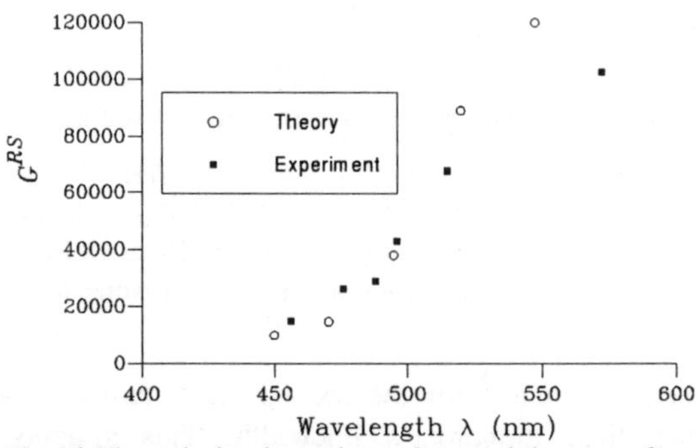

Fig. 16. Theoretical and experimental spectral (in terms of wavelength) dependencies of the enhancement coefficient G^{RS} for silver colloid clusters.

Equation (30) shows that the Raman scattering is enhanced for an adsorbed molecule by a factor of $G^{RS} \cong Q^3$. For noble metals in the red region of visible light, we have $Q \cong 30 - 100$, so the predicted enhancement is very large, $G^{RS} \cong 10^4 - 10^6$. This explains the range of enhancements found experimentally.[9] The comparison[22] of theoretical spectral dependence with the experiment is shown in Fig. 16. As we can see, the theory explains the experimental observations quite well.

Now, we will very briefly discuss coherent, or parametric, nonlinear photoprocesses. These processes are due to nonlinear wave mixing. One of the most interesting is frequency-degenerate four-wave mixing (a third-order process), responsible for such an interesting effect as phase conjugation. We have found the corresponding enhancement coefficient in Ref.10 in the binary approximation and Ref.31 numerically.

Physically, the enhancement of an nth -order parametric photoprocess can be estimated in the following simple way. The enhancement of the amplitude of each of the $n + 1$ participating photons (including the emitted photon) is $\cong Q$. This yields the enhancement of the amplitude of the process as $\cong Q^{n+1}$. For any coherent process, the amplitude, not probability, is the quantity to average. The averaging is done by multiplying by the fraction of resonant monomers, which is $\cong Q^{-1}$, leading to mean amplitude as $\cong Q^{n}$. Finally, the amplitude should be squared, yielding the enhancement coefficient $G^{(n)} \approx Q^{2n}$. More precise theory gives

$$G^{(n)} \approx Q^{2n} \left(X \operatorname{Im} \alpha \right)^2 . \tag{31}$$

For $n = 3$, this reduces to

$$G^{(3)} \approx \left(X/\delta \right)^6 \left(X \operatorname{Im} \alpha \right)^2 \tag{32}$$

in agreement with the corresponding result of Ref.31. The comparison of Eq. (32) to the numerical calculations shown in Fig. 17 supports this result.

A remarkable feature of Eq. (32) is that the predicted enhancement is quite large. For the realistic dielectric

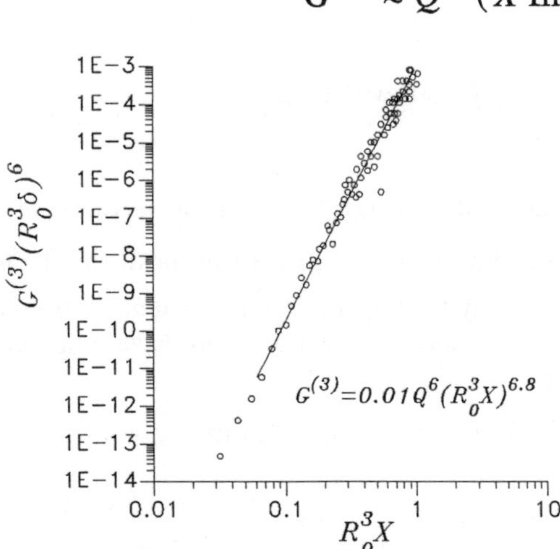

Fig. 17. Scaled enhancement coefficient for third-order degenerate parametric process, calculated for CCA.

parameters of silver in the visible range, $Q \cong 10-30$, and correspondingly $G^{(3)} \cong 10^4 - 10^7$. The experimental investigation[32] has indeed found a very strong enhancement for the phase conjugation, with $G^{(3)} \cong 10^6$, confirming the theoretical predictions.

The enhancement discussed above in this section is calculated for nonlinear-optical molecules adsorbed at the surface of monomers of a fractal cluster. Below we will consider another important system, namely Maxwell Garnett composite modeled as a RLG. It will be assumed for simplicity that only inclusion particles possess optical nonlinearity, while the host is optically linear. We will compute the nonlinear polarizability from the spectral theory and also compare it with the results of mean-field theory of Ref. 13.

For the linear polarization of exciting radiation, the third-order nonlinear (hyper)polarizability of an isotropic medium is completely characterized by one constant, $\chi^{(3)}$, that enters the expression for induction

$$\mathbf{D} = \varepsilon\mathbf{E} + \chi^{(3)}\left|\mathbf{E}\right|^2\mathbf{E}. \tag{33}$$

This hyperpolarizability for a composite, $\chi_c^{(3)}$, is expressed in terms of the corresponding hyperpolarizability for inclusion particles $\chi_p^{(3)}$ as (see Ref. 16 for the derivation of a similar formula)

$$\chi_c^{(3)} = \frac{1}{V\left|\mathbf{E}_0\right|^2\mathbf{E}_0^2}\int_V \chi_p^{(3)}\left|\mathbf{E}_1\right|^2\mathbf{E}_1^2 dV, \tag{34}$$

where $\mathbf{E}_1$ is a local (mesoscopic) field in the composite (i.e., a superposition of the average macroscopic field $\mathbf{E}_0$ and the dipolar fields of the inclusions), and V is a volume of the composite. Note that actually the integral in this equation is extended over the volume of the inclusions only, because the host does not have a nonlinear response.

The substitution of the local (mesoscopic) fields into Eq. (34) finally gives

$$\chi_c^{(3)} = \chi_p^{(3)} f |q|^2 q^2 \left\langle \left| \alpha_{\beta z}^{(i)} \right|^2 \left(\alpha_{\beta z}^{(i)} \right)^2 \right\rangle, \tag{35}$$

where f is a fill factor (a fraction of the total volume occupied by the inclusions) and

$$q = \frac{1}{R_m^3} \frac{3\varepsilon_h}{\varepsilon_p - \varepsilon_h}, \quad \alpha_{\beta z}^{(i)} = \sum_j \mathbf{G}_{i\beta, jz}, \tag{36}$$

where subscripts p and h denote the inclusion particles and the host, $\alpha_{\beta z}^{(i)}$ is a polarizability of an i th particle in the composite, $\mathbf{G}_{i\beta, jz}$ is the Green function (11) for the z polarization of the exciting light, and as above R_m is the radius of an inclusion particle. The expression of Eq. (35) should be compared to the corresponding result of a mean-field theory of Ref. 13,

$$\chi_c^{(3)} = \chi_p^{(3)} f |b|^2 b^2, \quad \text{where} \quad b = \frac{3\varepsilon_h}{(2 + f)\varepsilon_h + (1 - f)\varepsilon_p}. \tag{37}$$

We have calculated numerically the enhancement coefficient $g^{(3)} = \chi_c^{(3)} / \chi_p^{(3)}$ for the third-order susceptibility of a Maxwell Garnett composite in the RLG model using both the spectral theory (35) and the mean-field theory (37). The results are shown in Fig. 18. As one can immediately conclude, the mean-field theory completely fails to reproduce the results of the spectral theory that is exact within the framework of the dipole approximation. The underlying cause of this failure is in strong fluctuation and chaos of the eigenmodes and local fields discussed throughout this paper. The degree of the failure of the mean-field theory for nonlinear processes is much more dramatic than for linear polarizability (cp. Fig. 5 with Fig. 18). It is understandable, because nonlinear processes are much more sensitive to fluctuations of the local fields.

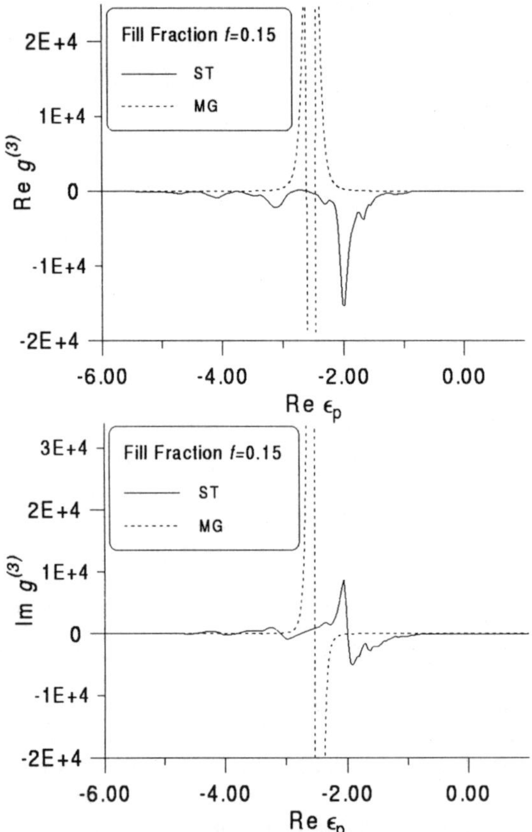

Fig. 18. Real (upper panel) and imaginary (lower panel) parts of the enhancement coefficient for RLG composite. The argument ε_p is shown relative to ε_h. The results of the present spectral theory are indicated by ST and those of the mean-field theory by MG (referring to Maxwell Garnett).

9. Concluding Remarks

We have considered a variety of the photoprocesses mediated by disordered clusters and composites. We employed models with both fractal (self-similar) and conventional geometries to simulate different existing systems.

The remarkable feature of the disordered clusters and composites is giant fluctuations of the local fields that brings about their enhanced nonlinear-optical responses. Namely, the enhancement is due not only to the high averaged value of the local fields, but principally due to their fluctuations in space, from one monomer (inclusion particle) to another. Nonlinearity, causing a higher power of the local fields to be an acting parameter, enhances the effects of these fluctuations. Therefore, the enhancements are found to increase dramatically with the order of nonlinearity.

The chaos of the local fields bears many similarities to quantum chaos, but differs due to the long range of the dipole interaction. Not only individual eigenmodes are chaotic, but also their spatial correlation factors. These chaos and fluctuations are responsible for the dramatic failure of the mean-field theory to describe the nonlinear polarizability of composites.

Finally, many of the effects predicted have been verified experimentally, and we have mentioned some of them. However, many very interesting effects are not discussed due to the space limitations. Among them, we recognize the enhanced laser production of plasma by colloidal-metal clusters.[33]

References

1. J.C. Maxwell Garnett, Philos. Trans. R. Soc. London **203**, 385 (1906).
2. H.A.Lorentz, *Theory of Electrons* (Dover, New York), 1952.
3. D.A.G.Bruggeman, Ann. Phys. (Leipzig), **24**, 636 (1935).
4. V. M. Shalaev and M. I. Stockman, ZhETF **92**, 509 (1987) [Translation: Sov. Phys. JETP **65**, 287 (1987)].
5. V. A. Markel, L. S. Muratov and M. I. Stockman, ZhETF **98**, 819 (1990) [translation: Sov. Phys. JETP **71**, 455 (1990)].
6. V. A. Markel, L. S. Muratov, M. I. Stockman, and T. F. George, Phys. Rev. B **43**, 8183 (1991).
7. M. I. Stockman, L. N. Pandey, L. S. Muratov and T. F. George, Phys. Rev. B **51**, 185 (1995).
8. M. I. Stockman, L. N. Pandey and T. F. George, Phys. Rev. B **53**, 2183 (1996).
9. M. Moskovits, Rev. Mod. Phys. **57**, 783 (1985).
10. A. V. Butenko, V. M. Shalaev and M. I. Stockman, ZhETF **94**, 107 (1988) [translation: Sov. Phys. JETP, **67**, 60 (1988)].
11. T.A.Witten and L.M.Sander, Phys. Rev. Lett. **47**, 1400 (1981).
12. G. L. Fischer, R. W. Boyd, R. J. Gehr, S. A. Jenekhe, J. A. Osaheni, J. E. Sipe and L. A. Wellerbrophy, Phys. Rev. Lett. **74**, 1871 (1995).
13. J. E. Sipe and R. W. Boyd, Phys. Rev. B **46**, 1614 (1992).
14. K. Ghosh and R. Fuchs, Phys. Rev. B **38**, 5222 (1988).
15. R. Fuchs and F. Claro, Phys. Rev. B **39**, 3875 (1989).
16. D.J.Bergman and D.Stroud, In: *Solid State Physics,* v.**46**, p.148 (Academic Press, Boston, 1992).
17. S. Alexander, Phys. Rev. B **40**, 7953 (1989).
18. V. M. Shalaev, R. Botet and A. V. Butenko, B **48**, 6662 (1993).
19. M. I. Stockman, L. N. Pandey, L. S. Muratov and T. F. George, Phys. Rev. Lett. **75**, 2450 (1995).
20. M.I.Stockman, L.N.Pandey, and T.F.George, Phys. Rev. B 53(5), 2183 (1996).
21. M. I. Stockman, L. N. Pandey, L. S. Muratov and T. F. George, Phys. Rev. Lett. **72**, 2486 (1994).

22. M.I.Stockman, V.M.Shalaev, M.Moskovits, R.Botet, and T.F.George, Phys. Rev. B **46**, 2821 (1992).

23. A. V. Karpov, A. K. Popov, S. G. Rautian, V. P. Safonov, V. V. Slabko, V. M. Shalaev and M. I. Stockman. , Pis'ma ZhETF **48**, 528 (1988) [Translation: JETP Lett. **48**, 571 (1988)].

24. Yu. E. Danilova, A. I. Plekhanov and V. P. Safonov, Physica A **185**, 61 (1992).

25. D. P. Tsai, J. Kovacs, Z. Wang, M. Moskovits, V.M.Shalaev, J.S.Suh and R. Botet, Phys. Rev. Lett. **72,** 4149, 1994.

26. F.J.Dyson, J. Math. Phys. (N.Y.) **3**, 1189 (1962).

27. M.V.Berry, J. Phys. A **10**, 2083 (1977).

28. M.I.Stockman, Phys. Rev. Lett. **79**, 4562 (1997).

29. M.I.Stockman, Phys. Rev. E **56**, 6494 (1997).

30. V.I.Fal'co and K.B.Efetov, Phys. Rev. Lett. **77**, 912 (1996).

31. V. M. Shalaev and M. I. Stockman, Physica A **185**, 181 (1992).

32. A. V. Butenko, P. A. Chubakov, Yu. E. Danilova, S. V. Karpov, A. K. Popov, S. G. Rautian, V. P. Safonov, V. V. Slabko, V. M. Shalaev and M. I. Stockman, Z.Phys. D **17**, 283 (1990).

33. M. M. Murnane, H. C. Kapteyn, S. P. Gordon, J. Bokor, E. N. Glytsis and R. Falcone, Appl. Phys. Lett. **62**, 1068 (1993).

ATOMIC VALENCES IN APERIODIC CRYSTALS STUDIED BY THE BOND VALENCE METHOD

SANDER VAN SMAALEN

Laboratory of Crystallography, University of Bayreuth,
D-95440 Bayreuth, Germany[a]

The bond valence method is a semi-empirical method for studying the strength of bonds and the atomic valences in crystals of inorganic compounds. A review is presented of the bond valence method, and of the possibilities to extent this method to aperiodic crystals. Applications to various compounds are given, including incommensurately modulated crystals and incommensurate composite crystals.

1   Introduction

The valence of an atom can be considered as the amount of chemical bonding this atom is involved in. Going back to the ideas of Pauling,[1,2] this amount of bonding is shared by the chemical bonds between the central atom and the atoms in its coordination polyhedron. In simple "well behaved" compounds (e.g. NaCl) the coordination of each atom is well defined, and a valence can be assigned to each bond, equal to the atomic valence divided by the coordination number.[1]

The bond valence method is an extension of these ideas, and it is based on two simple concepts.[3,4,5] Firstly, it is assumed that atoms have definite valences, and that these valences can be obtained as the sum of contributions of bond valences,

$$V_i = \sum_j \nu_{ij}. \tag{1}$$

V_i is the valence of atom i and ν_{ij} is the bond valence of the atom pair (i, j). Secondly, it is assumed that there is a definite relationship between the length d_{ij} of a bond and its bond valence, given by

$$\nu_{ij} = exp[(R_{ij} - d_{ij})/b]. \tag{2}$$

The bond valence parameter R_{ij} only depends on the chemical identity of the atoms. The parameter $b = 0.37$ Å is a universal constant.[6]

Extensive calculations have shown that a consistent set of bond valence parameters can be derived from the vast amount of structures in the inorganic structure database.[6,7,8] The parameters R_{ij} are transferable between

---

[a]E-mail: smash@uni-bayreuth.de

compounds and between crystallographically independent bonds within the same structure. Thus, once determined, they can be used to analyze other structures or test the feasibility of prospective structures for new compounds. Theoretically it can be shown, that Eqs.1 and 2 provide a reasonable description of the cohesive forces in inorganic solids, both for compounds with ionic bonds and for covalent compounds.[9]

Incommensurately modulated crystals and incommensurate composite compounds lack 3-dimensional (3D) translational symmetry.[10] They are characterized by the presence of an infinite number of different coordinations of the atoms, contrary to periodic crystals, where the number of different coordinations is restricted to the number of atoms in the asymmetric part of the unit cell. The bond valence method analyses coordination polyhedra, thus it can provide a useful tool to study incommensurate structures.[11,12] However, it is then to be established that it is meaningful to do so.

In Section 2 the essential structural features of aperiodic crystals will be introduced. The bond valence method for incommensurate crystals will be presented in Section 3. In the second half of this paper, a series of examples will be given. The main purpose is to show that the concept of atomic valence is still valid in aperiodic structures. Furthermore, it will be shown that the bond valence method can be used to distinguish different oxidation states of the same element; that it can be used to calculate the inter-layer interaction in inorganic misfit layer compounds; and that it can be used to study the varying valence state in charge density wave compounds.

2   Incommensurate Structures

In the present paper we will consider incommensurately modulated crystals and incommensurate intergrowth compounds. Together with quasicrystals, they represent the three types of aperiodic crystals presently known.

In first approximation, the structure of an incommensurately modulated crystal is periodic, and the coordinates of the atoms can be given with respect to the average lattice $\{\vec{a}_1, \vec{a}_2, \vec{a}_3\}$, according to

$$\bar{x}_i(\mu) = L_i + x_i^0(\mu), \tag{3}$$

for $i = 1, 2, 3$. L_i are integers representing the unit cell, and $x_i^0(\mu)$ are the coordinates of atom μ with respect to the unit cell.

The true positions of the atoms in the modulated structure can be given as the sum of the average position and a modulation function. The latter is periodic, with a period given by a modulation wavevector which is incommensurate with respect to the average lattice:

$$\vec{q}^{\,j} = \sigma_{j1}\vec{a}_1^{\,*} + \sigma_{j2}\vec{a}_2^{\,*} + \sigma_{j3}\vec{a}_3^{\,*}, \tag{4}$$

where $j = 1, \cdots, d$ allows for more than one modulation wavevector. The matrix σ contains the components of the modulation vectors with respect to the basis $(\vec{a}_1^{\,*}, \vec{a}_2^{\,*}, \vec{a}_3^{\,*})$ of the reciprocal lattice. Each row of the σ-matrix contains at least one irrational entry.

The so-called superspace description considers the $j = 1, \cdots, d$ arguments of the modulation functions as additional coordinates.[13] Restricting the description to a single modulation $(d = 1)$ then gives 4 coordinates characterizing the structure:

$$\bar{x}_4(\mu) = t + \vec{q}^{\,1} \cdot \left(\vec{L} + \vec{x}^{\,0}(\mu) \right) \tag{5}$$

$$x_i(\mu) = \bar{x}_i(\mu) + u_i^{\mu}(\bar{x}_4). \tag{6}$$

The parameter t can be interpreted as the initial phase of the modulation functions. It is also the section of (3+1)D superspace describing one particular representation of the structure in 3D physical space. The other symbols have the same meaning as in Eqs. 3 and 4. The periodicity of the modulation is given by $\vec{u}(\bar{x}_4) = \vec{u}(\bar{x}_4 + 1)$. Due to the incommensurateness, the same type of atom has different values for the modulation functions in different unit cells (different $\vec{L}$). Alternatively, these different positions can be described by different values for t. The periodicities of the modulation functions then determine that all different values for $\vec{x}(\mu)$ are contained in the range $0 \leq t < 1$. This property of the superspace description allows a concise representation of the displacements of the atoms out of their average positions as a continuous function of t. Only the arguments in the unit interval need to be considered. The same applies to functions of the atomic positions. As an example Fig.1a gives the distances between Nb atoms in NbTe$_4$ and the eight Te and two Nb atoms in the first coordination shell.

Incommensurate composite crystals require two lattices to describe their structures. They can be considered as the coherent intergrowth into a single thermodynamic phase of two incommensurately modulated structures. The average unit cell of one subsystem is incommensurate with the unit cell of the other subsystem. The interaction between the subsystems determines that each subsystem is modulated with periodicities given by the average unit cell of the other subsystem. The physical realization of such crystals can be envisaged in, for example, layered crystals, where layers of different chemical composition alternate, and each type of layer can have its own intra-layer lattice constants. All presently known substances have two reciprocal periodicities common to the

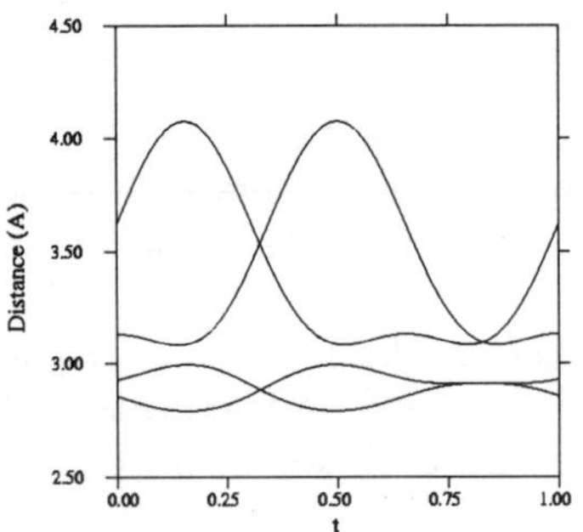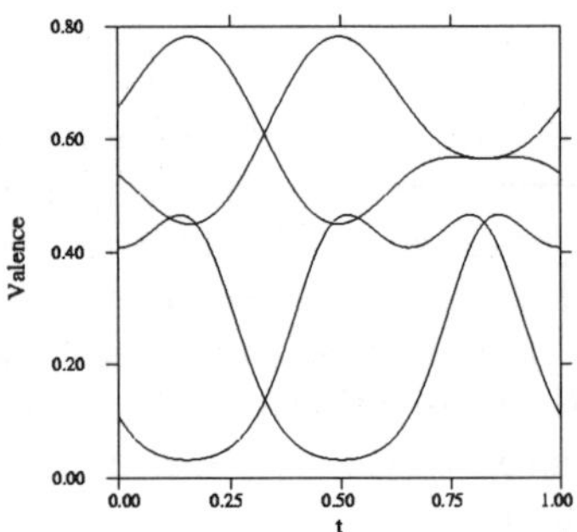

Figure 1: The coordination of Nb atoms in the incommensurately modulated structure of NbTe$_4$ as a function of the phase coordinate t. (a) Interatomic distances between Nb and eight coordinating Te and two Nb atoms. (b) The corresponding bond valences as a function of t.

two subsystems, and most of the intergrowth compounds can be characterized by a single additional period, i.e. they can be described in $(3+1)$D superspace.

The subsystems have incommensurately modulated structures, and the atomic coordinates can be described in the same way as for modulated crystals, whereby each subsystem ν has its own average lattice $\Lambda_\nu = \{\vec{a}_{\nu 1}, \vec{a}_{\nu 2}, \vec{a}_{\nu 3}\}$. Similarly, the coordinates Eqs. 3 through 6 will be given an additional subscript ν. However, this simple description does not give the relation between the structures of the two subsystems. That is provided by the fact, that the 3 average periodicities and the d modulation wavevectors of each subsystem can be obtained as integer linear combinations of a single set of $3 + d$ reciprocal vectors $M^* = \{\vec{a}_1^*, \vec{a}_2^*, \vec{a}_3^*, \vec{a}_4^*\}$. Each subsystem is characterized by an integer $(3 + d)x(3 + d)$ matrix W^ν to derive Λ_ν^* and $\vec{q}^\nu$:

$$\vec{a}_{\nu i}^* = \sum_{k=1}^{3+d} W_{ik}^\nu \vec{a}_k^* \tag{7}$$

$$\vec{q}^\nu = \sum_{k=1}^{3+d} W_{4k}^\nu \vec{a}_k^* \tag{8}$$

The matrices W^ν applied in direct space give the coordinates of the atoms in the different subsystems

$$\bar{x}_{\nu i}(\mu) \;=\; L_{\nu i} \;+\; x^0_{\nu i}(\mu) \;-\; W^\nu_{i4}t \tag{9}$$

$$\bar{x}_{\nu 4}(\mu) \;=\; W^\nu_{44}t \;+\; \vec{q}^{\,j}\cdot\left(\vec{L}_\nu + \vec{x}^{\,0}_\nu(\mu)\right) \tag{10}$$

$$x_{\nu i}(\mu) \;=\; \bar{x}_{\nu i}(\mu) \;+\; u^\mu_{\nu i}(\bar{x}_{\nu 4}). \tag{11}$$

It is noted that the phases of the modulation functions in both subsystem, as well as the average structure coordinates depend on a single phase-parameter t. It is the presence of one such parameter instead of different parameters, which represents the inter relation between the subsystems within one compound. Like for modulated crystals, atomic displacement and inter-atomic distances can now be displayed as a function of t. Complications arise due to the dependence of the basic coordinates on the parameter t, and the different periodicities of the modulation functions in the two subsystems. The distance between an atom of subsystem 1 and an atom of subsystem 2 becomes infinite for $t \to \pm\infty$. Periodicity is recovered, when we consider the t-dependence of the distances between one atom of subsystem 1 and all atoms of subsystem 2, and vice versa. As an example the structure of the inorganic misfit layer compound $(LaS)_{1.14}NbS_2$ is considered. One parabolic-like curve in Fig.2a represents the t-dependence of the distance of one pair of atoms. The collection of such curves for a central atom from subsystem 1 corresponds to a different periodicity in t (Fig.2b, as the collection of curves for a central atom from subsystem 2 (Fig.2a). Combining Fig.2a with the interatomic distances within the LaS subsystem then provide a complete description of the coordination of La; both functions have the same periodicity in t. Similarly Fig.2b combined with distances within the NbS_2 subsystem provides the complete description of the coordination of S of subsystem 1. It are these combinations, which can be used in the bond valence method.

3   The Bond Valence Method

The principles underlying the bond valence method have been presented in the Introduction. The definite bond valence / bond length relation (Eq. 2) can be applied directly to incommensurate structures. For a particular pair of atoms the bond valence becomes a function of t, and it can be displayed in the same way as the distances themselves. As an example, the counterparts of Figures 1a and 2a are given in Fig. 1b and Fig. 3, respectively.

A property of the t-plots is the correlation between different curves. If the coordination shell around a particular atom is considered, the distances computed at a single value for t correspond to this coordination shell somewhere

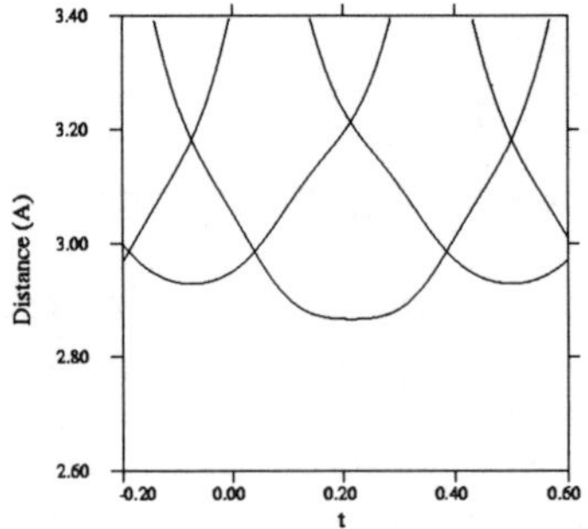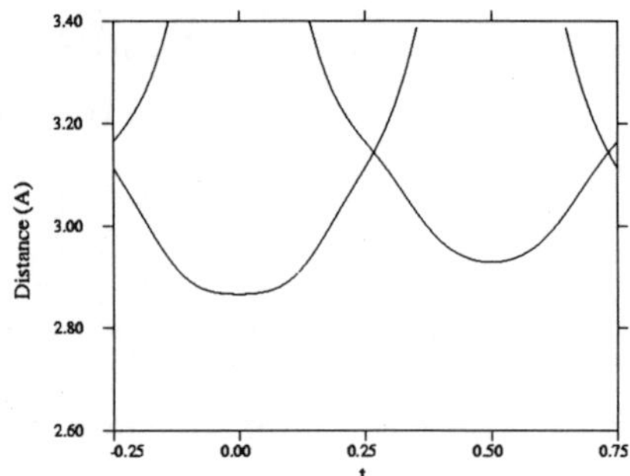

Figure 2: Interatomic distances as a function of the fourth coordinate t for the inorganic misfit layer compound $(LaS)_{1.14}NbS_2$. The distances are shown between La in subsystem 2 and S in subsystem 1 for (a) La as as the central atom with periodicity 0.57 in t; and (b) S as the central atom with periodicity 1 in t.

in the crystal. All possible coordinations are contained within one period on the t-axis, and the t-plots form a concise way to display the various environments of atoms, as occur throughout the structure. Considering all atoms in the coordination shell (e.g. Fig.1b), then allows the generalization of Eq.1 by adding the bond valences at constant t, and then displaying the result as a function of t (Fig.4a). Functions like these summarize the possible valences of a particular type of atom, as occur anywhere in the structure.

The valence of the central atom usually is obtained from the contributions of the bond valences towards atoms bonded to the central atom. The first step in the application of the bond valence method then is to determine which atoms belong to the first coordination shell, and, therefore, should be included in the bond valence sum (Eq.1). For many structures the first coordination shell can identified unambiguously, and it are these compounds which have been used to derive a consistent set of bond valence parameters R_{ij}.[6,7,8] For complicated periodic structures, it is difficult to determine the atoms that are to be included into the bond valence sum, and some arbitrariness is introduced into the method. For incommensurate composite crystals a first coordination shell is not defined. The environments of the symmetry equivalent copies of a particular central atom involve distances to other atoms which range

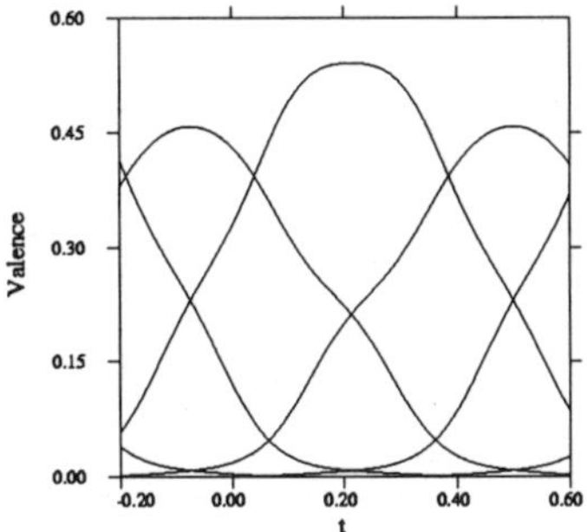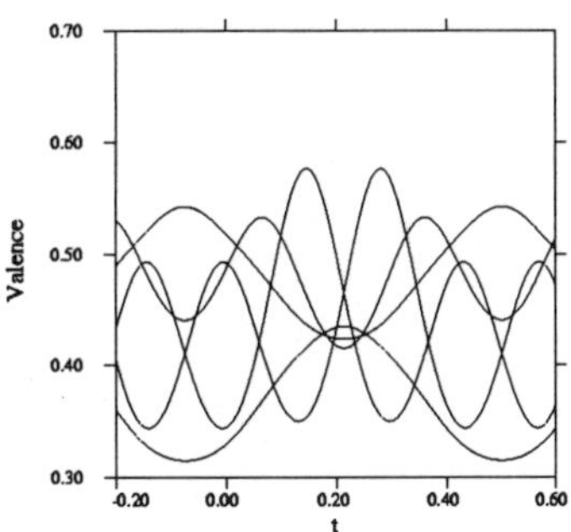

Figure 3: Coordination of La in the inorganic misfit layer compound $(LaS)_{1.14}NbS_2$. Bond valences are shown as a function of t for the S atoms around La. (a) For the S atoms from the NbS_2 subsystem. (b) For the S atoms from the LaS subsystem. (After Ref. 12.)

continuously from a certain minimum value to infinity (Fig.2). To resolve this problem it was proposed that all atoms in the crystal should be included in the sum Eq.1, and the observation was made that this represents only a small correction to the bond valence sums for the "well behaved" structures.[12] The rapid convergence of the sum determines that only atoms up to about 5 Å need to be included into Eq.1. Of course, a new set of bond valence parameters needs to be derived then, but it is a good first approximation to use the set of parameters derived with the conventional approach.

Theoretically, it can be questioned, whether atoms 5 Å or more apart do form a partial chemical bond, i.e. participate in defining the valences of these atoms. On the other hand, it is known that the decomposition of the crystal cohesion energy into contributions from chemical bonds is an approximation itself. Taking into account the (small) contributions from atoms farther apart then can be considered as a different approximation to the cohesion energy of the crystal.

With these modifications a consistent bond valence method has been obtained for incommensurate structures. However, questions about the transferability of bond valence parameters, the interpretation of the results, and the possible uses of the method must to be answered at new. The first application of the bond valence method is to predict bond lengths in inorganic compounds.[14,4,15] That can be done, when the bond valence method is extended by a third principle, known as the equal valence rule: [15] "Each atom

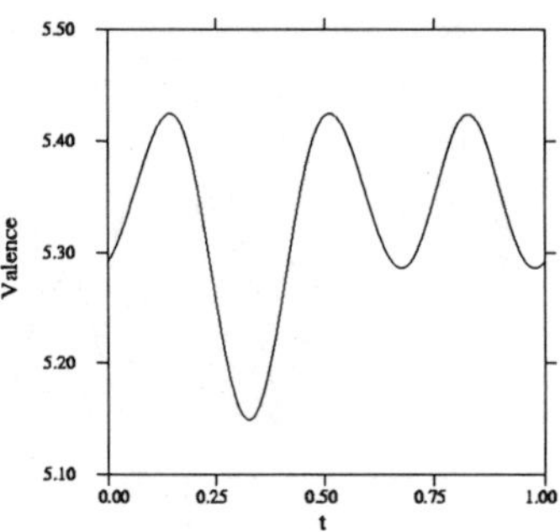
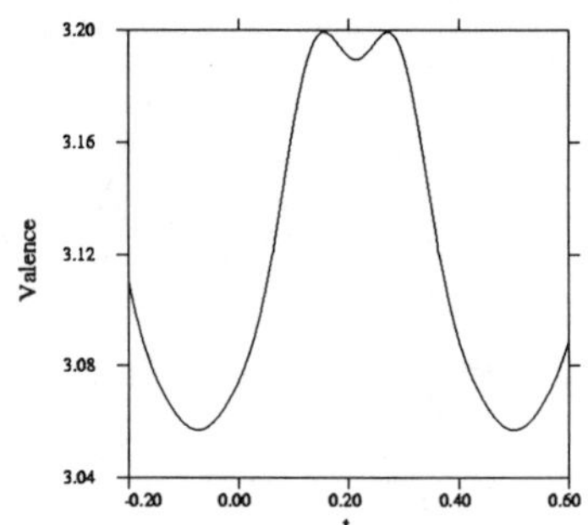

Figure 4: The valence of a central atom as a function of t, obtained by adding the bond valences from Fig.1 and Fig.3. (a) Valence of Nb in $NbTe_4$. (b) Valence of La in $(LaS)_{1.14}NbS_2$. (After Ref. 12.)

shares its valence as equally as possible among the bonds it forms." This principle works fine for the "well behaved" structures, but it essentially fails for more complicated structures.[15] The bond valence method in this form cannot predict the presence of bonds of different length between atoms of the same chemical type. Because, incommensurate structures always contain a range of values for the bonds between atoms of the same chemical type, it is not useful to try to predict bond length in these structures.

The second application of the bond valence method relies on Eqs.1 and 2 only, and it has been denoted as the description of structures. For a known structure, the bond valences and atomic valences can be computed from Eqs.1 and 2 combined with the set of bond valence parameters derived from the inorganic structure database.[6,7,8] The result can then be used in various ways:[4]

1. To check the accuracy of a structure;

2. As an aid in structure determination, e.g. to distinguish between atoms with similar scattering power but different valences, or to locate the most probable position of a light atom in an inorganic structure;

3. To distinguish between different oxidation states of transition metal atoms.

Part 1 is based on the principle that one valence value is expected for atoms of a particular element, and a deviation from this rule is taken as an indication for possible errors in the structure. For the examples presented here, it will be shown that in incommensurate crystals there is a tendency towards a constant valence for all atoms of one type throughout the crystal. To some extent, the limited quality of the description of the modulation functions will be responsible for the remaining variation, and it can be concluded that the constant valence rule also applies to aperiodic crystals. Examples will be given of the application of (2) and (3).

4   Inorganic Misfit Layer Compounds

Inorganic misfit layer compounds belong to the class of incommensurate composite crystals. They can be characterized by the alternate stacking of two types of layers of different chemical composition and with different structures (Fig.5).[16] The first subsystem ($\nu = 1$) comprises of layers TX_2, with T a transition metal atom Nb, Ta, V, Cr, or Ti, and with X equal to S or Se. For Nb and Ta, a single layer is isostructural to one layer of NbS_2 (M in trigonal prismatic coordination). The metal atom is in octahedral coordination for V, Cr and Ti. The second subsystem ($\nu = 2$) corresponds to layers MX, with M one of the p-block metals, Sn, Pb, Bi or Yb, or a rare earth metal. The structure corresponds to a two atom thick (1 0 0) slab of the NaCl-type structure. The periodicities parallel to the layers are independent for the two types of layers, resulting in an incommensurate composition $(MX)_x TX_2$. x depends on the particular compound, and usually is in the range $1.1 < x < 1.3$. Actually, one reciprocal axis parallel to the layers is common to the subsystems, as well as is the reciprocal axis $\vec{c}^*$ perpendicular to the layers. There is a single incommensurate direction (taken as the $\vec{a}$-axis), and all misfit layer compounds can be described in (3+1)D superspace.[17]

Apart from the fascinating non-periodic structural order, these compounds are interesting for their physical properties.[16] Metallic and semiconducting behavior have been observed, and some of the misfit layer compounds become superconducting at low temperatures. The MX_2 compounds have been studied extensively for their feasibility to intercalate atoms or molecules, and for the properties related to the low-dimensional character of the conduction band residing on the plane of M atoms. The misfit layer compounds provide a large range of compounds with different properties in which this intercalation can be analyzed. A substantial part of the studies on the misfit layer compounds have been aimed at elucidating the valence and charge state of the constituting atoms, as well as measuring and calculating their electronic band structures. Problems include the amount of charge transfer between the two types of layers; the possibility of vacancies in the sublattice of M atoms;[18] and the relation between valence of the atoms, their charge state, and the electrical properties

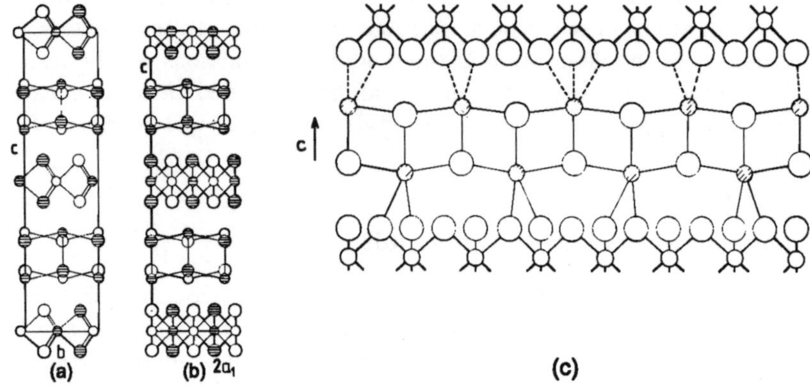

Figure 5: The basic structure of $(LaS)_{1.14}NbS_2$ in the projection along (a) the incommensurate axis; and (b) the commensurate in-plane axis. (c) is an extended view of (b) showing the variation in local environment of the La atoms. Small circles are metal atoms; large circles are sulfur atoms. Hatched and open circles differ by 0.5 in the projected coordinate.

of the misfit compounds. The bond valence method has provided part of the answer to these questions, and the results of the application of this method to misfit layer compounds will be reviewed here.

First the valences of the atoms are considered. As discussed in Section 3, the atomic valences can be obtained as a function of t, and all values possible in the structure are contained within one period on the t-axis. The result for the valence of La in $(LaS)_{1.14}NbS_2$ was given in Fig.4b,[12] and similar plots can be obtained for the other misfit layer compounds with T equal to a rare earth metal atom. Despite the fundamental aperiodic character, an effectively constant valence was found. Furthermore, the calculated value $(La^{3.1+})$ was only slightly larger than the expected value of 3. The small difference can be explained by the problem that re-normalized bond valence parameters were not available, and that the published values have been used (Section 3). These observations incited the conclusions that:

- the local environments in incommensurate crystals vary in such a way that there is a definite valence for each atom;

- the valence state of atoms is the same in aperiodic and periodic crystals.

A further understanding can be obtained by analyzing the individual bond valences. Several partial La to S bonds exist between the subsystems, and it was shown that the constant valence is achieved, because the variations of the valences of the different bonds compensate each other (Fig.3). Dependent on the position in the crystal (the value of t), the coordination of La can be described approximately as varying between 6 and 7, where the higher coordination number corresponds to weaker individual bonds.

Based on the idea that the bond valence is a quantitative measure for the amount of bonding between atoms, incommensurate composite crystals can be analyzed further by considering the partial sums of bond valences for bonds between the subsystems and for bonds within a single subsystem. Again concentrating on the valence of the M atom, the misfit compounds could be classified into three types (Table 1).[19] Compounds with M a rare earth atom have the valence 3 for M, which is composed of a partial valence of about 2 within the MX subsystem and a partial valence of about 1 due to bonding towards the other subsystem. For M is the p-block metal Sn or Pb the valence is slightly over 2, with only a weak inter-layer bonding. The p-block elements Bi and Sb again show weak inter-layer bonding. The total valence of about 3 is now achieved through M-M bonds within the MX subsystem, as does occur due to superstructure formation with respect to the rock salt type of structure. The importance of these metal bonds and the validity of the constant valence rule was demonstrated most spectacularly in $(SbS)_{1.15}(TiS_2)_n$.[20,21] In these compounds $(n = 1, 2)$ the MX subsystem exhibits an incommensurate super-structure with respect to the rock salt type of structure, and only a restricted set of harmonics of the occupation modulations and displacement modulations of Sb and S could be determined. Despite a coordination shell continuously varying between none and a full Sb-Sb bond, the valence of Sb was found to be close to 3, while the average valence of Sb was equal to 3.1 (Fig.6). A single valence state for Sb, while there are two different types of coordination, was recently confirmed experimentally by Mossbauer spectroscopy.[22] The valence of 3 for the rare earth atoms, and of 2 for Pb and Sn has been found experimentally by photoelectron spectroscopy (XPS and UPS), X-ray absorption spectroscopy (XAS), and electron energy loss spectroscopy (EELS). [18,23,24,25,26,27,28]

For periodic crystals, it has been argued that the valence obtained by the bond valence method include both ionic and covalent bonds. Whereas for ionic compounds the valence can be identified with the charge of the ions, it is the amount of bonding when applied to covalent compounds.

One property of the misfit compounds is charge transfer from M towards the d-band on the T atoms in the TX_2 layers. [16,18,19,24] However, this does not imply an ionic type of bonding between the layers. The bond valence method shows that the interlayer interaction is provided by the bonds between M and X of TX_2 (Table 1), which will be mainly of covalent character. The following model evolves for the interlayer bonding. Covalent bonds exist between M and X of TX_2. With M contributing 1 or 2 electrons and X contributing

Table 1: Average values for the valences of the metal atoms M in inorganic misfit layer compounds. The valence has contributions from bonds within its own subsystem and of bonds towards the other subsystem.

	Own subsystem	Other subsystem	Total
$(LaS)_{1.14}NbS_2$ [29]	2.20	0.93	3.13
$(PbS)_{1.18}TiS_2$ [30]	1.96	0.29	2.21
$(BiSe)_{1.09}TaSe_2$ [19]	2.74	0.39	3.13
$(SbS)_{1.15}TiS_2$ [20]	2.92	0.21	3.13
$(SbS)_{1.15}(TiS_2)_2$ [21]	2.86	0.25	3.11

2 electrons, the antibonding orbitals contain electrons too, and the resulting bonding interaction is weak. The bond strength is increased by charge transfer from the antibonding orbitals towards the d-band on the T atoms of the TX_2 subsystem. For M is a rare earth, strong interlayer bonding thus is obtained, while for the p-block elements only weak bonds exist, in full accordance with the results of the bond valence method.

5   Charge Density Wave Compounds

For the inorganic misfit layer compounds it was shown that the valences of the metal atoms have a definite value for each element, despite the dramatic variations of their coordination shell (Section 4). When one element is present in two different valences (e.g. like Fe in Fe_3O_4), the bond valence method can determine the correct valence of each atom in the unit cell. The question then arises, whether the bond valence method may be used to determine valence fluctuations in incommensurate crystals, where, in principle, the complete range of valence values between some minimum and some maximum value may be realized. The so-called Charge Density Wave (CDW) compounds form a class of materials on which these ideas can be tested.[31] The CDW state is characterized by two intimately connected components. The conduction electrons order into areas of higher and lower electron density according to a wave with a wavevector equal to twice the Fermi wavevector. The atoms are displaced according to a wave with the same wavelength.[32] It will be shown here that the bond valence method does not provide evidence that the modulation of the density of conducting electrons would correspond to valence fluctuations. On the contrary, the calculations give a strong indication that there is a tendency

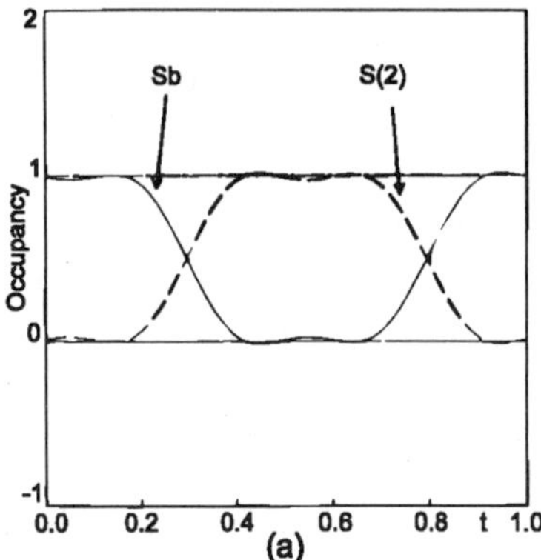
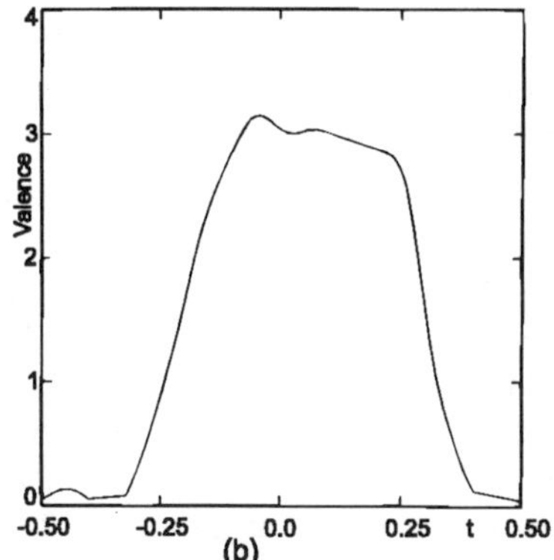

Figure 6: The inorganic misfit layer compound $(SbS)_{1.15}TiS_2$. (a) Occupation probability of Sb and S as a function of the phase parameter t_2 corresponding to the modulation within the SbS subsystem. (b) Valence of Sb multiplied by the value of the occupation modulation function. (From Ren et al.[20].)

of all atoms of the same type towards a constant valence.

5.1 $NbTe_4$

The high-temperature structure of $NbTe_4$ consists of chains of equally spaced Nb atoms parallel to the tetragonal axis. Each Nb atom is surrounded by eight Te atoms in a square anti-prismatic coordination.[33] Different columns are connected through Te - Te bonds, such that the formal valences are $Nb^{+4}(Te_2^{-2})_2$. On the chains of Nb atoms resides a partially filled quasi-one-dimensional electron band, which has been made responsible for the occurrence of a CDW state. At room temperature the structure is incommensurately modulated. The largest displacements describe a longitudinal wave along the chains of Nb atoms, with displacements of the Te atoms following those of the Nb atoms.[34] Bond valences were calculated as a function of t for the 8 coordinating Te atoms, and the 2 nearest Nb atoms around a single Nb atom.[10] It followed that the variation of the bond valence of a single bond ($\delta\nu = 0.33$) was larger than the variation of the valence of Nb ($\delta V = 0.29$), as was obtained by summing all 10 contributing bond valences (Figs. 1b and 4a). It was concluded that the modulation in $NbTe_4$ adapts itself in such a way, that the valence of Nb is constant throughout the structure.

5.2   AuTe$_2$

Valence fluctuations of Au were proposed as the origin for the incommensurately modulated structure of the mineral AuTe$_2$ (calaverite).[35] The basic structure is a distorted Cd(OH)$_2$ type structure, with Au in distorted octahedral coordination by Te. The largest modulation amplitude is 0.36 Å along $\vec{b}$ for Te. Despite this large amplitude, contributions of the second and third harmonic to the modulation functions were found to be small.

In the basic structure, Au is coordinated by 2 Te atoms at a distance of 2.67 Å and by 4 Te atoms at a distance of 2.97 Å. The modulation hardly affects the two short distances, while the four longer distances attain a variation with t between a minimum of 2.68 Å and a maximum of 3.23 Å. This was interpreted as the variation between a square planar coordination and a linear coordination. Comparing with other gold compounds, it was concluded that this large variation in local environment corresponded to a variation between Au^{3+} and Au^{1+}.[35]

The variation of the coordination of Au throughout the structure is better analyzed by the bond valence method than by studying inter atomic distances. Corroborating the interpretation of Schutte and De Boer,[35] at one extremum ($t = 0.25$) a square planar coordination is found, with four Au - Te bonds of valence of about $\nu \approx 0.5$ and the other two Te atoms only weakly interacting with $\nu \approx 0.1$ (Fig. 7a). However, a linear coordination is not found. At the other extremum ($t = 0$), two bonds are strong ($\nu \approx 0.5$), but the interaction with the other four Te atoms cannot be neglected. Each of these four bonds is about half the strength of the strongest bond. Because there are four weaker and only two stronger bonds, both types are of equal importance in determining the valence state of Au.

Evidence for valence fluctuations is not provided by the bond valence method. Adding up the contributions of the 6 Te atoms in the first coordination shell of Au, a valence is obtained which varies between 1.81 and 2.16. Again, it is found that the variation of the valence ($\delta V = 0.35$) is less than the variations of the bond valences ($\delta \nu = 0.37$) of each of the 4 longer Au - Te bonds. The structure is not accurate enough to determine whether the remaining variation is a real effect. In any case, the bond valence method does not give valences +1 and +3 for the Au atoms, but results in a much smaller remaining variation. Again, a tendency is found towards a constant valence for all atoms of the same type.

Calaverite exists for a certain range of compositions Au$_{1-x}$Ag$_x$Te$_2$, where silver replaces gold. The crystal analyzed by Schutte and De Boer[35] corresponded to $x = 0.09$. The occupation probability of Ag at the Au sites was

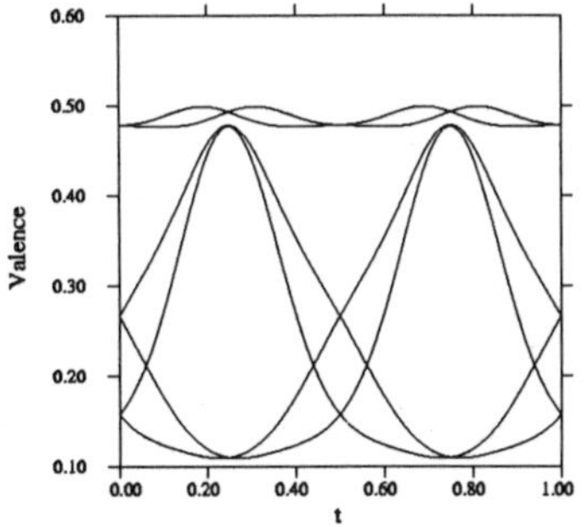 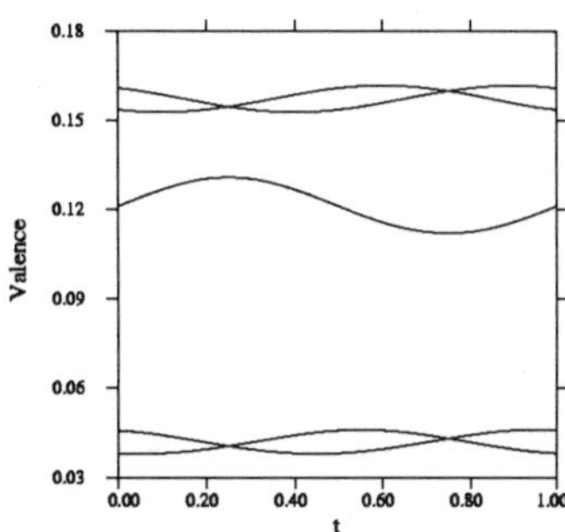

Figure 7: Bond valences as a function of t. (a) For the six bonds between Au and its coordinating Te atoms in $AuTe_2$. (b) For the ten bonds between K and its surrounding O in $K_{0.3}MoO_3$.

found to be modulated between $x = 0.18$ and $x = 0.0$. It was observed that this modulation correlates with the modulation of the local environment of the Au site, where minimum silver occupation was found for the sites with square planar coordination. This was taken as evidence for the presence of Au^{1+}, as Ag^{1+} would only substitute for Au^{1+}, and linear coordination is preferred by silver. The bond valence method shows that there is no site of linear coordination, but that it really should be considered as distorted octahedral coordination. Then the correlation with the modulation of the occupation probability cannot be explained by a straightforward argument. Additional complications arise, because the bond valence parameter for the Ag - Te pair ($R = 2.51$) is larger than for the Au - Te pair ($R = 2.41$).[7] With this larger parameter the valence of Ag at the "linearly coordinated" site is calculated even larger than 1.81 as was found for Au. Elucidation of the role of Ag could come from structure refinements of calaverite with varying amounts of silver.

5.3   *NbSe$_3$ and NiTa$_2$Se$_7$*

The basic structure of $NbSe_3$ is built of infinite chains of Nb atoms in triangular prismatic coordination by Se. The prisms are linked through shared triangular faces. Neighboring columns are displaced over one half prism length (one half of the b-axis), such that each Nb is in 8-fold coordination by Se. The room temperature structure of $NbSe_3$ is periodic, and it contains three independent columns (I, II, and III) with three independent Nb atoms (Nb1, Nb2, and Nb3, respectively).

At low temperatures, two independent CDW's are present. The structure determination showed that the so-called q^1-type modulation resides on the type-III columns, while the q^2-type modulation resides on the type-I columns.[36] Nb2 does not participate in the modulations. This relation between CDW's and basic structure was predicted by Wilson,[37] based on a model for the valences of the atoms and the resulting fillings of the quasi-1D electron bands. From the differences in their coordinations, it was derived that Nb1 and Nb3 should be less positive than Nb2, resulting in an empty 3d conduction band on the Nb2 chains and partially filled bands on the Nb1 chains and the Nb3 chains. The partially filled quasi-one-dimensional bands then give rise to the CDW formation on these columns.

The bond valence method applied to the non-modulated, room-temperature structure showed only a small difference between the three Nb atoms, with $V(\text{Nb3}) = V(\text{Nb1}) = 5.28$, and $V(\text{Nb2}) = 4.90$. The valence of Nb2 thus was found to be smaller than that of Nb3 and Nb1, contrary to the theoretical model, which predicted a more positive ionic charge for Nb2. It should be noted that this difference does not depend on the precise values of the bond valence parameters, because only Nb - Se bonds were taken into account, and because, due to symmetry, a possible contribution of Nb - Nb interactions is the same for all three chains.[38] This immediately poses the question about the meaning of valence. It certainly is likely that the three Nb atoms differ in the filling of the quasi-one-dimensional electron band, as follows from band structure calculations.[39] However, the lesser amount of conduction electrons on Nb2 apparently is overcompensated by more covalent bonding. The bond valence method adds up all types of bonding, whether it be covalent, ionic or metallic type of bonding.

The variation of the valence of Nb3 and Nb1 with the phase of the modulation was studied too. Like in the other examples, a strong tendency towards a constant valence is observed.

$NiTa_2Se_7$ has a CDW transition similar to $NbSe_3$, but with larger modulation amplitudes of the atoms. The same tendency towards a constant valence of all atoms was found in this compound.[40]

5.4   Blue bronze

Molybdenum bronzes, $A_xMo_yO_z$ (A = metal atom), have structures based on corner sharing and edge sharing MoO_6 octahedra, with the A atoms occupying the interstitial sites between corner sharing octahedra.[41] The blue bronze

Table 2: The bond valence method applied to the modulated structure of blue bronze. Given are the valences of the metal atoms averaged over t $(\bar{V})$; the coordination number; The difference between maximum and minimum valence (Max-Min); and the sum of variations of the individual bond valences $(\Sigma(\text{Max-Min})_{ij})$.

Atom	Coordination number	$\bar{V}$	Max-Min	$\Sigma(\text{Max-Min})_{ij}$
Mo1	6	5.88	0.00	0.24
Mo2	6	5.82	0.13	0.38
Mo3	6	5.84	0.10	0.50
K1	10	1.04	0.01	0.11
K2	10	1.16	0.02	0.06

$K_{0.3}MoO_3$ has a CDW transition at $T_{CDW} = 180$ K, towards an incommensurately modulated structure at low temperatures. The modulated structure was determined by Schutte and De Boer,[42] who also analyzed the bond strength according to Zachariasen for the Mo atoms in the basic structure. For the three crystallographically independent Mo atoms they found valences of 5.83, 5.69, and 5.65, respectively. These values were scaled such that their average value matched the formal valences of the atoms, as can derived directly from the composition: $K_{0.3}^{1+}Mo^{+5.7}O_3^{2-}$. It was proposed that the CDW reflects small valence fluctuations on the Mo atoms, between 5.6 and 5.8.[42]

Application of the bond valence method to the modulated structure again shows that the variations of the bond valences with t of the individual bonds compensate each other to a large extent (Table 2). Furthermore, the three crystallographically independent Mo atoms have almost the same valence values. Absolute values are slightly higher than expected, due to the fact that bond valence parameters for room temperature have been used, and distances tend to be smaller at lower temperature. The very small dependence of the valence of K1 on t is remarkable, despite 10 contributing partial bonds, each of which with a larger or smaller variation with t (Fig. 7). Schutte and De Boer proposed valence fluctuations for the atoms Mo2 and Mo3.[42] In view of the large uncertainties for the modulation parameters of the O atoms, and the observed compensating effect of the variations of the individual bond valences, this conclusion is not supported by the bond valence method. Whether or not the CDW causes a (small) variation of the valences of the atoms throughout the structure, awaits a more accurate description of the modulated structures.

6   Transition Metal Oxides

Compounds in the solid solution field $(Ta_2O_5)_{1-x}(WO_3)_x$ can be synthesized for a large range of x values. Detailed structure refinements have been reported for $x = 0.1$, $x = 0.14$, and $x = 0.267$.[43,44] Dependent on the value of x, these compounds are commensurate or incommensurate composite crystals. The first subsystem has composition $Ta_{1-x}W_xO$, with lattice parameters $a = 6.19$Å, $b = 3.67$Å, and $c = 3.89$Å (Orthorhombic Cmmm). The second subsystem comprises of O atoms only, with lattice parameters a, $b_O = 2.25$Å (for $x = 0.1$), and c. The b_O lattice parameter, and therefore the oxygen content, depends on the value of x.

Refinements for the compound with $x = 0.267$ showed a variation of the Ta/W occupancy ratio with t (Fig.8a), while the small difference in scattering power of Ta and W precluded an accurate determination of the substitutional modulations at smaller x values. Displacement modulation parameters could be determined for all atoms and compositions, while taking into account one set of parameters for the metal atom. The resulting variation of the valence of the metal atom site was large (Fig.8b; note that the bond valence parameters are equal for Ta - O and W - O bonds). This was explained by the different valences for Ta atoms (5+) and W atoms (6+): maximum valence for the metal site was found for those values of t where maximum occupancy for W occurred.[43,44] As the latter still reflects only 50% occupancy, disorder remains for the Ta/W substitution. It was proposed that there are oxygen vacancies around the Ta atoms on these sites, thus reducing the valences of Ta towards 5.

Unlike the other examples, a large residual variation of the valence of the metal sites with t was found. This variation could be explained by the incommensurate substitutional order of the Ta and W atoms, which have different valences.[43,44] Taking into account the variation of the occupational probability, it is again found that throughout the incommensurate structure the same atom has a single value for its valence.

7   Organometallic Crystals

Originally devised for inorganic compounds, the bond valence method can also be applied to organometallic compounds, in order to describe the interaction between the metal atom and ligand atoms of the type O, S or Se.

$(ET)(SCN)_2Hg_{0.776}$ is an incommensurate composite crystal. The modulated structure was determined by Coppens et al.[11] One subsystem is formed by the mercury atoms, and the other subsystem contains $(ET)(SCN)_2$. Primary contacts between the subsystems are between Hg atoms and the S atoms

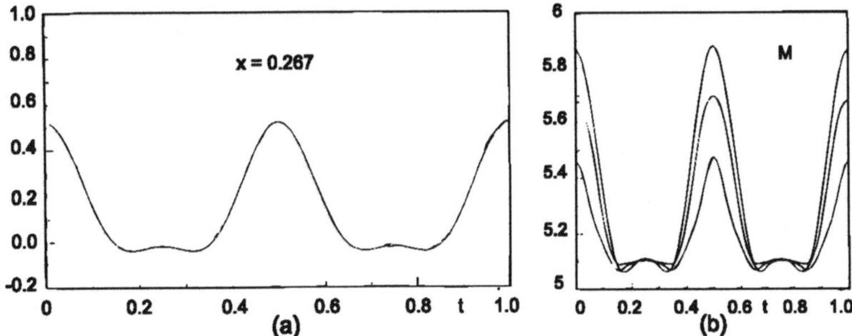

Figure 8: The composite crystal structures of $(Ta_2O_5)_{1-x}(WO_3)_x$. (a) Occupation probability of W for $x = 0.267$. (b) Valences of the M site for $x = 0.267$, $x = 0.14$, and $x = 0.1$, from top to bottom, respectively. (From Schmid et al.[44].)

of SCN. Taking into account these interactions, the valences of the Hg atoms were calculated as a function of t, both for the basic structure and for the structure including the modulation. The effect of the modulation was to reduce the variation of the valence from 0.95 to 0.57, and to reduce the average value from 3.48 to 2.2, close to the expected value of 2 for mercury (Fig.9a).[11]

(Perylene)Co(mnt)$_2$ has an incommensurately modulated structure. Relative large amplitudes were found for Co (0.8 Å) and for the S atoms of the maleonitrile-dithiolate (mnt) groups (0.6 Å).[45] The coordination of Co was found to vary between 6-fold and 5-fold. The corresponding valences did show an average value of about 3, but a variation of 0.6 remained. This was interpreted as an indication for a varying valence state of Co, in accordance with electronic band structure calculations.[46]

The bond valence calculations for both organometallic compounds showed a larger residual variation of the valence of the metal atom, than was determined for the incommensurate inorganic compounds. It is not possible to derive reliable conclusions from this observation, because the applicability of the bond valence method to organometallic compounds is not established. However, a tendency towards a constant valence for each type of atom is found again for these compounds.

8 Conclusions

Attempts have been made to employ the bond valence method to predict structures of inorganic compounds, and various degrees of success have been achieved.[14,4,15] It is inherent to this method, that bonds of different length for a single type of pair of atoms cannot be predicted correctly. A varying coordi-

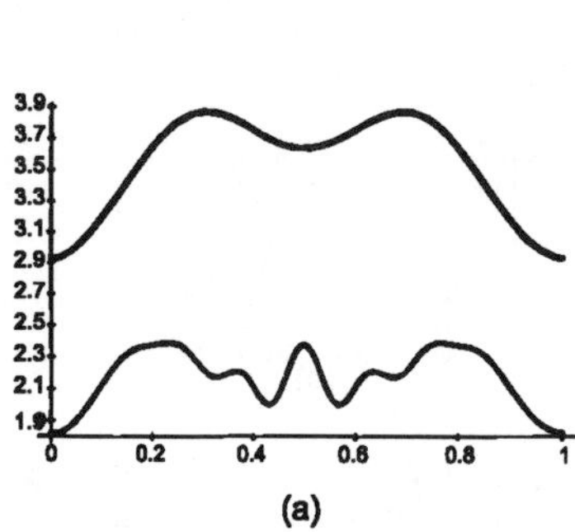
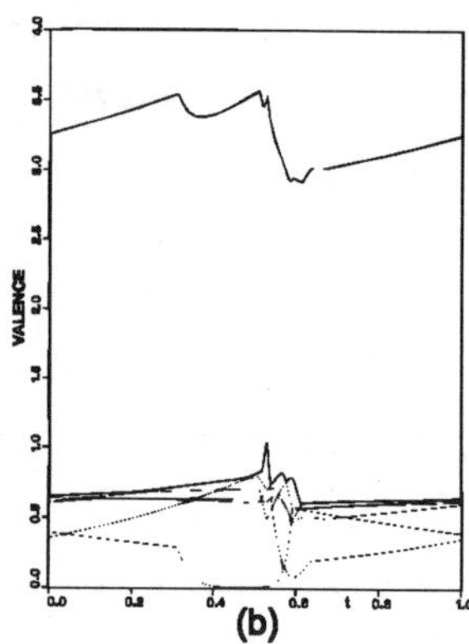

Figure 9: (a) The valence of Hg in $(ET)(SCN)_2Hg_{0.776}$ as a function of t for the basic structure (top) and for the modulated structure (bottom) (from Coppens et al.[11]); (b) Valence of Co and the 6 Co - S bond valences in $(Perylene)Co(mnt)_2$ as a function of t (from Lam et al.[45]).

nation shell is a essential characteristic of incommensurate crystals. Therefore, the bond valence method cannot be used to predict structures of incommensurate crystals. In particular, it cannot be used to predict whether a compound will crystallize with a periodic, a modulated, or a composite crystal structure.

Periodic and aperiodic crystals alike rely on the same types of interactions between atoms. Therefore, it is to be expected that the bond length / bond strength relation Eq.2 applies to both types of crystals, and no evidence of the contrary has been obtained. By analyzing a series of examples, it was shown that the constant valence rule applies equally well to aperiodic crystals (Sections 4 through 7). Generally, the modulations or the composite character do not correspond to a varying valence state, but a constant valence is achieved for each chemical element, despite large variations in the coordination shell. The limited accuracy of the modulation functions does not allow to determine whether the remaining variation is a structural effect or would disappear with a better description of the modulation functions.

When a variation of the valence with the incommensurate phase parameter t was found, it could be interpreted as due to a varying valence state of the atoms (Section 6).

References

1. L. Pauling, *J. Am. Chem. Soc.* **51**, 1010 (1928).
2. L. Pauling, *J. Am. Chem. Soc.* **69**, 542 (1947).

3. I.D. Brown, in *Structure and bonding in crystals II*, eds. M. O'Keeffe and A. Navrotsky (Academic Press, New York, 1981).

4. M. O'Keeffe, *Structure and Bonding* **71**, 161 (1989).

5. I.D. Brown, *Acta Crystallogr.* B **48**, 553 (1992).

6. I.D. Brown and D. Altermatt, *Acta Crystallogr.* B **41**, 244 (1985).

7. N.E. Brese and M. O'Keeffe, *Acta Crystallogr.* B **47**, 192 (1991).

8. M. O'Keeffe and N.E. Brese, *Acta Crystallogr.* B **48**, 152 (1992).

9. V.S. Urusov, *Acta Crystallogr.* B **51** 641 (1995).

10. S. van Smaalen, *Crystallogr. Rev.* **4**, 77 (1995).

11. P. Coppens *et al.*, *J. Am. Chem. Soc.* **113**, 5087 (1991).

12. S. van Smaalen, *Acta Crystallogr.* A **48**, 408 (1992).

13. T. Janssen *et al.*, in *International Tables for Crystallography Vol. C*, ed. A.J.C. Wilson (Kluwer Academic Publishers, Dordrecht, 1992).

14. I.D. Brown, *Acta Crystallogr.* B **33**, 1305 (1977).

15. I.D. Brown, *Acta Crystallogr.* B **53**, 381 (1997).

16. G.A. Wiegers, *Prog. Solid State Chem.* **24**, 1 (1996).

17. S. van Smaalen, *Mater. Sci. Forum* **100-101**, 173 (1992).

18. J. Rouxel *et al.*, *Inorg. Chem.* **33**, 3358 (1994).

19. V. Petricek *et al.*, *Acta Crystallogr.* B **49**, 258 (1993).

20. Y. Ren *et al.*, *Acta Crystallogr.* B **51**, 275 (1995).

21. Y. Ren *et al.*, *Acta Crystallogr.* B **52**, 389 (1996).

22. J.P. Espinos *et al.*, *Chem. Mater.* **9**, 1393 (1997).

23. Y. Ohno, *Phys. Rev.* B **44**, 1281 (1991).

24. Y. Ohno, *J. Phys.: Condens. Matter* **4**, 7815 (1992).

25. K. Suzuki, T. Enoki and S. Bandow, *Phys. Rev.* B **48**, 11077 (1993).

26. A.R.H.F. Ettema *et al.*, *Phys. Rev.* B **49**, 10585 (1994).

27. C.M. Fang *et al.*, *J. Phys. Chem. Solids* **58**, 1103 (1997).

28. L. Cario *et al.*, *Phys. Rev.* B **55**, 9409 (1997).

29. S. van Smaalen, *J. Phys.: Condens. Matter* **3**, 1247 (1991).

30. S. van Smaalen *et al.*, *Acta Crystallogr.* B **47**, 314 (1991).

31. C. Schlenker *et al.* (eds.) *Physics and chemistry of low-dimensional inorganic conductors* (Plenum Press, New York, 1996).

32. J.P. Pouget, in *Physics and chemistry of low-dimensional inorganic conductors*, eds. C. Schlenker *et al.* (Plenum Press, New York, 1996).

33. H. Bohm, *Z. Kristallogr.* **180**, 113 (1987).

34. S. van Smaalen *et al.*, *Acta Crystallogr.* B **42**, 43 (1986).

35. W.J. Schutte and J.L. de Boer, *Acta Crystallogr.* B **44**, 486 (1988).

36. S. van Smaalen *et al.*, *Phys. Rev.* B **45**, 3103 (1992).

37. J.A. Wilson, *Phys. Rev.* **19**, 6456 (1979).

38. S. van Smaalen, J.L. de Boer and P. Coppens, *J. Physique IV* **3**, 89

(1993).

39. J. Ren and M.H. Whangbo, *Phys. Rev.* B **46**, 4917 (1992).
40. S. van Smaalen, in *Physics and chemistry of low-dimensional inorganic conductors*, eds. C. Schlenker *et al.* (Plenum Press, New York, 1996).
41. M. Greenblatt, in *Physics and chemistry of low-dimensional inorganic conductors*, eds. C. Schlenker *et al.* (Plenum Press, New York, 1996).
42. W.J. Schutte and J.L. de Boer, *Acta Crystallogr.* B **49**, 579 (1993).
43. A.D. Rae *et al.*, *Acta Crystallogr.* B **51**, 708 (1995).
44. S. Schmid, K. Fuetterer and J.G. Thompson, *Acta Crystallogr.* B **52**, 223 (1996).
45. E.J.W. Lam *et al.*, *Acta Crystallogr.* B **51**, 779 (1995).
46. L.F. Veiros, M.J. Calhorda, and E. Canadell, *Inorg. Chem.* **33**, 4290 (1994)

SURFACE LIGHT-INDUCED DRIFT

MICHAEL A. VAKSMAN

Department of Chemistry, University of Wisconsin - Superior, Superior, Wisconsin 54880

1. Introduction

Better understanding of the mechanisms of gas-surface interactions is crucial both fundamentally and for a variety of applications (such as heterogeneous catalysis and sputtering). The study of molecule-surface interactions has undergone a remarkable increase within the last three decades, particularly owing to exploitation of powerful molecular-beam techniques [1]. This success has been augmented by the development of numerical methods combined with analytical studies [2,3]. Doppler measurements have been used [4] to obtain information concerning the velocity distribution of the scattered molecules, in parallel to time-of-flight measurements [5]. Different mechanisms of scattering have been studied [6], such as trapping-desorption, and direct inelastic scattering. A detailed study of direct inelastic scattering has been carried out by Hurst et al [6], with careful measurements of the accommodation of both tangential and normal velocities.

A number of papers are devoted to investigations of the adsorption-desorption process (cf., e.g., the review in Ref. [7]). The effect of rotational cooling (that is, the decrease of the mean rotational temperature of the scattered molecules compared to the surface temperature) has also been studied [8-12]. Chemical reactions on surfaces have been analyzed in a number of studies (see, e.g., Refs. [13,14]). According to Somorjai [14], rough surfaces are a preferred site of enhanced chemical activity. The "puzzle" of this unique chemical activity of rough surfaces is still an open subject of investigations.

Many studies on the state specificity of gas-surface interactions have been carried out so far. More particularly, it has been shown that the chemical activity of molecules can be enhanced by several orders of magnitude, using vibrational excitation (cf. the review in Ref. [15]). However, most papers on excited-molecule scattering have concentrated on experimental studies and calculations of vibrational deactivation in a gas-surface collision, or on the probability of molecular collisional excitation [16]. Only recently, studies [17-19,27,28] have appeared on the dependence of trapping and chemisorption on the vibrational and/or rotational state of the impinging molecules. All these studies describe and analyze the data obtained by molecular beam techniques. With this method, one often obtains the information on the state-dependence of scattering indirectly, by analyzing the data acquired by using the seeded beams at different nozzle temperatures [17,18]. Deconvoluting the data obtained by the molecular beam technique is often a difficult task [19].

There is another way to obtain similar data. Accommodation of the longitudinal momentum of a molecule in the collision with the surface is related mainly to its trapping and subsequent desorption (or chemisorption) [2]. Therefore, information on the state dependence of trapping or chemisorption can also be obtained by measuring the change in the accommodation coefficient of translational energy, due to radiative excitation preceding a surface collision. In other words, one can measure the fraction of parallel momentum of an **excited** molecule that is on average accommodated by the surface during molecule-surface collision, and same for a **non-excited** molecule. The question we ask is: what is *the difference* between these two numbers?

The phenomenon of Surface light-induced drift (SLID) can provide an answer. In the SLID studies (Fig. 1), a laser beam is traveling through a cell with a rarefied (Knudsen) gas, its frequency ω slightly tuned off a molecular frequency ω_0. Due to the Doppler effect, the laser excites only molecules with an x-projection of their velocity around $v_0 = (\omega-\omega_0)/k$ (where x is the direction of the laser beam, and **k** is the radiation wavevector). Consider the accommodation coefficient α describing the average attenuation of an incident molecule by the collision with the cell-wall surface (or, in other words, the transfer of parallel momentum to the wall): $\alpha \equiv |p_{xl}> - <p_{x\,f}>\,|\,/\,|<p_{x\,l}>|$, where $<p_{x\,l}>$ and $<p_{x\,f}>$ are the average longitudinal momenta of the impinging molecules before and after scattering by the surface respectively. Let the accommodation coefficient α be higher for molecules excited by the laser radiation than for non-excited ones. The molecules in the excited velocity group around v_0 will then be attenuated by the cell walls more efficiently than molecules in the opposing velocity group (around $-v_0$), thus giving rise to a net flux [20-22] and to a pressure gradient along the cell which can be monitored [64]. The magnitude of the resulting flux is proportional [20,22] to the difference $\Delta\alpha$ between the accommodation coefficients for the excited and the non-

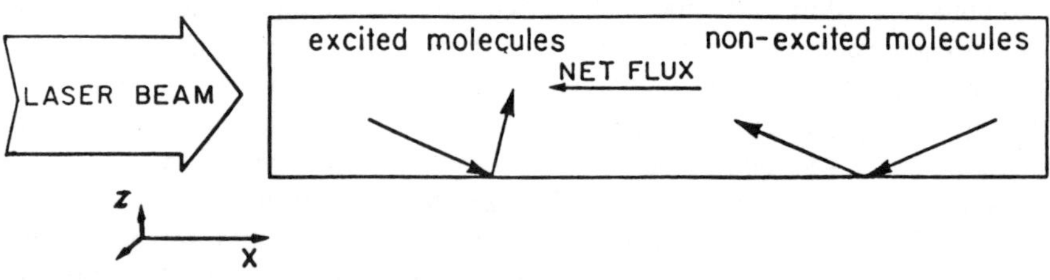

Fig. 1. The origin of SLID flux. Laser-radiation frequency is shifted to the blue ($\omega>\omega_0$); accommodation coefficient in the excited state is larger than in the ground state ($\Delta\alpha>0$).

excited molecules, and to the density n_e of excited molecules in the selected velocity interval:

$$j \approx -\Delta\alpha \; n_e \; \upsilon_0$$

As it can be seen from the above expression, the flux and the pressure drop resulting from SLID change their signs when the light frequency is swept within a narrow Doppler absorption line of the gas, which can assist in identifying SLID against the background of the masking phenomena. At present, development of the microscopic theory of the accommodation-coefficient state dependency is yet at the early stage [23]. Traditional methods (such as time-of-flight measurements) can hardly provide an applicable tool for such studies, since the change $\Delta\alpha$ in the magnitude of the accommodation coefficient often turns out to be quite low (about 0.1%; the highest value of $\Delta\alpha$ measured so far has been observed in the experiment [24] on SLID of $^{13}CH_3F$ molecules scattered on LiF(100) crystal surface, $\Delta\alpha$=1.4%). Let us remind in comparison, that the velocity spreads of approximately 10% are most typical for the modern molecular beam sources [3]; naturally, it would be hard to measure the above mentioned subtle differences in accommodation coefficients by molecular beam techniques. It is the phenomenon of SLID that has demonstrated the capacity of providing a method of high sensitivity for measurements of state-specificity of the accommodation coefficient α in molecule-surface interactions. In particular, SLID experiments have shown that the value of α is larger if the angular momentum of the resonant molecules is parallel to the surface ("cartwheels") than if perpendicular ("helicopters")[25]. Although the difference in the values of α was only of the order of 0.1 %, it was possible to register this subtle difference by using the SLID technique. On the other hand, at the present level of SLID theory, much of the available data on SLID basically allow yet for more than a single microscopic interpretation [26], [39]. Thus, additional studies on SLID are essential.

Besides being a powerful method of probing gas-surface interactions complimentary to molecular beams technique, SLID could also find a variety of applications in the areas where control is needed over the spatial distributions of rarefied gases. For example, one can eliminate certain component of an extremely rarefied gas mixture (where the particle mean free path greatly exceeds the size of the cell) from some region, without any changes in the concentrations of the other components. Such applications could become especially important in photochemistry and vacuum science. Another potential application of this effect is in isotope separations.

In this chapter we describe the studies of SLID that have been done so far for the cases of the Maxwell model of surface scattering, inelastic scattering, and scattering on rough surfaces. We also discuss the possibilities to use SLID for studying heterogeneous catalysis. We show that, contrary to an intuitive perception, SLID can be used to obtain the information on the *angular* dependence of the gas-surface scattering which is an important issue being widely discussed (see, in particular, the papers Refs.[19, 27-31]). We propose to use this idea to study the scaling laws for the angular dependence of surface scattering by the SLID technique. Finally, we briefly discuss other computational methods that are being developed for the SLID studies and outline some prospective directions for the future research. Although we attempted to review, to some extent, most of the publications that have appeared on SLID up to date, by no means this chapter is a complete bibliography source on the subject.

2. Surface light-induced drift in the Maxwell model for the gas - surface scattering

First, we consider SLID arising in the Maxwell model, where it is assumed that during the gas-surface collision a fraction $(1 - \alpha)$ of all the molecules are scattered in the specular fashion, while the remaining fraction α completely "forgets" the direction of incidence and is scattered in the diffuse fashion. We can calculate the speed of the SLID flux in a two-level approximation by solving a system of steady-state Maxwell-Boltzmann kinetic equations for the velocity distributions $f_e(\upsilon)$ of excited and $f_g(\upsilon)$ of non-excited gas-phase molecules (in the strong-collision limit):

$$\nu_e \; [N_e^{'} \; W(\upsilon) - f_e] \; + \; Q(\upsilon)(f_g - f_e) \; - \; \gamma f_e = 0 \qquad\qquad (1a,b)$$

$$\nu_g \; [N'_g \; W(\upsilon) - f_g] \; - \; Q(\upsilon)(f_g - f_e) \; + \; \gamma f_e = 0$$

Each of these equations consists of three terms: (a) a velocity-selective optical pumping term, with the velocity-dependent excitation rate $Q(\upsilon_x)$,

$$Q(\upsilon_x) \; \equiv \; \int BM(\omega) \; \{\Gamma^{2} \; / \; [\Gamma^{2} \; + \; (\omega - \omega_0 - k\upsilon)^2]\} \; d\omega$$

where B is the Einstein coefficient, $M(\omega)$ is the laser spectral intensity distribution, Γ is the homogeneous absorption linewidth, and γ is the decay rate of the excited state; (b) a damping term, depending on the decay rate of the excited state resulting from molecule-wall quenching collision and spontaneous emission; (c) a collision term depending on the free parameters N_e and N_g' having the meaning of effective molecular concentrations N_e and N_g (indices e and g correspond to the excited and to the non-excited resonant molecules respectively) and the rates of velocity-randomizing collisions v_e and v_g,

$$v_{e,g} = \alpha_{e,g}\ (v/d)\ \cos\theta$$

Here v is the magnitude of the molecule's velocity, θ is the angle between the normal to the surface and the direction of the particle velocity, and d is the distance between the walls. These equations are quite similar to those that can be used for analyzing the bulk LID [32],[33] and other kinetic phenomena arising in a gas under velocity-selective excitation [34,35]. It has been assumed in writing Eqs. (1) that the specular collisions with the cell walls do not alter the net velocity distributions of the molecules but just lead to their "symmetrization" in the direction orthogonal to the laser beam. After Eqs. (1) are solved, and the free parameters N_e' and N_g' are found using the self-consistency equations expressing the conservation of the number of particles in collisions, we determine the net velocity distribution $f(v) = f_e + f_g$ and the flux $j = \int v f(v) d^3 v$. In the case of rapid spontaneous decay of the excited state and excitation of molecules in a semi-infinite velocity interval $v_x > 0$, that is, $\kappa(v_x) \equiv Q(v_x)/\gamma = \kappa_0\,\theta(v_x)$, where $\theta(v_x)$ is Heaviside's unit function, we get

$$j = \sqrt{\frac{2}{\pi}}\ v_T\ \frac{\kappa_0\,(1-\beta)}{\kappa_0(\beta+3)+2}\ N =$$

$$= \sqrt{\frac{2}{\pi}}\ v_T\ (1-\beta)\ N_e, \tag{2}$$

where $\beta \equiv \alpha_e/\alpha_g$, N is the concentration of the gas, N_e is the concentration of the excited molecules, and v_T is the thermal velocity. In a closed cell the SLID flux must be compensated by a diffusive flux. Using the usual estimation for the diffusion coefficient of rarefied gas $D \sim d\, \upsilon_T$ [36], we obtain the resulting relative pressure drop along the cell [20-22]:

$$\frac{\Delta P}{P} \approx c\, \frac{j}{N\upsilon_T}\, \frac{L}{d}.$$

(3)

Here L is the length of the cell, $c \sim 1$ is a constant the precise value of which depends on the cell geometry [36]. The factor L/d of Eq. (3) above can be as large as 10^3 in real experimental conditions. SLID is thus exceedingly sensitive to changes in molecule-surface interactions with excitation. Even in cases when $\Delta\alpha$ is very small, substantial L/d can lead to a measurable pressure drop, as it happened in most of experiments by Hermans and co-workers in Leiden [37,23-25].

3. Surface light-induced drift arising in the case of inelastic scattering

In section two, we have considered the effect of SLID arising in a model where only specular or diffusive scattering are allowed. Another type of scattering, which is not a part of the Maxwell model, is direct inelastic (quasispecular) scattering [2]. Below we will consider, following Ref. [21], the effect of this latter type of scattering on SLID. We assume that after one act of quasispecular scattering the longitudinal components of the particle velocity change by $u_{e,g}$ and the normal component by $U_{e,g}$ respectively. We arrive at the following system of Boltzmann kinetic equations:

$$[\upsilon_e\, N_e'\, W(\upsilon_x) + (1-\eta)(\upsilon-\upsilon_e')(1-\alpha_e)\, f_e(\upsilon_x \pm u_e) - \upsilon f_e] +$$

$$+\, Q(\upsilon_x)(f_g - f_e) - \gamma f_e = 0,$$

(4a)

$$[\upsilon_g\, N_g'\, W(\upsilon_x) + \eta(\upsilon-\upsilon_{eg}')(1-\alpha_e)\, f_e(\upsilon_x \pm u_{eg}) +$$

$$(v - v'_g)(1 - \alpha_g) \, f_g(v_x \pm u_g) \, - v f_g] \, -$$

$$- Q(v_x)(f_g - f_e) \, + \gamma f_e \, = 0. \tag{4b}$$

Here the plus and minus signs in the arguments of velocity distribution functions correspond to the cases of $v_x > 0$ and $v_x < 0$, respectively;

$$v \equiv (v/d) \cos\theta, \quad v'_{e,g} \equiv (U_{e,g}/d) \cos\theta, \quad v'_{eg} \equiv (U_{eg}/d) \cos\theta,$$

u_{eg} and U_{eg} are the variations in the longitudinal and the normal components of the molecular velocity in the collisions where a molecule changes from the excited state to the ground state in the course of quasispecular collision with the wall; η is the probability of such a transition; other notations are the same as in section 2. For $u_{e,g} \ll v_T$ we can write

$$f_{e,g}(v_x \pm u_{e,g}) \approx f_{e,g}(v_x) \pm u_{e,g} \frac{\partial f_{e,g}}{\partial v_x},$$

and equations (4a,b) form a system of differential equations for f_e and f_g. Since, generally speaking, its solution is cumbersome, we will again consider the case of excitation of particles in a semi-infinite velocity interval $v_x > 0$, that is, $\kappa(v_x) \equiv Q(v_x)/\gamma = \kappa_0 \, \theta(v_x)$, where $\theta(v_x)$ is Heaviside's unit function. Qualitative considerations show that for $\alpha_{e,g} \sim 1$ one can assume that

$$\frac{\partial f_{e,g}}{\partial v_x} \approx \frac{\partial f_{e0,g0}}{\partial v_x},$$

to within a term of the order of $(u+U)/v_T$, for all $|v_x| \gtrsim u$, where f_{e0} and f_{g0} are the values of f_e and f_g as u and U tend to 0. Adding equations (4a) and (4b), for $\alpha_e = \alpha_g$ and $U \ll v_T$ we arrive at an algebraic equation expressing f in terms of f_{e0} and f_{g0} with accuracy of $[(u+U)/v_T]$ for all $|v_x| \gtrsim u$. In the calculation of the drift flux $j = \int v_x f\, dv_x$ the region $|v_x| \lesssim u \ll v_T$ provides only lower-order contributions. The values of f_{e0} and f_{g0} are found by using the self-consistency conditions $\int f_{e,g}\, dv_x = N_{e,g}$ and $\int St_{e,g}(v_x)\, dv_x = 0$, where $St_{e,g}(v_x)$ are the collision integrals [20]. We find

$$j \approx \frac{\kappa_0}{2\kappa_0+1} \frac{1-\alpha}{2\alpha} \theta_1 \{[(1-\eta)u_e + \eta u_{eg} - u_g] -$$

$$-[(1-\eta)U_e + \eta U_{eg} - U_g]\}\, N \tag{5}$$

where $N = N_e + N_g$ is the concentration of the gas and

$$\theta_1 \equiv \min(1, \gamma/\nu).$$

The factor θ_1 accounts for the decrease in the magnitude of the effect at $\gamma < \nu$. The resulting expression for the relative pressure drop $\Delta P/P$ in a cell with closed by lateral ends is still given by the expression Eq. (3).

Note that the terms in square brackets in equation (5) contribute to j (and correspondingly to $\Delta P/P$) with an opposite sign, because an increase in the normal component of the velocity leads to a decrease in the time of free flight of the molecule between the walls and, accordingly, to a decrease in its displacement in the longitudinal direction (x). By contrast, an increase in the x component of the velocity of the molecule obviously leads to an increase in such a displacement.

4. The study of anisotropy of gas-surface scattering using surface light-induced drift

Being a very sensitive tool for studying the gas-surface scattering, SLID apparently suffers from a major flaw in comparison with the molecular beam techniques. This apparent flaw stems from the very nature of SLID. Let us denote θ

the angle between the direction of the incidence of a molecule and the normal to the surface. It seems obvious that, since molecules that impinge on the surface at *all* possible θ's contribute to the SLID flux, the only information that the SLID data can provide is the value of $\Delta\alpha$ *averaged over all possible directions θ of incidence*, - unlike the molecular beams data.

It is interesting however that this apparent flaw of SLID does not exist [38]. Actually, one *can* obtain the angularly resolved information on the change of accommodation coefficient, $\Delta\alpha$ (θ), even though molecules from all θ's contribute to the SLID flux.

To gain a qualitative grasp of the effect, let us assume that $\Delta\alpha$ (θ) is anisotropic; for example,

$$\Delta\alpha \begin{cases} >0 & \textit{for the small } \theta's \textit{ (nearly normal incidence} \\ <0 & \textit{for the large } \theta's \textit{ (grazing-angle incidence} \end{cases} \qquad (6)$$

Since trapping is the main mechanism behind the accommodation of the longitudinal momentum [2], assumption (6) means that, considering the molecules whose directions of incidence are close to the surface normal, the trapping probability is greater for the excited molecules, while for the grazing angles of incidence the trapping probability is greater for the ground-state molecules. We will show now that the SLID velocity υ_{SLID} (and therefore the pressure drop ΔP along the cell) as a function of the detuning $\Omega \equiv \omega - \omega_0$ behaves quite differently than in the "isotropic" case

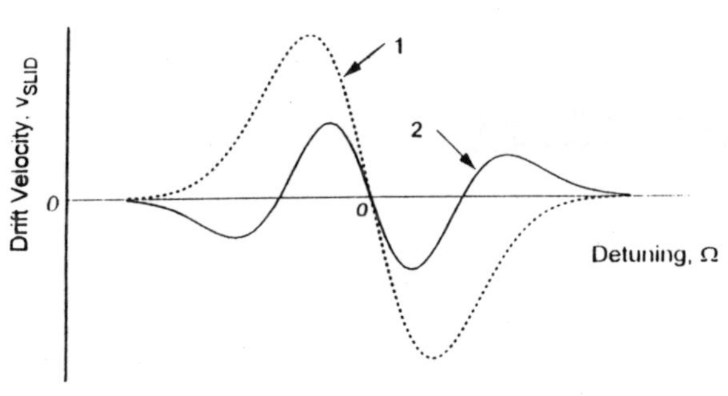

Fig. 2. The dependence of the SLID velocity upon the detuning $\Omega \equiv \omega - \omega_0$: 1- the isotropic scattering, 2 - the anisotropic scattering of Eq. (6).

when $\Delta\alpha$ $(\theta) = const$. In the isotropic situation [20-22,26,39-42], υ_{SLID} is zero at $\Omega = 0$ (since in this case $\upsilon_0=0$) and is very small far from the center of the resonant line (since there are very few absorbers far at the wings of the Maxwell distribution). Obviously, this dependency must be antisymmetric with respect to $\Omega = 0$. Thus, the resulting υ_{SLID} (Ω) dependency is a dispersion-like curve that has one zero, at $\Omega=0$ (dashed line in Fig. 2).

Let us consider now our angle-dependent example (Eq. (6)). We examine separately the situations arising at *two different* detunings Ω : (i) for a *small* detuning, $\Omega \approx k\Delta\upsilon_0$ (where $\Delta\upsilon_0$ is the width of the excited velocity interval which we assume to be much more narrow than the thermal velocity υ_T), and (ii) for a *large* detuning, say, $\Omega \approx 3k\upsilon_T$. In both of the cases (i) and (ii), the average *transverse* velocity component $<\upsilon_z^2>^{0.5}$ remains essentially unchanged [43]; in both cases, it is of the order of the thermal velocity υ_T. However, the value of the *longitudinal* component υ_x of the velocity is different in (i) and (ii): in the case (i), it is small $(0<\upsilon_x<\Delta\upsilon_0 \ll \upsilon_T)$, while in the case (ii), $\upsilon_x \approx 3\upsilon_T$. Thus, in the case (i) (Fig. 3(a)), preferentially the molecules that move *perpendicular* to the surface can get excited,

while in the case (ii) (Fig. 3(b)) primarily the molecules moving at *grazing* angles can get excited. It is clear from Eq. (6) and Figs. 3 (a,b) that the directions of the SLID fluxes in the cases (i) and (ii) should be opposite. The curve is reflected for negative detunings, and the resulting $\upsilon_{SLID}(\Omega)$ dependence should have **three** zeroes (the solid line in Fig. 2), unlike the "isotropic" case (the dashed line in Fig. 2). Thus, by monitoring the $\Delta P(\Omega)$ dependence and deconvoluting the resulting profile, one can study the *angular variation* of the state-dependence of the accommodation coefficient $\Delta\alpha(\theta)$.

We can calculate the magnitude of SLID flux by solving the same system (1a,b) of Boltzmann kinetic equations for the velocity distributions of excited, $f_e(\upsilon)$, and of non-excited, $f_g(\upsilon)$, gas-phase molecules. Unlike the previous cases

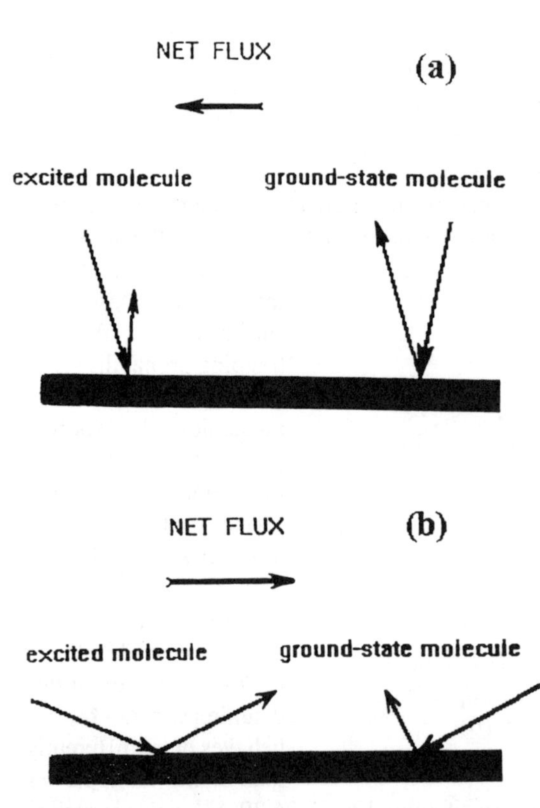

Fig. 3. SLID flux in the case when $\Delta\alpha$ is anisotropic and obeys Eq. (6): (a) Ω is small, (b) Ω is large.

though we will assume that the coefficients $\alpha_{e,g}$ (describing the fraction of the overall scattering that go along the trapping-desorption channel [2]) depend on the kinetic energy E_i of the impinging molecule and the angle of scattering θ, $\alpha_{e,g} = \alpha_{e,g}(E_i,\theta)$. After solving the Eqs. (1a,b) and obtaining the expressions for the total velocity distribution function $f(\upsilon) \equiv f_e(\upsilon) + f_g(\upsilon)$ and the SLID flux $\upsilon_{SLID} N \equiv \int \upsilon f(\upsilon) d^3\upsilon$, one can find the magnitude of the relative pressure drop along the closed cell:

$$\frac{\Delta P}{P} = -c\,\frac{\alpha}{2-\alpha}\,(N\upsilon_T)^{-1}\,\frac{L}{R}\,\times$$

$$\times \int_{-\infty}^{+\infty}\frac{\Delta\alpha(E_i,\ \theta)}{\alpha_g(E_i,\ \theta)}\,\upsilon\,\frac{Q(\upsilon_x)}{\nu_e+\gamma}\,(f_g - f_e)\,d^3\upsilon \tag{7}$$

Here α is an average value of the accommodation coefficient of the molecules, $c{\sim}1$ is a constant the precise value of which depends on the cell geometry [36]. The angular dependence of scattering enters Eq. (7) through $\alpha_{e,g}(E_i,\theta)$.

The scaling laws for the probabilities of trapping and chemisorption are important issues that recently have been frequently discussed [19,27-31]. These probabilities usually scale as $E_i \cos^n \theta$, where n can vary from $n=2$ ("normal energy scaling") to $n=0$ ("total energy scaling") [28]. As an example, we made calculations for different scaling laws of the excited molecules, $0 \leq n \leq 2$, using the formula (7) and assuming that the accommodation coefficient of the ground-state molecules obeys normal energy scaling. We have considered the case of small field saturation, $\kappa_0 \equiv Q(\upsilon_0)/\gamma \ll 1$ [42], $\Delta\upsilon_0 = 0.1\,\upsilon_T$, $\upsilon_T = 10^3$ m/s, $L/R = 10^4$ and $\nu_e \ll \gamma$. The results are presented in Fig. 4. One can see that as n changes from 2 to 0, the corresponding ΔP (Ω) curves change drastically. Thus, SLID data can be used for studying the state-dependence of scaling. Information on the dependence of scattering on the average energy of the impinging molecules can be obtained by varying the temperature of the cell. Deviation from the normal energy scaling indicates the importance of the parallel momentum in trapping [30,31] frequently caused by the corrugation of the surface potential. We would also like to stress the difference between the present effect and the earlier described effects [44,45] with the non-dispersive dependence of the magnitude of the effect on the detuning. Unlike the effects [44,45], such a non-dispersive dependence arises here because of the difference in *anisotropy* of scattering between the excited and non-excited molecules.

This study has shown that the SLID technique can be used to obtain independent results for the state dependency of $\alpha(\theta)$. The $\Delta\alpha(\theta)$-data have the significance on their own right, as they can assist in understanding the dependence of scattering and chemical reactivity of a molecule on its quantum state. On the other hand, if the value of $\alpha_g(\theta)$ is determined from an independent experiment, then using the results of this study one can find also the angular dependence for scattering of *excited* molecules $\alpha_e(\theta)$, thus providing the data on the angular dependence of the *individual* accommodation coefficients $\alpha_{e,g}(\theta)$.

There is one more advantage in using the described method. With the molecular beam method, one can draw conclusions about the differences in the scattering of excited and non-excited molecules by comparing the scattering diagrams with and without irradiation. If the spontaneous decay of the excited state is rapid (as for electronically excited molecules), then in order to enable the molecules to reach the surface prior to deexcitation, it may be necessary to irradiate a region directly adjacent to the wall, including the surface itself [22]. In this case the scattering diagram will be deformed by the changes in the states of both the impinging molecules and the surface (often the surface is covered by adsorbed molecules of the same gas). In the SLID method, on the other hand, the magnitude of the effect is proportional to the difference in scattering of the excited and non-excited *impinging* molecules and thus can still provide this information.

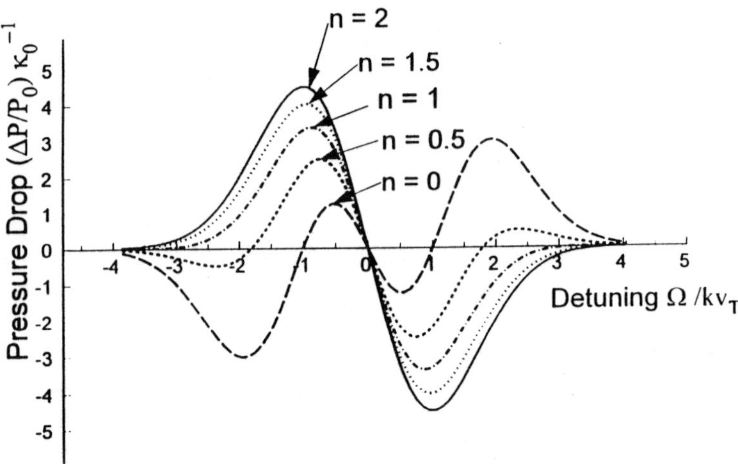

Fig. 4. Dependence of the relative pressure drop $\Delta P/P_0$ upon the detuning Ω for the different scalings of the excited-state accommodation coefficient α_e. In all cases $\alpha_g = 0.98 - (10^{-8}\,s^2\,m^{-2})\,\upsilon^2\,cos^2\,\theta$, $\alpha_e = 1.00 - (10^{-8}\,s^2\,m^{-2})\,\upsilon^2\,cos^n\,\theta$, $0 \le n \le 2$ (υ is the velocity of an impinging molecule). Notice the drastic change in the shape of the $\Delta P\,(\Omega)$ dependence as n varies between $n=0$ and $n=2$.

5. **Surface light-induced drift caused by roughness**

So far, we have considered SLID arising in cells with planar (smooth) walls.   However, surfaces of most of the cells are not flat but rough. We will show now that in a cell with rough walls SLID can arise even if all molecules are scattered in a locally diffuse manner. The origin of the drift in this case is illustrated in Fig. 5. Suppose the laser excites molecules in the velocity group near v_0, where $v_0 > 0$. Clearly, the molecules impinge mostly those facets of roughness which face the direction of incidence (those facets are labeled + in Fig. 5). At the same time, ground-state molecules scatter mainly on facets of the opposite orientation (facets labeled -). Let us denote by A the probability for a molecule to be adsorbed and subsequently travel along a few facets of roughness, and by a the probability for the molecule to be diffusively scattered without adsorption, or to be adsorbed and subsequently desorbed at the same facet. Clearly, the fraction a_e (a_g) of the incident excited and ground-state molecules correspondingly gives rise to backscattering, while the fraction A_e (A_g) of those molecules will "forget" the orientation of the facet where they have been adsorbed and desorb on the average in normal direction with respect to the x axis. Thus, at $A_e \neq A_g$ drift will occur [22,39], the direction of which will coincide with that of backscattering of molecules having smaller values of A. As one can see from comparison of the Figs. 1 and 5, the direction of drift in a cell with rough walls at $A_e < A_g$ is the same as in a cell with flat walls at $a_e > a_g$. The resulting change of the sign of the SLID flux stems from the observation that the simple rule "more adsorption - more accommodation" which holds for a flat surface, is broken in the case of rough one. Because of the backscattering, locally-diffusive scattering provides greater accommodation for the longitudinal momentum than

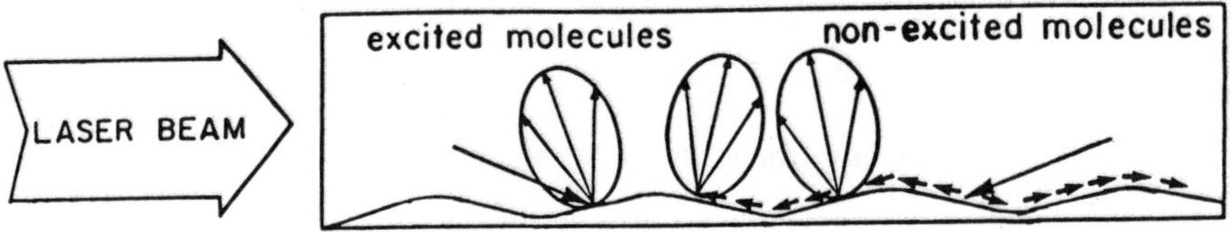

Fig. 5. Cell with rough walls: SLID arises even if the scattering of all molecules is diffusive (cf. Fig. 1). Here again laser-radiation frequency is shifted to the blue. We assume that adsorption probability is larger for the ground-state molecules, and that the adsorbed molecule during its surface diffusion passes several roughness facets prior to desorption.

scattering in which the adsorbed molecule travels over the surface before desorption.

As the direction of SLID can change due to roughness, one should first study the microscopic structure of the walls of the cell in order to draw any conclusions about the value and even the sign of the change of the coefficient of adsorption during excitation.

"Ordinary" mechanisms of SLID manifest themselves only in the case if an appreciable fraction of all molecules scatter in a locally-nondiffusive way with state-dependent scattering lobes. Considering that quite often [46] scattering is locally diffusive, one could be somewhat skeptical about the probability of the conditions for the "ordinary" SLID to be widely met. In contrast, the condition $A_e \neq A_g$ can be satisfied because of the state dependence of adsorption probability (or, in principle, of the length l_e (l_g) over which an adsorbed molecule travels before being desorbed). Since adsorption probability is generally accepted to be state dependent [2], the "roughness" mechanism provides a significant extension of real physical situations where SLID can occur.

The quantitative treatment of this effect [38] is somewhat similar to that used in Section 3 for calculating the magnitude of SLID caused by inelastic scattering. We restrict ourselves to a simple model of the regular zigzag-shaped corrugation and consider the case of a small angle of roughness ($\psi \ll 1$). In Fig. 5 notice the labeling of the facets of the two possible orientations as + and - . We do not take into consideration the double collisions of a molecule with the neighboring facets. Using the polar system for two-dimensional integration and denoting by ϕ the angle between the x axis and the direction of the velocity of the molecule, we obtain the following set of steady-state kinetic equations:

$$v a_e (1-\zeta)(n_e^+ W^+ + n_e^- W^-) + s_e [v^+ p^+ (f_e(\phi^+, \upsilon) +$$

$$+ v^- p^- (f_e(\phi^-, \upsilon)] - v f_e = -Q(\phi, \upsilon)(f_g - f_e) + \gamma f_e, \tag{8a}$$

$$v[A_g(n_g^+ + n_g^-) + A_e(n_e^+ + n_e^-)]W + v[a_g(n_g^+ W^+ + n_g^- W^-) + \zeta a_e(n_e^+ W^+ + n_e^- W^-)] +$$

$$+ s_g[v^+ p^+(\phi^+)\, f_g(\phi^+,\upsilon)+v^- p^-(\phi^-)\, f_g(\phi^-,\upsilon)]$$

$$= Q(\phi,\upsilon)(f_g - f_e) - \gamma f_e. \tag{8b}$$

The left-hand sides in (8a) and (8b) contain collision integrals, while in the right-hand sides the first term takes into consideration the molecule-field interaction, the second one describes spontaneous relaxation, with γ being its rate. Here $s_{e(g)}$ are the probabilities of specular scattering of molecules in corresponding states (so that $A+a+s=1$), $v(\phi,\upsilon) = \upsilon|sin\phi|/d = |\upsilon_z|/d$ is the inverse time of flight of a molecule between the walls, and d still the distance between the cell walls. We assumed here that all collisions of A type are accompanied by quenching, and ζ is a fraction of quenching collisions during scattering of a type. Further, $n_{e(g)}^+$ and $n_{e(g)}^-$ are the effective concentrations of particles scattered by the + and - facets, respectively,

$$n_{e,g}^{\pm} \equiv \int p^{\pm} f_{e,g}(\phi,\upsilon)\, d\phi\, d\upsilon, \tag{9}$$

$p^{\pm}(\phi)$ are the probabilities for the molecule to incident to + or - facets, respectively. An elementary geometrical analysis yields $p\pm = 0.5\,(1 \pm \psi\, ctg\, \phi)$ for $\psi \leq \phi \leq \pi-\psi$. Scattering at a very small angle $\phi < \psi$ or $\pi - \psi < \phi < \pi$, which often leads to double collisions with the neighboring facets, is neglected. $W(\upsilon) = (\alpha/\pi)exp(-\alpha\upsilon^2)$ is the Maxwellian, $W^{\pm}(\phi,\upsilon)$ are the distributions of the molecules being diffusively scattered by $\pm$ facets. As the probability for a particle to leave the surface after diffuse scattering is proportional to the cosine of the angle of its velocity to the surface normal (Knudsen cosine law), the following relation can be written for the ejection from those facets:

$$vW^{\pm} \sim \upsilon\, nW = \upsilon(sin\phi \mp \psi cos\phi)W.$$

Notice that $W^+ + W^- = 2W$, which is accounted for in the term that contains $A_{e,g}$. Also, $v^+ = v(\phi^+,\upsilon),\ \phi^+ = \phi-2\psi,\ v^- = v(\phi^-,\upsilon),\ \phi^- = \phi+2\psi$. One can see that specular scattering of a particle by the + facet leads to its turn over the angle $+2\psi$, while specular scattering by the - facet leads to the turn over the angle -2ψ, which is accounted for by the terms with shifted ϕ. Using the condition $\psi \ll 1$ we can retain only the first two terms in the Taylor expansion for $f_{e,g}(\phi \pm 2\psi, \upsilon)$ and thus with the help of Eq. (9) see that (8a) and (8b) are a set of integro-differential equations, solution of which is rather complicated. In order to simplify the treatment we consider a special case $s_e = s_g = 0$, when the "ordinary"SLID [20-22] is absent. We also neglect spontaneous relaxation the rate of which is exceedingly small for rovibrational transitions, and consider again the case of excitation of the half-infinite velocity interval $v_x > 0$, that is, $-\pi/2 < \phi < \pi/2$. We discuss the case of saturated absorption of radiation $Q \gg vT/d$ and in order to obtain $f_{e,g}$ restrict ourselves to the first-order approximation, substituting the zeroth-order approximation (that is, the value of $f_{e,g}$ at $\psi=0$) to calculate $n_{e,g}^{\pm}$. We obtain the following expression for the SLID flux $j \equiv \iint d\phi \, dv \, v \cos \phi \, f(\phi, v)$ (with $f(\phi, v) = f_e + f_g$ being the net velocity distribution) along the x-axis. As in many problems involving velocity-selective excitation, it may be convenient to express the result in terms of the absorbed radiation power density ΔI :

$$j \approx \frac{2}{\pi} \, (A_e - A_g) \, (\psi^2 \ln^2 \psi) \, \frac{\Delta I}{h\omega L} d. \tag{10}$$

Here we have considered the case of an optically thin system. The proportionality of the flux to ψ^2 is the result of multiplication of two quantities: (i) the difference of scattering probabilities on the $+$ and $-$ facets, and (ii) the asymmetry in the diffusive ejection from each facet, as both (i) and (ii) are proportional to ψ. Slight ($\sim\ln^2 \psi$) deviation from the ψ^2 is caused by the particles moving at tangent angles $\phi \ll 1$ for which the difference between p^+ and p^- is not small. Note that even in the case of excitation of the narrow velocity interval v_0 Eq. (10) can be used to estimate the flux which now is given by $(v_0/v_T) j$. Relative pressure drop $\Delta P/P$ can be calculated using the same formula (3). One can see that in order to obtain the values of the $\Delta P/P$ observed in many of the experiments on SLID [37,23,25], it is enough to assume the existence of roughness with $\psi^2 \approx 0.1$ when $s_e = s_g = 0$ and $\Delta A \equiv A_e - A_g = -\Delta a \sim -10^{-1} - -10^{-2}$. Of course, this does not exclude the possibility of an alternative interpretation [23].

6. Surface light-induced drift caused by heterogeneous chemical reactions

Rough surfaces are shown to be the site of enhanced chemical activity [14]. Moreover, the rate of surface chemical reaction can be increased several orders of magnitude by laser irradiation of the gas-phase molecules [15]. Such differences in the behavior of excited vs. nonexcited molecules can cause a drift of considerable magnitude, following a velocity-selective excitation. In order to provide a more quantitative description, the simple model of molecular scattering at rough surfaces proposed by Berman and Maegley [47] was adopted. It is assumed in this model that a fraction a of the incident molecules undergoes completely diffuse scattering with respect to the average plane of the surface, while the remaining fraction, b, undergoes direct back scattering (that is, the incident molecules constituting this fraction have the direction of their velocity reversed at scattering). In absence of chemical reactions, $a+b=1$. It is interesting that, despite its extreme simplicity, this model provides even slightly better agreement with the experimental data concerning rarefied gas flow through long capillaries than a model assuming locally-diffuse scattering [48] . As an extension of the Berman-Maegley model to reactive scattering, we expect that a fraction a_g (a_e) of the incident nonexcited (excited) particles undergoes diffuse scattering without chemical reactions. Similarly, a fraction b_g (b_e) undergoes back scattering without chemical reactions. Furthermore, a fraction a_{cg} (b_{cg}) $[a_{ce}(b_{ce})]$ undergoes diffuse scattering (back scattering) following a chemical reaction. Finally, let A and B be the fractions of products of chemical reaction which undergo diffuse scattering and back scattering, respectively, in their collisions with the surface. We assume here that (i) the reaction is not reversible, and that (ii) the products are not state-selective in their behavior. Thus,

$$a_g + b_g + a_{cg} + b_{cg} \;=\; a_e + b_e + a_{ce} + b_{ce} \;=\; A+B = 1.$$

Using this model, we get the following set of kinetic equations to describe the stationary, spatially-homogeneous state of the incident molecules and of the reaction products:

$$va_e(1-\zeta)N_eW+vb_e\,f_e(-\upsilon)-vf_e(\upsilon)+s(\upsilon)-\gamma f_e(\upsilon) \;=\; 0; \tag{11a}$$

$$va_gN_gW+va_e\zeta N_eW+vb_g\,f_g(-\upsilon)-vf_g(\upsilon)-s(\upsilon)+\gamma f_e(\upsilon)=0 \tag{11b}$$

$$n[va_{ce}N_eW+vb_{ce}\,f_e(-\upsilon)+va_{cg}N_gW+vb_{cg}f_g(-\upsilon)]+$$

$$+\;v_cAN_cW \;+\; v_cBf_c(-\upsilon) \;-\; v_c\,f_c(\upsilon) \;=\; 0; \tag{11c}$$

$$va_e(1-\zeta)N_eW+vb_e\, f_e(\upsilon)-vf_e(-\upsilon)+s(-\upsilon)-\gamma f_e(-\upsilon) = 0; \qquad (11d)$$

$$a_gN_gW+va_e\zeta N_eW+vb_g f_g(\upsilon)-vf_g(-\upsilon)-s(-\upsilon)+\gamma f_e(-\upsilon)=(\qquad (11e)$$

$$n[va_{ce}N_eW+vb_{ce}\, f_e(\upsilon)+va_{cg}N_gW+vb_{cg}f_g(\upsilon)]+$$

$$+ v_cAN_cW + v_cBf_c(\upsilon) - v_c\, f_c(-\upsilon) = 0. \qquad (11f)$$

Eqs. (11a)-(11c) describe the evolution of the velocity distribution functions of the nonexcited (f_g) and excited (f_e) particles, and of the reaction products (f_c). Since each velocity group (υ) is coupled to the opposite velocity group ($-\upsilon$) by back scattering, these equations must be complimented by the Eqs. (11d) - (11f) for the opposing velocity group. In Eqs. (11a)-(11f), $v\equiv\upsilon_T/d$ and $v_c\equiv\upsilon_{cT}/d$ are the mean rates of surface scattering for the resonant molecules and reaction products (c), respectively, where υ_T and υ_{cT} are the mean thermal velocities of the corresponding molecules, and d is the distance between the cell walls. We still consider a rarefied (Knudsen) gas, in which molecules collide with the surface rather than with each other. We also assume that some fraction, ζ, of the diffusive collisions of excited molecules is accompanied by quenching. Furthermore, n is the stoichiometric coefficient of the considered heterogeneous chemical reaction, and

$$s(\upsilon) = Q(\upsilon)\, [f_g(\upsilon) - f_e(\upsilon)]$$

is the velocity-selective laser excitation rate. The set of Eqs. (11a - f) is easy to solve in order to find the velocity distributions of the resonant molecules and the reaction products. In the usual way, these distribution functions can be used to find the drift fluxes of the resonant and the product molecules:

$$J = (b_g - b_e)\frac{1 - b_e - b_e b_g + \dfrac{\gamma}{\nu + \gamma}b_e(1 - b_g)}{(1 - b_g^2)[\gamma + \nu(1 - \phi b_e^2)]\hbar\omega}$$

$$\times \frac{dI}{dx}\upsilon_0, \tag{12a}$$

and

$$J_c = \frac{\xi}{(1 - B^2)(1 - b_g^2)[\gamma + \nu(1 - \phi b_e^2)]\hbar\omega}$$

$$\times\{(b_{cg} - b_{ce})(1 - b_g^2)(1 - B - \phi b_e + \phi B b_e)$$

$$+ (b_e - b_g)b_{cg}[1 - B - \phi b_e - b_g$$

$$+ B(b_g + \phi b_e(1 - b_g)) + \phi b_e b_g]\} \times \frac{dI}{dx}\upsilon_0, \tag{12b}$$

respectively. Notice that J and J_c can be non-vanishing even when $(b_g + b_{cg}) = (b_e + b_{ce})$, that is, when the total fraction of incident resonant molecules scattered in a non-diffuse way does not change upon excitation, contrary to situations studied previously (Sections 2-5). The magnitudes of J and J_c can yield interesting information concerning the heterogeneous reaction under consideration. For

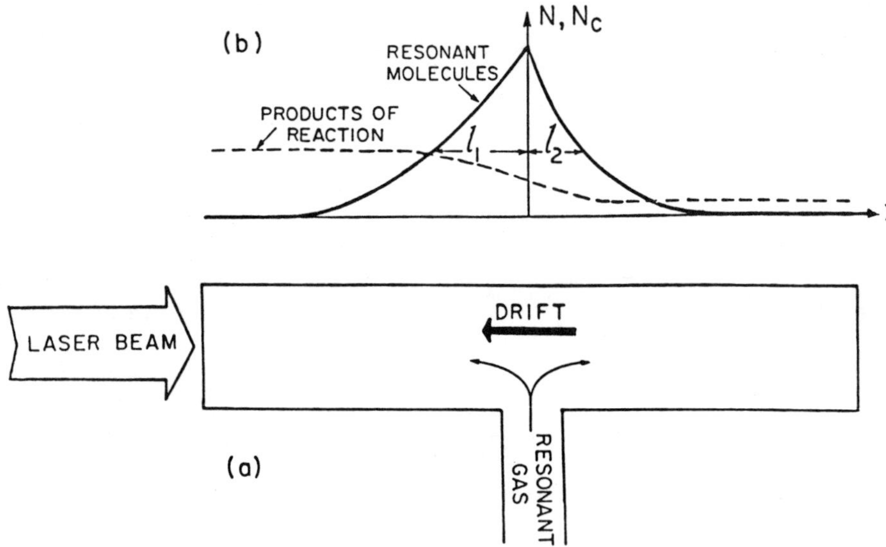

Fig. 6. (a) The form of the cell suggested for studying SLID in the presence of surface-enhanced chemical reactions. (b) A schematic distribution of the resonant molecules (solid line) and of reaction products (dashed line) along the cell.

example, if $J \neq 0$ it means that the reaction products were scattered without diffusing first over surface distances much exceeding the size of surface-roughness facets.

There remains, however, the question how to measure J and J_c experimentally. As one recalls from above, in a cell with closed ends the flux of molecules leads to a pressure gradient in the resonant gas along the cell, and the experimentally measured magnitude of the stationary pressure drop can be used in order to calculate the value of the SLID flux. However, in the case studied here such a simple way would be ineffective, since due to the chemical reaction one may not achieve a stationary molecular distribution. One could use a constant flux of resonant molecules injected into the cell from the lateral side, but then it would be necessary to register changes in the density gradients caused by switching on the laser light. These changes may be too small to detect on the background of fluctuations. Perhaps, the most advantageous scheme in this case would be a T-shaped enclosure (Fig. 6) like the one previously used in experiments on bulk light-induced drift of sodium vapor in helium.   The resonant gas penetrating the tube irradiated by the laser will preferably move in the direction of the drift flux (J). Therefore, the characteristic lengths, l_1 and l_2 , of the "wings" describing the resonant gas spatial distributions in the two opposing directions will differ from each other (Fig. 6 (b)). By measuring the value $l_2 - l_1$, one should be able to determine the value of J, and therefore calculate the magnitude of $(b_g - b_e)$. A similar analysis should be applicable to the determination of $(b_{cg} - b_{ce})$, by observing the spread of the reaction products. The spatial distribution of the resonant molecules can be derived by using an equation similar to the one used in the case of velocity-selective photodissociation and photoionization [49],

$$D\frac{d^2N}{dx^2} - \frac{dJ}{dx} - cN = 0, \tag{13}$$

which describes the conservation of the number of molecules in each layer along the x-axis.   In Eq.(13), $N = N_e + N_g$ is the (now inhomogeneous) concentration of resonant molecules, c is a constant describing the rate of chemical reaction, and D $\approx \upsilon_T d$ is the diffusion coefficient of the rarefied gas. We suppose that the concentration near the entrance point of the T-cell is constant, i.e. $N(x=0) = N_0$. Obviously, far from this point N should vanish due to the chemical reaction.   Eq. (13) then yields the following solution:

$$N(x) = \begin{cases} N_0 \, exp(x/l_1), & x<0, \\[2ex] N_0 \, exp(-x/l_2), & x>0, \end{cases} \tag{14}$$

where $l_1^{-1} \equiv R+F,$ $l_2^{-1} \equiv R-F,$ $F \equiv V/(2D),$ and $R \equiv (c/D + F^2)^{0.5}$, while $V \equiv J/N$ is the velocity of the light-induced flux. From Eq.(14) it follows immediately (see also Ref. [50]) that

$$V \approx \upsilon_T \left(\frac{d}{l_1}\right) \frac{l_2 - l_1}{l_2}. \tag{15}$$

Due to the factor d/l_1 [compare to the factor L/d of Eq. (3)], the suggested method may lead to the experimental observation of very small drift velocities V, and therefore to the determination of small values of $(b_g - b_e)$ through the use of Eq. (12). If, for example, $d/l_1=10^{-2}$, then V values as small as $10^{-3}\, v_T$ should lead to a comparatively large difference between the lengths of the right and left spreading wings in the spatial density distribution, i.e., $(l_2 - l_1)/l_2 \approx 10^{-1}$.

Both fluxes, J and J_c, change their sign under the small change (~ 100 MHZ) of laser frequency that would lead v_0 to change its sign. This fact can be used to detect the phenomenon discussed here on the background of parasitic ones.

Let us mention in conclusion that the multiphoton excitations, which happen to be a common source of laser-enhanced reactivity, can be treated by a straightforward modification of this model in a manner suggested elsewhere [49].

7. Surface light-induced drift of a dense gas

So far we have considered only the case of a rarefied (Knudsen) gas where the mean free path l is much larger than the cell radius R. What will happen if the gas density is raised to the point when l becomes much smaller than R? And what if we add a buffer gas to the resonant gas? The answers to these questions might have significant applications as in many of the real situations (isotope separation etc.) one often deals with denser gases and with gas mixtures, not a single-component rarefied gas.

First we consider the case of a dense *single-component* gas. The molecules interacting with the wall now lie in a thin layer of thickness $\sim l$ beside the wall. The problem thus gets the small parameter l/R. How would experimentally observable quantities depend on this parameter? In principle, derivation of exact result requires [22] solving a system of kinetic equations with appropriate boundary conditions in order to find the velocity distribution functions $f_{e,g}$ and thus the flux speed near the surface (see Section for details), and then the use of these results as boundary conditions for the gasdynamic equations. Because of the mathematical difficulties of this approach we will follow in our analysis the path described in our paper [22]. In our treatment, we neglect the "buffer" drift mechanisms [32,33,44]. We assume that no collisions between the molecules occur in a layer of gas of thickness l near the wall, while outside this layer the molecular velocity distribution in the volume, $f_{e,g}$, can be found from the kinetic equations similar to (1a,b):

$$v\,[N_e'\,W(v) - f_e] + Q(v)(f_g - f_e) - \gamma f_e = 0 \qquad (16\text{a,b})$$

$$v\,[N'_g\,W(v) - f_g] - Q(v)(f_g - f_e) + \gamma f_e = 0$$

where v is the rate of Maxwellizing collisions *in the bulk* and $W(v)$ is the Maxwellian with a shifted mean velocity V_s:

$$W(\upsilon) \equiv (\alpha/\pi)^{3/2} \exp[-\alpha(\upsilon-V_s)^2],$$

where $\alpha \equiv 0.5\,\upsilon_T^2$. In the steady state, the force exerted on the wall layer of gas vanish:

$$\alpha_e \int \upsilon_x \upsilon_\perp \, f_e d^3\upsilon \;+\; \alpha_g \int \upsilon_x \upsilon_\perp \, f_g d^3\upsilon \;=\; 0. \tag{17}$$

Here we again used the Maxwell model of surface scattering, with $\alpha_{e,g}$ being the coefficients of diffuse scattering. From Eqs. (16,17) we can find the value V_s of the gas velocity in the layer directly adjacent to the surface layer:

$$V_s = -\left[\, b \,\frac{\psi}{2(1+\psi\theta)+(1+\theta)/\kappa_0}\, \upsilon_0 \right] \times$$

$$\left[\, 1 + b \,\frac{\psi\theta}{2(1+\psi\theta)+(1+\theta)/\kappa_0} \right]^{-1}, \tag{18}$$

or

$$V_s = -b\frac{n}{N(\nu+\gamma)}\upsilon_0 \left[1+b\frac{n}{N(\nu+\gamma)}\frac{\nu}{\gamma}\right]^{-1}. \tag{19}$$

Here $b \equiv \alpha_e/\alpha_g - 1$; ψ is given by

$$\psi = (2\pi)^{0.5}\,\frac{\Delta\upsilon_0}{\upsilon_T}\,e^{-\alpha\upsilon_0^2}$$

$\theta \equiv v/\gamma$, n is the number of photons which are absorbed in the wall layer per unit volume per unit time, and

$$v_0 = \int v_x Q(\upsilon)(f_g - f_e)d^3\upsilon \left[\int Q(\upsilon)(f_g - f_e)d^3\upsilon\right]^{-1} = \phi v_T , \qquad (20)$$

where ϕ is a function introduced in ref. [51]. In a cell without ends or with infinitely remote ends, all of the gas is entrained in the motion at a velocity V_s by virtue of the viscosity. If the cell is bounded by lateral ends, the gas will accumulate at one of these ends until the drift is balanced by the oppositely directed convection caused by the pressure drop arising due to the arising density gradient. Taking the viscous friction into account in a force balance equation similar to Eq. (17) we find the following boundary conditions for the gas-dynamic equations:

$$V\big|_{r=R} + c(\partial V/\partial r)\big|_{r=R} = V_s , \qquad (21)$$

with

$$c \approx \sqrt{2\pi} l \left[\alpha_g + (\alpha_e - \alpha_g)\frac{n}{N(v+\gamma)}\frac{v}{\gamma}\right]^{-1} . \qquad (22)$$

Solving the Navier-Stokes equation for a long tube with boundary conditions (21) and $(\partial V/\partial r)_{r=0} = 0$ and neglecting compressibility, we find expressions for the concentration drop ΔN along the cell and for the Poiseuille-like velocity profile which is "shifted" in such a manner that the boundary velocity $V(R) = V_0 \neq 0$ and the total gas flow through the cross-section of the cell is zero:

$$\Delta N \approx 8\frac{V_0}{\upsilon_T}\frac{\eta L}{mN\upsilon_T R^2}N = 8\frac{V_0}{\upsilon_T}\frac{lL}{R^2}N, \qquad (23)$$

$$V(r) \approx V_0\left[2\left(\frac{r}{R}\right)^2 - 1\right], \quad V_0 \equiv V_s\left(1 + \frac{4c}{R}\right)^{-1} \qquad (24)$$

So far we have been discussing the case of a single-component dense gas. If buffer particles are added to the vessel, they will become entrained in the flow of the resonant molecules and will be slowed at the wall. The problem thus can be formulated in the following way: determine the state of a gas mixture in a cell if one of the components (the resonant component) perceives the side wall as moving, while the other (buffer) component perceives it to be at rest. Obviously, far from the wall the flux velocities of the resonant and the buffer components become equal. In contrast, in the near-the-wall layer with a thickness $\sim l$, the resonant molecules do not manage yet to undergo collisions with the buffer molecules, and the flux of the resonant molecules in this region remains fast. Detailed treatment [22] yields the following expression for the concentration drop along the cell valid in most experimentally interesting cases:

$$\Delta N \approx \frac{\Delta I}{\hbar\omega(\nu+\gamma)R}\frac{\upsilon_0}{\upsilon_T}A \sim \frac{V_0}{\upsilon_T}\frac{N}{R}L, \tag{25}$$

where $\Delta I \equiv n\,\hbar\omega L$ is the absorbed laser power density, and

$$A \equiv -\frac{N_b}{N+N_b}b\left\{\left[\frac{l_b}{l}+\frac{N_b}{N}\left(\frac{m_b}{m}\right)^{0.5}\right]\alpha_b+4\frac{\sqrt{2\pi\eta}}{m\upsilon_T NR}\right\}$$

$$\times\left[1+\frac{\alpha_b}{\alpha_g}\frac{N_b}{N}\left(\frac{m_b}{m}\right)^{0.5}+b\frac{n}{N(\nu+\gamma)}\frac{\nu}{\gamma}+4\frac{\sqrt{2\pi\eta}}{\alpha_g m\upsilon_T NR}\right]^{-1}. \tag{26}$$

Here m, N, l and m_b, N_b, l_b are the molecular masses, concentrations, and the mean free paths of the resonant and the buffer components correspondingly. One can see that, paradoxically enough, the magnitude of ΔN for a mixture can be much larger (by a factor R/l) than the value of ΔN for a single-component gas. The reason is that a compensation of the drift by the diffusion (as it happens in the multi-component mixture) requires a considerably higher value of ΔN than would be required for compensation by convective mixing. For the analysis of particular cases when Eq. (26) can be simplified the reader is referred to Ref. [22]. We will use elements of the described approach in part B of our analysis of SLID in porous media in the next section.

8. Surface light-induced drift in porous media

In this section, we will consider the new possibilities that arise for studies and applications of porous media using the effect of SLID. Diffusion and flow of gases in porous media have attracted considerable attention for many decades [52-55]. Interest in such research has been motivated by its many fields of application, such as surface catalysis, gas separations, soil physics, etc. Gas flows often arise here due to the difference in gas pressure or concentration between the opposing surfaces of the porous medium. If the gas flow in porous media is caused not by the concentration or pressure gradients but by SLID, then the flow of a gas becomes *selective*. Namely, one can force only the molecules of the resonant gas through the porous membrane, while leaving the other components untouched. It implies possible benefits in using the SLID for gas separations and other applications.

As the reader might recall from the discussion of Eq. (3), the magnitude of the relative pressure drop increases with the rise of the factor L/d. Obviously, L/d can be also large for microscopic capillaries, even at smaller L's. SLID is thus exceedingly sensitive to changes in molecule-surface interactions with excitation. Even in cases when $\Delta\alpha$ is very small, substantial L/d can lead to a measurable pressure drop. Indeed, as it was already mentioned above, in most of experiments [37,23-25], $\Delta\alpha$ measured in the experiments was as low as 0.1%. On the other hand for a given resonant gas, by both choosing an appropriate surface and its proper preparation, one could expect [22] to reach $\Delta\alpha \sim 1\text{-}10\%$, and thus we can have $\Delta P/P \gg 1$ (in the case where the laser intensity is sufficiently high, so that $(N_e/N)(v_0/v_T)(L/d) \gg 10 - 10^2$). It signifies the potential of SLID as a method for isotope separation. SLID through porous media might be advantageous also in order to operate at higher pressures, since the radius R of pores can be very small and the mean free path will exceed R even at elevated pressures. Also, due to a high surface/volume ratio, porous media can be especially favorable for studying the state-specificity of surface chemical reactions and catalysis.

Consider SLID in a transparent porous membrane that separates two macroscopic cells filled with the resonant gas irradiated by the laser. Many channels go from one end of the membrane to the opposite one, connecting the left cell to the right one. In turn each channel consists of many consecutive randomly oriented straight capillaries (pores). We will consider a simple model where the interconnectivity of different *channels* is not allowed. We also do not allow for the branching of channels. One should make a distinction between the initial flow profile which arises when the laser is switched on, and steady-state regime of flow which is set in the channels later, when the flux along each channel becomes constant throughout. When the laser is switched on, the initial speed of SLID (before pressure gradients build up) in each capillary depends upon the orientation of this capillary with respect to the laser beam; we denote the initial velocity in the

i-th capillary v_i. In what follows, SLID flux $\Pi_i = j_i s_i$ (where j_i is the flux density in the i-th capillary, and s_i is the capillary cross-section) is found in the regime of steady-state flow, where the value of the flux remains constant along the channel. This value, Π_i $(1 \le i \le n$, where n is the number of capillaries in a channel), is denoted below as simply Π.

We consider below two different regimes of the steady-state SLID flow. First we limit ourselves (as everywhere above except section 7) to the case of Knudsen diffusion, where the mean free path of a molecule l is much larger than the typical capillary radius R. The second regime of gas transfer occurs when $l \ll R$ and convection takes the place of diffusion as the mechanism of molecular drift [ ]. In both regimes, we neglect the adsorption of molecules by the capillary walls. We also limit the treatment to the case of media with the refractive index close to unity and neglect the possible distortion of the radiation wavefront and the scattering of light.

A. Rarefied (Knudsen) gas.

Similar to Section 2, we can calculate the initial speed of SLID in each capillary by solving a system of Maxwell-Boltzmann kinetic equations for the velocity distributions of excited, $f_e(v)$, and of non-excited, $f_g(v)$, gas-phase molecules in the strong-collision limit. We can determine then the net velocity distributions $f_i(v) = f_e + f_g$ and the flux velocities $v_i = j/N = N^{-1} \int v f_i(v) d^3 v$, where N is the concentration of the resonant gas.

It is useful to notice that, similar to the case of a single capillary parallel to the laser beam direction [22], the action of the walls of the pores on the irradiated gas is the same as in the situation when the walls of each capillary were moving in the direction of SLID flux with the velocity equal to v_i. Thus the problem of finding SLID flux Π in the regime of steady-state flow reduces to the following one: find the steady-state molecular flux in a channel consisting of n capillaries with the walls of each one moving with velocity v_i.

There are two contributions to the steady-state flux in each capillary: one, $v_i N$, is caused by the movement of the walls, and the second, $D_i \nabla N$, by diffusion, D_i being the diffusion coefficient for the i-th capillary. In the steady-state regime, if we neglect the channel branching, the flux along each capillary and along the whole channel does not vary. The first of these conditions, $v_i N - D_i \nabla N = const$, yields the spatial distribution of molecular concentration along the i-th capillary:

$$N(x_i) = A_i[exp(\alpha_i x_i) - 1] + N_{i-1}, \tag{27}$$

where A_i is a constant that can be found using the boundary conditions (see below), $\alpha_i = v_i/D_i$, x_i is the distance from the joint of $(i-1)$-th and i-th capillaries, and N_j is the concentration at the joint of the j-th and $(j+1)$-th capillaries. Equating then the concentrations at each of the joints, $N(x_i = L_i) = N(x_{i+1} = 0)$, where L_i is the length of the i-th capillary, we obtain the following set of equations:

$$A_1[\exp(\alpha_1 L_1)-1]-N_1 = -N_0,$$

$$A_2[\exp(\alpha_2 L_2)-1]+N_1-N_2=0,$$

$$A_j[\exp(\alpha_j L_j)-1]+N_{j-1}-N_j=0, \qquad (28)$$

$$\cdots\cdots$$

$$A_{n-1}[\exp(\alpha_{n-1} L_{n-1})-1]+N_{n-2}-N_{n-1}=0,$$

$$A_n[\exp(\alpha_n L_n)-1]+N_{n-1}=-N_r.$$

We assume here that the molecular concentrations at the edges of the membrane are equal to the concentrations, N_0 and N_r, in the left and the right cells correspondingly. Equating the fluxes at each of the joints, we get:

$$(s_1 v_1 - s_2 v_2)N_1 - s_1 v_1 A_1 \exp(\alpha_1 L_1) + s_2 v_2 A_2 = 0,$$

$$(s_2 v_2 - s_3 v_3)N_2 - s_2 v_2 A_2 \exp(\alpha_2 L_2) + s_3 v_3 A_3 = 0,$$

$$\cdots\cdots$$

$$(s_i v_i - s_{i+1} v_{i+1})N_i - s_i v_i A_i \exp(\alpha_i L_i) + s_{i+1} v_{i+1} \qquad (29)$$

$$(s_{n-1} v_{n-1} - s_n v_n)N_{n-1} - s_{n-1} v_{n-1} A_{n-1} \exp(\alpha_{n-1} L_{n-1}) + s_n v_n A_n = 0,$$

where $s_i = \pi R_i^2$ is the cross-section of the *i-th* capillary, R_i is its radius. We thus allow s_i to vary, as it is done in the converging-diverging pore model [53]. We have used a common estimation for the value of D_i, $D_i \sim R_i v_T$. Together, Eqns. (28) and (29) constitute a system of *(2n-1)* equations with *(2n-1)* unknowns, A_i *(1 ≤ i ≤ n) and N_j (1 ≤ j ≤ n-1)*, which has a unique solution for a non-zero determinant. Consider a further simplification yielded by treating the case when $max(\alpha_i L_i) \ll 1$, or $max(v_i L_i / R_i) \ll v_T$. In this case, the concentration profile along each pore is linear and the equality of fluxes in each capillary along the channel yields the following system of equations:

$$
\begin{pmatrix}
q_1-q_2-k_1-k_2 & -q_2+k_2 & 0 & 0 & \cdots & 0 & 0 & 0 & 0 \\
q_2+k_2 & q_2-q_3-k_2-k_3 & -q_3+k_3 & 0 & \cdots & 0 & 0 & 0 & 0 \\
0 & q_3+k_3 & q_3-q_4-k_3-k_4 & -q_4+k_4 & \cdots & 0 & 0 & 0 & 0 \\
 & & & & \cdots & & & & \\
0 & 0 & 0 & 0 & \cdots q_{n-3}+k_{n-3} & q_{n-3}-q_{n-2}-k_{n-3}-k_{n-2} & -q_{n-2}+k_{n-2} & 0 \\
0 & 0 & 0 & 0 & \cdots & 0 & q_{n-2}+k_{n-2} & q_{n-2}-q_{n-1}-k_{n-2}-k_{n-1} & -q_{n-1}+k_{n-1} \\
0 & 0 & 0 & 0 & \cdots & 0 & 0 & q_{n-1}+k_{n-1} & q_{n-1}-q_n-k_{n-1}-k_n
\end{pmatrix}
\begin{pmatrix} N_1 \\ N_2 \\ N_3 \\ \cdots \\ N_{n-3} \\ N_{n-2} \\ N_{n-1} \end{pmatrix}
=
\begin{pmatrix} -(q_1+k_1) \\ 0 \\ 0 \\ \cdots \\ 0 \\ 0 \\ (q_n-k_n)N_r \end{pmatrix} \qquad (30)
$$

Here $q_i=0.5s_iv_i(<L>/<R>)$, $k_i=s_iR_i/L_i$. $<R>$ and $<L>$ are the mean capillary radius and length correspondingly. We have rescaled the variables so that $v_i=v_i/v_T$, $L_i=L_i/<L>$, $R_i=R_i/<R>$, $s_i=s_i/(\pi<R>^2)$, and $N_i=N_i/N_0$. After the system (30) is solved for N_i, the rescaled steady-state flux through the channel Π can be easily calculated. This rescaled steady-state flux is defined as $\Pi\equiv\Pi/(\pi<R>^2\cdot v_T\cdot N_0)$, which is approximately equal to $\Pi\approx s_1[v_1(1+N_1)/2 - (R_1/L_1)(<R>/<L>)(N_1-1)]$.

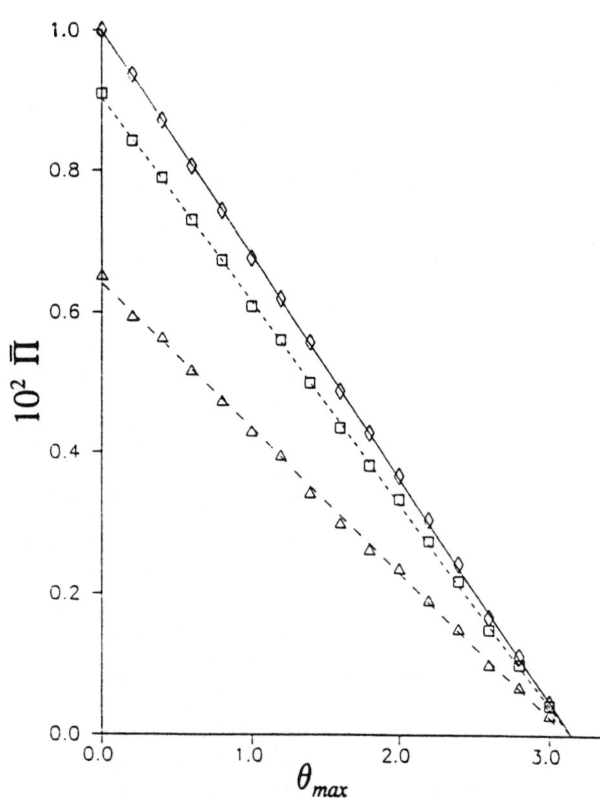

Fig. 7. Steady-state SLID flux vs. θ_{max}, where θ_{max} is the parameter describing the dispersion of the orientations of capillary pores; $\theta_{max}=0$ corresponds to a set of parallel capillaries (Fig. 9(a)), while $\theta_{max}=\pi$ corresponds to completely random orientations of the pores (Fig. 9 b).

As mentioned before, initial velocity of SLID flux in each capillary, v_i, depends upon its orientation with respect to the laser-beam direction. Initial SLID flux is maximal for the pores parallel to the laser beam, and zero for the pores perpendicular to it (where effective selectivity of excitation is zero). To simplify the analysis even further, we linearly approximate this dependence by $v_i = V_s[1-\theta_i/(\pi/2)]$, where θ_i is the direction between the orientations of the *i-th* capillary and of the laser beam and V_s is the initial velocity of SLID flux in a capillary parallel to the laser-beam direction ($\theta=0$). In our numerical analysis of the system (30), we allow θ_i to vary randomly between 0 and some θ_{max}. Similarly, the rescaled radius of the capillary varies randomly between r' and $2-r'$, and rescaled capillary length between L' and $2-L'$ (so

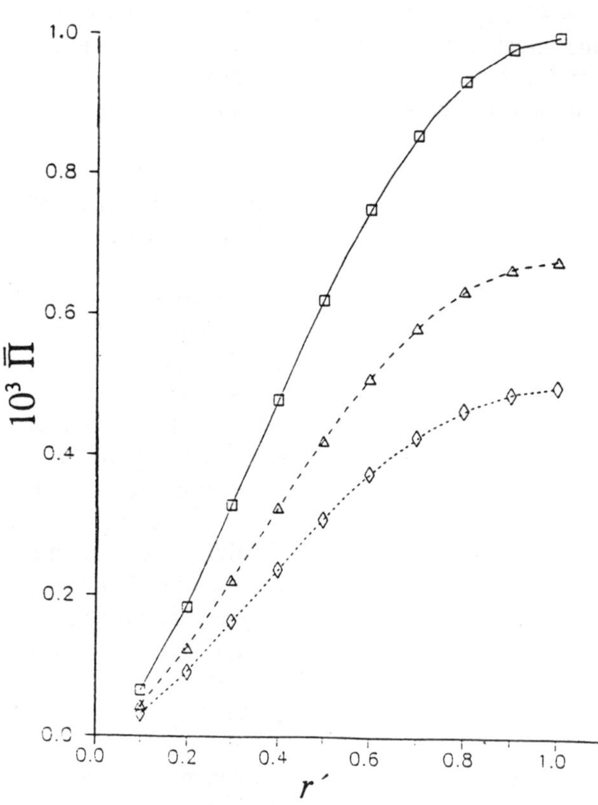

Fig. 8. Steady-state SLID flux vs. r', with parameter r' describing the dispersion of the pores' radii.

that the average values of R_i and L_i are both 1).

Some of our numerical results for the case $N_r = N_0$ are presented in Figs. 7 and 8. Fig. 7 represents the dependence on θ_{max} of the steady-state SLID flux, Π, through a channel consisting of $n = 1000$ capillaries (averaged over 100 runs). By virtue of using linear approximation for $v_i(\theta_i)$, the physical meaning of this result is readily apparent. Steady-state flux through the membrane turns to be equal to the average of the initial fluxes arising in the capillaries of different orientation. Case $\theta_{max} = 0$ corresponds to the porous medium consisting of a "bundle of capillary tubes" [54,55] (Fig. 9(a)) oriented parallel to the laser-beam direction; obviously the flux is maximal in this case. On the other hand, case $\theta_{max} = \pi$ corresponds to a situation when the interconnectivity of the pores in a hypothetical low-porosity medium is such that the path of the gas molecules inside this medium is a random walk (Fig. 9(b)). For this latter situation, simple considerations indicate that due to the counterbalancing of the oppositely directed SLID fluxes within each loop (Fig. 9(b)), resulting flux is drastically ($\propto n^{-0.5}$) decreased, as it is reflected in Fig. 7. Obviously, the parameter θ_{max} is closely associated with the porosity τ of the porous medium [52-54]. Therefore, strong dependence of the SLID flux upon θ_{max} can be used to determine τ.

Our calculations indicate that the average flux, Π, depends only slightly upon the dispersion of the capillary length L_i but quite strongly upon the

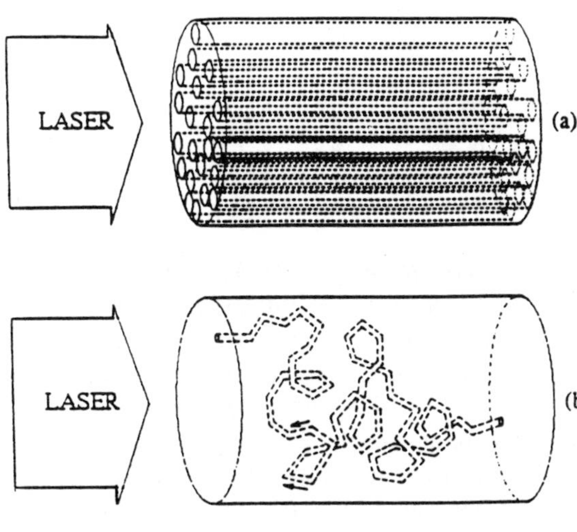

Fig. 9. SLID in porous media: steady-state SLID flux is sensitive to the texture of the medium. In (a) the porous media can be adequately represented as a set of parallel capillaries; in (b), interconnectivity of pores in a hypothetical low-porosity media is such that the path of the gas molecules inside this media is a random walk (only one channel is shown).

dispersion of the capillaries radii. The dependence of the average flux upon the parameter r' is shown in Fig. 8. Notice the monotonic increase of the flux as r' grows from 0 to 1, despite the fact that the average value of R_i remains unity throughout. Recall that the closer r' is to 1, the less is the variation of the rescaled capillary radii R_i (since $r' \leq R_i \leq 2-r'$). Thus, according to the data of Fig. 8, the less variation of R_i around 1 is allowed, the larger is $\prod$. It might be related to the 'bottleneck' effect, when the transmittance of the channel is determined by the transmittance of the most narrow links.

B. Dense gas.

Similar to the case $l \gg R$ considered above, in the case of a dense gas ($l \ll R$) the action of the walls of the pores on the irradiated gas is the same as the condition when the walls of each capillary were moving in the direction of SLID flux with the velocity equal to v_i. If, instead of a channel, we had a single capillary without ends or with infinitely remote ends, all of the gas would be drawn to the motion at a velocity $v = v_i$ by virtue of viscosity. In this case, the velocity of SLID flux may be not lower than for $l \gg R$. But if this single capillary would be bounded by the lateral ends, the gas would accumulate at one of these ends until the SLID flux were balanced by the oppositely directed convection caused by the pressure drop that would arise. Obviously, in this latter case, there is no net mass transfer through this capillary. The analysis presented below indicates that the situation realized in porous media is close to the case of an open-end capillary.

Following the treatment of the previous section, the case is considered when the frequencies of intermolecular bulk collisions are the same for the ground-state and excited molecules, so that the bulk light-induced phenomena do not occur. To find transverse velocity profile $V(r)$ in each capillary, we use the Navier-Stokes equation

$$\frac{\partial}{\partial r}(r\frac{\partial V}{\partial r})=\frac{\nabla N}{N}\frac{r}{l}\upsilon_T \tag{31}$$

with the following boundary conditions: $V(r=R_i) = \upsilon_i,\quad \partial V/\partial r(r=0) = 0$. A solution for (31) is readily obtained, $V(r) = \upsilon_i + [(r^2-R_i^2)/4l_i](\nabla N/N)\upsilon_T$. We find the net flux through capillary by integrating $V(r)$ over the cross-section. Equating fluxes in each pair of the adjacent capillaries, we arrive at a system of equations identical to Eq. (30) with $k_i = \pi R_i^4/(8l_i L_i)$, $l_i=l_i/<R>$ and same other notations. Thus, the results obtained in section 8A are applicable here.

Let us make an estimate. For the laser excitation in IR region with power density $P \sim 100\ W/cm^2$, and $\Delta\alpha \sim 0.1$ one can get $V_s \sim 10^2 \upsilon_T$ [22]. Taking $\theta_{max} \sim 1$, which corresponds to $\tau \sim 2$, porosity ~ 0.5 [52], $r'=0.5$, $N=10^{15}$ cm^{-3}, and $\upsilon_T = 10^5$ cm/s, one gets the SLID flux of about 10^{17} molecules per second for a membrane of 1 cm^2 area. If the membrane is 10 mm thick and $<R> \sim 10\mu$, then in accordance with Eq. (3) $\Delta P/P \sim 1$. For the volume 1 cm^3 for each of the two lateral cells, one needs to transfer $\sim 10^{15}$ molecules to reach the new equilibrium. Thus, time to reach equilibrium is about 10^{-2} s. Obviously, conditions arising for a real medium (adsorption, channel interconnectivity, dead-end loops, etc.) can significantly affect the above estimates and, in particular, increase the time needed to reach equilibrium.

The above estimation holds (at least in the case of Knudsen regime) also for the *mixture* of the resonant and non-resonant gases. Since there is no initial pressure drop along the membrane, there is no driving force to cause the non-resonant component(s) of the mixture to move [56]. It demonstrates once again the potential advantages of using SLID in the membrane separations in comparison with the traditional methods, where separation is achieved only due to the small differences in the diffusion rate for different components.

In conclusion, in this section we have analyzed here Surface light-induced drift in transparent porous media, which might be especially advantageous for the isotope separation and studying state-specific chemical reactions. In the simple model considered, we have calculated the steady-state SLID flux through the porous channels. The numerical results indicate that the magnitude of the steady-state SLID flux strongly depends upon the pores' texture. It means that SLID could be used to study this texture. Additional studies are necessary in order to analyze SLID in a real porous media, where models with internally connected pore segments (network models) [54,55] might be more appropriate. Optical properties of real porous media should be taken into account. More rigorous treatment of the angular dependence of v_i should be incorporated. We have completely neglected adsorption and surface diffusion of the resonant molecules along the pores; these effects can become especially important in the media with smaller pore sizes.

9. Other models and prospective directions

In the above treatment, in order to find velocity distributions $f_{e,g}$ we have been using a simple model where the scattering of the gas molecules by the surface has been incorporated into the collision kernel of the bulk kinetic equation, as if these were intermolecular collisions in the bulk. As it has already been mentioned in section 7 [22], an alternative approach would include solving a system of kinetic equations with only the bulk collisions accounted for by the collision kernels:

$$v_e \, [N_e \, W(v) - f_e] + Q(v)(f_g - f_e) - \gamma f_e = v_z \frac{\partial f_e}{\partial z} \qquad (32a)$$

$$v_g \, [N_g \, W(v) - f_g] - Q(v)(f_g - f_e) + \gamma f_e = v_z \frac{\partial f_g}{\partial z}, \qquad (32b)$$

while gas-surface collisions would then be taken into consideration by the appropriate boundary conditions (different for the excited and the ground-state molecules):

$$f_e^A(v_z > 0) = (1 - \alpha_e) \, f_e^A(v_z < 0) \qquad (33a)$$

$$f_g^A(v_z > 0) = (1 - \alpha_g) \, f_g^A(v_z < 0) \qquad (33b)$$

Here $v_{e,g}$ are now the frequencies of the *bulk* collisions; $f_{e,g}{}^A$ are the non-Maxwellian parts of the velocity distribution functions $f_{e,g}$, for which $\int f_{e,g}{}^A d^3 v = 0$; z is the direction of the surface normal, so that for the molecules approaching the wall $v_z < 0$, while for the scattered molecules $v_z > 0$. Boundary conditions (33a,b) describe the fact that in the Maxwell model of surface scattering, while the Maxwellian part remains Maxwellian after scattering, only the $(1 - \alpha_{e,g})$-*th* part of the non-Maxwellian $f_{e,g}{}^A$ doesn't vanish after scattering. Similar approach realized by Bazelyan and Kogan [57] and Kogan and Surdutovich [58] led to the prediction of a novel effect of the drift of a single-component gas arising due to the state dependency of the *bulk* collision frequencies. Notice that everywhere above we have neglected this state dependency; in this situation the exact solution yields results that differ from the presented ones only by a coefficient of the order of unity [59]. Chernyak, Vintovkina, and Chermyaninov [60] used similar approach to analyze a single-component mixture for a broad range of gas concentrations, from hydrodynamic to the Knudsen (free-molecular) regimes. Boltzmann collision integrals were not modelled by the strong-collision model (unlike what is represented in the Eqs. (1a,b)), and an integral-moment method based on the transformation of the integrodifferential kinetic equation for the velocity distribution function into a system of equations for its moments has been employed. The results are in a good agreement with the experimental data [23]. Zhdanov, Krylov, and Roldugin [61] proposed a theoretical treatment of light-induced kinetic effects in which the "bulk" light-induced drift and SLID are treated simultaneously by using an average of the system of linearized kinetic equations over the cell cross-section. This enables one to express the fluxes of interest averaged over the cross section in terms of several moments of the velocity distribution function at the cell wall. New terms related to the "wall" effects occur in the expression for the drift velocity of the resonant gas in the case when a buffer gas is present. The results agree qualitatively with the experimental dependence of the pressure drop along the cell vs. the initial pressure [62,63] for the range Kn $\equiv l/R \leq 0.25$. More satisfactory quantitative agreement in the results can probably be achieved [61] by taking into account a correction factor associated with the finite ratio of the homogeneous and Doppler line broadening, and also the variation in the radiation intensity over the cell cross-section.

It seems appropriate to discuss here also some of the prospective directions for the SLID-related research. At present, quantitative results on SLID with surface roughness [39] have been obtained only in a simple model of a regular zigzag-shaped corrugation having a small angle of roughness $\psi \ll 1$. When the roughness increases, its influence grows rapidly ($\sim \psi^2 \ln^2 \psi$), making the case $\psi \sim 1$ especially important for a quantitative analysis. This analysis can be performed by solving numerically the system of two non-linear integro-differential equations for the velocity distribution functions $f_{e,g}(v)$ similar to Eqs. (8a,b).

The only experiments on SLID performed so far have been concerned with molecular gases, with velocity-selective excitation of vibrotational transitions, which are obviously long-lived. No data are available yet on the variation of the momentum accommodation coefficient (α) with electronic excitation (visible, UV and VUV excitation). Being a very sensitive tool for measuring slight changes in α, SLID may be a favorable tool for obtaining such data. In the case of a short-lived electronic excitation, however, surface roughness can play also a negative role, destroying state-selectivity of excitation in the vicinity of the roughness facets. To perform experiments on SLID with short-lived electronic excitation (for example, SLID of K vapor), it becomes therefore important to study the influence on SLID of the light field distortion near the rough surface. In such an analysis, one can use the computer simulation approach developed for the studies of the scattering of light from one-dimensional rough surfaces [65].

Suppose one excites the walls of the cell by one short pulse (at some frequency $\tilde{\omega}_{wall}$ not necessarily close to $\tilde{\omega}_0$), and illuminates the near-surface region of the gas by another pulse (at the frequency $\tilde{\omega}$ close to $\tilde{\omega}_0$). Varying the duration of the pulses, frequencies and delay between pulses, one can obtain important information concerning the dynamics of scattering and adsorption-desorption processes. In addition, similar to the case of the bulk LID [66], pulse-periodic excitation will probably be the most energy-efficient way for SLID isotope separation in porous media. Quantitative analysis is necessary in order to optimize duration of pulses and their frequency. This analysis will require solving the system of kinetic equations for the gas-molecules velocity distributions in the gas phase and for adsorption-desorption phenomena on the surface.

If the cell is optically thick (i.e. the absorption length is much smaller than the cell length), then initially the laser radiation penetrates only a small fraction near the cell entrance and can SLID occur only in this illuminated region. If the SLID flux is directed from the laser, it will push the molecules away from the cell entrance, thus decreasing the concentration there and allowing the radiation to penetrate further into the bulk of the cell [67]. Often, besides the resonant gas, a buffer gas should be added to the cell, so that the mean free path l_f becomes much less than the cell diameter. Movement of the resonant gas in this case proceeds due to transverse diffusion of the resonant molecules toward the wall and a rapid near-wall SLID flux, resembling, in a sense, a worm's crawling ("light-induced worm") [68]. This effect is analogous to the experimentally observed phenomenon of optical piston [69,70] associated with the bulk mechanisms (LID). Light-induced "worm" provides another interesting direction of study. In particular, cases in which the directions of LID and SLID fluxes are anti-collinear, can be noteworthy. In this situation, one could expect existence of threshold values of $\Delta\alpha$, below which the "piston" can not arise. Determination of certain parameters of the optical piston, such as front steepness and velocity, can provide sufficient information to determine the state dependence of the surface accommodation coefficients vs. the surface coverage of the buffer-gas molecules. Quantitative analysis in this situation should include solving the system of equations describing light absorption along the cell and the kinetics of the gas mixture. Such an analysis is more complicated for SLID "piston" than in the case of a LID "piston" [69,70], where the problem can be reduced to one-dimensional. In some cases, numerical study will become necessary. Besides a simplified approach [22] to the kinetics of the dense gas interacting with the surface, a more rigorous, generalized Chapman-Enskog procedure developed [71] for the case of a pure resonant gas, could be useful in the proposed study.

Another scope of the prospective research is associated with the fact that SLID is influenced strongly by the features of surface coverage [72]. On the other hand, the surface coverage can change abruptly by adsorption or desorption at a certain value of the absorbing gas pressure [73]. Therefore, adsorption of resonant particles can provide a feedback between the concentration N and the SLID flux j. It seems interesting to analyze the possible results of this feedback on the stability of the state of the gas [73], and the existence of non-stationary regimes in SLID and other similar phenomena [74]. Again it will require solving the system of (coupled) equations describing kinetics of the gas phase and of adsorption/desorption processes at the cell-wall surface.

Although the subject of this chapter has been exclusively the theoretical and computational aspects of SLID, further developments definitely depend upon continued success of the experimental studies of the phenomenon; see papers by Hermans et al. [23-25] and the description of their most recent results [75] for details.

The above studies are related to gas scattering at the solid surfaces. SLID can also be used in the studies of gas-liquid scattering; the latter are yet at an early stage [76]. Another interesting extension of this subject and light-induced drift (LID) in general arises in applications to the solid-state physics; for the detailed description and bibliography the reader is referred to a recent paper by Shalaev et al. [77].

REFERENCES CITED

1. R.E. Smalley, D.H. Levy and L. Wharton, J. Chem. Phys. **64**, 3266 (1976).
2. J.A. Barker and D.J. Auerbach, Surf. Sci. Rep. **4**, 1 (1985).
3. C.R. Arumainayagam and R.J. Madix, Progress in Surf. Sci. **38**, 1 (1991).
4. D.S. King and R.R. Cavanagh, J. Chem. Phys. **76**, 5634 (1982).
5. D.J. Auerbach, C.A. Becker, J.P. Cowin, and L. Wharton, Appl. Phys. **14**, 141 (1977).
6. J.E. Hurst, L. Wharton, K.C. Janda and D.J. Auerbach, J. Chem. Phys. **78**, 1559 (1983).
7. G.P. Brivio and T.B. Grimley, Surf. Sci. Rep. **17**, 1 (1993).
8. R.R. Cavanagh and D.S. King, Phys. Rev. Lett. **47**, 1829 (1981).
9. M. Asscher, W.L. Guthrie, T.-H. Lin, and G.A. Somorjai, Phys. Rev. Lett. **49**, 76 (1982); J. Chem. Phys. **78**, 6992 (1983).
10. A. Modl, H. Robota, J. Segner, W. Vielhaber, M.C. Lin, and G. Ertl, J. Chem. Phys. **83**, 4800 (1985).
11. J.C. Tully and M. Cardillo, Science, **223**, 445 (1984).
12. A. Zangwill, *Physics at Surfaces*. Cambridge University Press, Cambridge, 1988, p. 392.
13. J.A. Serri, J.C. Tully, and M. Cardillo, J. Chem. Phys. **77**, 2185 (1982).
14. G.A. Somorjai, Surf. Sci. **242**, 481 (1991).
15. T.J. Chuang, Surf. Sci. Rep. **3**, 1 (1983).
16. P.L. Houston and R.P. Merrill, Chem. Rev. **88**, 657 (1988).

17.   H.F. Berger, M. Leisch, A. Winkler, and K.D. Rendulic, Chem. Phys. Lett. **175**, 425 (1990).

18.   H.F. Berger, E. Grosslinger, and K.D. Rendulic, Surf. Sci. **261**, 313 (1992).

19.   C.T. Rettner, D.J. Auerbach, and H.A. Michelsen, Phys. Rev. Lett. **68**, 1164 (1992).

20.   A.V. Ghiner, M.I. Stockmann, and M.A. Vaksman, Phys. Lett. **96A**, 79 (1983).

21.   M.A. Vaksman, Phys. Chem. & Mech. Surf. (Great Britain) **3**, 3222 (1985).

22.   M.A. Vaksman and A.V. Ghiner, Sov. Phys. JETP **62**, 23 (1985).

23.   R.W.M. Hoogeveen, G.J. van der Meer, and L.J.F. Hermans Phys. Rev. **A42**, 6471 (1990).

24.   G.J. van der Meer, B. Broers, R.W.M. Hoogeveen, and L.J.F. Hermans Physica **A 182**, 47 (1992).

25.   B. Broers, G.J. van der Meer, R.W.M. Hoogeveen, and L.J.F. Hermans J.Chem. Phys. **95**, 648 (1991).

26.   R.W.M. Hoogeveen, L.J.F. Hermans, V.D. Borman, and S. Yu. Krylov, Phys. Rev. **A42,** 6480 (1990).

27.   R.W. Verhoef, D. Kelly, C.B. Mullins, and W.H. Weinberg, Surf. Sci. **287/288**, 94 (1993).

28.   R.W. Verhoef, D. Kelly, C.B. Mullins, and W.H. Weinberg, Surf. Sci. **325**, 93 (1995).

29.   C.T. Rettner, H.A. Michelsen, and D.J. Auerbach, J. Chem. Phys. **102**, 4625 (1995).

30.   C.B. Mullins, C.T. Rettner, D.J. Auerbach, and W.H. Weinberg, Chem. Phys. Lett. **163**, 111 (1989).

31.   C.R. Arumainayagam, R.J. Madix, M.C. McMaster, V.M. Suzawa, and J.C. Tully, Surf. Sci. **226**,180 (1990).

32.   F.Kh. Gel'mukhanov and A.M. Shalagin, JETP Letters **29**, 711 (1979).

33.   A.L. Kaljazin and V.N. Sazonov, Sov. J. Quantum Electron **9**, 956 (1979).

34.   A.V. Ghiner, Optics Comm. **41**, 27 (1982).

35.   A.V. Ghiner and M.A. Vaksman, Phys. Lett. **100A**, 428 (1984).

36.   E.H. Kennard, Kinetic theory of gases. New York, London, McGraw-Hill Book Company, 1938, p. 306.

37.   R.W.M. Hoogeveen, R.J.C. Spreeuw, and L.J.F. Hermans, Phys. Rev. Lett. **59**, 447 (1987).

38.   M.A. Vaksman, Chem. Phys. Lett. **267**, 77 (1997).

39.   M.A. Vaksman, Phys. Rev. **A 44**, # 7, **R**4102 (1991).

40.   M.A. Vaksman and A. Ben-Reuven, Phys. Rev. **A 45**, # 11, 7883 (1992).

41.   M.A. Vaksman, Phys.Rev. **A 52**, 2179 (1995).

42.   M.A. Vaksman and Isabela Podgorski, Canadian Journ. Phys. **74**, 251 (1996).

43.   To be precise, the transverse velocity distribution does change slightly, but it brings only   higher-order contributions to the flux (see, for example, Refs. [34,35]).

44.   M.A. Vaksman, Sov. Phys. Tech. Phys. **29 (6)**, 681 (1984).

45.   G.J. van der Meer, J. Smeets, S.P. Pod'yachev, and L.J.F. Hermans, Phys. Rev. **A45**, R1303 (1992); F. Kh. Gel'mukhanov and A.I. Parkhomenko, Phys. Lett. **A162**, 45 (1992); F. Yahyaei-Moayyed and A.D. Streater, Phys. Rev. **A53**, 4331 (1996).

46.   M.A. Vaksman, Phys. Rev. **A 44,** 3125 (1991).

47.   A.S. Berman and W.J. Maegley, Phys. Fluids **15**, 772 (1972).

48.   F.O. Goodman, Phys. Fluids **A1**, 741 (1989).

49.   A.V. Ghiner and M.A. Vaksman, Opt. Commun. **41**, 263 (1982).

50.   S.N. Atutov, Phys. Lett. **A119**, 121 (1986).

51.   V.R. Mironenko and A.M. Shalagin, Izv. Akad. Nauk SSSR, Ser. Fiz. **45**, 995 (1981).

52.   P.C. Carman, *Flow of Gases Through Porous Media*. Academic Press (1956).

53.   G.R. Youngquist, Diffusion and Flow of Gases in Porous Solids.- In: *Flow Through Porous Media*. American Chemical Society Publications, Washington, D.C. (1970), p. 63.

54.   F.A.L. Dullien, *Porous Media: Fluid Transport and Porous Structure*. Academic Press (1992), pp. 167-313.

55.   E.A. Mason and A.P. Malinauskas, *Gas Transport in Porous Media: The Dusty-Gas Model*. Elsevier, Amsterdam-Oxford-NY (1983), pp. 66-67.

56.   In fact as the resonant component flows through the membrane, the non-resonant components will be forced in the *opposite* direction by the light-induced pressure drop ΔP arising between the macroscopic cells connected by the membrane. Clearly, it enhances the separation effect.

57.   A.E. Bazelyan and M.N. Kogan, Sov. Phys. Dokl. **34**, 770 (1989).

58.   M.N. Kogan and E.A. Surdutovich, Sov. Phys. JETP **71**, 62 (1990).

59.   We have obtained and studied the exact solution of kinetic equations with corresponding boundary conditions for the case of velocity-selective excitation prior to the publication of papers Refs. 57,58 [M.A. Vaksman and A.V. Ghiner, Report at the All-Union Seminar on Laser-Induced Surface Phenomena, Leningrad, 1986 (unpublished)].

60.   V.G. Chernyak, E.A. Vintovkina, and I.V. Chermyaninov, Sov. Phys. JETP **76**, 768 (1993).

61.   V.M.Zhdanov, A.A. Krylov, and V.I. Roldugin, Sov. Phys. JETP **78,**49 (1994).

62.   R.W.M. Hoogeveen, G.J. van der Meer, R.W.M. Hoogeveen, L.J.F. Hermans et al., Phys. Rev. **A39**, 5539 (1989).

63.   R.W.M. Hoogeveen, *Light-Induced Kinetic Effects in Molecular Gases,*

Huygens Laboratorium, Leiden (1990).

64. Levdanskii, Sov. Phys.-Tech. Phys. **28**, 518 (1983) has assessed independently the magnitude of the change in the capillaries' transmittance for the entering resonant molecules upon velocity-selective excitation.

65. A.A. Maradudin, T. Michel, A.R. McGurn, and E.R. Mendez. Ann. Phys. (N.Y.) **203**, 205 (1990).

66. A.K. Popov, V.M. Shalaev, and V.Z. Yakhnin, Sov. Phys. JETP **55**, 431 (1982).

67. M.A. Vaksman and <u>Ginger Musolf</u>. *Light-induced "Worm".* Talks presented at the 1996 Great Lakes College Chemistry Conference (March 16, 1996, Michigan State University, Lansing, Michigan) and the 28th Central Regional Meeting of American Chemical Society (June 9-12, 1996, Dayton, Ohio).

68. <u>M.A. Vaksman</u> and A. Ben-Reuven. Talk presented at the 12th International Vacuum Congress / 8th International Conference on Solid Surfaces, 12-16 October 1992, The Hague, The Netherlands. -Surf. Sci., **287/288**, 196 (1993).

69. G. Nienhuis, Phys. Rep. **138**, 152 (1986); Phys. Rev. **A40**, 269 (1989).

70. G. Nienhuis, Opt. Comm. **62**, 81 (1987); H.G.C. Werij and J.P. Woerdman, Phys. Rep. **169**, 145 (1988).

71. S.J. van Enk and G. Nienhuis, Physica **A179**, 199 (1991).

72. A. Thomy, X. Duval and J. Regnier, Surf. Sci. Rep. **1**, 1 (1981).

73. A.V. Ghiner and M.A. Vaksman, Phys. Rev. **A54**, 3270 (1996).

74. M.A. Vaksman, Phys. Rev. **A 48**, R26 (1993).

75. E.J. van Duijn, R. Nokhai, L.J.F. Hermans, A. Yu. Pankov, and S. Yu. Krylov, J. Chem. Phys. **107**, 3999 (1997).

76. M.E. Saecker et al., Science **252**, 1421 (1991).

77. V.M. Shalaev, C. Douketis, J.T. Stuckless, and M. Moskovits, Phys. Rev. **B53**, 11388 (1996).

THEORETICAL TREATMENT OF SURFACE ADSORBATES

László NÁNAI and Csaba BELEZNAI
Department of Experimental Physics, József Attila University,
Dóm tér 9, 6720 Szeged, Hungary
E-mail: nanai@physx.u-szeged.hu

Thomas F. GEORGE
Office of the Chancellor / Departments of Chemistry and Physics & Astronomy
University of Wisconsin-Stevens Point
Stevens Point, Wisconsin 54481-3897, USA
E-mail: tgeorge@uwsp.edu

Theoretical and computational methods used to describe chemisorption and adsorbate reactions on solids, mainly metallic surfaces, are summarized. The main focus is on electronic structural treatments of a one-electron nature, based on computational techniques in quantum chemistry. Different approaches are considered, such as *ab initio*, density functional, slab, atomic cluster and embedding methods, where the main goal is to describe the energetics of chemisorption and reactions of adsorbates on surfaces and structures with different bonding geometries at molecular and atomic levels.

1 General background

The mechanisms of surface reactions can be divided into two groups: Langmuir-Hinshelwood and Eley-Rideal-type reactions [1]. The first is based on the assumption that the surface species are in an energetic equilibrium with the surface; this mechanism includes the interaction between two adsorbates or between an adsorbed species and a vacant site. The Eley-Rideal mechanism considers reactions between gas-phase species and adsorbates, where products may remain adsorbed or desorb; the molecular-level description can be verified by experimentally-observable adsorbate-induced surface restructuring effects, where rearrangement of the surface atoms occurs around the adsorption site.

The general requirements for theories of surface processes are: (1) to be accurate enough to describe energetic species of molecular/atomic level reactions and the surface-adsorbate bond, and (2) to provide the appropriate framework to describe the different charge transfer and conduction processes within the system of interest. In order to evaluate the system size requirements with respect to the "energetics," one finds the following: a system with a small number of host atoms coupled with adatoms produces good results, while calculations of charge transfer processes require the inclusion of a large number of atoms (clusters) [2-6].

While the convergence situation is different from method to method the adsorbate-metal system generally requires a large number of atoms for convergence. Also, there is an important problem related to the cluster boundary, which might distort the periodicity of the electronic wavefunction in the vicinity of the adsorption site. Regarding the interaction between the adsorbate and metal's host atoms and the contribution of s, p and d orbitals, calculations show that even for transition metals, the contribution of s and p orbitals provide a major part of adsorption energy, with only minor contribution of the d orbitals.

The overall objective of various calculations, independent of the methods used, is to give information with respect to the:

> total energy as a function of adsorbate-surface circumstances.
> adsorption energy
> wavefunctions (electron and charge densities)
> work functions and other potentials
> equilibrium states with reaction ways and thermodynamic properties.

In Section 2 below, we provide a description of various theories and computational models developed for this task, and in Section 3 we give the summary and conclusions.

2   Theories and computations

Developing theories in order to understand the energetics, dynamics and other properties of adsorbate-surface systems is a task of wide range investigations from pure speculations to experimentally-based computer simulations. The theoretical studies deal mostly with problems of the determination of electronic structure and understanding adsorbate-surface interactions and reaction mechanisms at the molecular level. With respect to computational studies, one can distinguish between two different strategies based on the so-called cluster and slab calculations. The main advantages of cluster-type calculations are associated with the local character of the adsorbate-surface interaction and allow quantum chemistry calculations on atoms, fragments or more extended substrate parts. On the other hand, slab-type calculations incorporate more of the long-range effects, allowing investigations of overlayer and multilayer structures. Therefore, the slab calculations include adsorbate-adsorbate interactions at higher coverage, while the cluster calculations are more specific in the sense of isolated adsorbing species at low-coverage level.

The main requirement for the theories is their ability to answer the following questions [4,5,7-11]:

> adsorbate bonding peculiarities
> adsorbate structures

energetics
reaction heat and activation energies
nonlinearities arising in interaction phenomena
nonequilibrium geometries in potential surfaces
trapping and activation effects.

Methods developed to answer the above have a difficulty connected with the so-called size effects. In order to obtain an accurate description of the energetics and bonding of the adsorbate-surface system, small clusters (fewer than ten atoms) are sufficient, while for describing processes of conduction and charge transfer effects a large number of atoms has to be included in the calculations which must deal with the problems of convergence and boundary conditions [5,12]. Also, including the contribution of core-electron pseudopotentials for the electron correlation might become important. Nevertheless, we conclude that most of the adsorption energy in practically important applications comes from the s-p interactions.

2.1 *Vibrational relaxation of adsorbed particles related to periodic structures*

The localized vibrational relaxations of adsorbed particles on solids can be realized via interactions with phonons and conduction electrons. Therefore, a highly-accurate description (and computation) of periodic systems with one-, two- or three-dimensional treatments is highly desirable. Density functional and Hartree-Fock frameworks are used for approximate self-consistent field (SCF) calculations [1,9,12-15]. For the real (position) and k-space calculations, a special program named "CRYSTAL" has been developed and is available [8]. The density functional methods based on the Kohn-Sham approximation take into consideration numerical integration methods for the computation of Hamiltonian matrix elements. Numerical integration methods have been implemented based on the partitioning of space and application of product Gauss rules, where the approximation of integrals over the Brillouin zone and evaluation of the Coulomb potential via density fitting yield rapidly converging series with results of relatively high accuracy [11].

The vibrational relaxation of adsorbed atoms is usually described in the framework of:

- Density matrix theory and quantum Liouville equations
- Linear response theory, where the rate of quanta absorbed from an external field is taken as proportional to the Fourier transform of the correlation function of a physical variable relevant to the particular spectroscopic transition.
- The phonon mechanism of the relaxation described in terms of pair-potential methods, and the relaxation via electron-hole pairs described on the basis of the dipole mechanism [3].

- The adsorbate-substrate bond treated within the Anderson model, whose Hamiltonian consists of the adsorbate electron energy, the intra-atomic Coulomb repulsion, the metal energy levels and the one-electron exchange matrix elements of creation and annihilation operators.

2.2 Total-energy all-electron-density functional method

In the computational methods developed to determine the adsorbate electronic structure, an important technical problem is that numerous structural parameters are needed for calculations, which are not directly available from experiments. In order to avoid this problem, one can use the general principle of minimization of the total energy to determine the stability of a system. Density functional theories provide an excellent framework for obtaining the total energy from geometrical configurations. Local density particle equations help provide the total energy and related properties, as well as equilibrium phases, lattice constants and so forth. One of the main problems in computing the total energy of the system stems from the cancellation between the very large kinetic energy and potential energy for heavier atoms. A general approach using an all-electron total-energy calculation helps avoid this problem by introducing quantities obtained from self-consistent band structure calculations [16]. The total energy is constructed from the kinetic and exchange correlation terms. The nonrelativistic treatment includes the solution of the Poisson equation through solving the Dirichlet boundary-value problem. The exchange correlation energy is addressed by the local density approximation using the local-density full-potential linearized-augmented plane-wave method [17,18].

Finally, the total energy per unit cell can be expressed with easily achieved semiempirical functions. The use of experimentally-achieved parameters limits the accuracy of the method, which has been successfully exploited for a number of elements (e.g., Cs monolayer on metals and graphite [16]).

2.3 Cluster methods

The main idea in using cluster methods to represent the interaction between an adsorbate and a given surface is the local nature of such an interaction [6,13,19-20]. These methods are limited by the size of the clusters and basis sets, which might exceed the capability of computers. This approach leads to total energy calculations as a sum of many one-electron energies such as using a Hartree-Fock self-consistent field (SCF) approximation within the framework of *ab initio* techniques. Coupling the cluster model approach with SCF *ab initio* wavefunctions enables the determination of chemisorption energies. The configuration interaction (CI) method can be used to construct a basis set of ground and excited-state wavefunctions, where this method is able to include electron correlation effects which are not represented in the Hartree-Fock SCF approach. In both the SCF and CI cases, attention must be paid to inaccuracies in the calculated chemisorption energies, due to the approximate nature of SCF and CI wavefunctions and the possible dependence of the energies on cluster sizes. In any event, a more sophisticated wavefunction or enlarged cluster size might be used.

2.3.1 *Moderately-large-embedded cluster method (MLEC)*

As proved by many calculations, the cluster model approach is a theoretical approach for investigating the adsorption of isolated atoms and molecules on solid surfaces [4,19,21-24]. The theories are based on the assumption that the adsorbate-substrate interaction is short-ranged, consists of only a finite number of atoms, and uses only a locally-confined perturbation on the substrate. A key difficulty of such cluster methods is connected with the artificial boundary conditions and the discrete substrate density of states. The moderately-large-embedded-cluster model deals with the surface-adsorbate system connected with other surface regions. This model may use *ab initio* programs, but a special method called the Green's function formalism has also been developed [21,23]. A well-known program called "GAMESS" has been successfully used in related computations [B. R. Brooks, P. Saxe, W. D. Laidig and M. Dupuis, Program Library No. 481, Computer Center of the Institute for Molecular Science, 1981 – see the last reference in Ref. 23].

In Fig.1 taken from Ref. 23, we outline the general scheme of the MLEC model.

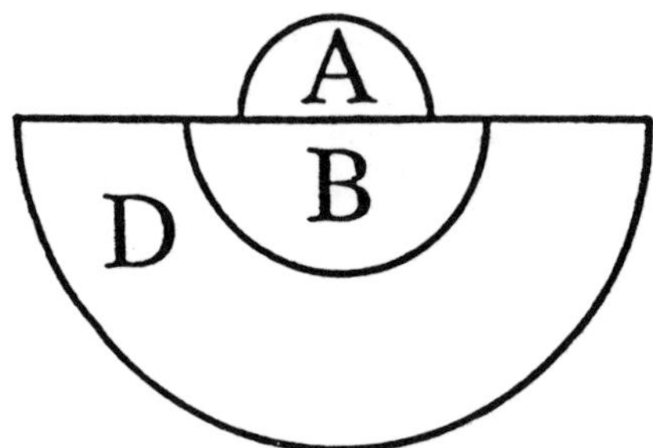

Figure 1.   Schematic representation of the embedded cluster model. *A* is the adsorbate, *B* the cluster, and *D* represents the solid.

The adsorbate is represented by A and the solid surface by B, belonging to the bulk D. Using the Green's function formalism, the perturbed coupled region can be coupled to its crystalline environment denoted by D. Parts A and B together form the "chemisorption cluster". Generally, part B is assumed to be chosen large enough in order to compress all effects of adsorption.

The MLEC formalism consists the following computational procedures:

- Self-consistent determination of the substrate electronic structure
- Construction of the substrate projected density of states and the Hamiltonian matrix in a localized basis
- Choice of the surface cluster collection of the corresponding submatrices and determination of the corrective term to the Hamiltonian.
- Self-consistent treatment of the adsorption problem.

This method has been successfully demonstrated for such a system as H and O adsorption on Li (001) [21]

2.3.2 Dipped adcluster method (DAM)

The conventional cluster model has achieved certain success in computational studies of the energetics of adsorbate-surface systems, but it mainly neglects electron transfer processes between the admolecule and surface (and also the bulk). An approach to help correct this involves an adcluster, consisting of an adatom (molecule) and cluster, which is "dipped" into the electron "background" of the substrate [24,25]. Equilibrium for electrons (and spins) is assumed to be established, where the energy $E(n)$ of the adcluster for n electrons transferred from the bulk is the chemical potential of the electrons at the solid surface. In the more sophisticated variants of the model, the electrostatic energy based on the image force has been used to fulfill requirements for the classical Poisson equations [24].

The total energy of the system can be written as

$$E = E_0 + E_1 , \tag{2.3.1}$$

where E_0 is the energy of just the adcluster, and E_1 is the electrostatic interaction energy between the adcluster and the bulk metal. The former can be computed by using molecular orbitals with n electrons and occupancy of m [25]. The latter can be calculated from a localized model, or from a delocalized model based on an image force method for a complete surface,

$$E_1 = E_{if} - E_{in} , \tag{2.3.2}$$

where E_{in} is the electrostatic energy for the holes located at some point outside of the cluster region, and E_{if} is the stabilization energy of the charge q of the adatom and the hole (see Fig. 2, taken from Ref. 24).

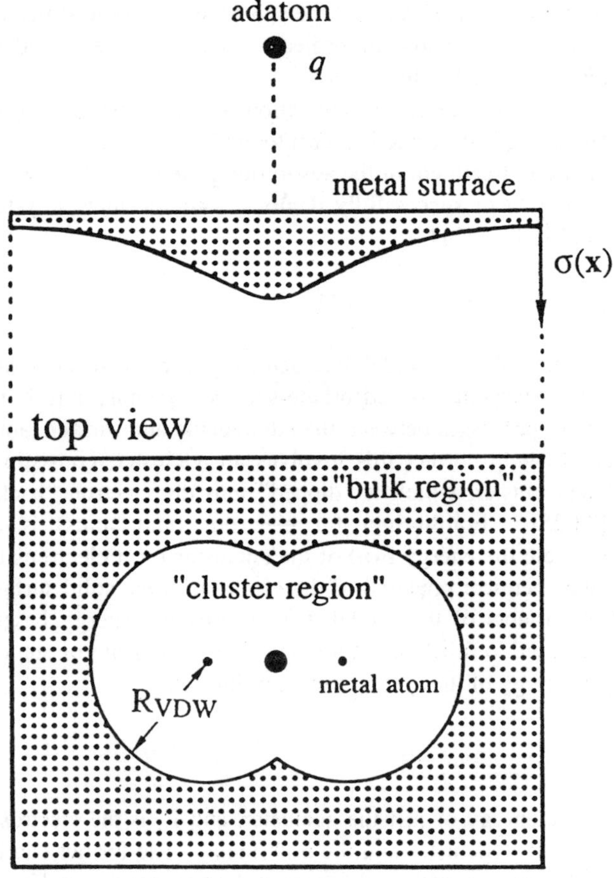

Figure 2. Schematic representation of the adatom and induced charge density on the surface. The cluster region is estimated from the van der Waals radius of the metal atom.

Let us sum the contributions of the charges q_i placed on the atoms of the cluster region, where i labels the atoms. The total charge density $\sigma(\vec{x})$ at the position $\vec{x}$ is

$$\sigma(\vec{x}) = \sum_i \sigma_i(\vec{x}) = -\sum_i \frac{q_i \left| \vec{a}_i - \vec{a}_i' \right|}{4\pi \left| \vec{a}_i - \vec{x} \right|^3} , \qquad (2.3.3)$$

where $\vec{a}_i$ is the position vector of atom i and $\vec{a}_i'$ is the position vector of the mirror image. The image force for atom i can be written as

$$\vec{f}_{if,i} = \sum_j \frac{q_i q_j (\vec{a}_j' - \vec{a}_i)}{\left| \vec{a}_i - \vec{a}_j' \right|^3} , \qquad (2.3.4)$$

and the total image force energy is

$$E_{if} = -\sum_i \sum_j \frac{q_i q_j}{2 \left| \vec{a}_i - \vec{a}_j' \right|} . \qquad (2.3.5)$$

We note that q_i and $\sigma(\vec{x})$ depend on each other and the calculations can be done only iteratively. DAM calculations have been succesfully performed for describing chemisorption of O_2 on Ag [25] and O_2 on Pd [24].

2.4 Embedded-atom method

This method has been developed on the basis of the quasiatom (or effective medium) theories [12,25,20], whereas the impurities are assumed to be locally uniform, or only slightly nonuniform, in this environment. Here each atom is considered to be an impurity embedded in a host consisting of all the other atoms. Therefore, the embedding energy is electron-density dependent, where the density of electrons is always definable, and the volume problem such as associated with pair-potential methods is circumvented [19]. According to Hohenberg and Kohn [26], the electron density uniquely specifies the potential, and thus the energy is a functional of the density. Therefore, the unperturbed host density is determined by its electron density. Introducing an impurity, the total potential will be a sum of the host and impurity potentials, set by the position and charge of the impurity nucleus,

$$E = \tilde{F}_{Z,R} [\rho_h(r)] , \qquad (2.4.1)$$

where $\rho_h(r)$ is the unperturbed host-electron density, while Z and R are the type (atomic number) and position of the impurity, respectively. We should point out that the self-consistency is contained implicitly in Eq. (2.4.1) through the argument of Z; the energy depends on the initial, unperturbed host electron density.

The total energy is

$$E_{tot} = \sum_i F_i(\rho_{h,i}) , \tag{2.4.2}$$

where F_i is the embedding energy, $\rho_{h,i}$ is the density of the host at R_i, but without atom i. We see in Eq. (2.4.1), in terms of the functional $\widetilde{F}$, the energy required to remove an atom from a solid leaving a vacancy is

$$E = \widetilde{F}_{atom}[\rho_{solid}(r)] , \tag{2.4.3}$$

where the "solid" here includes the vacancy. This energy can be obtained in the same fashion by viewing the solid with the vacancy as the impurity and a single atom host, such that

$$E = \widetilde{F}_{solid}[\rho_{atom}(r)] . \tag{2.4.4}$$

The energy to remove an atom from the solid is

$$E = \sum_i [F_i(\rho_i{}^*) - F_i(\rho_i)] , \tag{2.4.5}$$

where i runs over all atoms except the one removed, and $\rho_i{}^*$ is the density at atom i in the solid with the vacancy.

An important issue here is the core-core repulsion. Normally, atomic cores of the substrate repel atomic cores of the adsorbate. Introducing a correction for the core-core repulsion is generally not essential for flat metal surfaces, where typical adsorbates sit at positions sufficiently remote from surface atoms, so that the repulsion originating from the electron-electron interaction prevails over the core-core repulsion. However, the situation is dramatically different for the case of coadsorption phenomena, where the core of an adsorbed polarized atom can interact with other adsorbates. Also, core-core repulsion plays an important role in events where energetic atoms arrive at the surface such that the atomic cores can overlap. Moreover, for the situation of adsorption on stepped surfaces, the cores of surface atoms at the step positions might interact with other adsorbates. Since effective medium theories are based on the jellium model, these kinds of approximations do not consider the effects of the presence of individual surface atoms. For those situations where core-core interactions are significant, one can introduce a correction for the core-core repulsion,whereby the resulting energy is

$$E_{tot} = \sum_i F_i(\rho_{h,i}) + \frac{1}{2} \sum_{\substack{i,j \\ i \neq j}} \phi_{ij}(R_{ij}) \,, \tag{2.4.6}$$

where p_{ij} is the short-range pair potential for atoms i and j separated by the distance R_{ij}. Assuming also that the host density ($\rho_{h,i}$) might be approximated by a sum of atomic densities as

$$\rho_{h,i} = \sum_{j(j \neq i)} \rho_j^a(R_{ij}) \,, \tag{2.4.7}$$

where ρ_j^a is the atomic density at atom j, the total energy is then a simple function of the positions of the atoms.

The ground-state properties can be computed on the basis of Eq. (2.4.6). For equivalent atoms, we have

$$F = F_i \,, \quad \phi = \phi_{ij} \,, \quad \rho = \rho_j^a \,, \tag{2.4.8}$$

and at equilibrium we can write

$$\bar{\rho} = \sum_m \rho(a^m) \,, \quad \bar{\phi} = \sum_m \phi(a^m) \,, \tag{2.4.9}$$

where a^m is the distance to the mth neighbor. The lattice constant is given by the equilibrium condition

$$A_{ij} + F'(\bar{\rho})V_{ij} = 0 \,, \tag{2.4.10}$$

where

$$A_{ij} = \frac{1}{2} \sum_m \phi_m' \, a_i^m a_j^m / a^m \,, \tag{2.4.11}$$

$$V_{ij} = \sum_m \rho_m' \, a_i^m a_j^m / a^m \,. \tag{2.4.12}$$

Here a_i^m is the ith component of the position vector to the mth neighbor, and

$$\phi_m' = [d\phi(r)/dr]_{r=a^m} \; , \quad \rho_m' = [d\rho/dr]_{r=a^m} \; .$$

(2.4.13)

Analogous expressions are available for the elastic constants at equilibrium [10].

The sublimation energy is given by

$$E_S = - [F(\bar{\rho}) + \frac{1}{2}\bar{\phi}] \; .$$

(2.4.14)

The vacancy-formation energy is

$$E_{VF} = -\frac{1}{2}\bar{\phi} + \sum_m [F(\bar{\rho} - \rho_m) - F(\bar{\rho})] + E_{relax} \; ,$$

(2.4.15)

where E_{relax} takes into account the lattice relaxation around the vacancy.

The functions F and ϕ can be calculated by first-principles methods from the properties of the corresponding pure metals. What remains is the determination of the unlike atom-pair interaction, and in many cases we can write

$$\phi_{ij}(r) = Z_i(r)Z_j(r)/r \; .$$

(2.4.16)

Generally, for each element M one seeks the embedding energy $F_M(\rho)$ and effective charge $Z_M(r)$.

Most of the computations use minimizing-type procedures for the total energy with respect to the positions of the atoms. The forces are computed analytically from Eq. (2.4.6) as

$$\vec{f}_k = \sum_{j \neq k} (F_k'\rho_j' + F_j'\rho_k' + p_{jk}')\vec{r}_{jk} \; ,$$

2.4.17)

where $\vec{f}_k$ is the force on the kth atom, and $\vec{r}_{jk}$ is the unit vector between the jth and k^{th} atoms.

2.5 *Localized electronic subspace methods*

The computational strategies represented in the previous paragraphs are based on slab or cluster calculation methods. Both cluster and slab methods have their advantages and disadvantages, and they produce comparable results only for adsorbate-adsorbate interactions under similar conditions. The general question for the transition regime between the two methods relates to the convergence of the cluster calculation results connected with the size of cluster and the embedded region, e.g., the localization procedure. The cluster method consists of an alternative localization method based on the generation scheme of the cluster region basis set $\{\Phi^c\}$ with the orbital occupation numbers obtained from Hartree-Fock density matrix $\hat{P}^X$ computed for an extended system. After establishing the representation of that extended system, one partitions out a cluster region which is still linked in some way to the extended system [22,28,29]. This new cluster embedding procedure is able to utilize data from the CRYSTAL and GAMESS procedures mentioned earlier.

2.5.1 Localization

The localization procedure starts with an arbitrary partitioning of the extended basis set into sets associated with an initial cluster region $\{\Phi^C\}_i$ and the surrounding $\{\Phi^S\}$. After selecting the cluster region, the basis functions need to be transformed by mixing with functions from $\{\Phi^S\}$, such as

$$\Phi_i^C = \sum_\mu \chi_\mu C_{\mu i}^C , \tag{2.5.1}$$

where $\{\chi\}$ is the basis set with basis functions χ_μ ($C_{\mu i}^C$ reconstants). With similar expansion of Φ^S, one can determine the orbital occupation number

$$n_{ij} = C_i + \hat{S}\hat{P}^X\hat{S}\hat{C}_j , \tag{2.5.2}$$

where $\hat{P}^X$ is the density matrix for the extended substrate, and $\hat{S}$ is the overlap matrix between the χ_μ basis functions. The diagonal element n_{ii} is the occupation number of the cluster function Φ_i^C. One needs to compute n^{core} from the core density matrix $\hat{P}_{core}^X$, and thereafter one needs to diagonalize the cluster and the surrounding region n^{core} blocks. By carrying out the Jacobi 2×2 rotation between $\{\Phi^C\}_i$ and $\{\Phi^S\}$, one obtains appropriate core orbital occupation numbers. Then one transforms $\{\Phi^C\}_i$ and $\{\Phi^S\}$ so that the cluster-cluster and surrounding-surrounding block of $\hat{n}$ are diagonal. After that, one needs to maximize the off-diagonal cluster-surrounding $\tilde{n}_{ca}$ matrix elements by

$$\tilde{\Phi}_a^S = \sum_b \Phi_b^S t_{ba} \,, \tag{2.5.3}$$

where t_{ba} are obtained by diagonalizing

$$N_{ab} = \sum_c^{n_\delta} n_{ia} n_{ib} \,. \tag{2.5.4}$$

Here n_{ia} connect the $\{\Phi^c\}$ and $\{\Phi^s\}$ functions, and Eq. (2.5.4) is restricted to the n_δ partially-filled Φ_i^C with $\delta < n_{ii} < 2-\delta$, where the threshold δ determines whether Φ_i^C has a complete occupation.

2.5.2   *Energy calculations*

The determination of the interaction energy of an adsorbate with the extended substrate is based on the assumption that the adsorbate interacts directly with the cluster region but only indirectly with the surroundings. It can be computed by the partial self-consistent field (SCF) procedure based on the Hartree-Fock (HF) method including electron correlation effects. This calculation incorporates only contributions from the surroundings by using only the density $\hat{P}^X$ and Fock $\hat{F}^X$ computed for the base-extended substrate,

$$\hat{P}^X = \hat{P}^C + \hat{P}^S$$
$$\hat{F}^X = \hat{h} + \hat{G}^C + \hat{G}^S \tag{2.5.5}$$

where $\hat{h}$ are one-electron integrals, and $\hat{G}^C, \hat{G}^S$ are the two-electron integrals for the cluster and surroundings separately. The $\hat{P}^C$ are constructed as

$$P_{\mu\nu}^C = \sum_C^{\varepsilon_i < \varepsilon_F} 2 C_{\mu i} C\gamma i \,, \tag{2.5.6}$$

where ε_i are the eigenvalues of $\hat{F}^{CC}$, the cluster-cluster block, ε_F is the Fermi level, and $C_{\mu i}$ and $C_{\gamma i}$ are from the original $\{\chi_\mu\}$ basis. Similarly,

$$C^C_{T\mu\nu} = \sum_{\lambda\sigma} P^C_{\lambda\sigma} \left[\langle \mu\nu | \lambda\sigma \rangle - \frac{1}{2} \langle \mu\sigma | \lambda\nu \rangle \right]. \tag{2.5.7}$$

For the effective one-electron cluster-surrounding interactions, one writes

$$\hat{h}' = \hat{h} + \hat{G}^S \tag{2.5.8}$$

and

$$\hat{F}' = h' + \hat{G}^C , \tag{2.5.9}$$

where $\hat{F}'$ is the Fock matrix, $\hat{F}' = \begin{pmatrix} F^{CC} & F^{CS} \\ F^{SC} & F^{SS} \end{pmatrix}$. For the bare cluster energy, by

partitioning the HF energy, we get

$$E^C = \frac{1}{2} Tr\left[\hat{P}^C \left(\hat{h}' + \hat{F}' \right) \right] + V_{nn} , \tag{2.5.10}$$

where V_{nn} is an appropriate nuclear-nuclear repulsion term. For the cluster energy including the adsorbate, we derive

$$\tilde{E}^C = \frac{1}{2} Tr\left[\hat{P}^C \left(\tilde{h}' + \tilde{F}' \right) + Tr\hat{P}^S \Delta\hat{h} + V_{nn} + \Delta V_{nn} \right], \tag{2.5.11}$$

where $\Delta V_{nn} = \sum_B \dfrac{Z_A Z_B}{R_{AB}}$ is the adsorbate-base system nuclear-nuclear interaction,

and $\tilde{h}' = h' + \Delta\hat{h} + \Delta G^S$ is the effective cluster surroundings one-electron term. Here $\Delta\hat{h}$ is the extra electron-nuclear attraction,

$$\Delta h_{\lambda\sigma} = \langle \lambda | - \frac{Z_A}{R_{\lambda A}} | \sigma \rangle , \tag{2.5.12}$$

and $\tilde{F}' = \tilde{h}' + \tilde{G}^C$, the SCF energy based on an iteration of the HF calculation. Finally, for the adsorbate binding energy we get

$$E_B = \widetilde{E}^C - E^C - E_{ads} , \tag{2.5.13}$$

where E_{ads} is the HF energy for the free non-interacting adsorbate molecules.

3 Summary and Conclusions

After over a hundred years of work on adsorption phenomena, the main factors for determining the nature of the surface bond are well understood. The theoretical models described above are concerned mainly with metallic surfaces, due to the fact that most of the published work so far has been done on those materials. Since substrate properties influence significantly the character of the surface bond, one would expect that in the case of semiconductor surfaces, different computational approaches are needed. In fact, the methods presented above provide a good description for adsorption both on metallic and semiconductor surfaces; for example, slab calculations are widely used for describing adsorption phenomena on semiconductor surfaces.

On metals, computational results show that the adsorbate-surface chemical bond evokes a local perturbation of the electron density at the adsorption site. While this suggests that the bond is of local nature, a closer look reveals that there is a steady exchange of electrons between the bulk and the adsorbate. Hence, in the case of metals, considering the surface bonds as purely local in nature turns out to be incorrect, since the bonding involves the adsorbate and conduction band of the metal. In addition for transition metals, there is a contribution from the d bands to a slight extent, which is important to take into account in order to get a clear view on reactions on transition metal surfaces.

Adsorption on a semiconductor surface differs significantly from adsorption on a metallic surface. Since electrons in the semiconductor are less mobile, there is less electron exchange between the bulk and the surface bond, i.e., the bonds become more localized. They can be considered as simple covalent interactions, with eventual corrections because of crystal lattice strains or surface electric fields in the case of compound semiconductors.

Results from the effective medium model and density functional theory show that the bond strength is mainly governed by the electronegativity (work function) difference between the substrate and the adsorbate. This is an essential point, since electronegativity is a general surface feature which is not influenced by the adsorbate, and it allows us to predict whether bonding will be favored or not.

One has to keep in mind that all model calculations and all analytic schemes have limitations. The main drawback of semiempirical calculations is that they cannot be improved systematically. They depend very strongly on parametrization, where a chosen model Hamiltonian and the matrix elements will, by definition, have an essential influence on the computational results. Still, semiempirical calculations remain a useful tool for the physicist and chemist. *Ab initio* computations, by virtue of their physical completeness, are computationally very demanding, and therefore they are limited by the size of the adsorbate-substrate system.

The local nature of bonding is still not well understood. Local bonding can be essential in adsorption processes, where reactive intermediates are involved. In the case of atomic adsorption, the adsorbate forms a single bond with the surface. If adsorbate species form multiple bonds with the surface, the situation becomes less clear. In these events, localized interactions will dominate the adsorption process, and to understand this, a complete quantum mechanical treatment is needed. Certain methods, such as the slab and embedded cluster approaches, are capable of treating the problem of local bonding, although these methods are computationally very demanding.

The bonding of single atoms or simple molecules on metallic surfaces is largely understood. However, adsorption of complex species, reactive intermediates or the phenomenon of coadsorption still require more work in this field. Since these efforts involve a dramatic increase in the system sizes, methods with further computational shortcuts will be needed.

References

1. J. L. Whitten and H. Yang, *Surface Sci. Rep.* **218**, 55 (1996).
2. W. Ho, *Comments Condensed Mat. Phys.* **13**, 293 (1988).
3. V. P. Zhdanov, *Comments Condensed Mat. Phys.* **16**, 239 (1993).
4. J. D. Head and S. J. Silva, *J. Phys. Chem.* **104**, 3244 (1996).
5. A. Tomasulo and M. V. Ramakrishna, *J. Chem. Phys.* **105**, 10449 (1996).
6. A. Clotet, J. M. Ricart, J. Rubio and F. Illas, *Chem. Phys.* **177**, 61 (1993).
7. W. Steele, *Chem. Rev.* **93**, 2355 (1993).
8. C. Pisani, E. Apra and M. Causa, *Int. J. Quantum Chem.* **XXXVIII**, 395 (1990).
9. P. E. M. Siegbahn, L. G. M. Pettersson and U. Wahlgren, *J. Chem. Phys.* **94**, 4024 (1991).
10. V. Russier, D. R. Salahub and C. Mijoule, *Phys. Rev. B* **42**, 5046 (1990).

11. G. Te Velde and E. J. Baerends, *Phys. Rev. B* **44**, 7888 (1991).

12. J. L. Whitten, *Chem. Phys.* **177**, 387 (1993).

13. J. M. Ricart, A. Clotet, F. Illas and J. Rubio, *J. Chem. Phys.* **100**, 1988 (1994).

14. P. J. Freibelman, J. S. Nelson and G. L. Kellogg, *Phys. Rev. B* **49**, 10548 (1994).

15. John P. Perdew, *Phys. Rev. B* **33**, 8822 (1986).

16. M. Weinert, E. Wimmer and A. J. Freeman, *Phys. Rev. B* **26**, 4571 (1982).

17. E. Wimmer, H. Krakauer, M. Weinert and A. J. Freeman, *Phys. Rev. B* **24**, 864 (1981).

18. Peter J. Freibelman, *Phys. Rev. B* **49**, 14632 (1994).

19. M. S. Daw and M. I. Baskes, *Phys. Rev. B* **29**, 6443 (1976).

20. R. N. Barnett and U. Landman, *Phys. Rev. B* **48**, 2081 (1993).

21. S. Krüger and N. Rösch, *J. Phys. C* **6**, 8149 (1994).

22. P. S. Bagus, C. J. Nelin and C. W. Bauschlicher, *Phys. Rev. B* **28**, 5423 (1983).

23. Y. Fukunishi and H. Nakatsuji, *J. Chem. Phys.* **97**, 6535 (1992).

24. H. Nakatsuji, H. Nakai and Y. Fukunishi, *J. Chem. Phys.* **95**, 640 (1991).

25. H. Nakatsuji and H. Nakai, *J. Chem. Phys.* **98**, 2423 (1993).

26. P. Hohenberg and W. Kohn, *Phys. Rev. B* **136**, 864 (1964).

27. G. Te Welde and E. J. Baerends, *Chem. Phys.* **177**, 399 (1993).

28. L. G. M. Pettersson, H. Agren, O. Vahtras and V. Carravetta, *J. Chem. Phys.* **103**, 8713 (1995).

29. B. L. Maschhoff and J. P. Cowin, *J. Chem. Phys.* **101**, 8138 (1994).

PHASE CONJUGATION THROUGH FOUR-WAVE MIXING

Henk F. ARNOLDUS

Department of Physics and Astronomy, Mississippi State University,
P. O. Drawer 5167, Mississippi State, Mississippi 39762-5167, USA
E-mail: arnoldus@optics.ph.msstate.edu

Thomas F. GEORGE

Office of the Chancellor / Departments of Chemistry and Physics & Astronomy,
University of Wisconsin-Stevens Point, Stevens Point, Wisconsin 54481-3897, USA
E-mail: tgeorge@uwsp.edu

When a nonlinear crystal is illuminated by two strong counterpropagating laser beams, the medium acquires the potential for generating a phase-conjugated replica of a weak incident field through the mechanism of four-wave mixing. We have studied the configuration in which a planar slab of dielectric material is immersed in a strong light field, and where the incident radiation enters the medium as a plane wave at the boundary of the layer. For this geometry, Maxwell's equations can be solved analytically, and the solution is expressed in terms of Fresnel reflection and transmission coefficients, in analogy with the corresponding situation in linear optics. These Fresnel coefficients depend in a complicated way on the parameters of the material (dielectric constant, thickness, third-order susceptibility, and the intensity and polarization of the pump beams) and the properties of the incident field (frequency, angle of incidence and polarization). We show that such a pumped crystal acts indeed as a phase-conjugating device, and illustrate its capabilties and limitations with numerical examples.

1 Phase Conjugation

When a light beam carrying an image is distorted during propagation along its path, the distortions can be undone, in principle, by optical phase conjugation. The idea is to reverse the phase (spatial part) of the wave front, and then let the beam retrace its path. The process of wave front reversal is called phase conjugation, and a device that can accomplish this is termed a phase conjugator (PC) or phase-conjugating mirror. The group of Zel'dovich at the Lebedev Physical Institute in Moscow demonstrated for the first time experimentally wave front reversal by Brillouin scattering in a gas.[1] Shortly after that all kinds of different devices that act as a PC were developed.[2-6] In particular, four-wave mixing in photorefractive materials has been studied extensively.[7-14] It appears that the various practical methods for generating a phase-conjugated replica of a light wave rely on different physical mechanisms. This prohibits a uniform light wave rely on different physical mechanisms. This prohibits a uniform analsysis of the process of optical phase conjugation. Below we shall consider in detail four-wave mixing in a nonlinear crystal, a method first proposed in 1977.[15,16]

Wave front reversal is mathematically equivalent to complex conjugation of the phase of a wave (hence the name phase conjugation). In order to see this, let us consider an incident wave of the form

$$\mathbf{E}_{inc}(\mathbf{r},t) \propto e^{i\mathbf{k}\cdot\mathbf{r}}e^{-i\omega t} + c.c..$$ (1)

Complex conjugation of the phase gives

$$\mathbf{E}_{pc}(\mathbf{r},t) \propto (e^{i\mathbf{k}\cdot\mathbf{r}})^* \, e^{-i\omega t} + c.c.$$

$$= e^{-i\mathbf{k}\cdot\mathbf{r}}e^{-i\omega t} + c.c..$$ (2)

This is again a plane wave, but it has a wave vector -$\mathbf{k}$. Therefore, $\mathbf{E}_{pc}$ counterpropagates the incident wave, or we can say that it has a reversed wave front with respect to the incident wave. Comparison of Eqs. 1 and 2 shows that we can also write $\mathbf{E}_{pc}(\mathbf{r},t) = \mathbf{E}_{inc}(\mathbf{r},-t)$, so that phase conjugation is identical to time reversal. From this property it follows immediately that a PC can be used to undo distortions of wave fronts.[17] It also implies that a perfect PC can not exist, because that would violate causality.

2 Four-Wave Mixing

The electric field $\mathbf{E}(\mathbf{r},t)$ will be represented by its Fourier transform

$$\hat{\mathbf{E}}(\mathbf{r},\omega) = \int_{-\infty}^{\infty} dt \, \mathbf{E}(\mathbf{r},t) \, e^{i\omega t},$$ (3)

which is required to obey Maxwell's equation

$$\nabla \times (\nabla \times \hat{\mathbf{E}}(\mathbf{r},\omega)) - \frac{\omega^2}{c^2} \hat{\mathbf{E}}(\mathbf{r},\omega) = \mu_o \omega^2 \hat{\mathbf{P}}(\mathbf{r},\omega),$$ (4)

with $\hat{\mathbf{P}}$ the Fourier transform of the polarization of the medium. The lowest order nonlinearity in an isotropic inversion invariant medium is the third-order susceptibility, represented by $\chi^{(3)}$, a Cartesian tensor of rank three.[18,19] The corresponding nonlinear polarization is given by

$$\hat{\mathbf{P}}^{(3)}(\mathbf{r},\omega) = \frac{3\varepsilon_o}{4\pi^2} \int_{-\infty}^{\infty} d\omega_1 \int_{-\infty}^{\infty} d\omega_2 \int_{-\infty}^{\infty} d\omega_3\ \delta(\omega - \omega_1 - \omega_2 - \omega_3)$$

$$\times \chi^{(3)}_{xxyy}(\omega_1,\omega_2,\omega_3)\hat{\mathbf{E}}(\mathbf{r},\omega_1)[\hat{\mathbf{E}}(\mathbf{r},\omega_2)\cdot\hat{\mathbf{E}}(\mathbf{r},\omega_3)]. \tag{5}$$

Due to the delta function, three spectral components of the electric field at frequencies ω_1, ω_2 and ω_3 will create a polarization at frequency $\omega = \omega_1 + \omega_2 + \omega_3$. Since the polarization is a source term in Eq. 4, this will generate a fourth electric field at this sum frequency. When all three frequencies are equal, for instance, this gives rise to third harmonic generation. The frequency ω lies in the range $-\infty < \omega < \infty$, which includes negative frequencies as well. From $\mathbf{E}(\mathbf{r},t)^* = \mathbf{E}(\mathbf{r},t)$ it follows that

$$\hat{\mathbf{E}}(\mathbf{r},-\omega) = \hat{\mathbf{E}}(\mathbf{r},\omega)^*, \tag{6}$$

and therefore the field at negative frequencies is determined by the spectral component at $\omega > 0$. In linear optics, spectral components at different frequencies do not couple, which makes the appearance of negative frequency components trivial. In Eq. 1, for example, the second term (the c.c.) is the negative frequency part. For nonlinear media, Eq. 6 also holds, but the negative frequency fields can couple to the positive frequency components. We shall see below that this leads to the generation of phase-conjugated radiation.

The polarization in Eq. 4 also has a linear contribution, which can be accounted for in the usual way by the dielectric constant $\varepsilon(\omega)$. The equation for the electric field then becomes

$$\nabla \times (\nabla \times \hat{\mathbf{E}}(\mathbf{r},\omega)) - \varepsilon(\omega)\frac{\omega^2}{c^2}\hat{\mathbf{E}}(\mathbf{r},\omega) = \mu_o\omega^2\hat{\mathbf{P}}^{(3)}(\mathbf{r},\omega), \tag{7}$$

with the nonlinear polarization given by Eq. 5. The solution also determines the magnetic field, according to

$$\hat{\mathbf{B}}(\mathbf{r},\omega) = -\frac{i}{\omega}\nabla \times \hat{\mathbf{E}}(\mathbf{r},\omega). \tag{8}$$

3 PC Setup

We consider the configuration shown schematically in Fig. 1. A layer of nonlinear material with thickness L is illuminated uniformly by two strong counterpropagating laser beams, the pumps, which have a frequency $\overline{\omega}$. We take the top surface as the xy-plane and the z-axis as 'up'. The direction of the wave vector of the incident wave and the z-axis determine the plane of incidence. The pump beams are assumed to be linearly polarized into the z-direction. This is not essential, but the analysis below depends on the choice of polarization.

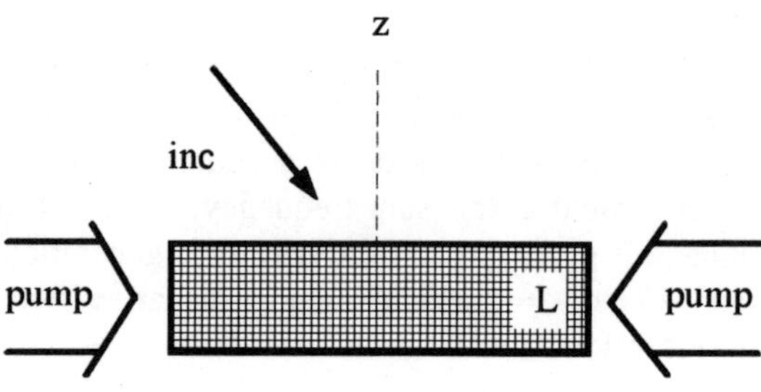

Figure 1: Diagram of the four-wave mixing configuration for phase conjugation of an incident field.

4 Weak Fields

The electric field in Eq. 5 is the total field inside the layer, e.g., the pumps, the incident field and any field generated during the interaction. For a linear medium the incident wave refracts at the surface z = 0, and there would already be a specular reflection of this refracted wave at the surface z = -L. The nonlinear interaction couples these waves to the pump fields, and couples the pumps among themselves. Let us take the propagation direction of the pumps as the y-direction, and split off the solution for the pump fields as

$$\hat{\mathbf{E}}(\mathbf{r},\omega) = \hat{\mathbf{E}}(\mathbf{r},\omega)' + \cos(\frac{\overline{\omega}\sqrt{\varepsilon}}{c}\,y)\mathbf{e}_z\{E_p\delta(\omega-\overline{\omega}) + E_p^*\delta(\omega+\overline{\omega})\}. \qquad (9)$$

Here, E_p is the complex amplitude of the standing-wave pump field in the medium, which is determined by the linear interaction only. It is the solution of Eq. 7 with the right-hand side set equal to zero, and subject to the appropriate boundary conditions at the sides of the crystal where the pump fields enter. We shall assume this E_p to be known. Due to interference between the pumps, there is a critical dependence on the length of the crystal (in the 'horizontal' direction in Fig. 1), and on the dielectric constant.

When we sustitute Eq. 9 three times into the right-hand side of Eq. 5 and carry out the integrations over the frequencies, we obtain an expression with 16 terms. All terms couple through various elements of the susceptibility tensor. Now we assume that the incident field is weak compared to the pump fields, with the result that we can neglect many terms. We shall also assume that the elements of $\chi^{(3)}$ at $\omega \approx \pm 3\overline{\omega}$ are small compared to the values for $\omega \approx \pm\overline{\omega}$. In this way, there is no coupling between the incident field and generated third harmonics. Finally, we shall adopt Kleinman's conjecture, which states that the elements of $\chi^{(3)}$ are approximately invariant under a permutation of frequency arguments.[20] Then the nonlinear interaction can be expressed in terms of the coupling parameter

$$\gamma \equiv \gamma_o e^{i\theta_p} \equiv \frac{3}{8\pi^2} E_p^2 \, \chi^{(3)}_{xxyy}(\overline{\omega}, \overline{\omega}, -\overline{\omega}). \tag{10}$$

The only remaining term in the nonlinear polarization can thus be written as

$$\hat{\mathbf{P}}^{(3)}(\mathbf{r},\omega) = \begin{cases} \gamma_o \varepsilon_o \wp\{2\hat{\mathbf{E}}(\mathbf{r},\omega)' + e^{i\theta_p}\hat{\mathbf{E}}(\mathbf{r},\omega-2\overline{\omega})'\} &, \quad \omega \approx \overline{\omega} \\[2mm] \gamma_o \varepsilon_o \wp\{2\hat{\mathbf{E}}(\mathbf{r},\omega)' + e^{-i\theta_p}\hat{\mathbf{E}}(\mathbf{r},\omega+2\overline{\omega})'\} &, \quad \omega \approx -\overline{\omega} \end{cases}, \tag{11}$$

where the operator $\wp$ accounts for the polarization of the pumps. For linear polarization in the z-direction, its action on an arbitrary vector $\mathbf{v}$ is defined as

$$\wp\mathbf{v} = \mathbf{v} + 2(\mathbf{e}_z \cdot \mathbf{v})\mathbf{e}_z. \tag{12}$$

Then Eq. 7 should hold for the weak field $\hat{\mathbf{E}}(\mathbf{r},\omega)'$ separately. From now on we shall drop the prime that indicates the weak field, since the pump fields only appear parametrically through the coupling constant γ.

The incident field can have an arbitrary spectral distribution. Let us consider a component at a given frequency ω_a, with $\omega_a \approx \omega$. If we set $\omega = \omega_a$ in the first line of Eq. 11, then we see that this component of the field couples to the component at frequency $\omega_b = \omega_a - 2\overline{\omega}$. This is a negative frequency component in the neighborhood of $-\overline{\omega}$. But then, if we set $\omega = \omega_b$ in the second line of Eq. 11, this component couples to $\omega_b + 2\overline{\omega}$, which is ω_a. Therefore, due to the form of the nonlinear polarization in this medium, frequencies only couple in pairs. Since positive and negative frequency components are related as in Eq. 6, we see that basically ω_a couples to $-\omega_b$, and they are related as $\omega_a + |\omega_b| = 2\overline{\omega}$, e.g., they are located symmetrically around the pump frequency. It will turn out to be advantageous to work with the negative frequency ω_b, rather than the positive one $-\omega_b$. Combining Eqs. 7 and 11 then yields

$$\nabla \times (\nabla \times \hat{\mathbf{E}}(\mathbf{r},\omega_a)) - \frac{\omega_a^2}{c^2}[\varepsilon(\omega_a) + 2\gamma_0 \wp]\hat{\mathbf{E}}(\mathbf{r},\omega_a) = \gamma \frac{\omega_a^2}{c^2}\wp\,\hat{\mathbf{E}}(\mathbf{r},\omega_b), \qquad (13)$$

$$\nabla \times (\nabla \times \hat{\mathbf{E}}(\mathbf{r},\omega_b)) - \frac{\omega_b^2}{c^2}[\varepsilon(\omega_b) + 2\gamma_0 \wp]\hat{\mathbf{E}}(\mathbf{r},\omega_b) = \gamma^* \frac{\omega_b^2}{c^2}\wp\,\hat{\mathbf{E}}(\mathbf{r},\omega_a), \qquad (14)$$

a linear set of coupled wave equations. We shall assume that the dielectric constant is real and does not vary much around $\overline{\omega}$. Then $\varepsilon(\omega_b) = \varepsilon(-\omega_b)^* \approx \varepsilon(\omega_a) \equiv \varepsilon$.

5 Dispersion

The linearity of the set suggests that it might admit plane wave solutions. We try

$$\hat{\mathbf{E}}(\mathbf{r},\omega_a) = \mathbf{a}e^{i\mathbf{k}\cdot\mathbf{r}} \quad , \quad \hat{\mathbf{E}}(\mathbf{r},\omega_b) = \mathbf{b}e^{i\mathbf{k}\cdot\mathbf{r}}, \qquad (15)$$

with vectors $\mathbf{a}$, $\mathbf{b}$ and $\mathbf{k}$ to be determined. Substitution gives

$$\kappa^2\mathbf{a} - (\boldsymbol{\kappa}\cdot\mathbf{a})\boldsymbol{\kappa} - (\varepsilon + 2\gamma_0 \wp)\mathbf{a} = \gamma\wp\mathbf{b}, \qquad (16)$$

$$\kappa^2\mathbf{b} - (\boldsymbol{\kappa}\cdot\mathbf{b})\boldsymbol{\kappa} - \rho^2(\varepsilon + 2\gamma_0 \wp)\mathbf{b} = \gamma^*\rho^2\wp\mathbf{a}, \qquad (17)$$

with $\boldsymbol{\kappa} = c\mathbf{k}/\omega_a$, the dimensionless wave vector, and $\rho = -\omega_b/\omega_a$, a dimensionless frequency parameter. Usually, electromagnetic waves are transverse, which would give here

$$\boldsymbol{\kappa}\cdot\mathbf{a} = \boldsymbol{\kappa}\cdot\mathbf{b} = 0. \qquad (18)$$

With the definition of $\wp$, Eq. 12, it then follows that

$$\mathbf{a} \cdot \mathbf{e}_z = \mathbf{b} \cdot \mathbf{e}_z = 0, \tag{19}$$

and this implies that the polarization vectors $\mathbf{a}$ and $\mathbf{b}$ are perpendicular to the plane of incidence. This is called s-polarization, and we shall write $\mathbf{e}_s$ for the corresponding unit vector. Combining everything then leads to

$$[\kappa^2 - (\varepsilon + 2\gamma_0)][\kappa^2 - \rho^2(\varepsilon + 2\gamma_0)] = \gamma_0^2 \rho^2, \tag{20}$$

an equation for the magnitude of the wave vector κ. This is the dispersion relation. It has two solutions for κ^2, given by

$$\left(\kappa^2\right)\binom{1}{2} = \frac{1}{2}(\varepsilon + 2\gamma_0)\left\{\rho^2 + 1 \mp \delta\sqrt{(\rho^2 - 1)^2 + \left(\frac{2\rho\gamma_0}{\varepsilon + 2\gamma_0}\right)^2}\right\}, \tag{21}$$

where $\delta = \mathrm{sgn}(\rho - 1)$. The two branches of the dispersion relation are defined such that

$$\kappa^{(1)} \to \sqrt{\varepsilon} \quad , \quad \kappa^{(2)} \to |\rho|\sqrt{\varepsilon}, \tag{22}$$

which gives for the corresponding wave numbers ($k = \omega_a \kappa / c$)

$$k^{(1)} \to \frac{\omega_a}{c}\sqrt{\varepsilon} \quad , \quad k^{(2)} \to \frac{|\omega_b|}{c}\sqrt{\varepsilon}. \tag{23}$$

These are the dispersion relations for an ω_a- and an ω_b-wave, respectively, in a dielectric (limit $\gamma \to 0$). The inclusion of δ in the solution given by Eq. 21 introduces a discontinuity in the dispersion relation at the resonance frequency $\overline{\omega}$, but in this way the separation between waves of type (1) and type (2) has a clear physical meaning. Figure 2 shows the dispersion relation for s-waves.

For the solutions κ of the dispersion relation, Eqs. 16 and 17 are dependent, and they imply a relation between vectors $\mathbf{a}$ and $\mathbf{b}$. Since the set is homogeneous, we can normalize one of these to unity. We shall take

$$\mathbf{a}_1 = \mathbf{b}_2 = \mathbf{e}_s, \tag{24}$$

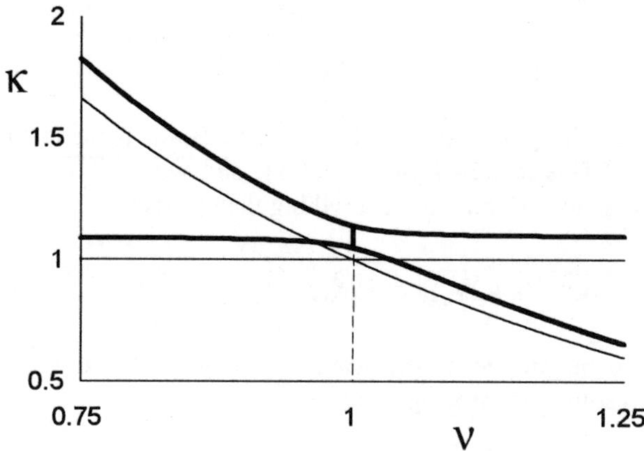

Figure 2: Dispersion relation for s-polarized waves in the nonlinear medium, for $\varepsilon = 1$. The variable on the horizontal axis is $v = \omega_a/\overline{\omega}$. There is a discontinuity at the resonance $v = 1$, such that the upper branch for $v < 1$ continues as the lower branch at $v > 1$. This corresponds to solution $\kappa^{(2)}$, and the other branch is $\kappa^{(1)}$. The thin lines are the dispersion curves for a linear medium. The curved one, close to the $\kappa^{(2)}$ branch, is the dispersion relation for an ω_b-wave, and the line $\kappa = 1$ is the dispersion for an ω_a-wave.

where the subscripts 1 and 2 refer to the corresponding solution of the dispersion relation. The other two must be proportional to $\mathbf{e}_s$, and we write

$$\mathbf{b}_1 = \zeta_1 \mathbf{e}_s \quad , \quad \mathbf{a}_2 = \zeta_2 \mathbf{e}_s . \tag{25}$$

From either Eq. 16 or 17 we then find

$$\zeta_1 = \frac{-2\gamma^* \rho^2}{(\varepsilon + 2\gamma_o)\{\rho^2 - 1 + \delta \sqrt{(\rho^2 - 1)^2 + \left(\dfrac{2\rho\gamma_o}{\varepsilon + 2\gamma_o}\right)^2}\}} , \tag{26}$$

$$\zeta_2 = -\frac{e^{2i\theta_p}}{\rho^2}\zeta_1 . \tag{27}$$

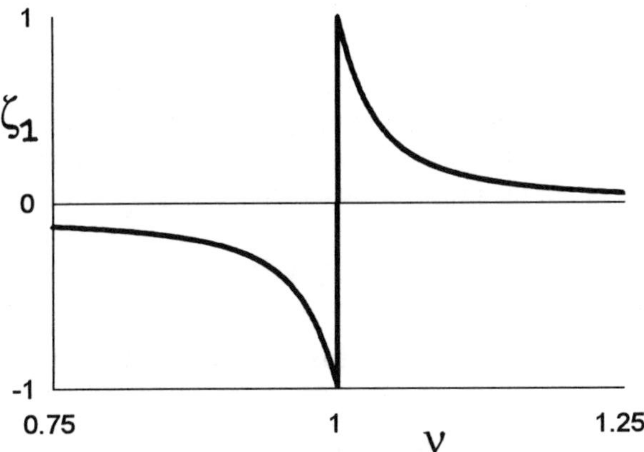

Figure 3: Parameter ζ_1 for s-waves as a function of $\nu = \omega_a/\overline{\omega}$.

These parameters have a strong resonant behavior near the pump frequency, as shown in Fig. 3. Far off resonance, both ζ_1 and ζ_2 go to zero, indicating again that solution 1 is essentially an ω_a-wave and solution 2 is an ω_b-wave. At resonance we have $|\zeta_1| = |\zeta_2| = 1$, and both components of the coupled set are excited equally.

The two fields from Eq. 15 are excited simultaneously in the medium as a set of coupled waves. They differ slightly in frequency (for $\omega_a \neq \overline{\omega}$), but have the same wave vector $\mathbf{k}$. Since $\omega_b < 0$, however, the ω_b-wave travels into the direction opposite to its wave vector. Therefore, the solution of Eq. 15 is a set of counterpropagating waves.

There is a second class of independent solutions, which have their polarization vectors $\mathbf{a}$ and $\mathbf{b}$ in the plane of incidence, and these are called p-waves. They can be derived in a similar way from Eqs. 16 and 17, but the formalism is more cumbersome. A particular interesting aspect of p-waves is that they are not transverse, e.g., Eq. 18 does not hold for these solutions. For the details of p-polarized solutions, we refer to Ref. 21.

6 Reflection at the Surface

Let the incident field be a monochromatic s-polarized plane wave of the form

$$\hat{E}(r,\omega_a)_{inc} = e_s\, e^{ik\cdot r},\tag{28}$$

with $k = \omega_a/c$. The direction of the wave vector $\mathbf{k}$ determines the angle of incidence, which is a free parameter. The wave vector $\mathbf{k}$ will be written as $\mathbf{k} = \mathbf{k}_\parallel + \mathbf{k}_\perp$, where the subscripts $\parallel$ and $\perp$ refer to the plane $z = 0$. The unit polarization vector $\mathbf{e}_s$ is perpendicular to the plane of incidence, and as phase convention we shall take this vector 'out of the paper' in the diagram of Fig. 1. Specifically,

$$e_s = (\mathbf{k}_\parallel \times e_z)/k_\parallel.\tag{29}$$

For later convenience we introduce the dimensionless vector $\boldsymbol{\kappa}_\parallel$, defined as

$$\boldsymbol{\kappa}_\parallel = \mathbf{k}_\parallel/k.\tag{30}$$

The length of this vector is

$$\kappa_\parallel = \sin\theta_i,\tag{31}$$

with θ_i the angle of incidence. The wave number $k = \omega_a/c$ is determined by the frequency ω_a, and the vector $\mathbf{k}_\parallel$ is determined by the angle of incidence. Then the z-component of the wave vector $\mathbf{k}$ follows from

$$k^2 = k_\parallel^2 + k_z^2,\tag{32}$$

and is therefore

$$k_z = -k\sqrt{1-\kappa_\parallel^2} = -k\cos\theta_i.\tag{33}$$

We have taken the solution with the negative sign, because this wave is traveling into the negative z-direction. It will turn out to be useful to define

$$\kappa_a = -\sqrt{1-\kappa_\parallel^2},\tag{34}$$

in terms of which the wave vector of the incident field can be written as

$$\mathbf{k} = k(\boldsymbol{\kappa}_\parallel + \kappa_a e_z).\tag{35}$$

Already for an ordinary dielectric, a specular wave will appear in the region $z > 0$. We shall call this the r-wave. Due to boundary conditions at the surface $z = 0$, this wave must have a wave vector with the same parallel component as the incident wave. On the other hand, the dispersion relation for this wave is $k_r = \omega_a/c$, so that $k_r = k$. The only possibility for the wave vector is therefore

$$\mathbf{k}_r = k(\boldsymbol{\kappa}_\| - \kappa_a \mathbf{e}_z), \tag{36}$$

and from this it follows that the angle of reflection is equal to the angle of incidence. Then, inside the medium the nonlinear interaction will couple any wave with frequency ω_a to waves with frequency ω_b. At the boundary $z = 0$, these waves will emanate from the medium and travel into the region $z > 0$. This is the phase-conjugated image of the incident field, and we shall call this the pc-wave. The dispersion relation for ω_b-waves in vacuum is

$$k_{pc} = -\omega_b / c = \rho k, \tag{37}$$

and due to the boundary conditions at $z = 0$ the corresponding wave vector must again have the same parallel component as the wave vector of the incident wave. Finally, the wave must propagate away from the medium, so that the z-component of this wave vector must be negative (because this negative frequency wave travels in the direction opposite to its wave vector). This uniquely determines the wave vector of the pc-wave as

$$\mathbf{k}_{pc} = k(\boldsymbol{\kappa}_\| + \kappa_b \mathbf{e}_z), \tag{38}$$

where we have set

$$\kappa_b = -\sqrt{\rho^2 - \kappa_\|^2}, \tag{39}$$

in analogy to Eq. 34.

The three possible waves in the region $z > 0$ are shown in Fig. 4. The pc-wave has a slight frequency shift with respect to the incident wave, and its frequency is $-\omega_b = 2\overline{\omega} - \omega_a$, which is $-\omega_b = \rho\omega_a$. This shift leads to interference fringes that have been observed experimentally.[22-23] The pc-wave almost counterpropagates the incident wave, but there is a small deviation in the angle of propagation. Let θ_{pc} be the angle of reflection. Then we have $\sin\theta_{pc} = k_\|/k_{pc}$, which is $\sin\theta_{pc} = \kappa_\|/\rho$. With Eq. 31 we then find

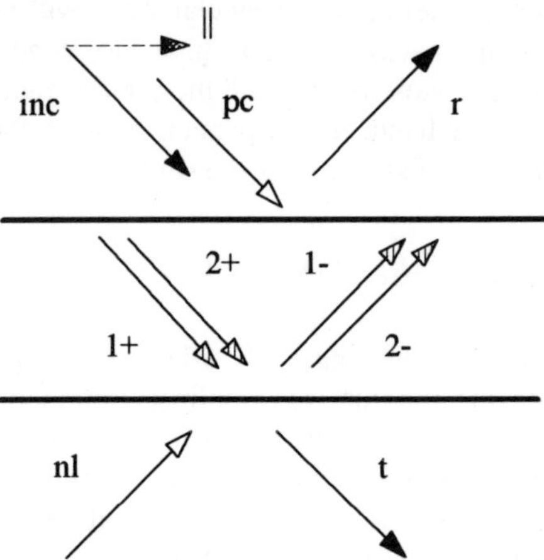

Figure 4: Illustration of the various wave vectors. Wave vectors with a solid arrow head (inc-, r- and t-waves) are waves with frequency ω_a, and they propagate in the direction of the wave vector. These are the waves that would appear for an ordinary dielectric medium. The waves indicated by a wave vector with a white arrow head (pc- and nl-waves) are ω_b-waves, and they propagate in the direction opposite the wave vector. The four wave vectors inside the medium have shaded arrow heads, indicating that they have an ω_a- and an ω_b-component. Therefore, each wave vector represents a set of two counterpropagating waves. The horizontal wave vector ($\|$) represents $\mathbf{k}_{\|}$, which is common to all wave vectors in this diagram.

$$\rho \sin \theta_{pc} = \sin \theta_i, \tag{40}$$

or equivalently,

$$-\omega_b \sin \theta_{pc} = \omega_a \sin \theta_i. \tag{41}$$

It is interesting to notice that this reflection law is independent of the details of the mechanism of four-wave mixing in the medium. We derived this from the fact that at the boundary the parallel component of the wave vectors has to be the same for all waves near the boundary, in order to satisfy Maxwell's equations.

7   Excitation of Plane Waves

At the boundary $z = -L$ every wave vector must also have the same parallel component as the incident wave. This implies that the only ω_a-wave in the region

$z < -L$ which travels away from the medium must have a wave vector equal to $\mathbf{k}$ of the incident wave. This transmitted wave will be called the t-wave, and we have

$$\mathbf{k}_t = \mathbf{k} = k(\boldsymbol{\kappa}_{\|} + \kappa_a \mathbf{e}_z). \tag{42}$$

Similarly, the only possible wave with frequency ω_b in $z < -L$ must have wave vector

$$\mathbf{k}_{nl} = k(\boldsymbol{\kappa}_{\|} - \kappa_b \mathbf{e}_z), \tag{43}$$

and the corresponding plane wave will be called the nl-wave (nonlinear wave).

Inside the medium, the dispersion relation determines the z-components of the possible waves, given $\mathbf{k}_{\|}$. Since there is no causality requirement in a layer with finite thickness, waves can propagate either in the positive or the negative z-direction. In addition, the dispersion relation has two solutions, given ω_a, and each solution corresponds to a set of two counterpropagating waves. This leaves a total of eight possible waves inside the medium, and an overall total of thirteen waves. This is shown in Fig. 4. For the four wave vectors we write

$$\mathbf{k}_1^{\pm} = k(\boldsymbol{\kappa}_{\|} \pm \kappa_\alpha \mathbf{e}_z), \tag{44}$$

$$\mathbf{k}_2^{\pm} = k(\boldsymbol{\kappa}_{\|} \pm \kappa_\beta \mathbf{e}_z), \tag{45}$$

with

$$\kappa_\alpha = -\sqrt{\left(\kappa^{(1)}\right)^2 - \kappa_{\|}^2}, \tag{46}$$

$$\kappa_\beta = -\sqrt{\left(\kappa^{(2)}\right)^2 - \kappa_{\|}^2}, \tag{47}$$

in terms of the solutions of the dispersion relation, as given by Eq. 21.

8 Fields

Each of the thirteen wave vectors corresponds to a plane wave of the type given by Eq. 28. Since the incident wave is s-polarized, all waves will be s-polarized. The incident wave is taken to have a unit amplitude, but the amplitudes of the other waves have yet to be determined. These amplitudes, relative to the incident waves, are the Fresnel coefficients. When we write R and P for the Fresnel coefficients of the r-wave and pc-wave, respectively, then the two frequency components of the field in $z > 0$ have the form

$$\hat{\mathbf{E}}(\mathbf{r},\omega_a) = \mathbf{e}_s \left\{ e^{i\mathbf{k}\cdot\mathbf{r}} + R e^{i\mathbf{k}_r\cdot\mathbf{r}} \right\}, \tag{48}$$

$$\hat{\mathbf{E}}(\mathbf{r},\omega_b) = \mathbf{e}_s\, P e^{i\mathbf{k}_{pc}\cdot\mathbf{r}}. \tag{49}$$

Using the fact that the wave vectors have the same parallel components, we can write Eq. 48 as

$$\hat{\mathbf{E}}(\mathbf{r},\omega_a) = \mathbf{e}_s\, e^{i\mathbf{k}_\parallel\cdot\mathbf{r}} \left\{ e^{ikz\kappa_a} + R e^{-ikz\kappa_a} \right\}, \tag{50}$$

and similarly for all other waves. It then follows immediately that for boundary conditions at $z = 0$ and $z = -L$, only the factor in curly brackets contributes. The exponentials with $\mathbf{k}_\parallel$ and the polarization vectors $\mathbf{e}_s$ cancel.

For the fields in $z < -L$ we write

$$\hat{\mathbf{E}}(\mathbf{r},\omega_a) = \mathbf{e}_s\, T e^{i\mathbf{k}_t\cdot\mathbf{r}}, \tag{51}$$

$$\hat{\mathbf{E}}(\mathbf{r},\omega_b) = \mathbf{e}_s\, N e^{i\mathbf{k}_{nl}\cdot\mathbf{r}}, \tag{52}$$

with T and N the Fresnel coefficients for the t-wave and nl-wave, respectively. Inside the medium we have eight waves, which appear in the form

$$\hat{\mathbf{E}}(\mathbf{r},\omega_a) = \mathbf{e}_s \left\{ Z_1^+ e^{i\mathbf{k}_1^+\cdot\mathbf{r}} + Z_1^- e^{i\mathbf{k}_1^-\cdot\mathbf{r}} + \zeta_2 \left(Z_2^+ e^{i\mathbf{k}_2^+\cdot\mathbf{r}} + Z_2^- e^{i\mathbf{k}_2^-\cdot\mathbf{r}} \right) \right\}, \tag{53}$$

$$\hat{\mathbf{E}}(\mathbf{r},\omega_b) = \mathbf{e}_s \left\{ \zeta_1 \left(Z_1^+ e^{i\mathbf{k}_1^+\cdot\mathbf{r}} + Z_1^- e^{i\mathbf{k}_1^-\cdot\mathbf{r}} \right) + Z_2^+ e^{i\mathbf{k}_2^+\cdot\mathbf{r}} + Z_2^- e^{i\mathbf{k}_2^-\cdot\mathbf{r}} \right\}, \tag{54}$$

with the Z's the four Fresnel coefficients.

The expressions for the fields above satisfy Maxwell's equations in the three regions of space. In addition, they must be connected across the two boundaries according to Maxwell's equations. This will yield a set of equations for the eight Fresnel coefficients, and the solution of this set then uniquely determines the fields everywhere.

9 Matching the Fields Across the Boundaries

The boundary conditions are

$$\hat{\mathbf{E}}_{\parallel}, \hat{\mathbf{B}}, \quad (\varepsilon\hat{\mathbf{E}} + \hat{\mathbf{P}}^{(3)}/\varepsilon_o)_{\perp} \quad \text{continuous.} \tag{55}$$

The $\hat{\mathbf{B}}$-field follows from the $\hat{\mathbf{E}}$-field with Eq. 8. These conditions have to hold both for ω_a and for ω_b, and both at $z = 0$ and $z = -L$. For s-waves, all fields are parallel to the surface, and therefore the last boundary condition holds automatically. The first two conditions then give a set of eight linear equations for the eight unknown Fresnel coefficients. It turns out that the set decouples into two sets of four equations. The equations for the Fresnel coefficients for the fields inside the medium can be written in matrix form as

$$F \begin{pmatrix} Z_1^+ \\ Z_1^- \\ Z_2^+ \\ Z_2^- \end{pmatrix} = \begin{pmatrix} 2\kappa_a \\ 0 \\ 0 \\ 0 \end{pmatrix}, \tag{56}$$

with F a 4x4 matrix. This matrix is explicitly

$$\begin{pmatrix} \kappa_a + \kappa_\alpha & \kappa_a - \kappa_\alpha & \zeta_2(\kappa_a + \kappa_\beta) & \zeta_2(\kappa_a - \kappa_\beta) \\ \zeta_1(\kappa_b - \kappa_\alpha) & \zeta_1(\kappa_b + \kappa_\alpha) & \kappa_b - \kappa_\beta & \kappa_b + \kappa_\beta \\ (\kappa_a - \kappa_\alpha)e^{-i\phi\alpha} & (\kappa_a + \kappa_\alpha)e^{i\phi\alpha} & \zeta_2(\kappa_a - \kappa_\beta)e^{-i\phi\beta} & \zeta_2(\kappa_a + \kappa_\beta)e^{i\phi\beta} \\ \zeta_1(\kappa_b + \kappa_\alpha)e^{-i\phi\alpha} & \zeta_1(\kappa_b - \kappa_\alpha)e^{i\phi\alpha} & (\kappa_b + \kappa_\beta)e^{-i\phi\beta} & (\kappa_b - \kappa_\beta)e^{i\phi\beta} \end{pmatrix}, \tag{57}$$

where we have introduced the phase angles

$$\phi_i = \ell\kappa_i \quad , \quad i = a, b, \alpha, \beta. \tag{58}$$

The parameter ℓ is the dimensionless layer thickness in units of $1/k$, e.g.,

$$\ell = kL = \frac{\omega_a L}{c}. \tag{59}$$

The matrix F is therefore determined by the dimensionless z-components κ_i of the various wave vectors, the two amplitude parameters ζ_1 and ζ_2 and the layer thickness parameter ℓ. After solving this set, the Fresnel coefficients for the fields outside the medium follow as

$$\begin{pmatrix} R \\ P \\ Te^{-i\phi_a} \\ Ne^{i\phi_b} \end{pmatrix} = \begin{pmatrix} 1 \\ 0 \\ 0 \\ 0 \end{pmatrix} + G \begin{pmatrix} Z_1^+ \\ Z_1^- \\ Z_2^+ \\ Z_2^- \end{pmatrix}, \tag{60}$$

with the matrix G given by

$$G = \begin{pmatrix} 1 & 1 & \zeta_2 & \zeta_2 \\ \zeta_1 & \zeta_1 & 1 & 1 \\ e^{-i\phi\alpha} & e^{i\phi\alpha} & \zeta_2 e^{-i\phi\beta} & \zeta_2 e^{i\phi\beta} \\ \zeta_1 e^{-i\phi\alpha} & \zeta_1 e^{i\phi\alpha} & e^{-i\phi\beta} & e^{i\phi\beta} \end{pmatrix}. \tag{61}$$

10   Phase-Conjugated Reflection

Equation 56 can be solved numerically for the four Fresnel coefficients for the waves inside the medium, and then Eq. 60 gives the remaining four coefficients. They depend on the frequency parameter $v = \omega_a/\overline{\omega}$, the dielectric constant ε of the medium, the nonlinear complex coupling parameter γ, the layer thickness ℓ and the angle of incidence, which comes in as $\kappa_\parallel = \sin\theta_i$.

The most interesting one is P, the complex amplitude of the pc-wave, with reference to an incident wave with unit amplitude. Figure 5 shows the typical frequency dependence of $|P|$. It has a strong resonance for $v = 1$, and there are many sidebands. The coupling parameter was taken as $\gamma = 0.01$, which is a small number, but nevertheless the peak at $v = 1$ is very pronounced. This indicates that the peak is indeed the result of a resonance phenomenon, just like in Fig. 3.

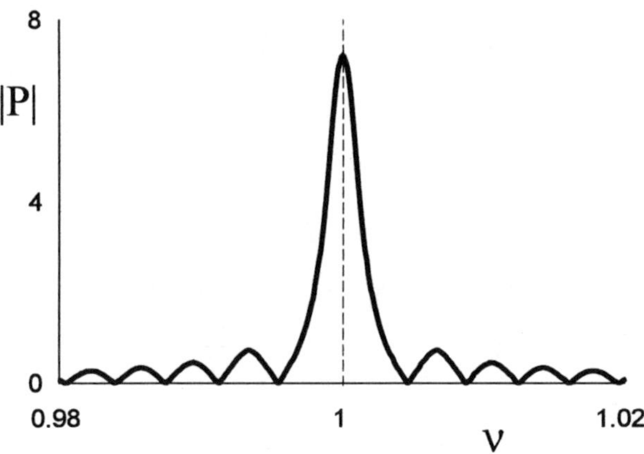

Figure 5: Absolute value of the Fresnel reflection coefficient for the pc-wave as a function of the frequency of the incident wave. The parameters are: $\varepsilon = 1$, $\gamma = 0.01$, $\ell = 660$ and the angle of incidence is 45°.

Another interesting feature is that the peak value exceeds unity, indicating amplification. For linear media this can never happen, due to energy conservation. Here, energy can be extracted from the pumps during the four-wave mixing. Sometimes this is called self-oscillation, and this has been observed experimentally.[24] The resonance condition is approximately given by $|\rho^2-1|\ll\gamma_o$, as can be seen from Eq. 26. This makes the width of the resonance approximately $\Delta v = \gamma_o/4$, in good agreement with Fig. 5.

Although Fig. 5 represents typical behavior, the Fresnel coefficients depend in a very intricate way on the various parameters. Figure 6 shows again $|P|$ for the same parameters as for Fig. 5, except that the layer thickness has been changed from $\ell = 660$ to $\ell = 942$. The behavior of $|P|$ is very similar, but now the value at resonance is a minimum. This is due to interference between waves propagating in the positive and the negative z-direction. Near resonance this creates a standing wave in the z-direction, but the interference can be constructive or destructive. This is the same effect as for a linear medium. Noticing the difference in scales in both figures, we see that this interference hardly affects the sidebands, although they have diminished a little. The Fresnel coefficient P also includes a phase shift, and in order to illustrate this we have graphed the result from Fig. 6 in the complex plane in Fig. 7. It is seen that the phase of P varies in a nontrivial way with the frequency. We recall that the frequency dependence is entirely geometric; we assume here that the dielectric constant is unity and that the

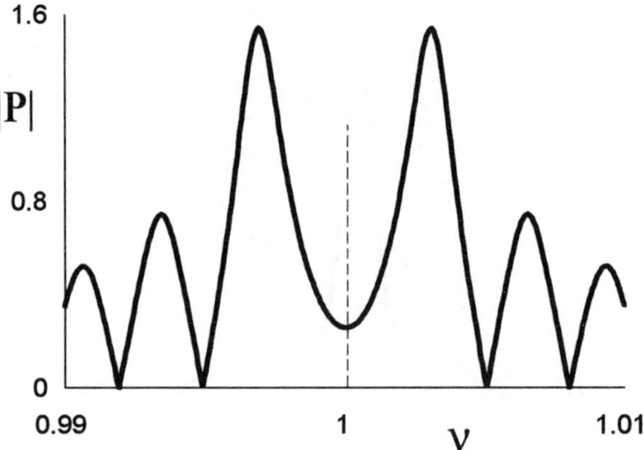

Figure 6: Same as Fig. 5, but now with $\ell = 942$.

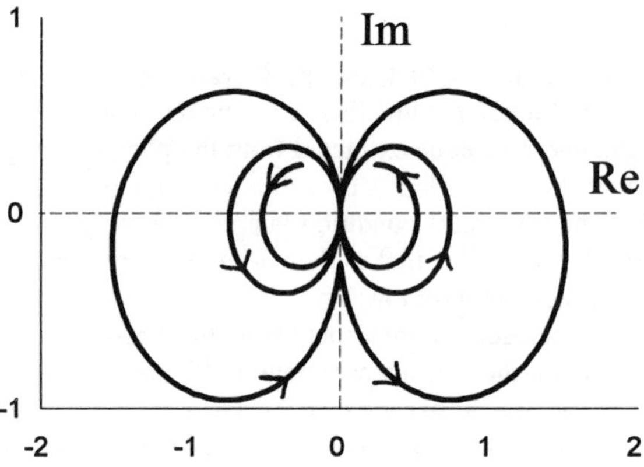

Figure 7: Fresnel coefficient P in the complex plane, for the same parameters as in Fig. 6. The arrow heads indicate the direction of increase of the frequency ν. For $\nu < 1$ the curve is entirely at the left side of the imaginary axis. The resonance frequency $\nu = 1$ corresponds to the point on the negative imaginary axis where the graph has a cusp. There the phase shift is $\pi/2$. Above resonance, the curve is at the right side of the imaginary axis. It is clear that each sideband in Fig. 6 corresponds to a loop in this graph, and we have P = 0 when the curve passes through the origin. For the situation of Fig. 5, where the resonance is a maximum, the point where the curve crosses the imaginary axis is much lower. This gives a smooth transition between the left and the right side, instead of a cusp like here.

susceptibility tensor is frequency independent over the range of interest (making γ independent of the frequency).

The polarization of the waves also affects the Fresnel coefficients considerably. Although the analysis for this more involved situation is not given here, we show in Fig. 8 the dependence of |P| on v for the same parameters as for Fig. 5. For s-polarization, Fig. 5., there is a pronounced maximum at resonance, but for p-polarization the value of |P| only has a local maximum, and there are two strong sidebands. A graph in the complex plane, not shown here, reveals that at resonance the curve has a discontinuity and jumps from one side of the imaginary axis to the other.

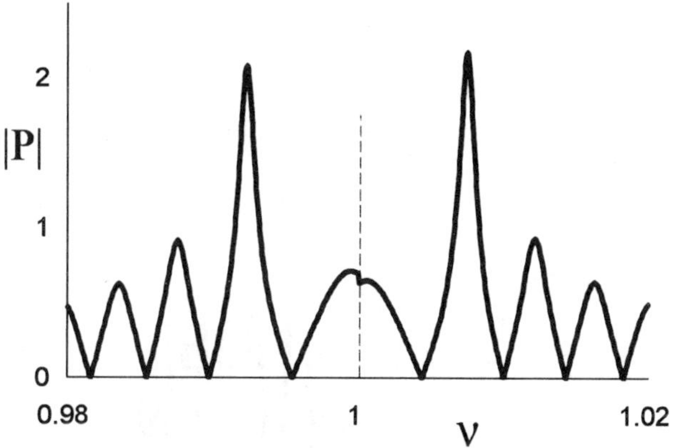

Figure 8: Fresnel coefficient |P| as a function of v for the same parameters as in Fig. 5, but now for p-polarization. The slight jump at resonance is probably a numerical instability.

11 Dependence on Other Parameters

The Fresnel coefficient |P| depends strongly on the angle of incidence, as shown in Figs. 9 and 10. For moderate values of the layer thickness ℓ, there is a peak close to grazing incidence, but for larger values of ℓ, the behavior is strongly oscillatory at large angles. This is due to the phase factors $\exp(\pm i\phi_\alpha)$ and $\exp(\pm i\phi_\beta)$ in the F and G matrices. The wave numbers κ_α and κ_β only have a modest dependence on θ_i, through $\kappa_\parallel = \sin\theta_i$ in Eqs. 46 and 47, but they are multiplied by ℓ in Eq. 58. Then a small change in θ_i leads to a large change in the phase, and this makes the exponentials oscillate very rapidly. Since the Fresnel

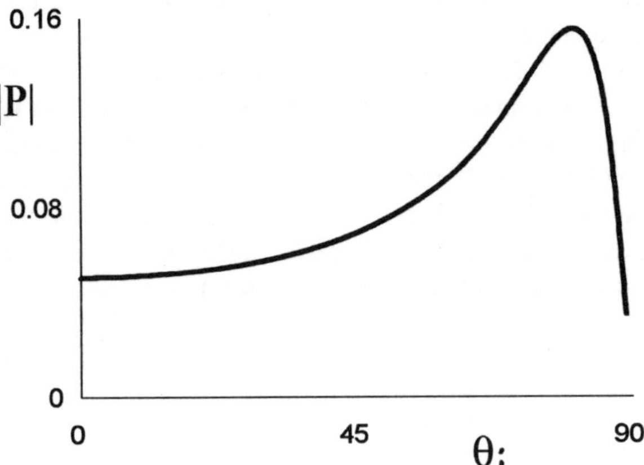

Figure 9: Fresnel coefficient $|P|$ as a function of the angle of incidence θ_i. The parameters are: $v = 0.99$, $\gamma = 0.01$, $\ell = 10$ and $\varepsilon = 1$.

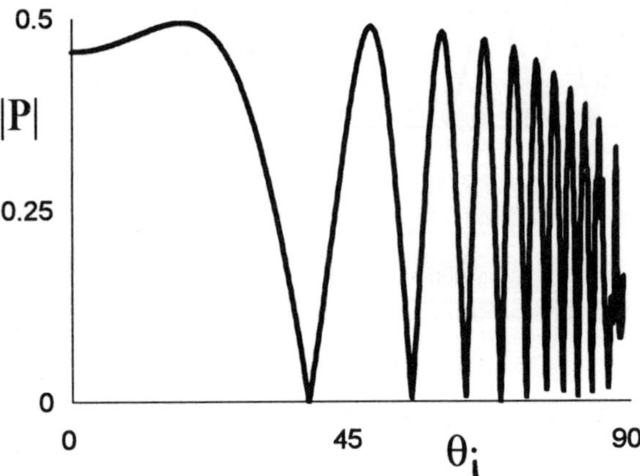

Figure 10: Same as Fig. 9, but now with $\ell = 660$.

coefficients depend in a complicated way on these phase factors, the resulting behavior can look chaotic in some instances.

A similar situation occurs when we consider the dependence on the dielectric constant ε. The parameters ζ_1 and ζ_2 in Eqs. 26 and 27 only depend mildly on ε, and so does the dispersion relation. This makes the wave numbers κ_α and κ_β vary

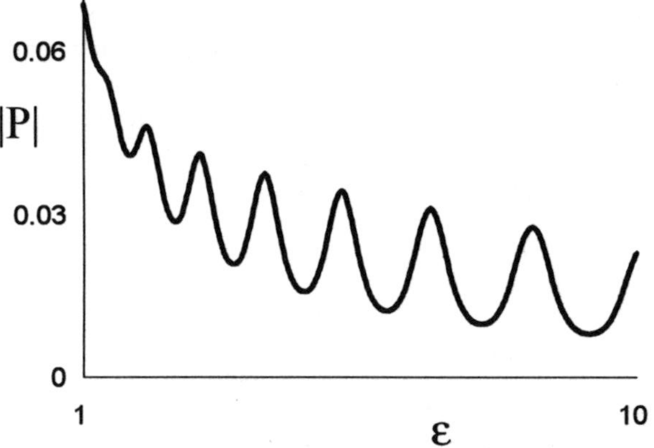

Figure 11: Fresnel coefficient |P| as a function of the dielectric constant ε, for $v = 0.99$, $\gamma = 0.01$, $\ell = 10$ and a 45° angle of incidence.

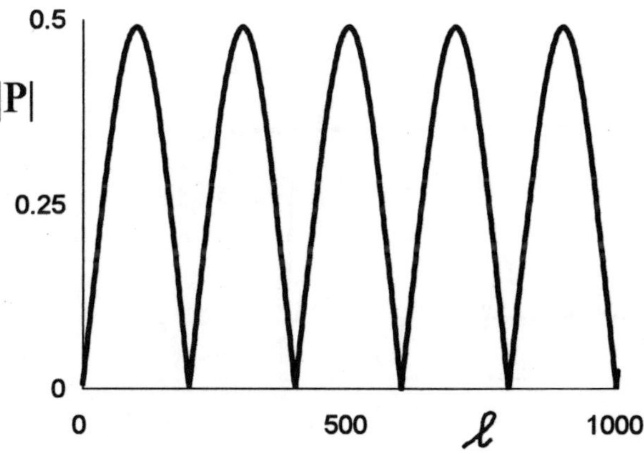

Figure 12: Absolute value of P as a function of ℓ for $v = 0.99$, $\gamma = 0.01$, $\varepsilon = 1$ and $\theta_i = 45^\circ$.

slightly with ε. But then for relatively large ℓ the phase factors become very oscillatory, leading to wild looking behavior of the Fresnel coefficients. This is illustrated in Fig. 11. For even larger ℓ, the oscillations increase rapidly.

For large ℓ, the Fresnel coefficients appear to have an irregular behavior when seen as a function of the various parameters. The dependence on ℓ itself, however, is very simple. The phases in Eq. 58 are proportional to ℓ, which makes the phase factors, the exponentials, vary sinusoidally. So even though a large value of ℓ leads to some unpredicatable behavior, the dependence on ℓ for large ℓ is smooth. This is shown in Fig. 12. It is seen that for periodic values of ℓ, the pc-wave disappears. Since ℓ is the layer thickness in units of a wavelength (apart from a factor of 2π), it follows that the material has to be manufactured with an accuracy of the order of a few wavelengths.

12 Specular Reflection

A dielectric with $\varepsilon = 1$ is completely transparent, and that gives $R = 0$. For the nonlinear medium under consideration, even for $\varepsilon = 1$ there is still a coupling to the specular wave. The nonlinear interaction does not only generate a phase-conjugated image of the incident field, but also a specular wave. This is illustrated in Fig. 13. Interesting to see is that $|R|$ has an off-resonance peak for this particular choice of parameters.

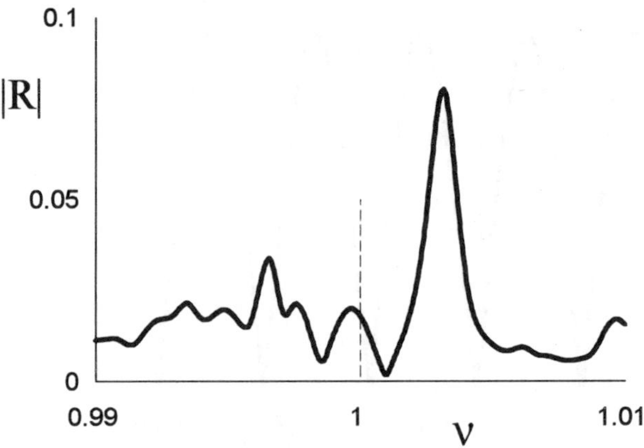

Figure 13: Graph of the Fresnel coefficient for specular reflection as a function of the frequency ν. The parameters are: $\varepsilon = 1$, $\gamma = 0.01$, $\ell = 942$ and $\theta_i = 45^{\circ}$.

13 Conclusions

When a nonlinear medium is pumped by two strong counterpropagating laser beams, the material acts as a phase conjugator for weak incident radiation. Even though the nonlinearity is very small (the third-order susceptibility $\chi^{(3)} \sim 10^{-22}$ m^2/V^2), it leads to a coupling parameter γ given by Eq. 10. The $\chi^{(3)}$ is multiplied by the square of the electric field of the pump beams, yielding a γ of the order of 10^{-2} or smaller. This is still a small number compared to $\chi^{(1)} = \varepsilon - 1$ which accounts for the linear interaction. We have shown, however, that due to the resonant nature of the interaction, a substantial signal can be generated, and even amplification can occur (Fig. 5). The bandwidth of the response is large, and approximately given by $\Delta\omega_a = \gamma_o \overline{\omega}$. The Fresnel reflection coefficient for the phase-conjugated image of the incident field depends in a complicated way on the frequency, the angle of incidence, the dielectric constant of the material, the parameter γ, the layer thickness and the polarization of the incident wave. We have illustrated graphically several features of the reflection coefficient. Especially interesting is that |P| can actually have a minimum at resonance (Fig. 6), indicating the delicate dependence on the various parameters.

References

1. B. Ya. Zel'dovich, V. I. Popovichev, V. V. Ragul'skii and F. S. Faizullov, *JETP Lett.* **15**, 109 (1972).
2. O. Yu. Nosach, V. I. Popovichev, V. V. Ragul'skii and F. S. Faizullov, *JETP Lett.* **16**, 435 (1972).
3. B. Ya. Zel'dovich, N. A. Mel'nikov, N. F. Pilipetskii and V. V. Ragul'skii, *JETP Lett.* **25**, 36 (1977).
4. D. M. Bloom and G. C. Bjorklund, *Appl. Phys. Lett.* **31**, 592 (1977).
5. S. M. Jensen and R. W. Hellwarth, *Appl. Phys. Lett.* **32**, 166 (1978).
6. P. V. Avizonis, F. A. Hopf, W. D. Bomberger, S. F. Jacobs, A. Tomita and K. H. Womack, *Appl. Phys. Lett.* **31**, 435 (1977).
7. J. Feinberg, *Opt. Lett.* **8**, 480 (1983).
8. M. Cronin-Golomb, B. Fisher, J. O. White and A. Yariv, *J. Quantum Electron.* **QE-20**, 12 (1984).
9. G. Salamo, M. J. Miller, W. W. Clark III, G. L. Wood and E. J. Sharp, *Opt. Commun.* **59**, 417 (1986).
10. P. M. Peterson and P. M. Johansen, *Opt. Lett.* **13**, 45 (1988).
11. K. Sayano, G. A. Rakuljic and A. Yariv, *Opt. Lett.* **13**, 143 (1988).
12. H. Kong, C. Lin, A. M. Biernacki and M. Cronin-Golomb, *Opt. Lett.* **13**, 324 (1988).
13. M. R. Belic, *Phys. Rev. A* **37**, 1809 (1988).
14. J. Goltz, C. Denz and T. Tschudi, *Opt. Commun.* **68**, 453 (1988).
15. R. W. Hellwarth, *J. Opt. Soc. Am.* **67**, 1 (1977).
16. A. Yariv and D. M. Pepper, *Opt. Lett.* **1**, 16 (1977).

17. B. Ya. Zel'dovich, N. F. Pilipetskii and V. V. Shkunov, *Principles of Phase Conjugation* (Springer, Berlin, 1985).
18. Y. R. Shen, *The Principles of Nonlinear Optics* (Wiley, New York, 1984).
19. M. Schubert and B. Wilhelmi, *Nonlinear Optics and Quantum Electronics* (Wiley, New York, 1986).
20. D. A. Kleinman, *Phys. Rev.* **126**, 1977 (1962).
21. H. F. Arnoldus and T. F. George, *Phys. Rev. A* **51**, 4250 (1995).
22. E. Wolf, L. Mandel, R. W. Boyd, T. M. Habashy and M. Nieto-Vesperinas, *J. Opt. Soc. Am. B* **4**, 1260 (1987).
23. A. A. Jacobs, W. R. Tompkin, R. W. Boyd and E. Wolf, *J. Opt. Soc. Am. B* **4**, 1266 (1987).
24. D. M. Pepper, D. Fekete and A. Yariv, *Appl. Phys. Lett.* **33**, 41 (1978).

ELECTROMAGNETIC PROPAGATORS IN MICRO- AND MESOSCOPIC OPTICS

OLE KELLER

Institute of Physics, Aalborg University,
Pontoppidanstræde 103, DK-9220 Aalborg Øst, Denmark
E-mail: okeller@physics.auc.dk

Fundamental aspects of semiclassical (field unquantized) electrodynamics are studied starting from a propagator point of view. If the seat of the transverse electromagnetic field is assumed to be the transverse atomic current densities the associated propagator is isotropic and only events on the light cone are correlated. If instead, the atomic current densities themselves are considered as seat for the transverse field this is composed of an attached part (the self-field) and a detached part propagating outwards from the atom at the speed of light. In the far field zone the propagating field correlates only events lying on the light cone, but in the near field also space-like couplings are present. The propagator formalism is used to investigate the spatial confinement of light in mesoscopic systems where the electrons are subjected to one-, two-, or three-dimensional spatial confinement.

1 Introduction

One often seeks to describe the optical bulk properties of condensed-matter media on the basis of macroscopic electrodynamics, i.e. by means of the macroscopic Maxwell equations and relevant constitutive equations. If the prevailing electromagnetic field is of moderate strength the material response is linear in the field, and in textbooks usually characterized by the (linear) susceptibility, permeability and conductivity tensors. Technically, one may join the susceptibility and conductivity into a single quantity, the macroscopic (M) relative dielectric tensor, $\overleftrightarrow{\epsilon}_{\mathrm{M}}$. In the macroscopic approach the constitutive equations are most often considered as point relations. This, for instance means that the macroscopic electric displacement field, $\vec{D}_{\mathrm{M}}(\vec{r}, t)$, in the space point $\vec{r}$ only depends on the macroscopic electric field $\vec{E}_{\mathrm{M}}(\vec{r}, t')$ at the same point. Although the assumption of spatial locality in the field-matter interaction usually is correct from a macroscopic point of view there exist electrodynamic phenomena where a macroscopic approach can be established only if the spatial-locality approximation is abandoned. In this respect, the optical activity in dielectrics and semiconductors, the anomalous skin effect in metals, and the far-infrared properties of superconductors are prominent examples. While a spatially local theory thus often works well in macroscopic optics, it is certainly needed to incorporate nonlocality in time in the overwhelming majority of macroscopic calculations. The macroscopic dielectric tensor hence is a function of two time coordinates, i.e. $\overleftrightarrow{\epsilon}_{\mathrm{M}} = \overleftrightarrow{\epsilon}_{\mathrm{M}}(\vec{r}, t, t')$, and the associated linear constitutive equation has the form $\vec{D}_{\mathrm{M}}(\vec{r}, t) = \epsilon_0 \int \overleftrightarrow{\epsilon}_{\mathrm{M}}(\vec{r}, t, t') \cdot \vec{E}_{\mathrm{M}}(\vec{r}, t')dt'$. The presence of an $\vec{r}$-dependence in $\overleftrightarrow{\epsilon}_{\mathrm{M}}$ allows one to deal with macroscopically inhomogeneous media. If the system in consideration does not change its (electrodynamic) properties with time, the dielectric tensor can only depend on the time difference $t-t'$. The resulting constitutive equation $\vec{D}_{\mathrm{M}}(\vec{r}, t) = \epsilon_0 \int \overleftrightarrow{\epsilon}_{\mathrm{M}}(\vec{r}, t - t') \cdot \vec{E}_{\mathrm{M}}(\vec{r}, t')dt'$ therefore simplifies to the local form $\vec{D}_{\mathrm{M}}(\vec{r}; \omega) = \epsilon_0 \overleftrightarrow{\epsilon}_{\mathrm{M}}(\vec{r}; \omega) \cdot \vec{E}_{\mathrm{M}}(\vec{r}; \omega)$ in the frequency (ω) domain. Nonlocality in time thus is synonymous with frequency dispersion in media exhibiting time-invariant properties. In a macroscopically homogeneous and isotropic medium the spatially local electrodynamic response hence may be characterized by the dielectric constant $\epsilon_{\mathrm{M}}(\omega)$ and the permeability $\mu_{\mathrm{M}}(\omega)$. To calculate these quantities one needs an equation of motion for the relevant charged particles in the medium studied. Such an equation is usually of the harmonic oscillator type with the crucial quantities determined by Schrödinger-equation dynamics. Even though the particle dynamics is described by the Schrödinger equation, the resulting theory essentially still is a macroscopic one from an optical point of view as long as spatially nonlocal effects are excluded. Including anharmonicity in the oscillator motion nonlinear optical bulk effects of various kinds may be incorporated in the framework of a macroscopic approach, as well. If different media are brought in contact interface regions occur, and in macroscopic optics the effect of an interface is felt by imposing a set of jump (boundary) conditions on the electromagnetic field. If no essential charge and/or current (polarization) densities are created by the electromagnetic field prevailing in the interface region, which usually has a width much less than the electromagnetic wavelength(s) of interest, the jump conditions are trivial and do not affect the macroscopic analysis in any deep manner. If essential field induced charge and current densities do occur the jump conditions are crucial and the adequate ones are difficult to obtain without resorting to a microscopic theory for the optical response of the interface region.

In the last decade the scientific interest in quantum-well, quantum-wire and quantum-dot structures has increased along with the improved ability to tailor these in a controllable manner for specific purposes.[1,2] The one-, two- or three-dimensional spatial confinement of the mobile particles (electrons) in these mesoscopic structures leads to an essential discretization of the stationary-state energy-level spectrum. This in turn implies that macro- and mesoscopic samples made of the same material may exhibit very different optical properties. Since mesoscopic media still contains many atoms the optical properties of these samples in general deviate appreciably from those of the parent atoms (molecules) constituting the medium. The very fact that the linear extension of a mesoscopic medium is much smaller than the wavelength of light in one, two or all three of the characteristic space coordinates makes it doubtful to rely on a macroscopic theory for the optical properties.

The need for going beyond the macroscopic field theory also appears in the context of optical investigations of surfaces and interfaces.[3-7] In these cases the transition layer thicknesses are typically of the order of a few atomic monolayers and thus much smaller than the wavelength of light. Attempts to analyze both the linear and nonlinear optical response of such transition layers by means of a macroscopic refractive index (and a geometrical thickness) in most cases lead to erroneous conclusions. Also the optical response of monolayer thin molecular layers cannot be understood on the basis of macroscopic field electrodynamics.

In the traditional optical studies of quantum wells, surfaces, interfaces, thin molecular layers, etc. one has a depth resolution on the atomic length scale, whereas the lateral resolution is only macroscopic. Using near field optical techniques it has, however, become possible in recent years to improve the lateral resolution so that structures of subwavelength extension in the lateral direction can be investigated.[8,9] To understand the electrodynamics in these structures it is certainly necessary in many cases to go beyond a macroscopic approach.

In the wake of the above-mentioned considerations it appears natural to begin theoretical studies of mesoscopic media from the microscopic Maxwell equations.[10] In these equations the particle dynamics enters in terms of the microscopic current $(\vec{J}(\vec{r},t))$ and charge $(\rho(\vec{r},t))$ densities, between which a constraint, namely the equation of charge continuity, exists. A closed set of equations for the description of the microscopic electrodynamics is obtained by adding to the microscopic Maxwell equations the many-body Schrödinger equation. In linear response theory the Schrödinger equation results in a constitutive equation of the form $\vec{J}(\vec{r};\omega) = \int \overleftrightarrow{\sigma}_{\mathrm{MB}}(\vec{r},\vec{r}';\omega)\cdot[\vec{E}_{\mathrm{T}}(\vec{r}';\omega)+\vec{E}_{\mathrm{L}}^{\mathrm{ext}}(\vec{r}';\omega)]d^3r'$ in the frequency representation provided the mesoscopic system exhibits translational invariant properties in time in the absence of the external (ext) field. The microscopic many-body (MB) conductivity tensor $\overleftrightarrow{\sigma}_{\mathrm{MB}}(\vec{r},\vec{r}';\omega)$, which is a function of two space coordinates $\vec{r}$ and $\vec{r}'$ relates the microscopic current density at $\vec{r}$ to the sum of the transverse (T) part of the local field $(\vec{E}_{\mathrm{T}}(\vec{r}';\omega))$ and the longitudinal (L) part of the external field $(\vec{E}_{\mathrm{L}}^{\mathrm{ext}}(\vec{r}';\omega))$ at $\vec{r}'$. Once it comes to a practical calculation of the microscopic electrodynamics, effective one-electron schemes of one sort or another are forced upon us. In particular a constitutive equation of the form $\vec{J}(\vec{r};\omega) = \int \overleftrightarrow{\sigma}(\vec{r},\vec{r}';\omega)\cdot\vec{E}(\vec{r}';\omega)d^3r'$, where $\vec{E}(\vec{r}';\omega)$ is the total local field prevailing in $\vec{r}'$, has been popular in condensed matter physics. Instead of using the microscopic one-body conductivity tensor $\overleftrightarrow{\sigma}(\vec{r},\vec{r}';\omega)$ one may formulate the linear response theory in terms of the microscopic dielectric tensor $\overleftrightarrow{\epsilon}(\vec{r},\vec{r}\,';\omega)$ and the microscopic permeability tensor $\overleftrightarrow{\mu}(\vec{r},\vec{r}\,';\omega)$. These two tensors in general have no unique physical meaning, and within certain limits different choices leading to equivalent physical results are possible. As an example, the possible choice $\overleftrightarrow{\mu}(\vec{r},\vec{r}\,';\omega) = \delta(\vec{r}-\vec{r}\,')\overleftrightarrow{U}$ transfers all magnetic properties to the dielectric tensor, which in turn now is related to the conductivity tensor by $\overleftrightarrow{\epsilon}(\vec{r},\vec{r}\,';\omega) = \overleftrightarrow{U}\delta(\vec{r}-\vec{r}\,') + i\overleftrightarrow{\sigma}(\vec{r},\vec{r}\,';\omega)/(\epsilon_0\omega)$.

In two recent review articles (Refs. 11 and 12) various fundamental aspects of the microscopic electrodynamic theory of mesoscopic media have been treated, and the application of the theory in specific cases discussed. Also a comprehensive list of references to relevant literature can be found in these articles. The microscopic theory may be formulated in different equivalent manners provided rigour is kept. The intuitive pictures one obtains in the various formulations may look quite different, and therefore a reformulation of the basic problem in a new picture in a sense often allows one to gain additional or complementary insight.

In the above-mentioned review articles the microscopic theory was based on an electromagnetic propagator picture which in an intuitively appealing manner emphasizes that a part of the microscopic transverse local field propagates away (becomes detached) from the microscopic atomic sources of the medium under study at the speed of light, and that another part of the transverse field and the total longitudinal field stay attached to the microscopic sources. Both parts and the interplay between them are of importance for our understanding of the microscopic electrodynamics of the new mesoscopic media (materials) and of optics in systems with confined electron dynamics as such. In the present article the electromagnetic propagator picture is further developed paying particular attention to the propagator itself, and I aim at a description which I hope supplements the previous review articles of mine on local-field electrodynamics in a natural and fruitful manner.

In section 2, we begin the description of the space-time electrodynamics in the propagator picture by separating the transverse and longitudinal vector-field dynamics. The retarded and advanced electromagnetic propagators associated with a transverse current-density source are introduced next. The advanced interactions are eliminated by means of the principle of causality, but only after the response-theory approximation has been adopted. As long as the transverse part of the particle current density is considered as the seat for the generation of the electromagnetic field, the field is only different from zero on the light cone. The retarded propagator is isotropic and exhibits a characteristic R^{-1}-distance dependence only. It is common in the literature to take the particle current density itself as the generator of the transverse electromagnetic field, and if this equivalent point of view is adopted, the associated transverse electromagnetic propagator turns out to possess a number of more complicated features than the aforementioned one. In particular both the advanced and retarded parts of this new propagator have far field (R^{-1}), middle field (R^{-2}), and near field (R^{-3}) parts in the space-frequency representation. Each of these parts are analyzed separately in the space-time domain, and it is demonstrated that the middle field part of the propagator is absent in this domain. The retarded far field response is different from zero only on the light cone, whereas the near field term contributes only to space-like events. Both disturbances propagate outwards from the source point at the speed of light. In the description where the particle current density distribution itself is considered the source for the electromagnetic field, a transverse self-field appears, in addition. A longitudinal self-field derived from the particle position variables themselves is always present. The electrodynamics of the two self-fields is analyzed in some details. Starting from a delta function current density source one is led to the standard (textbook) propagator. We show that this propagator exhibits a number of unacceptable properties. We end section 2 by discussing the relation between the standard formulation of electrodynamics and the propagator picture. In section 3, we analyze the transverse and longitudinal current densities from the point of view of quantum physics, and we discuss the associated electrodyanmics. Among other things, we study the asymptotic behaviour of the transverse and longitudinal current densities as well as the microscopic current flowing when an electron undergoes a transition between two bound states. Also the particular nature of the spin current density is described. In the electrodynamics of mesoscopic media not only the spatial confinement of the electrons plays a role, but also the localization of the electromagnetic field is of crucial importance, and in section 4 these two confinements are studied and compared. In a study of three-dimensional spatial confinement, we reconsider the electric point dipole model, discuss the confinement of the electromagnetic self-field, and bridge the gap to the radiation reaction phenomenon in semiclassical electrodynamics. Two-dimensional spatial confinement is examplified via a study of both the self-field and the retarded field of a mesoscopic (quantum) wire. The one-dimensional confinement found in quantum well systems is here related to an analysis of the commonly used non-retarded approach, but also the retarded quantum well propagator is discussed. Finally, the transverse and longitudinal delta functions appropriate for studies of quantum well systems excited by a single plane wave are introduced.

2   Space-time electrodynamics from a propagator point of view

2.1   *Transverse and longitudinal vector-field dynamics*

As a starting point for the present description we take the microscopic Maxwell equations in the time (t)-space $(\vec{r})$ domain, i.e.

$$\vec{\nabla} \times \vec{E}(\vec{r},t) = -\frac{\partial \vec{B}(\vec{r},t)}{\partial t}, \tag{1}$$

$$\vec{\nabla} \times \vec{B}(\vec{r},t) = \mu_0 \vec{J}(\vec{r},t) + \frac{1}{c_0^2}\frac{\partial \vec{E}(\vec{r},t)}{\partial t}, \tag{2}$$

$$\vec{\nabla} \cdot \vec{E}(\vec{r},t) = \frac{1}{\epsilon_0}\rho(\vec{r},t), \tag{3}$$

$$\vec{\nabla} \cdot \vec{B}(\vec{r},t) = 0, \tag{4}$$

where $\vec{E}(\vec{r},t)$ and $\vec{B}(\vec{r},t)$ are the prevailing electric and magnetic fields, and $\rho(\vec{r},t)$ and $\vec{J}(\vec{r},t)$ are the microscopic charge and current densities associated with the distribution and movement of the particles of the matter field. A Fourier integral transformation to the frequency (ω) domain brings Eqs. (1)–(4) to the form

$$\vec{\nabla} \times \vec{E}(\vec{r};\omega) = i\omega \vec{B}(\vec{r};\omega), \tag{5}$$

$$\vec{\nabla} \times \vec{B}(\vec{r};\omega) = \mu_0 \vec{J}(\vec{r};\omega) - \frac{i\omega}{c_0^2}\vec{E}(\vec{r};\omega), \tag{6}$$

$$\vec{\nabla} \cdot \vec{E}(\vec{r};\omega) = \frac{1}{\epsilon_0}\rho(\vec{r};\omega), \tag{7}$$

$$\vec{\nabla} \cdot \vec{B}(\vec{r},t) = 0. \tag{8}$$

Let us now divide the fields and current density into their transverse (T) and longitudinal (L) parts. By definition, a transverse vector field $\vec{V}_T(\vec{r})$ is a vector field satisfying the condition $\vec{\nabla} \cdot \vec{V}_T(\vec{r}) = 0$, and a longitudinal vector field $\vec{V}_L(\vec{r})$ is characterized by $\vec{\nabla} \times \vec{V}_L(\vec{r}) = \vec{0}$. The names transverse and longitudinal have an obvious geometrical meaning in the wave-vector $(\vec{q})$ domain. Thus, for a transverse vector field, $\vec{V}_T(\vec{q})$ is perpendicular to $\vec{q}$, i.e. $\vec{q} \cdot \vec{V}_T(\vec{q}) = 0$, and for a longitudinal vector field, $\vec{V}_L(\vec{q})$ is parallel to $\vec{q}$, $\vec{q} \times \vec{V}_L(\vec{q}) = \vec{0}$. Eq. (8) shows that the magnetic field only has a transverse part, i.e.

$$\vec{B}(\vec{r};\omega) = \vec{B}_T(\vec{r};\omega). \tag{9}$$

The electric field

$$\vec{E}(\vec{r};\omega) = \vec{E}_T(\vec{r};\omega) + \vec{E}_L(\vec{r};\omega), \tag{10}$$

and current density

$$\vec{J}(\vec{r};\omega) = \vec{J}_{\mathrm{T}}(\vec{r};\omega) + \vec{J}_{\mathrm{L}}(\vec{r};\omega), \tag{11}$$

on the other hand in general have both T and L components. Although the decomposition of a vector field into transverse and longitudinal components is not relativistically invariant, the T-L division in Eqs. (10) and (11) is convenient for our discussion of the spatial confinemenet of electromagnetic fields, and as such in low-energy physics where the particle dynamics is treated nonrelativistically. Also when it comes to a canonical quantization of the electromagnetic field the present approach is more adequate for the analysis than the manifestly covariant formulation in which the longitudinal and scalar photons have to be eliminated. Though the magnetic field is not a relativistically invariant quantity, the form invariance of the microscopic Maxwell equations guarantees that this field is transverse in all Lorentzian frames, cf. Eq. (8). In a given Lorentzian frame the division of a vector field $\vec{V}(\vec{r})$ into transverse and longitudinal parts is unique, viz.

$$\vec{V}_{\mathrm{T}}(\vec{r}) = \vec{\nabla} \times \left[\vec{\nabla} \times \int \frac{\vec{V}(\vec{r}')d^3r'}{4\pi|\vec{r}-\vec{r}'|} \right], \tag{12}$$

and

$$\vec{V}_{\mathrm{L}}(\vec{r}) = -\vec{\nabla} \left[\vec{\nabla} \cdot \int \frac{\vec{V}(\vec{r}')d^3r'}{4\pi|\vec{r}-\vec{r}'|} \right]. \tag{13}$$

It appears from Eq. (12) [or (13)] that the relationsship between $\vec{V}(\vec{r})$ and $\vec{V}_{\mathrm{T}}(\vec{r})$ [or $\vec{V}_{\mathrm{L}}(\vec{r})$] is nonlocal, which means that the value of $V_{\mathrm{T}}(\vec{r})$ [or $V_{\mathrm{L}}(\vec{r})$] at space point $\vec{r}$ depends on the values $V(\vec{r}')$ of the total vector field at all other points $\vec{r}'$. In the wave-vector domain the relations between the L and T parts of the vector field and the field itself are local, as one readily sees from the fact that $\vec{V}_{\mathrm{L}}(\vec{q}) = \vec{q}\vec{q} \cdot \vec{V}(\vec{q})/q^2$ and $\vec{V}_{\mathrm{T}}(\vec{q}) = (\vec{U} - \vec{q}\vec{q}/q^2) \cdot \vec{V}(\vec{q})$, $\vec{U}$ being the unit tensor and $q = |\vec{q}|$. The nonlocal effects introduced in real space by the divisions in Eqs. (10) and (11) are important for the spatial confinement problem for the electromagnetic field, as we shall realize later on. The T-L division splits the microscopic Maxwell equations in Eqs. (5)–(8) into a set

$$\vec{\nabla} \times \vec{E}_{\mathrm{T}}(\vec{r};\omega) = i\omega\vec{B}(\vec{r};\omega), \tag{14}$$

$$\vec{\nabla} \times \vec{B}(\vec{r};\omega) = \mu_0\vec{J}_{\mathrm{T}}(\vec{r};\omega) - \frac{i\omega}{c_0^2}\vec{E}_{\mathrm{T}}(\vec{r};\omega), \tag{15}$$

which describes the transverse electrodynamics, and a set

$$\vec{J}_{\mathrm{L}}(\vec{r};\omega) = i\epsilon_0\omega\vec{E}_{\mathrm{L}}(\vec{r};\omega), \tag{16}$$

$$\vec{\nabla} \cdot \vec{E}_{\mathrm{L}}(\vec{r}; \omega) = \frac{1}{\epsilon_0} \rho(\vec{r}; \omega), \tag{17}$$

accounting for the longitudinal dynamics. By combining Eqs. (16) and (17) one may obtain the local charge conservation condition $\vec{\nabla} \cdot \vec{J}_{\mathrm{L}}(\vec{r}; \omega) = i\omega\rho(\vec{r}; \omega)$.

By eliminating the magnetic field among Eqs. (14) and (15) it appears that the transverse electric field obeys the inhomogeneous Helmholtz equation

$$\left(\nabla^2 + q_0^2\right) \vec{E}_{\mathrm{T}}(\vec{r}; \omega) = -i\mu_0\omega \vec{J}_{\mathrm{T}}(\vec{r}; \omega), \tag{18}$$

where $q_0 = \omega/c_0$. A transformation of Eq. (18) to the time domain gives us the following wave equation for $\vec{E}_{\mathrm{T}}(\vec{r}, t)$ in the source field of the transverse current density:

$$\left(\nabla^2 - \frac{1}{c_0^2}\frac{\partial^2}{\partial t^2}\right) \vec{E}_{\mathrm{T}}(\vec{r}, t) = \mu_0 \frac{\partial \vec{J}_{\mathrm{T}}(\vec{r}, t)}{\partial t}. \tag{19}$$

In the time-space domain the relation between $\vec{E}_{\mathrm{L}}(\vec{r}, t)$ and $\vec{J}_{\mathrm{L}}(\vec{r}, t)$ takes the form

$$\frac{\partial \vec{E}_{\mathrm{L}}(\vec{r}, t)}{\partial t} = -\frac{1}{\epsilon_0} \vec{J}_{\mathrm{L}}(\vec{r}, t). \tag{20}$$

The electrodynamics of the system in consideration thus is governed by Eqs. (19) and (20). These equations are manifestly coupled since the transverse and longitudinal current densities are (unique) projections of the total current density, $\vec{J}(\vec{r}, t)$, cf. Eqs. (12) and (13). Indirectly the equations are coupled such that the source currents are driven in a selfconsistent manner by the prevailing (local) field. In passing, it is important to note that the results above are independent of the choice of gauge since they have been derived directly from the microscopic Maxwell euqations for $\vec{E}$ and $\vec{B}$ without reference to the vector and scalar potentials.

2.2   Retarded and advanced electromagnetic propagators associated with a transverse current-density source

Let us now return to the inhomogeneous Helmholtz equation for the transverse electric field, Eq. (18). From a mathematical point of view the general solution to this equation may be written in the form

$$\vec{E}_{\mathrm{T}}(\vec{r}; \omega) = \vec{\mathcal{E}}_{\mathrm{T}}(\vec{r}; \omega) - i\mu_0\omega \int \left[c^{\mathrm{R}} \overleftrightarrow{d}_0^{\mathrm{R}}(R; \omega) + c^{\mathrm{A}} \overleftrightarrow{d}_0^{\mathrm{A}}(R; \omega)\right] \cdot \vec{J}_{\mathrm{T}}(\vec{r}'; \omega)d^3r',$$

$$\tag{21}$$

where

$$\vec{\vec{d}}_0^{\,\mathrm{R}}(R;\omega) = -\frac{e^{iq_0R}}{4\pi R}\vec{\vec{U}},$$

$$(22)$$

$$\vec{\vec{d}}_0^{\,\mathrm{A}}(R;\omega) = -\frac{e^{-iq_0R}}{4\pi R}\vec{\vec{U}},$$

$$(23)$$

and $c^{\mathrm{R}} + c^{\mathrm{A}} = 1$, with $R = |\vec{R}|$ and $\vec{R} = \vec{r} - \vec{r}'$. The first term, $\vec{\mathcal{E}}(\vec{r};\omega)$, on the right hand side of Eq. (21) represents the general solution to the homogeneous Helmholtz equation, and as indicated by the subscript, $\vec{\mathcal{E}}_{\mathrm{T}}(\vec{r};\omega)$ has to be a transverse vector field. The second term is a particular solution to the inhomogeneous Helmholtz equation. The different choices which can be adopted for the particular solution are parametrized via the constants c^{R} and c^{A} which fulfil the constraint $c^{\mathrm{R}} + c^{\mathrm{A}} = 1$. The minus signs appearing in Eqs. (22) and (23) originate in a convention, and the (superfluous) unit tensors $(\vec{\vec{U}})$ are adequate for the subsequent discussion. With the help of the expansion(s)

$$\delta\left(\frac{R}{c_0} \mp \tau\right) = \frac{1}{2\pi}\int_{-\infty}^{\infty} e^{\pm i\frac{\omega}{c_0}R}e^{-i\omega\tau}\,d\omega,$$

$$(24)$$

of the Dirac delta function, δ, it is realized that Eq. (21) can be written in the form

$$\vec{E}_{\mathrm{T}}(\vec{r},t) = \vec{\mathcal{E}}_{\mathrm{T}}(\vec{r},t) + \mu_0\int\left[c^{\mathrm{R}}\,\vec{\vec{\delta}}_0^{\,\mathrm{R}}(R,\tau) + c^{\mathrm{A}}\,\vec{\vec{\delta}}_0^{\,\mathrm{A}}(R,\tau)\right]\cdot\frac{\partial\vec{J}_{\mathrm{T}}(\vec{r}',t')}{\partial t'}d^3r'dt',$$

$$(25)$$

with

$$\vec{\vec{\delta}}_0^{\,\mathrm{R}}(R,\tau) = -\frac{1}{4\pi R}\delta\left(\frac{R}{c_0} - \tau\right)\vec{\vec{U}},$$

$$(26)$$

$$\vec{\vec{\delta}}_0^{\,\mathrm{A}}(R,\tau) = -\frac{1}{4\pi R}\delta\left(\frac{R}{c_0} + \tau\right)\vec{\vec{U}},$$

$$(27)$$

and $\tau = t - t'$, in the time-space domain. The result in Eq. (25) thus represents the general solution to the wave equation for the transverse electric field, Eq. (19).

In classical electrodynamics the seat for all electromagnetic fields is the matter field. This means that if the transverse current density in Eq. (25) is thought of as comprising the transverse flow of all charges (in the universe), $\vec{J}_{\mathrm{T}}(\vec{r},t) = \vec{J}_{\mathrm{T}}^{\mathrm{all}}(\vec{r},t)$, the $\vec{\mathcal{E}}_{\mathrm{T}}(\vec{r},t)$-term must be zero. Hence,

$$\vec{E}_{\mathrm{T}}(\vec{r},t) = \mu_0\int\left[c^{\mathrm{R}}\,\vec{\vec{d}}_0^{\,\mathrm{R}}(R,\tau) + c^{\mathrm{A}}\,\vec{\vec{d}}_0^{\,\mathrm{A}}(R,\tau)\right]\cdot\frac{\partial\vec{J}_{\mathrm{T}}^{\mathrm{all}}(\vec{r}',t')}{\partial t'}d^3r'dt'.$$

$$(28)$$

In quantum electrodynamics, the ground state of the quantum field has a nonzero absolute energy, and the variances of the electric ($\vec{E}_{\mathrm{T}}$) and magnetic ($\vec{B}$) fields are nonvanishing, and therefore the homogeneous part of the solution to the inhomogeneous Helmholtz equation among the operators of the electrodynamic and matter fields cannot be omitted. Let us consider next a simpler situation where one approximately can replace $\vec{J}_{\mathrm{T}}^{\mathrm{all}}(\vec{r}, t)$ by two spatially separated flows, $\vec{J}_{\mathrm{T}}^{(1)}$ and $\vec{J}_{\mathrm{T}}^{(2)}$, and neglect all of the remaining contributions to $\vec{J}_{\mathrm{T}}^{\mathrm{all}}(\vec{r}, t)$. In this case we have

$$
\vec{E}_{\mathrm{T}}(\vec{r}, t) = \mu_0 \int_{V^{(1)}} \left[c^{\mathrm{R}} \overset{\leftrightarrow}{d}_0^{\mathrm{R}}(R, \tau) + c^{\mathrm{A}} \overset{\leftrightarrow}{d}_0^{\mathrm{A}}(R, \tau) \right] \cdot \frac{\partial \vec{J}_{\mathrm{T}}^{(1)}(\vec{r}\,', t')}{\partial t'} d^3 r' dt'
$$

$$
+ \mu_0 \int_{V^{(2)}} \left[c^{\mathrm{R}} \overset{\leftrightarrow}{d}_0^{\mathrm{R}}(R, \tau) + c^{\mathrm{A}} \overset{\leftrightarrow}{d}_0^{\mathrm{A}}(R, \tau) \right] \cdot \frac{\partial \vec{J}_{\mathrm{T}}^{(2)}(\vec{r}\,', t')}{\partial t'} d^3 r' dt', \tag{29}
$$

where $V^{(1)}$ and $V^{(2)}$ denote the effective volumes of the two flows. Usually, the two flows $\vec{J}_{\mathrm{T}}^{(1)}$ and $\vec{J}_{\mathrm{T}}^{(2)}$ have to be determined in a selfconsistent manner since the transverse radiation field provides an electromagnetic coupling between them. It may happen that the coupling almost is a one-way coupling from say $\vec{J}_{\mathrm{T}}^{(1)}$ to $\vec{J}_{\mathrm{T}}^{(2)}$. In such cases the so-called response theory may be helpful in analyzing the electrodynamics. In response theory one assumes a one-way interaction only, in the above case from $\vec{J}_{\mathrm{T}}^{(1)}$ to $\vec{J}_{\mathrm{T}}^{(2)}$. This means that the dynamics of the particles in volume $V^{(1)}$ evolves independently of the particle motion in $V^{(2)}$. The response-theory approximation is important from an experimental point of view, and one might be tempted to claim that all experimental studies in one way or another are based on response theory. In the framework of response theory the electrodynamics of system (2) thus is driven by the particle motion in system (1).

A further reduction of Eq. (29) is obtained in the basis of the principle of causality. This principle is based on the apparently never violated experimental observation that the response of the particles in system (2) always is delayed (retarded) in time with respect to the onset of the charge motion in the driving system (1), i.e. $\tau = t - t' > 0$. This means that in Eq. (29) the quantity $\overset{\leftrightarrow}{d}_0^{\mathrm{A}}(|\vec{r}^{(2)} - \vec{r}^{(1)}|, \tau)$ must vanish, $\vec{r}^{(1)}$ and $\vec{r}^{(2)}$ being space coordinates in system (1) and (2), respectively. If this observation is generalized by postulating a time delayed interaction between the transverse current-density flows at two arbitrary space-time points, $(\vec{r}^{(1)}, t^{(1)})$ and $(\vec{r}^{(2)}, t^{(2)})$, we arrive at the principle of causality. In the case of transverse electrodynamics treated here all so-called advanced interactions disappear. Analytically, we achieve this by setting $c^{\mathrm{A}} = 0$ (and hence $c^{\mathrm{R}} = 1$). In turn this means that the advanced (A) electromagnetic propagator $\overset{\leftrightarrow}{d}_0^{\mathrm{A}}(R, \tau)$ do not enter the formalism.

In the spirit of the previous discussion, let us name the (source) field stemming from the $\vec{J}_{\mathrm{T}}^{(1)}$-distribution the transverse external (ext) field, $\vec{E}_{\mathrm{T}}^{\mathrm{ext}}(\vec{r},t)$. In the framework of response theory the external field is a prescribed quantity. The particles of system (2) constitutes the matter particles under study, and in the following we omit for brevity the superscript (2) on the transverse current density, i.e. $\vec{J}_{\mathrm{T}}^{(2)}(\vec{r},t) = \vec{J}_{\mathrm{T}}(\vec{r},t)$, and on other relevant quantities of system (2). Invoking also the principle of causality, Eq. (29) hence takes the form

$$\vec{E}_{\mathrm{T}}(\vec{r},t) = \vec{E}_{\mathrm{T}}^{\mathrm{ext}}(\vec{r},t) + \mu_0 \int \overleftrightarrow{d}_0^{\mathrm{R}}(R,\tau) \cdot \frac{\partial \vec{J}_{\mathrm{T}}(\vec{r}\,',t')}{\partial t'} d^3 r' dt', \qquad (30)$$

where the so-called retarded (R) dyadic electromagnetic propagator (Green's function) $\overleftrightarrow{d}_0^{\mathrm{R}}(R,\tau)$ is given by Eq. (27). This Green's function shows that the propagation characteristic of a transverse electromagnetic disturbance from a transverse current-density source point located at $(\vec{r}\,',t')$ is very simple: (i) the field is only different from zero on the light cone $|\vec{r} - \vec{r}\,'| = c_0(t - t')$, (ii) the propagation is isotropic (the unit tensor describes the tensor properties of $\overleftrightarrow{d}_0^{\mathrm{R}}(R,\tau)$), the disturbance (iii) propagates with the vacuum speed of light (c_0) and (iv) exhibits a R^{-1} distance dependence.

2.3 Longitudinal response

It appears from Eq. (20) that the longitudinal electric field is different from zero only in space points where the longitudinal current density $\vec{J}_{\mathrm{L}}(\vec{r},t)$ ($= \vec{J}_{\mathrm{L}}^{\mathrm{all}}(\vec{r},t)$) is nonvanishing, but depends on $\vec{J}_{\mathrm{L}}$ not only at time t but also at earlier times. The fact that the longitudinal electric field at $\vec{r}$ only originates in the longitudinal current density at the same point means that no propagation effects at distance appear, and hence no light velocity. In the case where we are dealing with two separated longitudinal current density flows, $\vec{J}_{\mathrm{L}}^{(1)}(\vec{r},t)$ (source) and $\vec{J}_{\mathrm{L}}^{(2)}(\vec{r},t)$ (system), the longitudinal field in the system, $\vec{E}_{\mathrm{L}}^{(2)}(\vec{r},t) \equiv \vec{E}_{\mathrm{L}}(\vec{r},t)$, only depends on $\vec{J}_{\mathrm{L}}^{(2)}(\vec{r},t) \equiv \vec{J}_{\mathrm{L}}(\vec{r},t)$, the explicit relation being

$$\vec{E}_{\mathrm{L}}(\vec{r},t) = \vec{E}_{\mathrm{L}}(\vec{r},t_0) - \frac{1}{\epsilon_0} \int_{t_0}^{t} \vec{J}_{\mathrm{L}}(\vec{r},t')dt', \qquad (31)$$

where $\vec{E}_{\mathrm{L}}(\vec{r},t_0)$ is the longitudinal field at a reference time t_0.

Figure 1: Schematic illustration meant to emphasize the difference between the spatial confinement of the electrons (black domain) and the attached field (hatched domain) of an atomic source. The electrons are confined to a region where the relevant transition current densities are different from zero. The black domain indicates the $1s \leftrightarrow 2p_z$ electron confinement in a hydrogen atom, and the extension of this domain is of the order of the Bohr radius, a_0. The attached field (self-field) is localized in a region which coincides with the transverse (or longitudinal) transition current density. The extension of the attached field is given by the asymptotic form $(a_0/R)^3$, R being the distance from the nucleus.

2.4 Retarded and advanced transverse propagators. Near, middle, and far field dynamics

In section 2.2, we considered the transverse part of the prevailing current density $\vec{J}_{\mathrm{T}}(\vec{r},t)$ as the source for the generation of the retarded (or possibly advanced) transversely polarized electromagnetic field, and we found that the associated propagators $\vec{d}_0^{\mathrm{R}}(R,\tau)$ and $\vec{d}_0^{\mathrm{A}}(R,\tau)$ were isotropic, had a simple R^{-1} distance dependence, and were different from zero only on the retarded and advanced light cones. It is possible instead, however, to consider the current density itself, $\vec{J}(\vec{r},t)$, as the generator of the transverse electromagnetic field, and in fact, this point of view is the most common one in the literature. It seems natural to adopt the last point of view since a quantum mechanical calculation leads directly to $\vec{J}(\vec{r},t)$, whereas the determination of $\vec{J}_{\mathrm{T}}(\vec{r},t)$ is more difficult. For the electrodynamics of atoms, molecules and mesoscopic particles the common approach also has the advantage that the related source region, $\vec{J}(\vec{r},t)$, in general is much stronger localized in space than $\vec{J}_{\mathrm{T}}(\vec{r},t)$ (Fig. 1). When switching from the approach described in section 2.2 to the standard formulation there is a price to pay, however, because the associated electromagnetic propagator has a quite complicated structure, as we shall see below.

Finite-c_0 propagator electrodynamics

To turn from the picture where the seat of the transverse electromagnetic field is identified with the transverse current density distribution, $\vec{J}_{\mathrm{T}}(\vec{r},t)$, to a picture in which only the current density itself, $\vec{J}(\vec{r},t)$ acts as the seat, we start from the formal expression

$$\vec{J}_{\mathrm{T}}(\vec{r},t) = \int \vec{\vec{\delta}}_{\mathrm{T}}(\vec{r}-\vec{r}\,') \cdot \vec{J}(\vec{r}\,',t)d^3r'. \tag{32}$$

The spatially nonlocal relation between the transverse part of a vector field and the vector field itself, already mentioned in the context of Eq. (12), is here expressed in terms of the so-called transverse (T) delta function, $\vec{\vec{\delta}}_{\mathrm{T}}(\vec{r}-\vec{r}\,')$, which is a dyad projecting out the transverse part of a vector field from the total vector field. For time-dependent vector fields, $\vec{\vec{\delta}}_{\mathrm{T}}(\vec{r}-\vec{r}\,')$ relates the relevant quantities at a given time (locality in time). By inserting Eq. (32) into Eq. (30) and using Eq. (26), one obtains

$$\vec{E}_{\mathrm{T}}(\vec{r},t) = \vec{E}_{\mathrm{T}}^{\mathrm{ext}}(\vec{r},t) + \mu_0 \int \vec{\vec{G}}_0^{\mathrm{T}}(\vec{r}-\vec{r}\,',t-t') \cdot \frac{\partial \vec{J}(\vec{r}\,',t')}{\partial t'} d^3r'dt', \tag{33}$$

with

$$\vec{\vec{G}}_0^{\mathrm{T}}(\vec{r}-\vec{r}\,',t-t') = -\frac{1}{4\pi} \int_{-\infty}^{\infty} \frac{1}{|\vec{r}-\vec{r}\,''|} \delta\left(\frac{|\vec{r}-\vec{r}\,''|}{c_0} - t + t'\right) \vec{\vec{\delta}}_{\mathrm{T}}(\vec{r}\,''-\vec{r}\,')d^3r''. \tag{34}$$

In writing the expression for $\vec{\vec{G}}_0^{\mathrm{T}}(\vec{r}-\vec{r}\,',t-t')$ it has explicitly been stressed that the $\vec{r}\,''$-integration runs over the entire space. The tensor $\vec{\vec{G}}_0^{\mathrm{T}}(\vec{r}-\vec{r}\,',t-t')$ is a dyadic propagator which relates the transverse electric field at $(\vec{r},t)$ to the time-derivative of the total current density at $(\vec{r}\,',t')$. To underline that this propagator projects out the transverse field from the current density itself it has been given a superscript T.

To obtain an explicit expression for $\vec{\vec{G}}_0^{\mathrm{T}}$ we have to carry out the integration over $\vec{r}\,''$ in Eq. (34). Since $\vec{\vec{\delta}}_{\mathrm{T}}(\vec{r}\,''-\vec{r}\,')$ is singular at $\vec{r}\,''=\vec{r}\,'$, and $|\vec{r}-\vec{r}\,''|^{-1}$ goes to infinity for $\vec{r}\,''=\vec{r}$, one has to be extremely careful when seeking to evaluate the integral in Eq. (34) for $\vec{r}=\vec{r}\,''$. When $\vec{r}=\vec{r}\,'$, the field observation point in Eq. (33) coincides with a source point, and one obtains a so-called self-field contribution to $\vec{E}_{\mathrm{T}}(\vec{r},t)$. The self-field effect is independent of the vacuum velocity of light, and I shall undertake a discussion of the self-field dynamics in section 2.5. To study the finite-c_0 electrodynamics (retarded or advanced) it is adequate at this stage to introduce the (undamped) plane-wave expansion

$$\vec{\vec{\delta}}_{\mathrm{T}}(\vec{r}\,''-\vec{r}\,') = (2\pi)^{-3} \int_{-\infty}^{\infty} (\vec{\vec{U}} - \vec{e}_{\vec{q}}\vec{e}_{\vec{q}})\, e^{i\vec{q}\cdot(\vec{r}\,''-\vec{r}\,')}d^3q, \qquad \vec{r}\,'' \neq \vec{r}\,', \tag{35}$$

with $\vec{e}_{\vec{q}} = \vec{q}/q$, for the transverse delta function. As indicated, this expansion is valid only for $\vec{r}'' \neq \vec{r}'$. To extend this type of expansion to the singular point, a specifically defined procedure must be devised. The transverse propagator obtained by leaving out self-field effects I denote by $\overleftrightarrow{D}_0^{\mathrm{T}}(\vec{r}-\vec{r}',t-t') \equiv \overleftrightarrow{D}_0^{\mathrm{T}}(\vec{R},\tau)$. The integral expression for this propagator is found by inserting Eq. (35) into Eq. (34). Using the abbreviation $\vec{R}_0 = \vec{r} - \vec{r}''$ (and as before $\vec{R} = \vec{r} - \vec{r}'$ and $\tau = t - t'$) one has

$$\overleftrightarrow{D}_0^{\mathrm{T}}(\vec{R},\tau) = -\frac{1}{32\pi^4} \int_{-\infty}^{\infty} \frac{1}{R_0}\delta\left(\frac{R_0}{c_0} - \tau\right)(\overleftrightarrow{U} - \vec{e}_{\vec{q}}\vec{e}_{\vec{q}})\, e^{i\vec{q}\cdot(\vec{R}-\vec{R}_0)}d^3R_0 d^3q. \quad (36)$$

The integration over the $\vec{R}_0$- and $\vec{q}$-spaces can be carried out using spherical coordinates in both spaces. For the sake of the subsequent analysis I emphasize a few important intermediate results below. Performing thus first the integration over the $\vec{R}_0$-coordinates one gets

$$\overleftrightarrow{D}_0^{\mathrm{T}}(\vec{R},\tau) = \frac{ic_0}{16\pi^3} \int_{-\infty}^{\infty} \frac{1}{q}(\overleftrightarrow{U} - \vec{e}_{\vec{q}}\vec{e}_{\vec{q}})\left(e^{iqc_0\tau} - e^{-iqc_0\tau}\right) e^{i\vec{q}\cdot\vec{R}}d^3q. \quad (37)$$

As the second step spherical coordinates in $\vec{q}$-space are introduced, as mentioned above. By placing the polar axis along the $\vec{R}$-direction and carrying out subsequently the angular integrations we obtain

$$\overleftrightarrow{D}_0^{\mathrm{T}}(\vec{R},\tau) = \overleftrightarrow{D}_{\mathrm{F}}(\vec{R},\tau) + \overleftrightarrow{D}_{\mathrm{M}}(\vec{R},\tau) + \overleftrightarrow{D}_{\mathrm{N}}(\vec{R},\tau), \quad (38)$$

where

$$\overleftrightarrow{D}_{\mathrm{F}}(\vec{R},\tau) = \frac{c_0}{8\pi^2 R}(\overleftrightarrow{U} - \vec{e}_{\vec{R}}\vec{e}_{\vec{R}}) \int_0^{\infty} \left(e^{iqR} - e^{-iqR}\right)\left(e^{iqc_0\tau} - e^{-iqc_0\tau}\right)dq, \quad (39)$$

with $\vec{e}_{\vec{R}} = \vec{R}/R$,

$$\overleftrightarrow{D}_{\mathrm{M}}(\vec{R},\tau) = \frac{ic_0}{8\pi^2 R^2}(\overleftrightarrow{U} - 3\vec{e}_{\vec{R}}\vec{e}_{\vec{R}}) \int_0^{\infty} \left(e^{iqR} + e^{-iqR}\right)\left(e^{iqc_0\tau} - e^{-iqc_0\tau}\right)\frac{dq}{q}, \quad (40)$$

and

$$\overleftrightarrow{D}_{\mathrm{N}}(\vec{R},\tau) = -\frac{c_0}{8\pi^2 R^3}(\overleftrightarrow{U} - 3\vec{e}_{\vec{R}}\vec{e}_{\vec{R}}) \int_0^{\infty} \left(e^{iqR} - e^{-iqR}\right)\left(e^{iqc_0\tau} - e^{-iqc_0\tau}\right)\frac{dq}{q^2}. \quad (41)$$

The result in Eqs. (38)–(41) presents the propagator $\overleftrightarrow{D}_0^{\mathrm{T}}(\vec{R},\tau)$ in terms of an expansion of outgoing and ingoing spherical waves with wave numbers (q)

ranging from zero to infinity. The first term, $\vec{\vec{D}}_{\mathrm{F}}(\vec{R},\tau)$, gives the far (F) field contribution to the propagator because the factor in front of the integral in Eq. (39) is proportional to R^{-1}. In comparison with this the middle (M) field (in Eq. (40)) and near (N) field (in Eq. (41)) terms vary with R^{-2} and R^{-3}, respectively. In the subsequent three subsections we shall analyze each of these three terms (F, M, N) separately.

Transverse far field propagator

Since the outgoing and ingoing spherical waves belonging to the various wave numbers all have the same amplitude one expects that the transverse far field propagator is composed of delta function contributions, and a straight forward integration (see the appendix) also shows that

$$\vec{\vec{D}}_{\mathrm{F}}(\vec{R},\tau) = \vec{\vec{D}}_{\mathrm{F}}^{\mathrm{R}}(\vec{R},\tau) + \vec{\vec{D}}_{\mathrm{F}}^{\mathrm{A}}(\vec{R},\tau), \tag{42}$$

where

$$\vec{\vec{D}}_{\mathrm{F}}^{\mathrm{R}}(\vec{R},\tau) = -\frac{1}{4\pi R}\delta\left(\frac{R}{c_0} - \tau\right)\left(\vec{\vec{U}} - \vec{e}_{\vec{R}}\vec{e}_{\vec{R}}\right), \tag{43}$$

and

$$\vec{\vec{D}}_{\mathrm{F}}^{\mathrm{A}}(\vec{R},\tau) = \frac{1}{4\pi R}\delta\left(\frac{R}{c_0} + \tau\right)\left(\vec{\vec{U}} - \vec{e}_{\vec{R}}\vec{e}_{\vec{R}}\right), \tag{44}$$

are retarded (R) and advanced (A) contributions, respectively. The first term is different from zero only on the retarded light cone, $|\vec{r} - \vec{r}'| = c_0(t - t')$, and the second term vanishes outside the advanced light cone, $|\vec{r} - \vec{r}'| = c_0(t' - t)$. In section 2.2, I adopted the principle of causality when relating the transverse electric field at a given space-time point $(\vec{r}, t)$ to a transverse current density source located at $(\vec{r}', t')$. Since the relation between $\vec{J}_{\mathrm{T}}$ and $\vec{J}$ is local in time, this forces us to keep only the retarded part $\vec{\vec{D}}_{\mathrm{F}}^{\mathrm{R}}(\vec{R},\tau)$ of the far field propagator $\vec{\vec{D}}_{\mathrm{F}}(\vec{R},\tau)$. Likewise, if we had started postulating the principle of causality to hold between the transverse electric field, $\vec{E}_{\mathrm{T}}(\vec{r}, t)$, generated at $(\vec{r}, t)$ by the total current density, $\vec{J}(\vec{r}', t')$, at $(\vec{r}', t')$, we had been forced to accept a causal connection between $\vec{E}_{\mathrm{T}}(\vec{r}, t)$ and $\vec{J}_{\mathrm{T}}(\vec{r}', t')$. By comparing Eqs. (26) and (43) it appears that the propagators $\vec{d}_0^{\mathrm{R}}(R,\tau)$ and $\vec{\vec{D}}_{\mathrm{F}}^{\mathrm{R}}(\vec{R},\tau)$ only differ in their tensorial properties. The $\vec{\vec{d}}_0^{\mathrm{R}}(R,\tau)$-propagator thus is isotropic $(\vec{\vec{U}})$, as one might have expected since it relates two transverse (divergence-free) vector fields, and the $\vec{\vec{D}}_{\mathrm{F}}^{\mathrm{R}}(\vec{R},\tau)$-propagator is transverse $(\vec{\vec{U}} - \vec{e}_{\vec{R}}\vec{e}_{\vec{R}})$ since it in a given direction of observation $(\vec{e}_{\vec{R}})$ gives a (far) field which is perpendicular to $\vec{e}_{\vec{R}}$. One might have foreseen also the transverse character of $\vec{\vec{D}}_{\mathrm{F}}^{\mathrm{R}}(\vec{R},\tau)$, because this propagator gives the relation between a transverse vector field (far field of $\vec{E}_{\mathrm{T}}(\vec{r}, t)$) and a vector field ($\vec{J}(\vec{r}', t')$) having both transverse and longitudinal components. If one compares Eqs. (27) and (44), it appears that the advanced propagators $\vec{d}_0^{\mathrm{A}}(R, \tau)$ and $\vec{\vec{D}}_{\mathrm{F}}^{\mathrm{A}}(\vec{R}, \tau)$, apart from an extra factor of minus one, also only deviate in their tensorial (directional) properties.

In the context of classical optics, which is synonymous with far field optics, one might thus either use a $\vec{J}$-source formalism based on anisotropic $(\vec{\vec{U}} - \vec{e}_{\vec{R}}\vec{e}_{\vec{R}})$ propagators or equivalently a $\vec{J}_{\mathrm{T}}$-source description with isotropic $(\vec{\vec{U}})$ propagators.

The middle field propagator, a spurious quantity?

By carrying out the integration over q in Eq. (40), as indicated in the appendix, one obtains the following expression for the middle field part of the transverse electromagnetic propagator:

$$\vec{\vec{D}}_{\mathrm{M}}(\vec{R}, \tau) = \frac{c_0}{8\pi R^2} \left[\mathrm{sgn}\left(\frac{R}{c_0} - \tau\right) - \mathrm{sgn}\left(\frac{R}{c_0} + \tau\right) \right] (\vec{\vec{U}} - 3\vec{e}_{\vec{R}}\vec{e}_{\vec{R}}), \quad (45)$$

where the function $\mathrm{sgn}\, x$ equals $+1$ for $x > 0$ and -1 for $x < 0$. As it stands, $\vec{\vec{D}}_{\mathrm{M}}(\vec{R}, \tau)$ is composed of an outgoing part, proportional to $\mathrm{sgn}\,(R/c_0 - \tau)$, and an ingoing part proportional to $\mathrm{sgn}\,(R/c_0 + \tau)$, but in contrast to the situation with the far field propagator these two parts are not retarded and advanced, respectively. The two disturbances do propagate with the vacuum speed of light but they are not causal and non-causal as one would prefer. It is possible, of course, to establish an equivalent expression for the middle field propagator in which $\vec{\vec{D}}_{\mathrm{M}}(\vec{R}, \tau)$ still is divided into outgoing and ingoing parts which, however, now are causal and non-causal, respectively. Hence, it seems natural to take

$$\vec{\vec{D}}_{\mathrm{M}}(\vec{R}, \tau) = \vec{\vec{D}}_{\mathrm{M}}^{\mathrm{R}}(\vec{R}, \tau) + \vec{\vec{D}}_{\mathrm{M}}^{\mathrm{A}}(\vec{R}, \tau), \quad (46)$$

where

$$\vec{\vec{D}}_{\mathrm{M}}^{\mathrm{R}}(\vec{R}, \tau) = \frac{c_0}{4\pi R^2} \left[\Theta\left(\frac{R}{c_0} - \tau\right) - 1 \right] (\vec{\vec{U}} - 3\vec{e}_{\vec{R}}\vec{e}_{\vec{R}}) \quad (47)$$

is the retarded (and outgoing) part, and

$$\vec{\vec{D}}_{\mathrm{M}}^{\mathrm{A}}(\vec{R}, \tau) = \frac{c_0}{4\pi R^2} \left[1 - \Theta\left(\frac{R}{c_0} + \tau\right) \right] (\vec{\vec{U}} - 3\vec{e}_{\vec{R}}\vec{e}_{\vec{R}}) \quad (48)$$

is the advanced (and ingoing) part. The Θ-function appearing in the two equations above is the Heaviside unit step function, given by $\Theta(x) = 1$ for $x > 0$ and $\Theta(x) = 0$ for $x < 0$. It appears from Eq. (47) that only time-like events, i.e.

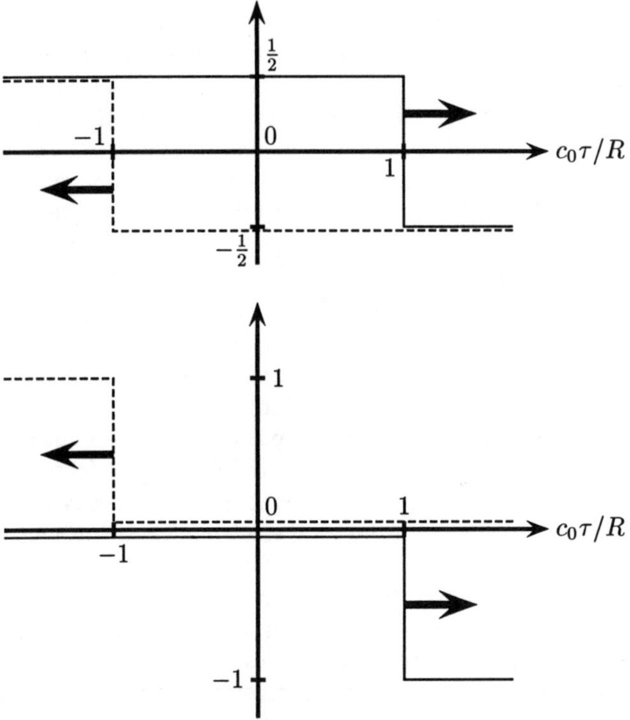

Figure 2: Schematic diagrams showing the time-space dynamics of the transverse middle field propagator. This propagator is proportional to $\frac{1}{2}[\mathrm{sgn}\,(\frac{R}{c_0} - \tau) - \mathrm{sgn}\,(\frac{R}{c_0} + \tau)]$, and in the upper figure is shown the outgoing $[\frac{1}{2}\mathrm{sgn}\,(\frac{R}{c_0} - \tau)]$ (full line) and ingoing $[-\frac{1}{2}\mathrm{sgn}\,(\frac{R}{c_0} + \tau)]$ (broken line) parts as a function of $c_0\tau/R$. The two parts are not causal and non-causal, respectively. In the lower figure is shown the division of the middle field propagator into an outgoing and retarded part $\Theta(\frac{R}{c_0} - \tau) - 1$ (full line) and an ingoing and advanced part $1 - \Theta(\frac{R}{c_0} + \tau)$ (broken line). The two contributions in the lower figure are different from zero only for time-like events.

those satisfying the inequality $|\vec{r} - \vec{r}\,'| < c_0(t - t')$, are coupled by the retarded part of the middle field propagator. In order that the advanced middle field propagator be different from zero, one must require that $|\vec{r} - \vec{r}\,'| < c_0(t' - t)$, and again only time-like (and non-causal) events are coupled. In the expression for $\overleftrightarrow{D}_{\mathrm{M}}(\vec{R}, \tau)$ given in Eq. (45), in both the outgoing and ingoing parts time-like ($|\vec{r} - \vec{r}\,'| < c_0|t - t'|$) and space-like ($|\vec{r} - \vec{r}\,'| > c_0|t - t'|$) couplings occur. The above-mentioned results are illustrated in Fig. 2. The presence of also space-like events in the separate ingoing and outgoing contributions to $\ddot{D}_{\mathrm{M}}(\vec{R}, \tau)$ is not a physically meaningless feature in itself as we shall realize when we analyse the quantum mechanics of the transverse (and longitudinal) current densities of our massive particle field. In fact, once one is dealing with (quantum) physical phenomena outside the light cone space- and time-like events seems to be equally acceptable. On the retarded and advanced light cones ($|\vec{r}-\vec{r}'| = c_0|t-t'|$), the delta functions character of the far field propagator makes it clear that here only far field phenomena need attention (exist). Though the division of the middle field propagator given in Eqs. (46)–(48) has a number of pleasant features compared to the split of $\ddot{D}_{\mathrm{M}}(\vec{R}, \tau)$ given in Eq. (45) it has at least one drawback: the retarded and advanced parts do not separately exist in the frequency (ω) domain. In contrast the outgoing but non-causal part of Eq. (45) does have a frequency domain representative and the same holds for the ingoing part (see also the discussion in section 2.6). Based on the considerations put forward above altogether it appears tempting for me to claim that the middle field propagator taken alone from a physical [(quantum) electrodynamic] point of view is a spurious quantity. One might of course always argue that only the sum of the retarded part of the total electromagnetic propagator $\ddot{D}_0(R, \tau)$, given in Eq. (38), and the self-field terms (to be discussed in section 2.5) has a physical meaning, and even if this is our point of view it is needed to stress that the propagator formalism of quantum electrodynamics is just one among the physically equivalent (acceptable) ones. To shed some additional light on the physical picture the propagator description offers us, we now turn our attention towards the remaining part of the finite-c_0 propagator.

The near field propagator, pleasant and unpleasant properties

Following the integration procedure sketched in the appendix, the near field propagator, $\ddot{D}_{\mathrm{N}}(\vec{R}, \tau)$, attains a form

$$\ddot{D}_{\mathrm{N}}(\vec{R}, \tau) = \frac{c_0^2}{8\pi R^3}\left(\left| \frac{R}{c_0} + \tau \right| - \left| \frac{R}{c_0} - \tau \right| \right)(\ddot{U} - 3\vec{e}_{\vec{R}}\vec{e}_{\vec{R}}), \qquad (49)$$

which written equivalently term by term as follows:

$$\ddot{D}_{\mathrm{N}}(\vec{R}, \tau) = \frac{c_0}{8\pi R^2}\left[\left(1 + \frac{c_0\tau}{R}\right)\mathrm{sgn}\left(\frac{R}{c_0} + \tau \right) - \left(1 - \frac{c_0\tau}{R}\right)\mathrm{sgn}\left(\frac{R}{c_0} - \tau \right) \right]$$
$$\times (\ddot{U} - 3\vec{e}_{\vec{R}}\vec{e}_{\vec{R}}), \qquad (50)$$

brings us to a mathematical form closer in look to the one given in Eq. (45) for the middle field propagator. As it was the case with the middle field propagator in its original form the presence of the sign functions $\mathrm{sgn}\,(R/c_0 \mp \tau)$ indicates that the outgoing and ingoing portions of $\vec{\vec{D}}_{\mathrm{N}}(\vec{R}, \tau)$ do not satisfy the pure conditions, causality and non-causality, respectively. This unpleasant feature is easily removed, and we hence obtains the division

$$\vec{\vec{D}}_{\mathrm{N}}(\vec{R}, \tau) = \vec{\vec{D}}_{\mathrm{N}}^{\mathrm{R}}(\vec{R}, \tau) + \vec{\vec{D}}_{\mathrm{N}}^{\mathrm{A}}(\vec{R}, \tau), \tag{51}$$

where

$$\vec{\vec{D}}_{\mathrm{N}}^{\mathrm{R}}(\vec{R}, \tau) = \frac{c_0}{4\pi R^2}\left[1 + \left(\frac{c_0\tau}{R} - 1\right)\Theta\left(\frac{R}{c_0} - \tau\right)\right]\Theta(\tau)\,(\vec{\vec{U}} - 3\vec{e}_{\vec{R}}\vec{e}_{\vec{R}}) \tag{52}$$

is the retarded part of the near field propagator, and

$$\vec{\vec{D}}_{\mathrm{N}}^{\mathrm{A}}(\vec{R}, \tau) = \frac{c_0}{4\pi R^2}\left[\left(\frac{c_0\tau}{R} + 1\right)\Theta\left(\frac{R}{c_0} + \tau\right) - 1\right]\Theta(-\tau)\,(\vec{\vec{U}} - 3\vec{e}_{\vec{R}}\vec{e}_{\vec{R}}) \tag{53}$$

is its advanced part. It readily appears from Eq. (52) that the retarded near field propagator may couple both time-like ($|\vec{r} - \vec{r}\,'| < c_0(t - t')$) and space-like ($|\vec{r} - \vec{r}\,'| > c_0(t - t')$) events in contrast to the retarded part of the middle field propagator which can intermediate interaction only between time-like events. Also the advanced part of $\vec{\vec{D}}_{\mathrm{N}}(\vec{R}, \tau)$ is different from zero for both space-like ($|\vec{r} - \vec{r}\,'| > c_0(t' - t)$) and time-like ($|\vec{r} - \vec{r}\,'| < c_0(t' - t)$) events, and this is again different from the middle field case, where the advanced propagator can provide time-like couplings, only (see Fig. 3 and compare also with Fig. 2). Despite their obvious advantages, the propagators $\vec{\vec{D}}_{\mathrm{N}}^{\mathrm{R}}(\vec{R}, \tau)$ and $\vec{\vec{D}}_{\mathrm{N}}^{\mathrm{A}}(\vec{R}, \tau)$ suffer from the weakness that they have no frequency domain representatives. The ingoing and outgoing parts of Eq. (50) do exist separately in the frequency domain, but have as we have seen above an unpleasant causality (non-causality) problem. So, altogether as in the case of the middle field propagator, the near field propagator can scarcely be considered to possess meaningful physical properties in itself.

A better propagator: the sum of the middle and near field propagators

Without performing any calculations, the fact that the middle field ($\vec{\vec{D}}_{\mathrm{M}}(\vec{R}, \tau)$) and near field ($\vec{\vec{D}}_{\mathrm{N}}(\vec{R}, \tau)$) propagators in Eqs. (40) and (41) have the same tensorial form, viz. $\vec{\vec{U}} - 3\vec{e}_{\vec{R}}\vec{e}_{\vec{R}}$, indicates a possible close relation between the two. This indication readily appears to be fruitful if one adds the two retarded parts (Eqs. (47) and (52)), and the two advanced parts (Eqs. (48) and (53)). This immediately gives for the R-part

$$\vec{\vec{D}}_{\mathrm{M}}^{\mathrm{R}}(\vec{R}, \tau) + \vec{\vec{D}}_{\mathrm{N}}^{\mathrm{R}}(\vec{R}, \tau) = \frac{c_0^2\tau}{4\pi R^3}\Theta(\tau)\Theta\left(\frac{R}{c_0} - \tau\right)(\vec{\vec{U}} - 3\vec{e}_{\vec{R}}\vec{e}_{\vec{R}}), \tag{54}$$

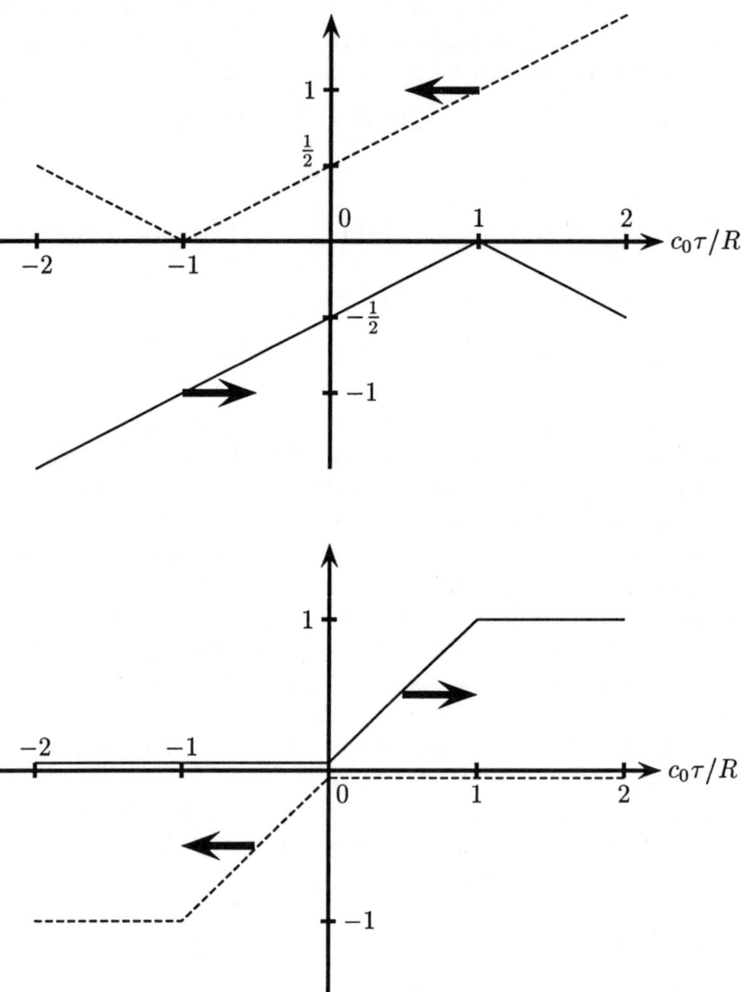

Figure 3: Schematic diagrams showing the time-space dynamics of the transverse near field propagator. This propagator is proportional to $\frac{1}{2}[(1+\frac{c_0\tau}{R})\mathrm{sgn}\,(\frac{R}{c_0}+\tau)-(1-\frac{c_0\tau}{R})\mathrm{sgn}\,(\frac{R}{c_0}-\tau)]$, and in the upper figure is shown the outgoing $[-\frac{1}{2}(1-\frac{c_0\tau}{R})\mathrm{sgn}\,(\frac{R}{c_0}-\tau)]$ (full line) and ingoing $[\frac{1}{2}(1+\frac{c_0\tau}{R})\mathrm{sgn}\,(\frac{R}{c_0}+\tau)]$ (broken line) parts as a function of $c_0\tau/R$. The two parts do not satisfy the criteria of causality and non-causality. In the lower figure is shown the division of the near field propagator into an outgoing retarded part, $[1+(\frac{c_0\tau}{R}-1)\Theta(\frac{R}{c_0}-\tau)]\Theta(\tau)$ (full line) and an ingoing advanced part $[(\frac{c_0\tau}{R}+1)\Theta(\frac{R}{c_0}+\tau)-1]\Theta(-\tau)$ (broken line). The two contributions connect both time- and space-like events.

using the identity $\Theta(R/c_0 - \tau) - 1 = \Theta(\tau)[\Theta(R/c_0 - \tau) - 1]$. A brief look at Eq. (54) shows that we seem to have obtained a more acceptable propagator by addition of the two propagators $\vec{\vec{D}}_{\mathrm{M}}^{\mathrm{R}}(\vec{R},\tau)$ and $\vec{\vec{D}}_{\mathrm{N}}^{\mathrm{R}}(\vec{R},\tau)$, which according to the analysis presented in the two previous subsections, both contains unpleasant features. The outgoing propagator in Eq. (54) thus satisfies the principle of causality as one would like, and due to the fact that it is only different from zero (and finite) in the time interval $0 < \tau < R/c_0$ (space-like event) it certainly can be transformed to the frequency domain. It is worth emphasizing that the time-like part of the near field coupling completely cancels the time-like middle field coupling, so that what is left in Eq. (54) is the space-like part of the retarded near field propagator. The distance dependence of the sum propagator $\vec{\vec{D}}_{\mathrm{M}}^{\mathrm{R}}(\vec{R},\tau) + \vec{\vec{D}}_{\mathrm{N}}^{\mathrm{R}}(\vec{R},\tau)$ is of the near field R^{-3}-type, as one might have anticipated, and at a given observation point it grows linearly in time, i.e. proportional to τ, until $\tau = R/c_0$. At this time the far field delta-function pulse passes the point of observation and destroys the all space-like near field coupling. As indicated previously it seems that quantum physics allows space-like couplings in nature. I shall return to this aspect after having introduced the self-field coupling in section 2.5, and again in relation to the quantum theory of the transverse current density.

By adding the advanced parts of the middle and near field propagators one obtains

$$\vec{\vec{D}}_{\mathrm{M}}^{\mathrm{A}}(\vec{R},\tau) + \vec{\vec{D}}_{\mathrm{N}}^{\mathrm{A}}(\vec{R},\tau) = \frac{c_0^2 \tau}{4\pi R^3}\Theta(-\tau)\Theta\left(\frac{R}{c_0} + \tau\right)\left(\vec{\vec{U}} - 3\vec{e}_{\vec{R}}\vec{e}_{\vec{R}}\right), \qquad (55)$$

after having used the identity $1 - \Theta(R/c_0 + \tau) = \Theta(-\tau)[1 - \Theta(R/c_0 + \tau)]$. The reader may easily convince herself that the sum propagator is different from zero only for space-like events $(|\vec{r} - \vec{r}'| > c_0(t' - t))$ lying in front of the advanced light cone, $|\vec{r} - \vec{r}'| = c_0(t' - t)$. The time-like part of $\vec{\vec{D}}_{\mathrm{N}}^{\mathrm{A}}(\vec{R},\tau)$ has also cancelled the middle field propagator, $\vec{\vec{D}}_{\mathrm{M}}^{\mathrm{A}}(\vec{R},\tau)$, and left us with a near field distance dependence, R^{-3}. The near field coupling changes linearly with τ (which here is negative) backwards in time from zero at $\tau = 0$ to the minimum value $-c_0(\vec{\vec{U}} - 3\vec{e}_{\vec{R}}\vec{e}_{\vec{R}})/(4\pi R^2)$ attained at $\tau = -R/c_0$. A schematic illustration of the time dependence of the sum of the middle and near field propagators is presented in Fig. 4.

The synthesized transverse propagator in time and space. The retarded transverse electric field

Starting from the integral representation for the non-singular part of the transverse propagator, given in Eq. (36) we have carried out a systematic analysis

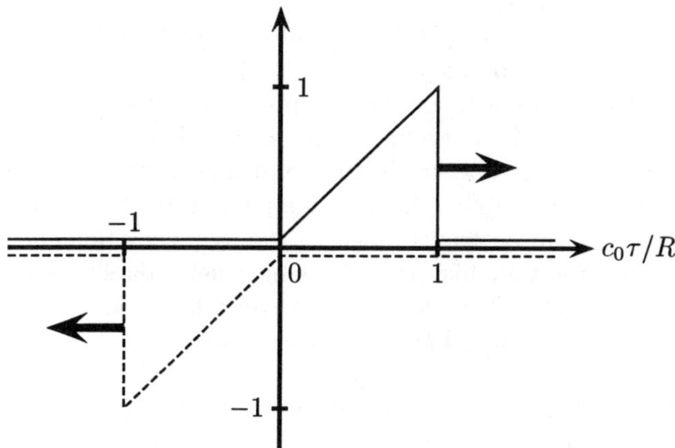

Figure 4: Schematic diagram illustrating the time-space dynamics hidden in the sum of the transverse middle and near field propagators. The sum consists of retarded and advanced parts. The full line shows the retarded part, which is proportional to $\frac{c_0\tau}{R}\Theta(\tau)\Theta(\frac{R}{c_0}-\tau)$, as a function of $c_0\tau/R$, and the broken line the advanced part, $\frac{c_0\tau}{R}\Theta(-\tau)\Theta(\frac{R}{c_0}+\tau)$. Note that the retarded part couples space-like events only.

of its different ingredients, stressing the physical picture. By an addition of the results given in Eqs. (43) and (54) it appears that the retarded part of $\overset{\leftrightarrow}{D}_0^{\mathrm{T}}(\vec{R},\tau)$, i.e.

$$\overset{\leftrightarrow}{D}_0^{\mathrm{R}}(\vec{R},\tau) = \overset{\leftrightarrow}{D}_{\mathrm{F}}^{\mathrm{R}}(\vec{R},\tau) + \overset{\leftrightarrow}{D}_{\mathrm{M}}^{\mathrm{R}}(\vec{R},\tau) + \overset{\leftrightarrow}{D}_{\mathrm{N}}^{\mathrm{R}}(\vec{R},\tau) \tag{56}$$

may be written explicitly in the following dyadic form:

$$\overset{\leftrightarrow}{D}_0^{\mathrm{R}}(\vec{R},\tau) = -\frac{1}{4\pi R}\delta\left(\frac{R}{c_0}-\tau\right)\left(\overset{\leftrightarrow}{U} - \vec{e}_{\vec{R}}\vec{e}_{\vec{R}}\right)$$
$$+\frac{c_0^2\tau}{4\pi R^3}\Theta(\tau)\Theta\left(\frac{R}{c_0}-\tau\right)\left(\overset{\leftrightarrow}{U} - 3\vec{e}_{\vec{R}}\vec{e}_{\vec{R}}\right), \tag{57}$$

where the first so-called far field term is different from zero only on the retarded light cone, and the second term, which might be called the near field term, contributes only to space-like events. Though different from zero in the entire space lying in front of the light cone (see Fig. 5), the near field disturbance travels with the vacuum speed of light, as one might have anticipated. At a fixed time delay τ (less than R/c_0), the near field term in Eq. (57) certainly has the characteristic R^{-3} distance dependence in front of the light cone. However, if one focuses the attention on the maximum value of the near field term at a

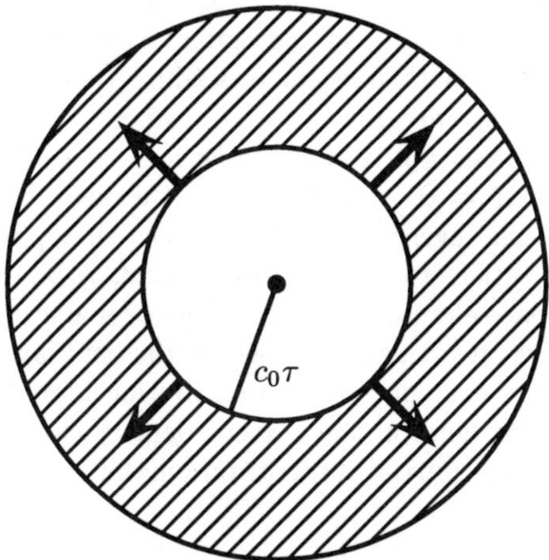

Figure 5: Schematic illustration showing the retarded electrodynamic response in the near field zone ($\sim R^{-3}$) of a decaying atom. In the time interval $\tau = t - t'$ the field has reached a distance $|\vec{r} - \vec{r}'| = c_0\tau$ away from the atom. Not only events on the light cone ($|\vec{r} - \vec{r}'| = c_0\tau$) are coupled, also space-like correlations are needed in a description where the current density of the atom (and not only the transverse part of the current density) acts as the source. The possible space-like observation points are located in the hatched domain in front of the light pulse. In the domain behind the light pulse no coupling exists, and this "hole" in the coupling prevents that a singularity of the field appears in the source point.

given distance R, this is given by $c_0/(4\pi R^2)$ and thus follows a R^{-2} distance dependence! At the various distances the maximum values are of course not obtained at the same delay time. In fact, the maximum value at a given point is reached just when the far field light pulse passes this point, i.e. at $\tau = \tau_{\mathrm{max}} = R/c_0$.

By combining Eqs. (33) and (57) it appears that the retarded part of the transverse electric field generated by the current density $\vec{J}(\vec{r}', t')$, given by

$$\vec{E}_{\mathrm{T}}^{\mathrm{R}}(\vec{r}, t) = \mu_0 \int \overleftrightarrow{D}_0^{\mathrm{R}}(\vec{r} - \vec{r}', t - t') \cdot \frac{\partial \vec{J}(\vec{r}', t')}{\partial t'} d^3 r' dt' \tag{58}$$

in integral form, can be written explicitly as

$$\vec{E}_{\mathrm{T}}^{\mathrm{R}}(\vec{r}, t) = -\frac{\mu_0}{4\pi} \int_{-\infty}^{\infty} \frac{1}{R} \left(\overleftrightarrow{U} - \vec{e}_{\vec{R}}\vec{e}_{\vec{R}} \right) \cdot \frac{\partial \vec{J}(\vec{r}', t - R/c_0)}{\partial t} d^3 r'$$

$$+ \frac{1}{4\pi\epsilon_0} \int_{-\infty}^{\infty} \frac{1}{R^3} \left(\overleftrightarrow{U} - 3\vec{e}_{\vec{R}}\vec{e}_{\vec{R}} \right) \cdot \left\{ \int_{t-\frac{R}{c_0}}^{t} (t - t') \frac{\partial \vec{J}(\vec{r}', t')}{\partial t'} dt' \right\} d^3 r'. \tag{59}$$

In the near field term of Eq. (59), the integration over $\vec{r}\,'$-space at first sight seems to involve an integration over a singularity of the unpleasant $|\vec{r} - \vec{r}\,'|^{-3}$-type. Unless a specific mathematical procedure (principal value integration e.g.) is adopted such an integral is not convergent. This situation of conditionally convergence does not appear, however, since the integral over t' in Eq. (59) is zero for $R = 0$. The fact that the near field term is different from zero for space-like events only thus makes the space integration absolutely convergent, a satisfactory conclusion indeed. A schematic illustration of the retarded time-space electrodynamics in the propagator picture is shown in Fig. 6, and a comparison with the case where the field is assumed to be generated by the transverse current density distribution is made. The dependence of the retarded electric field on the current density can be stated a bit more explicit by performing a partial integration on the time integral of Eq. (59). Doing this, one obtains

$$
\begin{aligned}
\vec{E}_{\mathrm{T}}^{\mathrm{R}}(\vec{r}, t) = {}&- \frac{\mu_0}{4\pi} \int_{-\infty}^{\infty} \frac{1}{R} \left(\overleftrightarrow{U} - \vec{e}_{\vec{R}} \vec{e}_{\vec{R}} \right) \cdot \frac{\partial \vec{J}(\vec{r}\,', t - R/c_0)}{\partial t} d^3 r' \\
&- \frac{1}{4\pi\epsilon_0 c_0} \int_{-\infty}^{\infty} \frac{1}{R^2} \left(\overleftrightarrow{U} - 3\vec{e}_{\vec{R}} \vec{e}_{\vec{R}} \right) \cdot \vec{J}(\vec{r}\,', t - R/c_0) d^3 r' \\
&+ \frac{1}{4\pi\epsilon_0} \int_{-\infty}^{\infty} \frac{1}{R^3} \left(\overleftrightarrow{U} - 3\vec{e}_{\vec{R}} \vec{e}_{\vec{R}} \right) \cdot \left\{ \int_{t-\frac{R}{c_0}}^{t} \vec{J}(\vec{r}\,', t') dt' \right\} d^3 r'. \quad (60)
\end{aligned}
$$

The result in Eq. (60) shows that the transverse retarded field consists of a far field term proportional to $\partial \vec{J}/\partial t$ taken at the retarded time, $t - R/c_0$, a middle field term proportional to $\vec{J}$ at the retarded time, and a near field term proportional to the integral of $\vec{J}$ between $t - R/c_0$ and t.

In most cases, the time development of the current density dynamics is so slow that one would automatically neglect the retardation in the near field zone. In the propagator picture this means that $\overleftrightarrow{D}_{\mathrm{M}}^{\mathrm{R}}(\vec{R}, \tau) + \overleftrightarrow{D}_{\mathrm{N}}^{\mathrm{R}}(\vec{R}, \tau)$ in Eq. (54) is replaced by

$$
\overleftrightarrow{D}_{\mathrm{M}}^{\mathrm{R}}(\vec{R}, \tau) + \overleftrightarrow{D}_{\mathrm{N}}^{\mathrm{R}}(\vec{R}, \tau) \approx \frac{c_0^2 \delta(\tau)}{4\pi R^3} \left(\overleftrightarrow{U} - 3\vec{e}_{\vec{R}} \vec{e}_{\vec{R}} \right) \int_{-\infty}^{\infty} \tau \Theta(\tau) \Theta \left(\frac{R}{c_0} - \tau \right) dt', \quad (61)
$$

which upon integration becomes

$$
\overleftrightarrow{D}_{\mathrm{M}}^{\mathrm{R}}(\vec{R}, \tau) + \overleftrightarrow{D}_{\mathrm{N}}^{\mathrm{R}}(\vec{R}, \tau) \approx \frac{\delta(\tau)}{8\pi R} \left(\overleftrightarrow{U} - 3\vec{e}_{\vec{R}} \vec{e}_{\vec{R}} \right). \quad (62)
$$

It is remarkable that the assumption of slowly varying current density dynamics to lowest order in the light retardation (neglect of retardation in the near field

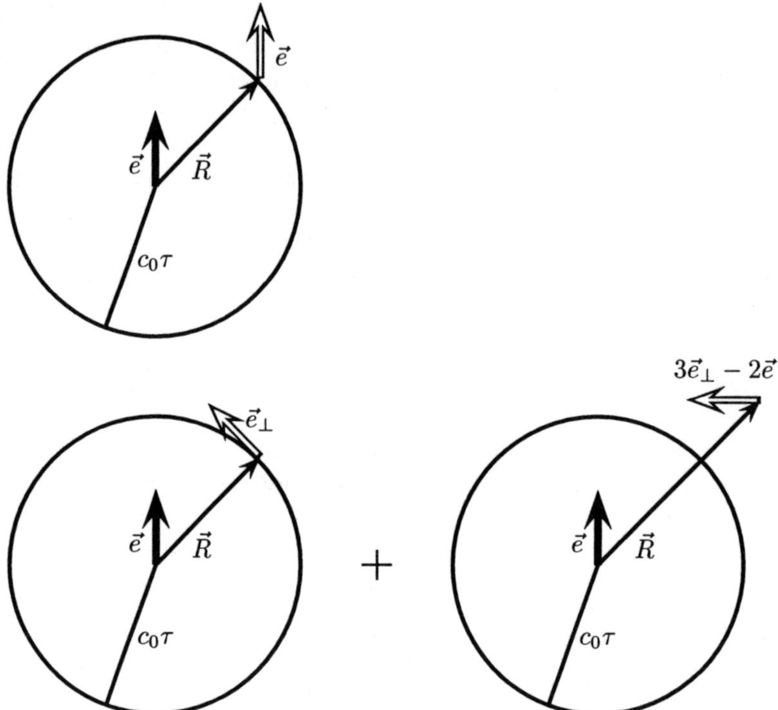

Figure 6: Schematic figures showing characteristics of an electromagnetic field, $\vec{E}_\mathrm{T}^\mathrm{RET}$, emanating from a source point of (i) a transverse current density distribution, $\vec{J}_\mathrm{T}$, (top figure), and (ii) a current density distribution, $\vec{J} = \vec{J}_\mathrm{T} + \vec{J}_\mathrm{L}$ (bottom figures). For a "$\vec{J}_\mathrm{T}$-source" the field is different from zero only on the spherical shell $|\vec{k}| = c_o\tau$, and $\vec{E}_\mathrm{T}^\mathrm{RET}$ and $\partial \vec{J}_\mathrm{T}/\partial t$ are polarized in the same direction (given by the unit vector $\vec{e}$ in the figure). For a "$\vec{J}$-source", both light-shell (left bottom figure) and space-like (right bottom figure) couplings appear. On the light shell the field is polarized in the (transverse) direction given by the vector $\vec{e}_\perp = (\overleftrightarrow{U} - \vec{e}_{\vec{R}}\vec{e}_{\vec{R}}) \cdot \vec{e}$, where $\vec{e}_{\vec{R}} = \vec{R}/R$. The space-like field is polarized in the direction $3\vec{e}_\perp - 2\vec{e}$.

zone) has turned the near field propagator part into a (time independent) far-field-like part. By using the approximate result in Eq. (62) the near field part of the retarded electric field becomes

$$\mu_0 \int \left[\overleftrightarrow{D}_\mathrm{M}^\mathrm{R}(\vec{R},\tau) + \overleftrightarrow{D}_\mathrm{N}^\mathrm{R}(\vec{R},\tau) \right] \cdot \frac{\partial \vec{J}(\vec{r}',t')}{\partial t'} dt' d^3 r' \approx$$

$$\frac{\mu_0}{8\pi} \int_{-\infty}^{\infty} \frac{1}{R} \left(\overleftrightarrow{U} - 3\vec{e}_{\vec{R}}\vec{e}_{\vec{R}} \right) \cdot \frac{\partial \vec{J}(\vec{r}',t)}{\partial t} d^3 r'. \tag{63}$$

Before ending this section it is worthwhile to consider the non-singular part of the source term $\partial \vec{J}(\vec{r}\,',t')/\partial t'$ starting from the advanced isotropic propagator $\overleftrightarrow{d}_0^{\mathrm{A}}(R,\tau)$ given in Eq. (27). The integral representation for the non-singular part of the propagator is in this case given by the expression one obtains by replacing τ with $-\tau$ on the right hand side of Eq. (36). If this is done, and the integration over $\vec{R}_0$-space performed, one obtains the result given in Eq. (37) merely with τ replaced by $-\tau$. Since the new factor $e^{iqc_0(-\tau)} - e^{-iqc_0(-\tau)}$ just equals -1 times the old factor $e^{iqc_0\tau} - e^{-iqc_0\tau}$, it readily appears that the new non-singular propagator belonging to $\overleftrightarrow{d}_0^{\mathrm{A}}(R,\tau)$ is

$$-\frac{1}{32\pi^4} \int_{-\infty}^{\infty} \frac{1}{R_0}\delta\left(\frac{R}{c_0}+\tau\right)\left(\overleftrightarrow{U} - \vec{e}_{\vec{q}}\vec{e}_{\vec{q}}\right)e^{i\vec{q}\cdot(\vec{R}-\vec{R}_0)}d^3R_0 d^3q = -\overleftrightarrow{D}_0^{\mathrm{R}}(\vec{R},\tau).$$

$$(64)$$

Apart from a factor of -1 the two non-singular anisotropic propagators one obtains starting from $\overleftrightarrow{d}_0^{\mathrm{R}}(R,\tau)$ and $\overleftrightarrow{d}_0^{\mathrm{A}}(R,\tau)$ are identical [compare e.g. Eqs. (36) and (64)].

2.5   Self-field dynamics

In the previous section we have analysed on the basis of a propagator description the time-space electrodynamics which relates in a retarded fashion a current density source $\vec{J}(\vec{r}\,',t')$ located at the space-time point $(\vec{r}\,',t')$ to the transverse field $\vec{E}_{\mathrm{T}}(\vec{r},t)$, created at the space-time point $(\vec{r},t)$. Only the situation where the space points $\vec{r}$ and $\vec{r}\,'$ did not coincide was studied, and in all cases the speed of light entered the analysis.

Electrodynamic phenomena which are independent of c_0 do seem to exist, however. From a heuristic point of view this is clear provided one accepts that electromagnetic fields may be generated in a space point $\vec{r}$ from a current density source located at the same point. Let us now study the so-called self-field effects, in which the speed of light does not enter, in a more systematic manner. Starting from the division of the electric field into its transverse and longitudinal parts in a given inertial frame, it follows directly from the microscopic Maxwell equations that a longitudinal current density located at $\vec{r}$ gives rise to a longitudinal electric field at the same point, and only at this point, cf. Eq. (31). Although strictly local in space, the relation between $\vec{E}_{\mathrm{L}}$ and $\vec{J}_{\mathrm{L}}$ is not local in time. It is interesting to note here that although the relation between $\vec{E}_{\mathrm{L}}$ and $\vec{J}_{\mathrm{L}}$ is spatially local, the relation between $\vec{E}_{\mathrm{L}}$ and the total current density $\vec{J}$ is spatially nonlocal. This is obvious from the fact that the connection between $\vec{J}$ and $\vec{J}_{\mathrm{L}}$ is a nonlocal one in space, cf. Eq. (13). To improve our understanding of the longitudinal self-field dynamics, it is adequate to write the relation between $\vec{J}_{\mathrm{L}}$ and $\vec{J}$ in a slightly different form, namely

$$\vec{J}_{\mathrm{L}} = \frac{1}{3}\vec{J}(\vec{r},t) - \mathrm{PV}\int \vec{\nabla}\vec{\nabla}\cdot\left(\frac{\vec{J}(\vec{r}',t)}{4\pi|\vec{r}-\vec{r}'|}\right)d^3r', \tag{65}$$

where PV means the principal value of the subsequent volume integral. By inserting Eq. (65) into Eq. (31) it appears that the longitudinal field at $\vec{r}$ has a contribution from the total current density at the same point, viz. $-(3\epsilon_0)^{-1}\int_{t_0}^{t}\vec{J}(\vec{r},t')dt'$. This contribution we name the longitudinal contact field. The remaining part of Eq. (65) gives rise to a spatially nonlocal contribution to $\vec{E}_{\mathrm{L}}$. In spite of this, the speed of light does not enter. I shall return to this point when discussing the quantum physics of the transverse and longitudinal current densities in section 3. Also the transverse electric field has a self-field (SF) contribution $\vec{E}_{\mathrm{T}}^{\mathrm{SF}}(\vec{r},t)$, which, however, cannot be calculated so easily from the microscopic Maxwell equations as the longitudinal self-field. A rigorous calculation leads to the following result:[11,12]

$$\vec{E}_{\mathrm{T}}^{\mathrm{SF}}(\vec{r},t) = \vec{E}_{\mathrm{T}}^{\mathrm{SF}}(\vec{r},t_0) - \frac{1}{3\epsilon_0}\int_{t_0}^{t}\vec{J}_{\mathrm{T}}(\vec{r},t')dt', \tag{66}$$

t_0 being a reference time. By a comparison of Eqs. (31) and (66) it is seen that the basic difference is the factor $1/3$ entering in front of the integral in Eq. (66). By writing the relation between $\vec{J}_{\mathrm{T}}$ and $\vec{J}$ as

$$\vec{J}_{\mathrm{T}}(\vec{r},t) = \frac{2}{3}\vec{J}(\vec{r},t) + \mathrm{PV}\int \vec{\nabla}\times\left[\vec{\nabla}\times\left(\frac{\vec{J}(\vec{r}',t)}{4\pi|\vec{r}-\vec{r}'|}\right)\right]d^3r', \tag{67}$$

it appears that the relation between $\vec{E}_{\mathrm{T}}^{\mathrm{SF}}$ and $\vec{J}$ consists of a so-called transverse contact term, $-2(9\epsilon_0)^{-1}\int_{t_0}^{t}\vec{J}(\vec{r},t')dt'$, which gives a spatially local relation between the transverse self-field and the full current density, and a spatially nonlocal term stemming from the PV-integral of Eq. (67). Also here the nonlocal part is non-retarded. One might be tempted to believe that the relation between the total electric self-field $\vec{E}^{\mathrm{SF}}(\vec{r},t) = \vec{E}_{\mathrm{L}}(\vec{r},t) + \vec{E}_{\mathrm{T}}^{\mathrm{SF}}(\vec{r},t)$ and the total current density $\vec{J}(\vec{r},t)$ would be strictly local in space so that no electrodynamic effects occur over finite distances unless the vacuum velocity of light is involved. The factor of $1/3$ appearing in Eq. (66) tells us that this is not true, however. Thus by adding Eqs. (31) and (66) with the expressions for $\vec{J}_{\mathrm{L}}$ (Eq. (65)) and $\vec{J}_{\mathrm{T}}$ (Eq. (67)) inserted one gets

$$\vec{E}^{\mathrm{SF}}(\vec{r},t) = \vec{E}^{\mathrm{SF}}(\vec{r},t_0) - \frac{5}{9\epsilon_0}\int_{t_0}^{t}\vec{J}(\vec{r},t')dt' + \frac{2}{3\epsilon_0}\int_{t_0}^{t}\vec{J}(\vec{r};t')dt', \tag{68}$$

where

$$\vec{J}(\vec{r},t') = \mathrm{PV} \int \vec{\nabla} \times \left[\vec{\nabla} \times \left(\frac{\vec{J}(\vec{r}',t')}{4\pi|\vec{r} - \vec{r}'|} \right) \right] d^3r', \tag{69}$$

and it appears that the nonlocal contribution (the last term on the right hand side of Eq. (68)) in general is different from zero.

2.6   *Retarded transverse propagator in the space-frequency domain*

For many purposes it is advantageous to describe the electrodynamic behaviour of the system in consideration in the space-frequency representation. Let us therefore consider the form the retarded part of the transverse propagator takes in this representation. The frequency analysis also has the bonus that it allows us to compare our rigorous $\vec{D}_0^{\mathrm{R}}$-propagator with the standard one usually used in textbook expositions and often in the research literature. A Fourier integral transformation of the retarded far field propagator, $\vec{D}_{\mathrm{F}}^{\mathrm{R}}(\vec{R},\tau)$, readily gives

$$\begin{aligned}
\vec{D}_{\mathrm{F}}^{\mathrm{R}}(\vec{R};\omega) &= -\frac{1}{4\pi R} \left(\overleftrightarrow{U} - \vec{e}_{\vec{R}}\vec{e}_{\vec{R}} \right) \int_{-\infty}^{\infty} \delta\left(\frac{R}{c_0} - \tau \right) e^{i\omega\tau} d\tau \\
&= \frac{q_0}{4\pi i} \frac{1}{iq_0 R} \left(\overleftrightarrow{U} - \vec{e}_{\vec{R}}\vec{e}_{\vec{R}} \right) e^{iq_0 R},
\end{aligned} \tag{70}$$

where, as before, $q_0 = \omega/c_0$. I have written the factor in front of the tensor in Eq. (70) in such a manner that the characteristic combination $iq_0 R$ occurs. Though the retarded middle ($\vec{D}_{\mathrm{M}}^{\mathrm{R}}(\vec{R},\tau)$) and near ($\vec{D}_{\mathrm{N}}^{\mathrm{R}}(\vec{R},\tau)$) field terms do not have their own frequency representations, the sum $\vec{D}_{\mathrm{M+N}}^{\mathrm{R}}(\vec{R},\tau) = \vec{D}_{\mathrm{M}}^{\mathrm{R}}(\vec{R},\tau) + \vec{D}_{\mathrm{N}}^{\mathrm{R}}(\vec{R},\tau)$ of the two does, and an elementary calculation leads to

$$\begin{aligned}
\vec{D}_{\mathrm{M+N}}^{\mathrm{R}}(\vec{R};\omega) &= \frac{c_0^2}{4\pi R^3} \left(\overleftrightarrow{U} - 3\vec{e}_{\vec{R}}\vec{e}_{\vec{R}} \right) \int_{-\infty}^{\infty} \tau\Theta(\tau)\Theta\left(\frac{R}{c_0} - \tau \right) e^{i\omega\tau} d\tau \\
&= \frac{q_0}{4\pi i} \left(\overleftrightarrow{U} - 3\vec{e}_{\vec{R}}\vec{e}_{\vec{R}} \right) \left[-\frac{1}{(iq_0 R)^2} e^{iq_0 R} + \frac{1}{(iq_0 R)^3} \left(e^{iq_0 R} - 1 \right) \right].
\end{aligned} \tag{71}$$

By addition of the results in Eqs. (70) and (71), it appears that the retarded transverse propagator is given by

$$\vec{D}_0^{\mathrm{R}}(\vec{R};\omega) = \frac{q_0}{4\pi i} \left\{ \left(\overleftrightarrow{U} - \vec{e}_{\vec{R}}\vec{e}_{\vec{R}} \right) \frac{e^{iq_0 R}}{iq_0 R} + \left(\overleftrightarrow{U} - 3\vec{e}_{\vec{R}}\vec{e}_{\vec{R}} \right) \left[-\frac{e^{iq_0 R}}{(iq_0 R)^2} + \frac{e^{iq_0 R} - 1}{(iq_0 R)^3} \right] \right\} \tag{72}$$

in the space-frequency domain. In this domain, the propagator exhibits the characteristic far $[\sim (iq_0 R)^{-1}]$, middle $[\sim (iq_0 R)^{-2}]$, and near $[\sim (iq_0 R)^{-3}]$ field terms.

To describe the radiation from a so-called electric point dipole it is commonly seen in the literature that the "textbook (tex)" propagator (valid only for $R \neq 0$)[13]

$$\vec{\vec{D}}_0^{\text{tex}}(\vec{R};\omega) =$$
$$\frac{q_0}{4\pi i}\left\{\left(\vec{\vec{U}} - \vec{e}_{\vec{R}}\vec{e}_{\vec{R}}\right)\frac{e^{iq_0 R}}{iq_0 R} + \left(\vec{\vec{U}} - 3\vec{e}_{\vec{R}}\vec{e}_{\vec{R}}\right)\left[-\frac{1}{(iq_0 R)^2} + \frac{1}{(iq_0 R)^3}\right]e^{iq_0 R}\right\} \quad (73)$$

is used. The $\vec{\vec{D}}_0^{\text{R}}$- and $\vec{\vec{D}}_0^{\text{tex}}$-propagators have the same far and middle field terms, but their near field parts are different, in the sense that the factor $e^{iq_0 R} - 1$ appearing in $\vec{\vec{D}}_0^{\text{R}}(\vec{R};\omega)$ is replaced by a factor $e^{iq_0 R}$ in $\vec{\vec{D}}_0^{\text{tex}}(\vec{R};\omega)$. This has the dramatic consequence that the textbook propagator does not exist in the space-time domain, a sad situation for this propagator indeed! A hint for understanding the physical difference between the two propagators, and the problem attached to the use of the textbook propagator in studies of radiation effects in the middle and near field zones, may be obtained by investigation of their respective properties in the limit where $R \to 0$. The asymptotic form of $\mu_0 \vec{\vec{D}}_0^{\text{R}}(\vec{R};\omega)$ thus is

$$\mu_0 \vec{\vec{D}}_0^{\text{R}}(\vec{R};\omega)\Big|_{R \to 0} = -\frac{1}{c_0^2}\frac{\vec{\vec{U}} - \vec{e}_{\vec{R}}\vec{e}_{\vec{R}}}{4\pi\epsilon_0 R} + \frac{1}{c_0^2}\frac{\vec{\vec{U}} - 3\vec{e}_{\vec{R}}\vec{e}_{\vec{R}}}{8\pi\epsilon_0 R} = -\frac{1}{c_0^2}\frac{\vec{\vec{U}} + \vec{e}_{\vec{R}}\vec{e}_{\vec{R}}}{8\pi\epsilon_0 R}, \quad (74)$$

where the first term in the expression in the middle is the far field contribution (unmodified for $R \to 0$) and the second term is the middle field plus the far field contributions. Our rigorous retarded and transverse propagator thus only exhibits an acceptable R^{-1}-singularity at $R = 0$, and having multiplied the propagator by μ_0 (as one must do as a step to determine the associated electric field) it is obvious that $\mu_0 \vec{\vec{D}}_0^{\text{R}}(\vec{R} \to \vec{0};\omega)$ is zero in the limit $c_0 \to \infty$. This is the expected behaviour for a propagator describing a retarded electromagnetic response. By comparison, the textbook propagator (multiplied by μ_0) has the asymptotic form

$$\mu_0 \vec{\vec{D}}_0^{\text{tex}}(\vec{R};\omega)\Big|_{R \to 0} = \frac{1}{\omega^2}\frac{\vec{\vec{U}} - 3\vec{e}_{\vec{R}}\vec{e}_{\vec{R}}}{4\pi\epsilon_0 R^3}. \quad (75)$$

The textbook propagator thus is dominated by the near field term, and it therefore exhibits the problematic R^{-3}-singularity at $R = 0$, but may be even worse, the quantity $\mu_0 \vec{\vec{D}}_0^{\text{tex}}(\vec{R} \to \vec{0};\omega)$ is finite for $c_0 \to \infty$ (it is independent of c_0). On top of the above-mentioned unpleasant properties the quantity in Eq. (75) diverges in the long-wavelength limit, whereas $\mu_0 \vec{\vec{D}}_0^{\text{R}}(\vec{R} \to \vec{0};\omega)$ in Eq. (74) is independent of ω. The quantity in Eq. (75) originates in the near field term of Eq. (73), and precisely by subtracting the limit $(R \to 0)$ form of the near field term from the propagator $\vec{\vec{D}}_0^{\text{tex}}(\vec{R};\omega)$ we obtain the rigorous $\vec{\vec{D}}_0^{\text{R}}(\vec{R};\omega)$-propagator.

Since the textbook propagator in a strict manner comes out from the (microscopic) Maxwell equations with current density support of delta-function character,[13] one might suspect "some reality" in the near field part of the textbook propagator also, and since this propagator apparently has a non-retarded part, we expect to find the roots of the (non-retarded) near field part of $\vec{\vec{D}}_0^{\text{tex}}(\vec{R};\omega)$ in the self-field dynamics. To verify this expectation, it is necessary to consider the self-field dynamics in the space-frequency domain. For this purpose, we start from Eq. (20), and the one obtained by taking the time derivative of Eq. (66), i.e.

$$\frac{\partial \vec{E}_{\text{T}}^{\text{SF}}(\vec{r},t)}{\partial t} = -\frac{1}{3\epsilon_0}\vec{J}_{\text{T}}(\vec{r},t). \tag{76}$$

In the space-frequency representation these equations are

$$\vec{E}_{\text{L}}(\vec{r};\omega) = \frac{1}{i\epsilon_0\omega}\vec{J}_{\text{L}}(\vec{r};\omega), \tag{77}$$

and

$$\vec{E}_{\text{T}}^{\text{SF}}(\vec{r};\omega) = \frac{1}{3i\epsilon_0\omega}\vec{J}_{\text{T}}(\vec{r};\omega), \tag{78}$$

respectively. Using the so-called longitudinal delta function, $\vec{\vec{\delta}}_{\text{L}}(\vec{r}-\vec{r}')$, which is a tensor quantity, defined operationally by an integral relation of the form (here with $\vec{J}$ and $\vec{J}_{\text{L}}$ used as vector fields)

$$\vec{J}_{\text{L}}(\vec{r},t) = \int \vec{\vec{\delta}}_{\text{L}}(\vec{r}-\vec{r}') \cdot \vec{J}(\vec{r}',t)d^3r', \tag{79}$$

and the analogous equation for the transverse delta function $\vec{\vec{\delta}}_{\text{T}}(\vec{r}-\vec{r}')$, see Eq. (32), Eqs. (77) and (78) can be written in the "propagator" forms

$$\vec{E}_{\text{L}}(\vec{r};\omega) = -i\mu_0\omega \int \vec{\vec{g}}_{\text{L}}(\vec{r}-\vec{r}';\omega) \cdot \vec{J}(\vec{r}';\omega)d^3r', \tag{80}$$

and

$$\vec{E}_{\text{T}}^{\text{SF}}(\vec{r};\omega) = -i\mu_0\omega \int \vec{\vec{g}}_{\text{T}}(\vec{r}-\vec{r}';\omega) \cdot \vec{J}(\vec{r}';\omega)d^3r', \tag{81}$$

where

$$\vec{\vec{g}}_{\text{L}}(\vec{r}-\vec{r}';\omega) = \left(\frac{c_0}{\omega}\right)^2 \vec{\vec{\delta}}_{\text{L}}(\vec{r}-\vec{r}') \tag{82}$$

is called the longitudinal self-field propagator in a somewhat misleading notation (no propagation effects are involved), and

$$\vec{\vec{g}}_{\mathrm{T}}(\vec{r}-\vec{r}\,';\omega) = \frac{1}{3}\left(\frac{c_0}{\omega}\right)^2 \vec{\vec{\delta}}_{\mathrm{T}}(\vec{r}-\vec{r}\,') \tag{83}$$

is named (just as badly) the transverse self-field propagator. The two self-field propagators do not exist in the space-time domain. The relation to the expression in Eq. (75) becomes clear if one uses the following explicit forms of the longitudinal and transverse delta functions:

$$\vec{\vec{\delta}}_{\mathrm{L}}(\vec{R}) = \frac{1}{3}\delta(\vec{R})\vec{\vec{U}} + \frac{\vec{\vec{U}} - 3\vec{e}_{\vec{R}}\vec{e}_{\vec{R}}}{4\pi R^3}, \tag{84}$$

and

$$\vec{\vec{\delta}}_{\mathrm{T}}(\vec{R}) = \frac{2}{3}\delta(\vec{R})\vec{\vec{U}} - \frac{\vec{\vec{U}} - 3\vec{e}_{\vec{R}}\vec{e}_{\vec{R}}}{4\pi R^3}. \tag{85}$$

The forms in Eqs. (84) and (85) only go along with a calculation of the self-field if spherical contraction is made around the point $\vec{r}\,' = \vec{r}$ ($\vec{R} = \vec{0}$). For a point source located at $\vec{r}\,' = \vec{0}$, only values of $\vec{R}$ different from zero can be used (are needed), since the self-field diverges at the source for a point source. This means that

$$\mu_0\vec{\vec{g}}_{\mathrm{L}}(\vec{R};\omega) = -3\mu_0\vec{\vec{g}}_{\mathrm{T}}(\vec{R};\omega) = \frac{1}{\omega^2}\frac{\vec{\vec{U}} - 3\vec{e}_{\vec{R}}\vec{e}_{\vec{R}}}{4\pi\epsilon_0 R^3}. \tag{86}$$

In comparison with Eq. (75), we thus have found

$$\vec{\vec{D}}_0^{\mathrm{tex}}(\vec{R};\omega)\Big|_{R\to 0} = \vec{\vec{g}}_{\mathrm{L}}(\vec{R};\omega) \ (= -3\vec{\vec{g}}_{\mathrm{T}}(\vec{R};\omega)). \tag{87}$$

The textbook propagator thus includes a non-propagating part, which in fact is identical to the longitudinal self-field propagator, and therefore $\vec{\vec{D}}_0^{\mathrm{tex}}(\vec{R};\omega)$ describes not only retarded electrodynamic phenomena but also a part of the self-field dynamics. Transverse self-field effects and contact phenomena are not incorporated in $\vec{\vec{D}}_0^{\mathrm{tex}}(\vec{R};\omega)$, however, but the most worrying aspect is associated with the fact that this propagator mixes retarded and non-retarded physical effects. In conclusion, we have thus obtained the following relation between the two propagators in Eqs. (72) and (73):

$$\vec{\vec{D}}_0^{\mathrm{tex}}(\vec{R};\omega) = \vec{\vec{D}}_0^{\mathrm{R}}(\vec{R};\omega) + \vec{\vec{g}}_{\mathrm{L}}(\vec{R};\omega), \qquad \vec{R} \neq 0. \tag{88}$$

In the space-frequency domain the previous analysis thus has demonstrated that electrodynamic phenomena, which obey the principle of causality, may be studied starting from the compact propagator description

$$\vec{E}(\vec{r};\omega) = \vec{E}_{\mathrm{T}}^{\mathrm{ext}}(\vec{r};\omega) - i\mu_0\omega \int \overleftrightarrow{G}_0(\vec{r} - \vec{r}';\omega) \cdot \vec{J}(\vec{r}';\omega)d^3r', \qquad (89)$$

provided there are no external current density sources inside the transverse (or equivalently longitudinal) current density domain of the system under study. The vacuum propagator

$$\overleftrightarrow{G}_0(\vec{R};\omega) = \overleftrightarrow{G}^{\mathrm{T}}(\vec{R};\omega) + \overleftrightarrow{g}_{\mathrm{L}}(\vec{R};\omega), \qquad (90)$$

is composed of a longitudinal self-field part, $\overleftrightarrow{g}_{\mathrm{L}}(\vec{R};\omega)$, and a transverse part

$$\overleftrightarrow{G}^{\mathrm{T}}(\vec{R};\omega) = \overleftrightarrow{D}_0^{\mathrm{R}}(\vec{R};\omega) + \overleftrightarrow{g}_{\mathrm{T}}(\vec{R};\omega), \qquad (91)$$

consisting of the sum of a transverse self-field part and a retarded part accounting for speed-of-light physics.

2.7   Relation to the standard formulation

Introductory remarks

It is of interest to investigate how the propagator description of the transverse and longitudinal electrodynamics discussed in the previous sections is related to the standard description commonly used. In the standard approach one starts from the Lorentz gauge for the vector, $\vec{A}(\vec{r},t)$, and scalar, $\phi(\vec{r},t)$, potentials and derives hereafter form-identical wave equations for $\phi(\vec{r},t)$ and the Cartesian components of $\vec{A}(\vec{r},t)$. The driving terms in these equations are the total (all) charge density, $\rho^{\mathrm{all}}(\vec{r},t)$, and the three Cartesian components of the total (all) current density, $\vec{J}^{\mathrm{all}}(\vec{r},t)$, and the solutions of the wave equations express in integral form the vector and scalar potentials as functions of the retarded current and charge densities, provided the principle of causality is adopted. Having obtained $\vec{A}(\vec{r},t)$ and $\phi(\vec{r},t)$, the total electric field $\vec{E}(\vec{r},t) = -\partial\vec{A}(\vec{r},t)/\partial t - \vec{\nabla}\phi(\vec{r},t)$ is readily calculated. In a compact form, classical electrodynamics thus is based on the following standard formula[14]

$$\vec{E}(\vec{r},t) = -\frac{\partial}{\partial t}\left[\frac{1}{4\pi\epsilon_0 c_0^2}\int \vec{J}^{\mathrm{all}}\left(\vec{r}',t - \frac{|\vec{r} - \vec{r}'|}{c_0}\right)\frac{d^3r'}{|\vec{r} - \vec{r}'|}\right]$$
$$-\vec{\nabla}\left[\frac{1}{4\pi\epsilon_0}\int \rho^{\mathrm{all}}\left(\vec{r}',t - \frac{|\vec{r} - \vec{r}'|}{c_0}\right)\frac{d^3r'}{|\vec{r} - \vec{r}'|}\right]. \qquad (92)$$

The expression in Eq. (92) shows that electrodynamic interactions over finite distances, i.e. for $|\vec{r} - \vec{r}'| \neq 0$, are always retarded in the sense that in order to obtain the electric field $\vec{E}(\vec{r}, t)$ at a given space-time observation point $(\vec{r}, t)$ one has to know the current density, $\vec{J}^{\mathrm{all}}(\vec{r}', t - R/c_0)$, and the charge density, $\rho^{\mathrm{all}}(\vec{r}', t - R/c_0)$, at the various $\vec{r}'$ space points at the earlier time $t' = t - R/c_0$. Though the time delay in the interaction certainly is characterized by the vacuum speed of light, it is not particularly easy to see from Eq. (92) how the propagation of the interaction takes place in time and space. For this purpose a propagator formulation is useful, *eo ipso*. In the standard description one is forced to consider both the charge and current density distributions as separate sources. From the equation of continuity

$$\vec{\nabla} \cdot \vec{J}^{\mathrm{all}}(\vec{r}, t) + \frac{\partial}{\partial t}\rho^{\mathrm{all}}(\vec{r}, t) = 0 \tag{93}$$

we know, however, that the dynamics of the two source fields are coupled, and therefore one would like to eliminate one of these particle source fields, say $\rho^{\mathrm{all}}(\vec{r}', t - R/c_0)$, from the formulation. This is also done in the propagator notation, where only the quantity $\partial \vec{J}(\vec{r}', t')/\partial t'$ acts as the source term. Furthermore, it does not appear so easy to extract the self-field contribution to the electrodynamics from the standard formula in Eq. (92). One might have hoped that a $c_0 \to \infty$-procedure would allow one to determine the self-field. Such a procedure implies, however, that one first performs the integrations over $\vec{r}'$-space, then carries out the partial derivatives ($\vec{\nabla}$, $\partial/\partial t$), and finally takes the limit $c_0 \to \infty$ in Eq. (92). This is certainly not easy unless the source distributions $\vec{J}(\vec{r}', t = R/c_0)$ and $\rho(\vec{r}', t - R/c_0)$ are extremely simple.

From the differential equations for the transverse and longitudinal fields to the standard integral formulation

Albeit it is easiest to derive the standard formula in Eq. (92) via the Lorentz gauge auxilary condition, it is instructive to see how this formula can be obtained starting from the differential equations for transverse (Eq. (19)) and longitudinal (Eq. (20)) electric fields. Hence, by differentiating Eq. (20) with respect to time we get

$$-\frac{1}{c_0^2}\frac{\partial^2 \vec{E}_{\mathrm{L}}(\vec{r}, t)}{\partial t^2} = \mu_0 \frac{\partial}{\partial t}\vec{J}_{\mathrm{L}}(\vec{r}, t), \tag{94}$$

and from the identity $\vec{0} = \vec{\nabla} \times (\vec{\nabla} \times \vec{E}_{\mathrm{L}}(\vec{r}, t)) = (\vec{\nabla}\vec{\nabla} - \overleftrightarrow{U}\nabla^2) \cdot \vec{E}_{\mathrm{L}}(\vec{r}, t)$ one obtains upon use of Eq. (17)

$$\nabla^2 \vec{E}_{\mathrm{L}}(\vec{r}, t) = \frac{1}{\epsilon_0}\vec{\nabla}\rho(\vec{r}, t). \tag{95}$$

By addition of Eqs. (94) and (95) one finds the following "wave equation" for the longitudinal part of the electric field:

$$\left(\nabla^2 - \frac{1}{c_0^2}\frac{\partial^2}{\partial t^2}\right)\vec{E}_{\rm L}(\vec{r},t) = \mu_0\frac{\partial}{\partial t}\vec{J}_{\rm L}(\vec{r},t) + \frac{1}{\epsilon_0}\vec{\nabla}\rho(\vec{r},t). \tag{96}$$

If one next divides $\vec{E}_{\rm L}(\vec{r},t)$ into two parts

$$\vec{E}_{\rm L}(\vec{r},t) = \vec{E}_{\rm L}^{(1)}(\vec{r},t) + \vec{E}_{\rm L}^{(2)}(\vec{r},t), \tag{97}$$

and requires that $\vec{E}_{\rm L}^{(1)}(\vec{r},t)$ satisfies the wave equation

$$\left(\nabla^2 - \frac{1}{c_0^2}\frac{\partial^2}{\partial t^2}\right)\vec{E}_{\rm L}^{(1)}(\vec{r},t) = \mu_0\frac{\partial}{\partial t}\vec{J}_{\rm L}(\vec{r},t), \tag{98}$$

the other part must fulfil

$$\left(\nabla^2 - \frac{1}{c_0^2}\frac{\partial^2}{\partial t^2}\right)\vec{E}_{\rm L}^{(2)}(\vec{r},t) = \frac{1}{\epsilon_0}\vec{\nabla}\rho(\vec{r},t). \tag{99}$$

By addition of Eqs. (19) and (98) one now obtains

$$\left(\nabla^2 - \frac{1}{c_0^2}\frac{\partial^2}{\partial t^2}\right)\left(\vec{E}_{\rm T}(\vec{r},t) + \vec{E}_{\rm L}^{(1)}(\vec{r},t)\right) = \mu_0\frac{\partial}{\partial t}\vec{J}(\vec{r},t). \tag{100}$$

By introducing a scalar potential $\phi^{(2)}(\vec{r},t)$ via $\vec{E}_{\rm L}^{(2)}(\vec{r},t) = -\vec{\nabla}\phi^{(2)}(\vec{r},t)$, it appears from Eq. (99) that $\phi^{(2)}(\vec{r},t)$ satisfies the wave equation

$$\left(\nabla^2 - \frac{1}{c_0^2}\frac{\partial^2}{\partial t^2}\right)\phi^{(2)}(\vec{r},t) = -\frac{1}{\epsilon_0}\rho(\vec{r},t). \tag{101}$$

The vector potential $\vec{A}^{(2)}(\vec{r},t)$ associated with $\phi^{(2)}(\vec{r},t)$ is given via

$$\vec{E}_{\rm T}(\vec{r},t) + \vec{E}_{\rm L}^{(1)}(\vec{r},t) = -\frac{\partial}{\partial t}\vec{A}^{(2)}(\vec{r},t), \tag{102}$$

and has hence both transverse and longitudinal parts. Introduction of $\vec{A}^{(2)}(\vec{r},t)$ reduces Eq. (100) to

$$\left(\nabla^2 - \frac{1}{c_0^2}\frac{\partial^2}{\partial t^2}\right)\vec{A}^{(2)}(\vec{r},t) = \mu_0\vec{J}(\vec{r},t). \tag{103}$$

By solving Eqs. (101) and (103) one obtains the standard integral formula in Eq. (92) for the total electric field

$$\vec{E}(\vec{r},t) = \vec{E}_{\mathrm{T}}(\vec{r},t) + \vec{E}_{\mathrm{L}}^{(1)}(\vec{r},t) - \vec{\nabla}\phi^{(2)}(\vec{r},t). \tag{104}$$

The quantities $\vec{J}(\vec{r},t)$ and $\rho(\vec{r},t)$ in Eqs. (101) and (103) represent the total current and charge densities, and in Eq. (92) a superscript "all" has been added to these for the sake of the discussion in the subsequent subsection.

The route from the standard formulation to the propagator description

With the standard formula in Eq. (92) as starting point, let us now see how the propagator picture of electrodynamics may be derived. Let us set out with a Fourier transformation of Eq. (92) to the frequency domain. This gives

$$\vec{E}(\vec{r};\omega) = \frac{i}{4\pi\epsilon_0 c_0^2} \int \omega \vec{J}^{\mathrm{all}}(\vec{r}';\omega) e^{i\frac{\omega}{c_0}|\vec{r}-\vec{r}'|} \frac{d^3 r'}{|\vec{r}-\vec{r}'|}$$
$$- \frac{1}{4\pi\epsilon_0} \vec{\nabla} \int \rho^{\mathrm{all}}(\vec{r}';\omega) e^{i\frac{\omega}{c_0}|\vec{r}-\vec{r}'|} \frac{d^3 r'}{|\vec{r}-\vec{r}'|}. \tag{105}$$

By introducing the function

$$g_0(|\vec{r}-\vec{r}'|;\omega) = \frac{e^{iq_0|\vec{r}-\vec{r}'|}}{4\pi|\vec{r}-\vec{r}'|}, \tag{106}$$

with $q_0 = \omega/c_0$, using the equation of continuity, $i\omega\rho^{\mathrm{all}}(\vec{r}';\omega) = \vec{\nabla}' \cdot \vec{J}^{\mathrm{all}}(\vec{r}';\omega)$, and the division $\vec{J}^{\mathrm{all}}(\vec{r}';\omega) = \vec{J}_{\mathrm{T}}^{\mathrm{all}}(\vec{r}';\omega) + \vec{J}_{\mathrm{L}}^{\mathrm{all}}(\vec{r}';\omega)$, Eq. (105) can be written in the form

$$\vec{E}(\vec{r};\omega) =$$
$$i\mu_0\omega \int g_0(|\vec{r}-\vec{r}'|;\omega)\vec{J}_{\mathrm{T}}^{\mathrm{all}}(\vec{r}';\omega)d^3 r' + i\mu_0\omega \int g_0(|\vec{r}-\vec{r}'|;\omega)\vec{J}_{\mathrm{L}}^{\mathrm{all}}(\vec{r}';\omega)d^3 r'$$
$$+ \frac{i}{\epsilon_0\omega} \int \left[\vec{\nabla}g_0(|\vec{r}-\vec{r}'|;\omega)\right] \vec{\nabla}' \cdot \vec{J}_{\mathrm{L}}^{\mathrm{all}}(\vec{r}';\omega)d^3 r'. \tag{107}$$

In passing, please note the close relation between g_0 and $\overleftrightarrow{d}_0^{\mathrm{R}}$ (Eq. (22)). Next, I draw specific attention to the particles which constitute the (sub)system under study, cf. the discussion in section 2.2, and as before the current and charge densities of the subsystem are denoted by $\vec{J}(\vec{r}';\omega) = \vec{J}_{\mathrm{T}}(\vec{r}';\omega) + \vec{J}_{\mathrm{L}}(\vec{r}';\omega)$, and $\rho(\vec{r}';\omega)$. Again I assume that the external charges are so far away from our system that the longitudinal (transverse) current densities do not overlap

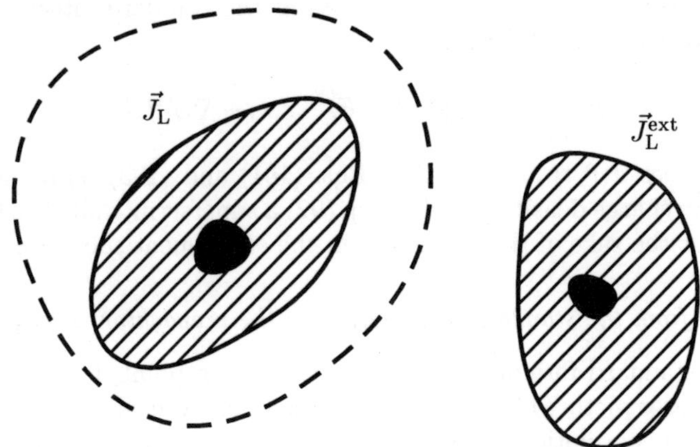

Figure 7: Schematic illustration of non-overlapping longitudinal current density distributions of a subsystem under study ($\vec{J}_{\mathrm{L}}$) and the so-called external source ($\vec{J}_{\mathrm{L}}^{\mathrm{ext}}$) driving the particles of the subsystem. To determine the electric field inside the subsystem a partial integration over a domain (indicated by the dashed line contour) enclosing the $\vec{J}_{\mathrm{L}}$-domain but not any part of the $\vec{J}_{\mathrm{L}}^{\mathrm{ext}}$-domain has to be carried out. The domains occupied by the current densities themselves, i.e. $\vec{J}$ and $\vec{J}^{\mathrm{ext}}$, are indicated in black.

(see Fig. 7). This in turn breaks the integrals in Eq. (107) into separate pieces over the $\vec{J}_{\mathrm{L}}$- ($\vec{J}_{\mathrm{T}}$-) and $\vec{J}_{\mathrm{L}}^{\mathrm{ext}}$- ($\vec{J}_{\mathrm{T}}^{\mathrm{ext}}$-) domains. Let us then consider the part of the last term in Eq. (107) which extends over the $\vec{J}_{\mathrm{L}}$-domain. Utilizing that $\vec{\nabla} g_0(|\vec{r} - \vec{r}'|; \omega) = -\vec{\nabla}' g_0(|\vec{r} - \vec{r}'|; \omega)$ and performing afterwards a partial integration over a domain enclosing the $\vec{J}_{\mathrm{L}}$-domain (Fig. 7), but not any part of the $\vec{J}_{\mathrm{L}}^{\mathrm{ext}}$-domain, we obtain

$$\int \left[\vec{\nabla} g_0(|\vec{r} - \vec{r}'|; \omega) \right] \vec{\nabla}' \cdot \vec{J}_{\mathrm{L}}(\vec{r}'; \omega) d^3 r' =$$

$$\int \left[\vec{\nabla}' \cdot \vec{\nabla}' g_0(|\vec{r} - \vec{r}'|; \omega) \right] \vec{J}_{\mathrm{L}}(\vec{r}'; \omega) d^3 r'. \qquad (108)$$

Since $g_0(|\vec{r} - \vec{r}'|; \omega)$ satisfies the differential equation

$$\left[(\nabla')^2 + q_0^2 \right] g_0(|\vec{r} - \vec{r}'|; \omega) = -\delta(\vec{r} - \vec{r}') \qquad (109)$$

we then have

$$\int \left[\vec{\nabla} g_0(|\vec{r} - \vec{r}'|; \omega) \right] \vec{\nabla}' \cdot \vec{J}_{\mathrm{L}}(\vec{r}'; \omega) d^3 r' =$$

$$-\vec{J}_{\mathrm{L}}(\vec{r}; \omega) - \left(\frac{\omega}{c_0} \right)^2 \int g_0(|\vec{r} - \vec{r}'|; \omega) \vec{J}_{\mathrm{L}}(\vec{r}'; \omega) d^3 r'. \qquad (110)$$

By inserting the result in Eq. (110) into Eq. (107), it is realized that for observation points inside the system under study one has

$$\vec{E}(\vec{r};\omega) = i\mu_0\omega \int g_0(|\vec{r}-\vec{r}'|;\omega)\vec{J}_{\mathrm{T}}^{\mathrm{all}}(\vec{r}';\omega)d^3r' + \frac{1}{i\epsilon_0\omega}\vec{J}_{\mathrm{L}}(\vec{r};\omega). \qquad (111)$$

In Eq. (111), one recognizes the last term as the longitudinal part of the self-field. In Ref. 11 is demonstrated that the first term on the right hand side of Eq. (111), which in fact gives the transverse part of the electric field, can be rewritten in the following form:

$$i\mu_0\omega \int g_0(|\vec{r}-\vec{r}'|;\omega)\vec{J}_{\mathrm{T}}^{\mathrm{all}}(\vec{r}';\omega)d^3r' =$$

$$\vec{E}_{\mathrm{T}}^{\mathrm{ext}}(\vec{r};\omega) + \frac{1}{3i\epsilon_0\omega}\vec{J}_{\mathrm{T}}(\vec{r};\omega) - i\mu_0\omega \int \vec{\vec{D}}_0^{\mathrm{R}}(\vec{r}-\vec{r}';\omega)\cdot\vec{J}(\vec{r}';\omega)d^3r'. \qquad (112)$$

By inserting Eq. (112) into Eq. (111), we finally have

$$\vec{E}(\vec{r};\omega) = \frac{1}{3i\epsilon_0\omega}\left(\vec{J}_{\mathrm{T}}(\vec{r};\omega) + 3\vec{J}_{\mathrm{L}}(\vec{r};\omega)\right) - i\mu_0\omega \int \vec{\vec{D}}_0^{\mathrm{R}}(\vec{r}-\vec{r}';\omega)\cdot\vec{J}(\vec{r}';\omega)d^3r',$$

$$(113)$$

i.e. in a slightly different form precisely Eq. (89). A route from the standard description of electrodynamics to the propagator formulation hence has been established.

3 Transverse and longitudinal current densities in quantum physics and the associated electrodynamics

In order to determine the local electromagnetic field from the microscopic Maxwell-Lorentz equations it is necessary either to have a preknowledge of the microscopic current density or to be able to establish a relation between the field induced current density and the prevailing electromagnetic field which thereafter allows one to setup a self-consistent loop equation for the electromagnetic field. If one prefers to formulate the Maxwell-Lorentz field dynamics in the transverse propagator picture discussed in section 2 it is necessary also to calculate the transverse and longitudinal parts of the prevailing current density, cf. e.g. Eq. (113). In atomic (molecular), mesoscopic, and condensed matter physics it is necessary in most cases to use quantum mechanics to calculate the current density distributions. In this section, we shall not only present and discuss the quantum theory for the material current densities, but in the wake also undertake an analysis of the self-field dynamics originating in these. Though a many-body approach certainly can be established, I shall limit myself here to considerations based on the one-particle picture.

3.1   *One-body current density, a reminder*

Starting from the classical expression, $-e\vec{v}\delta(\vec{r}-\vec{r}_e)$, for the current density of a point particle of charge $-e$ located at the space point $\vec{r}_e$, and moving with the instantaneous velocity $\vec{v}$, the Hamiltonian formulation of classical mechanics naturally leads to the following expression for the one-electron current density operator in quantum mechanics:[14]

$$\vec{j} = -\frac{e}{2m}\left[\vec{\pi}\delta(\vec{r}-\vec{r}_e) + \delta(\vec{r}-\vec{r}_e)\vec{\pi}\right], \tag{114}$$

where

$$\vec{\pi} = \vec{p} + e\vec{A}(\vec{r}_e, t) \tag{115}$$

is the so-called mechanical momentum operator, $\vec{p} = (\hbar/i)\vec{\nabla}$ being the canonical momentum operator. The electron charge and mass are respectively $-e$ and m, and the particle is subjected to the self-consistently determined vector potential $\vec{A}(\vec{r}_e, t)$. The form in Eq. (114) is symmetrized to guarantee that the current density operator is Hermitian. In Eq. (114), the spin part of the current density operator was left out, but it will be studied in section 3.4.

Let us assume now that the electron is in the pure state $|\Psi\rangle(t)$ which in the $\vec{r}_e$-representation is named $\Psi(\vec{r}_e, t) = \langle\vec{r}_e|\Psi\rangle(t)$. In this state the quantum mechanical mean value, $\vec{J}(\vec{r}, t)$, of the current density operator at the space-time point $(\vec{r}, t)$ is given by

$$\vec{J}(\vec{r}, t) = \frac{e\hbar}{2im}\left(\Psi(\vec{r}, t)\vec{\nabla}\Psi^*(\vec{r}, t) - \Psi^*(\vec{r}, t)\vec{\nabla}\Psi(\vec{r}, t)\right) - \frac{e^2}{m}\vec{A}(\vec{r}, t)\,|\Psi(\vec{r}, t)|^2. \tag{116}$$

Lack of information usually prevents us from knowing the state of the electron precisely, and we are therefore often forced to work with the so-called mixed states. If the characteristic electron is in a mixed state described by the density matrix operator ρ, the (measurable) current density becomes

$$\vec{J}(\vec{r}, t) = \mathrm{Tr}\{\rho\vec{j}\}, \tag{117}$$

where $\mathrm{Tr}\{\cdots\}$ means the trace of $\{\cdots\}$. Returning now to the expression in Eq. (116) for the current density in the case of a pure state, it appears that this has a term which depends on the (chosen) vector potential, $\vec{A}(\vec{r}, t)$. Although the longitudinal part of the vector potential is not gauge invariant, the current density does not depend on the gauge chosen, as must be required. A change of gauge for the vector and scalar potential in quantum mechanics is accompanied by a unitary transformation of the wave function, and without the term $-(e^2/m)\vec{A}|\Psi|^2$ the expression for $\vec{J}(\vec{r},t)$ in Eq. (116) would not be gauge independent. To demonstrate that $\vec{J}(\vec{r},t)$ is gauge invariant let us assume that we initially have chosen a gauge where the vector and scalar potentials are $\vec{A}_{\text{OLD}}(\vec{r},t)$ and $\phi_{\text{OLD}}(\vec{r},t)$, respectively, and the wave function is $\Psi_{\text{OLD}}(\vec{r},t)$. Using the (arbitrary) gauge function $\chi(\vec{r},t)$ the OLD quantities are transformed into the NEW ones

$$\vec{A}_{\text{NEW}}(\vec{r},t) = \vec{A}_{\text{OLD}}(\vec{r},t) + \vec{\nabla}\chi(\vec{r},t), \tag{118}$$

$$\phi_{\text{NEW}}(\vec{r},t) = \phi_{\text{OLD}}(\vec{r},t) - \frac{\partial\chi(\vec{r},t)}{\partial t}, \tag{119}$$

$$\Psi_{\text{NEW}}(\vec{r},t) = \exp\left(-i\frac{e}{\hbar}\chi(\vec{r},t)\right)\Psi_{\text{OLD}}(\vec{r},t). \tag{120}$$

If one in the expression

$$\vec{J}_{\text{OLD}}(\vec{r},t) = \frac{e\hbar}{2im}\left(\Psi_{\text{OLD}}(\vec{r},t)\vec{\nabla}\Psi^*_{\text{OLD}}(\vec{r},t) - \Psi^*_{\text{OLD}}(\vec{r},t)\vec{\nabla}\Psi_{\text{OLD}}(\vec{r},t)\right)$$
$$-\frac{e^2}{m}\vec{A}_{\text{OLD}}(\vec{r},t)\left|\Psi_{\text{OLD}}(\vec{r},t)\right|^2 \tag{121}$$

for the current density in the OLD gauge eliminates $\vec{A}_{\text{OLD}}$, ϕ_{OLD}, and Ψ_{OLD} by means of Eqs. (118)–(120), one finds

$$\vec{J}_{\text{OLD}}(\vec{r},t) = \vec{J}_{\text{NEW}}(\vec{r},t), \tag{122}$$

where

$$\vec{J}_{\text{NEW}}(\vec{r},t) = \frac{e\hbar}{2im}\left(\Psi_{\text{NEW}}(\vec{r},t)\vec{\nabla}\Psi^*_{\text{NEW}}(\vec{r},t) - \Psi^*_{\text{NEW}}(\vec{r},t)\vec{\nabla}\Psi_{\text{NEW}}(\vec{r},t)\right)$$
$$-\frac{e^2}{m}\vec{A}_{\text{NEW}}(\vec{r},t)\left|\Psi_{\text{NEW}}(\vec{r},t)\right|^2. \tag{123}$$

The gauge function $\chi(\vec{r},t)$ thus automatically is eliminated, and Eq. (122) proves the gauge invariance of the current density. The transverse part of the vector potential, $\vec{A}_{\text{T}}(\vec{r},t)$, is gauge invariant, and in the Coulomb gauge where the auxilary condition is $\vec{\nabla}\cdot\vec{A}(\vec{r},t) = 0$, the longitudinal part of the vector potential is zero so that $\vec{A}_{\text{T}}(\vec{r},t)$ appears in Eq. (116).

3.2 *Asymptotic behaviour of the transverse and longitudinal current density*

Let us imagine that we study the electrodynamics of a spatially strongly localized charge density distribution, say an atom, a molecule or a mesoscopic particle. For such a case the spatial extension of the one-electron wave function is determined by an essentially exponentially decaying tail, cf. e.g. the wave functions of the stationary states of hydrogen. Since the current density $\vec{J}(\vec{r},t)$ only is different from zero in regions of space where $\Psi(\vec{r},t)$ is nonzero, this means that $\vec{J}(\vec{r},t)$ decays rapidly ($\sim$exponentially) with distance as one moves away from the central part of the charge distribution.

To determine the asymptotic behaviour of the transverse and longitudinal parts of the current density we rewrite Eqs. (65) and (67) in the forms

$$\vec{J}_{\mathrm{L}}(\vec{r},t) = \frac{1}{3}\vec{J}(\vec{r},t) - \vec{\mathcal{J}}(\vec{r},t), \tag{124}$$

and

$$\vec{J}_{\mathrm{T}}(\vec{r},t) = \frac{2}{3}\vec{J}(\vec{r},t) + \vec{\mathcal{J}}(\vec{r},t), \tag{125}$$

where $\mathcal{J}$, defined in Eq. (69), is given by

$$\vec{\mathcal{J}}(\vec{r},t) = \mathrm{PV}\int \frac{1}{4\pi|\vec{r}-\vec{r}'|^3}\left(3\vec{e}_{\vec{r}-\vec{r}'}\vec{e}_{\vec{r}-\vec{r}'} - \vec{\vec{U}}\right)\cdot\vec{J}(\vec{r}',t)d^3r', \tag{126}$$

after carrying out the $\vec{\nabla}\times(\vec{\nabla}\times(\cdots))$-operation. Far away from the current density distribution Eq. (126) takes the asymptotic ($|\vec{r}|\to\infty$) form

$$\vec{\mathcal{J}}^{\infty}(\vec{r},t) = \frac{1}{4\pi r^3}\left(3\vec{e}_{\vec{r}}\vec{e}_{\vec{r}} - \vec{\vec{U}}\right)\cdot\int \vec{J}(\vec{r}',t)d^3r', \tag{127}$$

where $\vec{e}_{\vec{r}} = \vec{r}/r$, provided the integral $\int \vec{J}d^3r'$ does not vanish. If this should happen one just proceeds to the next term in the multipole expansion. Using the standard abbreviation

$$\langle\mathcal{O}\rangle(t) = \int \Psi^*(\vec{r},t)\mathcal{O}(\vec{r},t)\Psi(\vec{r},t)d^3r \tag{128}$$

for the quantum mechanical mean value of an operator $\mathcal{O}$, it is easy to show by means of Eq. (116) that

$$\int \vec{J}(\vec{r}',t)d^3r' = -\frac{e}{m}\langle\vec{p}+e\vec{A}\rangle(t). \tag{129}$$

The common asymptotic behaviour of the transverse and longitudinal current densities therefore is given by

$$\vec{J}_{\mathrm{T}}^{\infty}(\vec{r},t) = -\vec{J}_{\mathrm{L}}^{\infty}(\vec{r},t) = \frac{e}{4\pi m r^3}\left(\vec{\vec{U}} - 3\vec{e}_{\vec{r}}\vec{e}_{\vec{r}}\right)\cdot\langle\vec{\pi}\rangle(t) \tag{130}$$

in terms of the mean value of the mechanical momentum operator. It appears from Eq. (130) that $\vec{J}_\mathrm{T}$ and $\vec{J}_\mathrm{L}$ decay asymptotically as r^3, and thus have a much larger spatial extension than $\vec{J}$.

The asymptotic behaviour of the transverse and longitudinal self-fields may now be obtained via Eqs. (20) and (76), or equivalently Eqs. (31) and (66). Illustrative expressions for these quantities can be obtained by using the fact that the mean values of the mechanical momentum and position operators are related via

$$\langle \vec{\pi} \rangle(t) = m \frac{d}{dt} \langle \vec{r} \rangle(t). \tag{131}$$

By combining Eqs. (20), (130), and (131) one obtains the following expression for the time derivative of the asymptotic longitudinal self-field:

$$\frac{\partial}{\partial t} \vec{E}_\mathrm{L}^\infty(\vec{r}, t) = \frac{e}{4\pi\epsilon_0 r^3} \left(\overset{\leftrightarrow}{U} - 3\vec{e}_{\vec{r}}\vec{e}_{\vec{r}} \right) \cdot \frac{d}{dt} \langle \vec{r} \rangle(t). \tag{132}$$

By introduction of the electric dipole moment operator

$$\vec{d} = -e\vec{r}, \tag{133}$$

one gets upon integration of Eq. (132)

$$\vec{E}_\mathrm{L}^\infty(\vec{r}, t) = \frac{1}{4\pi\epsilon_0 r^3} \left(3\vec{e}_{\vec{r}}\vec{e}_{\vec{r}} - \overset{\leftrightarrow}{U} \right) \cdot \langle \vec{d} \rangle(t), \tag{134}$$

under the assumption that $\vec{E}_\mathrm{L}^\infty(\vec{r}, t)$ vanishes when the dipole moment $\langle \vec{d} \rangle(t)$ is zero. A close correspondence between the quantum mechanical expression for the asymptotic longitudinal self-field and the textbook formula for the quasi-static near field from a classical electric point dipole of moment $\vec{d} = -e\vec{r}$ now has emerged. This analogy one might have anticipated. In view of Eqs. (130) and (76) the corresponding expression for the transverse asymptotic self-field is of course

$$\vec{E}_\mathrm{T}^{\mathrm{SF},\infty}(\vec{r}, t) = \frac{1}{3} \frac{1}{4\pi\epsilon_0 r^3} \left(\overset{\leftrightarrow}{U} - 3\vec{e}_{\vec{r}}\vec{e}_{\vec{r}} \right) \cdot \langle \vec{d} \rangle(t). \tag{135}$$

The total self-field, i.e. the sum of $\vec{E}_\mathrm{T}^{\mathrm{SF}}$ and $\vec{E}_\mathrm{L}$, thus is spread over a region which extension is characterized by a r^{-3}-tail.

3.3 *Microscopic current flows in transitions between bound states*

Let us now briefly consider the self-field electrodynamics associated with induced electronic transitions between the various bound stationary states of a microscopic or mesoscopic particle. Although the general many-body approach conceptually is straight forward to establish I also here limit myself to the one-electron approximation for heuristic purposes, and deal with pure quantum states only. By assuming that the excitation probability amplitudes to the unbound states of the system are sufficiently small, an arbitrary pure state $|\Psi\rangle(t)$ can be expanded in terms of a (almost) complete orthonormal set of bound eigenfunctions, $|\psi_i\rangle$, for the field unperturbed Hamilton operator, i.e. in the $\vec{r}$-representation $[\psi_i(\vec{r}) = \langle\vec{r}|\psi_i\rangle,\ \Psi(\vec{r},t) = \langle\vec{r}|\psi\rangle(t)]$

$$\Psi(\vec{r},t) = \sum_i c_i(t)\psi_i(\vec{r}), \tag{136}$$

where $c_i(t) = \langle\psi_i|\Psi\rangle(t)$. In most bound-state transition cases one neglects the (small) term in the current density which is proportional to the vector potential. To study for instance diamagnetic effects in atoms (molecules) and small particles the $\vec{A}$-dependent term in Eq. (116) must be kept. In fact, this term in $\vec{J}(\vec{r},t)$ is often called the diamagnetic term. By inserting the expansion in Eq. (136) into Eq. (116) and omitting the diamagnetic contribution, one gets

$$\vec{J}(\vec{r},t) = \sum_{i,j} c_i(t)c_j^*(t)\vec{j}_{ij}(\vec{r}), \tag{137}$$

where

$$\vec{j}_{ij}(\vec{r}) = \frac{e\hbar}{2im}\left(\psi_i(\vec{r})\vec{\nabla}\psi_j^*(\vec{r}) - \psi_j^*(\vec{r})\vec{\nabla}\psi_i(\vec{r})\right) \tag{138}$$

is the so-called transition current density associated with a single-electron excitation from state i to state j. In the absence of dc vector potentials (giving rise e.g. to the Aharanov-Bohm effect) all the wave functions may be chosen to be real, and therefore $\vec{j}_{ii} = \vec{0}$ for all i. The stationary states thus carry no current in this kind of expansion. Only transition between different stationary states thus need to be considered. We already know of course that stationary states cannot be observed, only changes in such states may lead to detectable effects. To determine the transverse,

$$\vec{J}_{\mathrm{T}}(\vec{r},t) = \sum_{i,j\neq i} c_i(t)c_j^*(t)\vec{j}_{ij}^{\mathrm{T}}(\vec{r}), \tag{139}$$

and longitudinal,

$$\vec{J}_{\mathrm{L}}(\vec{r},t) = \sum_{i,j\neq i} c_i(t)c_j^*(t)\vec{j}_{ij}^{\mathrm{L}}(\vec{r}), \tag{140}$$

current densities one hence needs to calculate the transverse $(\vec{j}_{ij}^{\mathrm{T}})$ and longitudinal $(\vec{j}_{ij}^{\mathrm{L}})$ parts of the relevant transition current densities. The result of such a calculation is presented for the hydrogen $1s$-$2p_z$ transition in Ref. 15.

3.4   One-body spin current density

So far, the elctron spin has been neglected. To incorporate electrodynamic phenomena stemming from the spin dynamics in the propagator picture, it is adequate to take as a starting point the one-electron spin current density operator

$$\vec{j}_{\vec{\sigma}} = \frac{e}{2im}\left[\delta(\vec{r} - \vec{r}_e)\vec{p} \times \vec{\sigma} - \vec{p} \times \vec{\sigma}\delta(\vec{r} - \vec{r}_e)\right], \tag{141}$$

where $\vec{\sigma} = (\sigma_x, \sigma_y, \sigma_z)$ is the Pauli spin operator. In the two-component spin space of a spin-$\frac{1}{2}$ particle the three Cartesian components of the spin operator may be represented by the Pauli spin matrices

$$\sigma_x = \frac{\hbar}{2}\begin{pmatrix} 0 & 1 \\ 1 & 0 \end{pmatrix}, \tag{142}$$

$$\sigma_y = \frac{\hbar}{2}\begin{pmatrix} 0 & -i \\ i & 0 \end{pmatrix}, \tag{143}$$

$$\sigma_z = \frac{\hbar}{2}\begin{pmatrix} 1 & 0 \\ 0 & -1 \end{pmatrix}. \tag{144}$$

Once the electron spin is introduced the possible state vectors of the electron are vectors in the tensor product space of space and spin states. To determine the current density originating in the electron spin it is convenient to expand the quantum states in the $\vec{r}$-representation and the (s^2, s_z)-basis. Assuming once more that the electron is in a pure quantum state, the wave function of the electron is described by the two-component spinor

$$\begin{pmatrix} \langle \vec{r}; \uparrow | \Psi \rangle \\ \langle \vec{r}; \downarrow | \Psi \rangle \end{pmatrix}(t) = \begin{pmatrix} \Psi_{\uparrow}(\vec{r}, t) \\ \Psi_{\downarrow}(\vec{r}, t) \end{pmatrix}. \tag{145}$$

The spin current density, denoted in the following by $\vec{J}^{\text{SPIN}}(\vec{r}, t)$, is obtained as the quantum mechanical mean value of the spin current density operator in the two-component spinor state of Eq. (145), i.e.

$$\vec{J}^{\text{SPIN}}(\vec{r}, t) = \frac{ie}{2m}\int \left(\Psi_{\uparrow}^*(\vec{r}_e, t), \Psi_{\downarrow}^*(\vec{r}_e, t)\right)\left[\vec{p} \times \vec{\sigma}\delta(\vec{r} - \vec{r}_e)\right.$$
$$\left. -\delta(\vec{r} - \vec{r}_e)\vec{p} \times \vec{\sigma}\right]\begin{pmatrix} \Psi_{\uparrow}(\vec{r}_e, t) \\ \Psi_{\downarrow}(\vec{r}_e, t) \end{pmatrix} d^3r_e, \tag{146}$$

cf. Eq. (128). By utilizing the hermiticity of the electron momentum operator, a partial integration on the last term in Eq. (146), and the definition of the

Dirac delta function to carry out the integrations over $\vec{r}_e$-space, one obtains a spin current density

$$\vec{J}^{\text{SPIN}}(\vec{r},t) = \frac{ie}{2m}\left[\vec{p}\,\Psi_\uparrow(\vec{r},t),\vec{p}\,\Psi_\downarrow(\vec{r},t)\right]^* \times \vec{\sigma}\left(\begin{array}{c}\Psi_\uparrow(\vec{r},t)\\\Psi_\downarrow(\vec{r},t)\end{array}\right) + \text{c.c.}\tag{147}$$

Upon insertion of the Pauli spin matrices (Eqs. (142)–(144)), the spin contribution to the current density finally becomes

$$\vec{J}^{\text{SPIN}}(\vec{r},t) = \frac{e\hbar}{\sqrt{2}m}\left(\vec{e}_+ \times \vec{\nabla}\left[\Psi_\uparrow(\vec{r},t)\Psi_\downarrow^*(\vec{r},t)\right] + \vec{e}_- \times \vec{\nabla}\left[\Psi_\downarrow(\vec{r},t)\Psi_\uparrow^*(\vec{r},t)\right]\right.$$
$$\left.+\frac{\vec{e}_z}{\sqrt{2}} \times \vec{\nabla}\left[\Psi_\uparrow(\vec{r},t)\Psi_\uparrow^*(\vec{r},t) - \Psi_\downarrow(\vec{r},t)\Psi_\downarrow^*(\vec{r},t)\right]\right),\tag{148}$$

where $\vec{e}_z$ is a unit vector along the spin quantization axis (the z-axis), and $\vec{e}_\pm = (\vec{e}_x \pm i\vec{e}_y)/\sqrt{2}$ are complex unit vectors. It appears from Eq. (148) that $\vec{\nabla}\cdot\vec{J}^{\text{SPIN}}(\vec{r},t) = 0$, and this means that the spin current density is transverse, i.e.

$$\vec{J}^{\text{SPIN}}(\vec{r},t) = \vec{J}_{\text{T}}^{\text{SPIN}}(\vec{r},t).\tag{149}$$

Consequently, there are no longitudinal fields associated with the spin electrodynamics. This fact may be not so surprising from the point of view of quantum electrodynamics. Remembering that if one uses the economic Coulomb gauge, the redundant longitudinal field variables are eliminated in favour of the dynamical particle position variables, $\{\vec{r}_e\}$, and are therefore linked to the Hilbert subspace of the spatial degrees of freedom, and hence not to the spin subspace.

In spinor dynamics the spatial part of the current density, denoted by $\vec{J}^{\text{SPACE}}(\vec{r},t)$ below, has spin-up ($\uparrow$) and spin-down ($\downarrow$) parts, i.e.

$$\vec{J}^{\text{SPACE}}(\vec{r},t) = \vec{J}_\uparrow^{\text{SPACE}}(\vec{r},t) + \vec{J}_\downarrow^{\text{SPACE}}(\vec{r},t),\tag{150}$$

and the explicit expressions for these are

$$\vec{J}_{\uparrow(\downarrow)}^{\text{SPACE}}(\vec{r},t) = \frac{e\hbar}{2im}\left[\Psi_{\uparrow(\downarrow)}(\vec{r},t)\vec{\nabla}\Psi_{\uparrow(\downarrow)}^*(\vec{r},t) - \Psi_{\uparrow(\downarrow)}^*(\vec{r},t)\vec{\nabla}\Psi_{\uparrow(\downarrow)}(\vec{r},t)\right]$$
$$-\frac{e^2}{m}\left|\Psi_{\uparrow(\downarrow)}(\vec{r},t)\right|^2 \vec{A}(\vec{r},t).\tag{151}$$

Taking into account the electron spin we have hence realized that a single electron in the pure state given in Eq. (145) is accompanied by an electromagnetic field

$$\vec{E}(\vec{r};\omega) = \vec{E}_{\mathrm{T}}^{\mathrm{ext}}(\vec{r};\omega) + \frac{1}{i\epsilon_0\omega}\left(\vec{J}_{\uparrow,\mathrm{L}}^{\mathrm{SPACE}}(\vec{r};\omega) + \vec{J}_{\downarrow,\mathrm{L}}^{\mathrm{SPACE}}(\vec{r};\omega)\right)$$

$$+\frac{1}{3i\epsilon_0\omega}\left(\vec{J}_{\uparrow,\mathrm{T}}^{\mathrm{SPACE}}(\vec{r};\omega) + \vec{J}_{\downarrow,\mathrm{T}}^{\mathrm{SPACE}}(\vec{r};\omega) + \vec{J}^{\mathrm{SPIN}}(\vec{r};\omega)\right)$$

$$-i\mu_0\omega\int\vec{\vec{D}}_0^{\mathrm{R}}(\vec{r}-\vec{r}\,';\omega)\cdot\left(\vec{J}_{\uparrow}^{\mathrm{SPACE}}(\vec{r};\omega) + \vec{J}_{\downarrow}^{\mathrm{SPACE}}(\vec{r};\omega) + \vec{J}^{\mathrm{SPIN}}(\vec{r};\omega)\right)d^3r',$$

$$(152)$$

in the space-frequency domain. The quantities $\vec{J}_{\uparrow(\downarrow),\mathrm{T}}^{\mathrm{SPACE}}(\vec{r},t)$ and $\vec{J}_{\uparrow(\downarrow),\mathrm{L}}^{\mathrm{SPACE}}(\vec{r},t)$ denote the transverse and longitudinal parts of the spin-up ($\uparrow$) or spin-down ($\downarrow$) spatial current density.

Before finishing this section let us imagine a situation in which only the spin state is allowed to change in time. The associated pure state spinor thus is given by the space and spin uncorrelated ansatz

$$\begin{pmatrix}\Psi_{\uparrow}(\vec{r},t)\\\Psi_{\downarrow}(\vec{r},t)\end{pmatrix} = \psi(\vec{r})\begin{pmatrix}\uparrow(t)\\\downarrow(t)\end{pmatrix}, \qquad (153)$$

where $\psi(\vec{r})$ is the (time independent) wave function of the stationary state chosen. Freezing the spatial degrees of freedom one might be tempted to believe that the particle dynamics is completely confined to the spin subspace and in consequence of this no retarded electromagnetic fields nor transverse self-fields would occur. The spatial current densities $\vec{J}_{\uparrow(\downarrow)}^{\mathrm{SPACE}}(\vec{r},t)$ are zero, of course, but it appears from Eq. (148) that a spin current density does exist. By inserting Eq. (153) into Eq. (148) one obtains the following spin current density:

$$\vec{J}^{\mathrm{SPIN}}(\vec{r},t) = \frac{e\hbar}{\sqrt{2}m}\left[\uparrow(t)\downarrow^*(t)\vec{e}_+ + \downarrow(t)\uparrow^*(t)\vec{e}_-\right.$$
$$\left.+\left(|\uparrow(t)|^2 - |\downarrow(t)|^2\right)\vec{e}_z\right]\times\vec{\nabla}|\psi(\vec{r})|^2. \qquad (154)$$

To change the electron spin between two eigenstates [say $\left(\begin{smallmatrix}1\\0\end{smallmatrix}\right)$ and $\left(\begin{smallmatrix}0\\1\end{smallmatrix}\right)$] in spin space a current must flow in the direct $\vec{r}$-space, even though we imagine a situation in which the spatial part of the electron wave function has been frozen (see also Fig. 8). Since the spin current density is transverse, this time dependent current flow is not accompanied by any charge modulations, i.e.

$$\rho^{\mathrm{SPIN}}(\vec{r},t) = 0, \qquad (155)$$

identically.

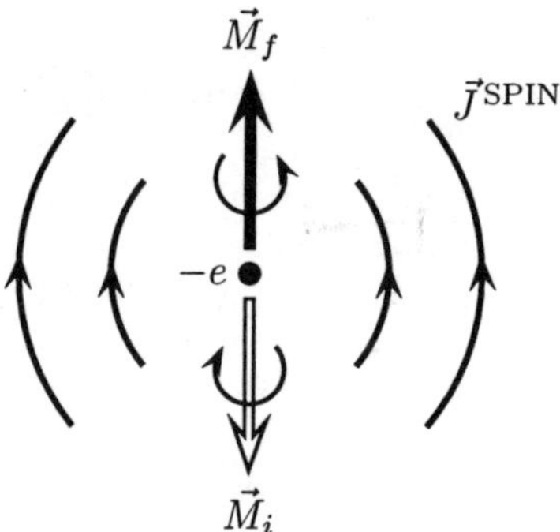

Figure 8: In order to change the internal magnetic moment of an electron (of charge $-e$) from the initial value $\vec{M}_i$ to the finale value $\vec{M}_f$, a (spin) current of density $\vec{J}^{\mathrm{SPIN}}$ must flow in direct space in the vicinity of this point particle, even if one imagine that the spatial part of the electron wave function does not change.

4   Spatial confinement of light and electrons

The propagator picture of (quantum) electrodynamics is useful in studies of the optical properties of mesoscopic media like quantum dots, quantum wires and rings, and quantum wells. In such media the spatial confinement of the electrons in one, two or three dimensions is accompanied by essential quantizations of the particle motion in certain directions and this in turn leads to a number of new and interesting electronic properties of the above-mentioned structures. When an electromagnetic field interacts with a mesoscopic medium the induced current densities necessarily exhibit particular quantum size effects, and these are therefore in a manifest manner present in the matter-attached part of the local field, i.e. in the self-field. The spatial confinement of the electron distribution, which roughly speaking can be characterized by the extension of the induced current density in space, consequently is accompanied by a spatial confinement of the induced electromagnetic field. Since the self-field is different from zero only in space regions where the longitudinal (or equivalently transverse) part of the induced current density does not vanish it is obvious that one should be careful to distinguish between the confinement regions for electrons and light in mesoscopic electrodynamics. Since the retarded part of the electromagnetic field generated by the motion of the electrons of the mesoscopic system under study is locked to the self-field in the transverse (longitudinal) current density region, also the field radiated from the mesoscopic medium may exhibit the fingerprint of the particular particle confinement present in these new mesoscopic materials. The local field prevailing in a mesoscopic medium depends in a crucial manner on the geometry of the medium, and this implies that the local field problem, at best, can only be solved if the artificial mesoscopic material has a particularly simple form. A number of theoretical issues associated with the local field problem in mesoscopic media have been treated in two recent review articles of mine (Refs. 11 and 12), and in these a comprehensive list of references to the remaining literature can be found. In the reviews also specific studies of the linear and nonlinear optical properties of quantum wells, dots (small particles), and rings are described. In the following I shall therefore limit the discussion to a few aspects not hitherto emphasized or unified.

4.1 Three-dimensional spatial confinement

Electric point dipole model

To describe the transverse and longitudinal electrodynamics associated with a localized but otherwise arbitrary current density source distribution $(\vec{J})$ one needs the transverse $(\vec{J}_{\mathrm{T}})$ and longitudinal $(\vec{J}_{\mathrm{L}})$ parts of $\vec{J}$. In principle, these can be obtained from Eqs. (124) and (125), but in practice it is often difficult to determine the nonlocal $\vec{J}(\vec{r}, t)$ contribution from the principal value integral in Eq. (126). In cases where one is interested in studying the electrodynamic interaction between localized charge distributions (atoms, molecules, mesoscopic particles) not too close to each other, it may be sufficient to retain only the asymptotic part $\vec{J}^{\infty}(\vec{r}, t)$ of $\vec{J}(\vec{r}, t)$. It appears from Eq. (127) that this part can be obtained by replacing the true current density distribution $\vec{J}(\vec{r}, t)$ with the delta function distribution

$$\vec{J}^{\mathrm{ED}}(\vec{r}, t) = -\frac{e}{m}\langle\vec{\pi}\rangle(t)\delta(\vec{r}) \tag{156}$$

The superscript ED indicates that $\vec{J}^{\mathrm{ED}}(\vec{r}, t)$ is the current density associated with a so-called electric dipole, here placed at $\vec{r} = \vec{0}$. In a consistent manner, the longitudinal and transverse parts of the electric dipole current density must be calculated from the relations

$$\vec{J}_{\mathrm{L}}^{\mathrm{ED}}(\vec{r}, t) = \int \overleftrightarrow{\delta}_{\mathrm{L}}(\vec{r} - \vec{r}\,') \cdot \vec{J}^{\mathrm{ED}}(\vec{r}\,', t) d^3 r', \tag{157}$$

$$\vec{J}_{\mathrm{T}}^{\mathrm{ED}}(\vec{r}, t) = \int \overleftrightarrow{\delta}_{\mathrm{T}}(\vec{r} - \vec{r}\,') \cdot \vec{J}^{\mathrm{ED}}(\vec{r}\,', t) d^3 r', \tag{158}$$

and thus become

$$\vec{J}_{\mathrm{L}}^{\mathrm{ED}}(\vec{r}, t) = -\frac{e}{m}\left(\frac{1}{3}\delta(r)\overleftrightarrow{U} + \frac{\overleftrightarrow{U} - 3\vec{e}_{\vec{r}}\vec{e}_{\vec{r}}}{4\pi r^3}\right) \cdot \langle\vec{\pi}\rangle(t), \tag{159}$$

and

$$\vec{J}_T^{ED}(\vec{r}, t) = -\frac{e}{m}\left(\frac{2}{3}\delta(r)\overleftrightarrow{U} - \frac{\overleftrightarrow{U} - 3\vec{e}_{\vec{r}}\vec{e}_{\vec{r}}}{4\pi r^3}\right) \cdot \langle\vec{\pi}\rangle(t). \tag{160}$$

The self-field of the electric dipole, $\vec{E}_{ED}^{SF}(\vec{r}, t)$, now can be determined using Eqs. (31), (66), (131), and (133). Hence,

$$\vec{E}_{ED}^{SF}(\vec{r}, t) = -\frac{1}{\epsilon_0}\left(\frac{5}{9}\delta(r)\overleftrightarrow{U} + \frac{2}{3}\frac{\overleftrightarrow{U} - 3\vec{e}_{\vec{r}}\vec{e}_{\vec{r}}}{4\pi r^3}\right) \cdot \langle\vec{d}\rangle(t). \tag{161}$$

If we had retained only the longitudinal part of the electric-dipole current density when calculating the electric self-field the factors 5/9 and 2/3 in Eq. (161) would have been replaced by 1/3 and 1, respectively, and the textbook result for the quasi-static longitudinal field from an electric point dipole would emerge, see e.g. Ref. 13. The delta function term, $(5/9)\delta(r)\overleftrightarrow{U}$, appearing in Eq. (161) does not contribute to the self-field away from the site of the dipole, but nevertheless it is kept here, because it allows one to incorporate contact-interaction effects in the framework of the electric point dipole model, albeit in a crude manner. The retarded (R) and transverse (T) part of the electric field from the electric point dipole, here denoted by $\vec{E}_{T,ED}^{R}(\vec{r}, t)$, may readily be obtained from Eq. (60) inserting $\vec{J}(\vec{r}, t) = \vec{J}^{ED}(\vec{r}, t) = \delta(\vec{r})d/dt\langle\vec{d}\rangle(t)$, and $\partial/\partial t\vec{J}(\vec{r}, t) = \partial/\partial t\vec{J}^{ED}(\vec{r}, t) = \delta(\vec{r})d^2/dt^2\langle\vec{d}\rangle(t)$. This gives

$$\vec{E}_{T,ED}^{R}(\vec{r}, t) = -\frac{\mu_0}{4\pi r}\left(\overleftrightarrow{U} - \vec{e}_{\vec{r}}\vec{e}_{\vec{r}}\right) \cdot \frac{d^2}{dt^2}\langle\vec{d}\rangle\left(t - \frac{r}{c_0}\right)$$
$$-\frac{1}{4\pi\epsilon_0 c_0 r^2}\left(\overleftrightarrow{U} - 3\vec{e}_{\vec{r}}\vec{e}_{\vec{r}}\right) \cdot \left[\frac{d}{dt}\langle\vec{d}\rangle\left(t - \frac{r}{c_0}\right) - \frac{c_0}{r}\left(\langle\vec{d}\rangle(t) - \langle\vec{d}\rangle\left(t - \frac{r}{c_0}\right)\right)\right]. \tag{162}$$

In the near field zone (or equivalently the self-field zone), where the retardation times are small, the expression in Eq. (162) can be simplified by expanding the quantities $\langle\vec{d}\rangle(t - \frac{r}{c_0})$, $d/dt\langle\vec{d}\rangle(t - \frac{r}{c_0})$, and $d^2/dt^2\langle\vec{d}\rangle(t - \frac{r}{c_0})$ in Taylor series around the time t. By neglecting fourth and higher order time derivatives of $\langle\vec{d}\rangle(t)$, one gets after a straightforward calculation

$$\vec{E}_{T,ED}^{R}(\vec{r}, t|\text{NEAR FIELD ZONE}) \approx$$
$$-\frac{\mu_0}{8\pi r}\left(\overleftrightarrow{U} + \vec{e}_{\vec{r}}\vec{e}_{\vec{r}}\right) \cdot \frac{d^2}{dt^2}\langle\vec{d}\rangle(t) + \frac{1}{6\pi\epsilon_0 c_0^3}\frac{d^3}{dt^3}\langle\vec{d}\rangle(t). \tag{163}$$

It appears from Eq. (163) that the retarded field in the near field region of the electric point dipole consists of a term varying with distance as r^{-1} and a distance independent term. If we had retained higher order time derivatives in $\langle \vec{d} \rangle(t)$, we would readily have seen that the factor in front of the n-th order derivatives $d^n \langle \vec{d} \rangle(t)/dt^n$, is proportional to r^{n-3}. For $n \geq 4$, the corresponding terms thus vanish in the limit $r \to 0$. If the point dipole emits a quasi-monochromatic field, the quantities $\langle \vec{d} \rangle(t)$ and $d^2 \langle \vec{d} \rangle(t)/dt^2$ will be essentially in phase, and therefore one would expect in such cases that the part of total near field which is in phase with $\langle \vec{d} \rangle(t)$ is dominated by the self-field part near the charge distribution itself, since this diverges as r^{-3} when $r \to 0$, whereas the retarded part goes as r^{-1} (compare the two terms for $q_0 r \ll 1$). To study quasi-monochromatic near field interactions in a localized charge density distribution it is thus expected to be sufficient to take

$$\vec{E}_{\mathrm{ED}}^{\mathrm{SF}}(\vec{r}, t) + \vec{E}_{\mathrm{T,ED}}^{\mathrm{R}}(\vec{r}, t | \text{NEAR FIELD ZONE}) \approx$$

$$-\frac{5}{9\epsilon_0}\delta(r)\langle \vec{d} \rangle(t) - \frac{1}{6\pi\epsilon_0 r^3}(\vec{\vec{U}} - 3\vec{e}_{\vec{r}}\vec{e}_{\vec{r}}) \cdot \langle \vec{d} \rangle(t) + \frac{1}{6\pi\epsilon_0 c_0^3}\frac{d^3}{dt^3}\langle \vec{d} \rangle(t), \quad (164)$$

in the framework of the electric point dipole model. If we abandon the point particle model the contact term and the remaining part of the self-field term (varying as r^{-3}) in Eq. (164) would be changed in a significant manner, whereas one would expect the out-of-phase retarded term to be essentially unchanged due to the fact that it is distance independent. I shall return to this point when discussing the radiation reaction phenomenon.

On the spatial confinement of light

The retarded (and transverse) electromagnetic field generated by a source distribution $\partial \vec{J}(\vec{r}', t')/\partial t'$ is given by the integral expression in Eq. (58). It spreads out from each source point with the speed of light, as described by the retarded transverse propagator in Eq. (57). Let us imagine that we follow the wavelet emitted from a given source point when the emission started. Is it then correct to claim that the electromagnetic field generated by the source initially, i.e. at the time immediately after the emission started (with zero amplitude), was localized to the region in space where $\partial \vec{J}(\vec{r}', t')/\partial t'$ is different from zero? If yes, one would in principle be able to compress electromagnetic fields to spatial regions as small as the electron sources themselves. For an atomic source the light thus could have been compressed to atomic dimensions. The answer to the aforementioned question is in general *no*, the reason being that we have not accounted for the self-field yet. The retarded field in a sense is allowed to be detached (radiated) from the source, whereas the self-field inevitably always stays attached to the source current density distribution. The spatial confinement of the total self-field is readily determined in the space-frequency representation. Thus, by means of Eqs. (80)–(83) one gets

$$\vec{E}^{\mathrm{SF}}(\vec{r};\omega) = \vec{E}^{\mathrm{SF}}_{\mathrm{T}}(\vec{r};\omega) + \vec{E}_{\mathrm{L}}(\vec{r};\omega) = \frac{1}{3i\epsilon_0\omega}\left(\vec{J}_{\mathrm{T}}(\vec{r};\omega) + 3\vec{J}_{\mathrm{L}}(\vec{r};\omega)\right), \quad (165)$$

or equivalently

$$\vec{E}^{\mathrm{SF}}(\vec{r};\omega) = \frac{1}{3i\epsilon_0\omega}\left(\vec{J}(\vec{r};\omega) + 2\vec{J}_{\mathrm{L}}(\vec{r};\omega)\right). \quad (166)$$

The self-field hence is not limited to regions where $\vec{J}(\vec{r};\omega)$ effectively is different from zero. If the electric dipole moment of the source current density is different from zero the self-field spreads over a region characterized by the r^{-3} tail, cf. Eqs. (134) and (135). If the dipole moment of $\vec{J}(\vec{r},t)$ is zero the compression becomes better. A spatial localization of the electromagnetic field just as strong as for the electrons of the source may be achieved in relation to a pure spin transition, where $\vec{J}(\vec{r},t) = \vec{J}^{\mathrm{SPIN}}(\vec{r},t) = \vec{J}^{\mathrm{SPIN}}_{\mathrm{T}}(\vec{r},t)$, and consequently $\vec{J}^{\mathrm{SPIN}}_{\mathrm{L}}(\vec{r},t) = 0$. The self-field, $\vec{E}^{\mathrm{SF}}_{\mathrm{SPIN}}(\vec{r},t)$, accompanying such a transition is given by

$$\vec{E}^{\mathrm{SF}}_{\mathrm{SPIN}}(\vec{r};\omega) = \frac{1}{3i\epsilon_0\omega}\vec{J}(\vec{r};\omega) \quad (167)$$

in the space-frequency domain, and therefore, the field and electrons are equally well localized in space. The general considerations above also make sense in relation to the retarded field, since equivalently one may consider the spatial region where the transverse part of the current density is different from zero as the source region for the retarded field, cf. the analysis presented in sections 2.2 and 2.4.

The velocity of light does not enter the self-field dynamics, and the correlations between the self-fields in different space-time points are not mediated by the propagation of light signals, but by the time development of the wave function of the source particles, i.e. in the nonrelativistic regime by the Schrödinger (or Pauli) equation. Let us for a moment consider the electrodynamics associated with a single transition between two bound states, denoted by indices 1 and 2. If we choose the involved energy eigenstate wave functions to be real, so that the two transition current densities satisfy $\vec{j}_{21}(\vec{r}) = \vec{j}^{\,*}_{12}(\vec{r}) = -\vec{j}_{12}(\vec{r})$, the source current density is given by

$$\vec{J}(\vec{r},t) = 2\mathrm{Im}\{c_1(t)c^*_2(t)\}\vec{j}_{12}(\vec{r}). \quad (168)$$

Without knowing the explicit time behaviour of c_1 and c_2 (if needed it may in a number of important cases be calculated from the well known two-level model of semiclassical dynamics), it appears from Eq. (168) that the current density is a product of a time-only dependent factor, $2\text{Im}\{c_1(t)c_2^*(t)\}$, and a factor, $\vec{j}_{12}(\vec{r})$, which depends only on the space coordinates. The "standing wave character" of the source current density tells us that the associated self-field

$$\vec{E}^{\text{SF}}(\vec{r}, t) = -\frac{2}{3\epsilon_0}\left(\vec{j}_{12}^{\text{T}}(\vec{r}) + 3\vec{j}_{12}^{\text{L}}(\vec{r})\right)\int_0^t \text{Im}\{c_1(t')c_2^*(t')\}dt', \tag{169}$$

where $\vec{j}_{12}^{\text{T}}(\vec{r})$ and $\vec{j}_{12}^{\text{L}}(\vec{r})$ are the transverse and longitudinal parts of $\vec{j}_{12}(\vec{r})$, develops in the same standing wave manner from the time $t = 0$, where the process was assumed to begin. The spatial confinement of the field in the initial phase of the transition thus is given by the extension of the transverse (or longitudinal) part of the transition current density $\vec{j}_{12}$. Later on the confinement becomes poorer due to the outwards propagation of the detached field. Based on the considerations above one can conclude that it is the nonlocal nature of quantum mechanics which sets the limit for the degree of spatial field confinement one can hope to achieve. If it is necessary to take into account more than two bound states, the conclusion is still the same, since the associated current density in Eq. (137) now is just a sum of terms (a term for each pair of states) where each term is a product of time and space dependent factors. In the many cases where the electric dipole moment of the current density does not vanish, the field confinement is determined by the transverse (longitudinal) part of the associated asymptotic current density, $\mathcal{J}^\infty(\vec{r}, t)$. Since the asymptotic current density has a standing wave character (see Eq. (127) or Eq. (130)) it is obvious once more that the nonlocal time-space quantum correlations set a limit for the degree of spatial field confinement.

Radiation reaction

We now consider the retarded field inside or at distances so close to the current density source distribution of the microscopic or mesoscopic object under study that retardation effects may be treated in a perturbative manner. Starting from the definition

$$\vec{J}(\vec{r}, t) = \frac{\partial \vec{P}(\vec{r}, t)}{\partial t} + \vec{\nabla} \times \vec{M}(\vec{r}, t), \tag{170}$$

in which the microscopic polarization, $\vec{P}(\vec{r}, t)$, and magnetization, $\vec{M}(\vec{r}, t)$, vectors have been introduced in an indirect way from a split of the microscopic current density, we use the fact that the transverse parts of $\vec{P}(\vec{r}, t)$ and $\vec{M}(\vec{r}, t)$ are not uniquely defined to make the choice $\vec{M}_{\text{T}}(\vec{r}, t) = \vec{0}$ for the transverse part of $\vec{M}(\vec{r}, t)$. The longitudinal part of $\vec{M}(\vec{r}, t)$ is irrelevant since only $\vec{\nabla} \times \vec{M}(\vec{r}, t)$

appears in Eq. (170). With this choice Eq. (170) becomes

$$\vec{J}(\vec{r},t) = \frac{\partial \vec{P}(\vec{r},t)}{\partial t}. \tag{171}$$

A detailed discussion of the freedom of choice allowed by the "weak" definition in Eq. (170) may be found in Ref. 16. By inserting Eq. (171) into Eq. (60), and using the Taylor series expansions

$$\frac{\partial}{\partial t}\vec{J}\left(\vec{r}',t-\frac{R}{c_0}\right) = \frac{\partial^2}{\partial t^2}\vec{P}(\vec{r}',t) - \frac{R}{c_0}\frac{\partial^3}{\partial t^3}\vec{P}(\vec{r}',t) + \cdots, \tag{172}$$

$$\vec{J}\left(\vec{r}',t-\frac{R}{c_0}\right) = \frac{\partial}{\partial t}\vec{P}(\vec{r}',t) - \frac{R}{c_0}\frac{\partial^2}{\partial t^2}\vec{P}(\vec{r}',t) + \frac{R^2}{2c_0^2}\frac{\partial^3}{\partial t^3}\vec{P}(\vec{r}',t) - \cdots, \tag{173}$$

and

$$\int_{t-\frac{R}{c_0}}^{t} \vec{J}(\vec{r}',t')\,dt' = \frac{R}{c_0}\frac{\partial}{\partial t}\vec{P}(\vec{r}',t) - \frac{R^2}{2c_0^2}\frac{\partial^2}{\partial t^2}\vec{P}(\vec{r}',t) + \frac{R^3}{6c_0^3}\frac{\partial^3}{\partial t^3}\vec{P}(\vec{r}',t) - \cdots, \tag{174}$$

one obtains

$$\vec{E}_{\mathrm{T}}^{\mathrm{R}}(\vec{r},t) = -\frac{\mu_0}{8\pi}\frac{\partial^2}{\partial t^2}\int_{-\infty}^{\infty}\frac{1}{R}\left(\vec{\vec{U}} + \vec{e}_{\vec{R}}\vec{e}_{\vec{R}}\right)\cdot\vec{P}(\vec{r}',t)d^3r'$$

$$+\frac{\mu_0}{6\pi c_0}\frac{d^3}{dt^3}\int_{-\infty}^{\infty}\vec{P}(\vec{r}',t)d^3r' + \cdots. \tag{175}$$

Neglecting fourth and higher order time derivatives of the polarization, the retarded field is given by

$$\vec{E}_{\mathrm{T}}^{\mathrm{R}}(\vec{r},t) =$$
$$-\frac{\mu_0}{8\pi}\frac{\partial}{\partial t}\int_{-\infty}^{\infty}\frac{1}{R}\left(\vec{\vec{U}} + \vec{e}_{\vec{R}}\vec{e}_{\vec{R}}\right)\cdot\vec{J}(\vec{r}',t)d^3r' + \frac{1}{6\pi\epsilon_0 c_0^3}\frac{d^2}{dt^2}\int_{-\infty}^{\infty}\vec{J}(\vec{r}',t)d^3r' \tag{176}$$

in terms of the microscopic current density. The second term in Eq. (176) is the well known[14] radiation reaction (RR) field, $\vec{E}_{\mathrm{RR}}(t)$, which in terms of the quantum mechanical mean value of the electric dipole moment operator can be written

$$\vec{E}_{\mathrm{RR}}(t) = \frac{1}{6\pi\epsilon_0 c_0^3}\frac{d^3}{dt^3}\langle\vec{d}\rangle(t). \tag{177}$$

For a well localized particle of charge $-e$ following a classical trajectory $\vec{\mathcal{R}}(t)$, the radiation reaction field is given by $\vec{E}_{\mathrm{RR}}(t) = -e(d^3\vec{\mathcal{R}}(t)/dt^3)/(6\pi\epsilon_0 c_0^3)$, cf.

e.g. Ref. 14. If the particle undergoes an oscillatory motion of frequency ω, $-e\vec{E}_{\mathrm{RR}}(t)$ describes a frictional force, which damps the motion of the particle. The first term in Eq. (176) for the retarded field usually is small in comparison with the self-field, and therefore the total field is approximately

$$\vec{E}(\vec{r},t) = \vec{E}^{\mathrm{SF}}(\vec{r},t) + \vec{E}_{\mathrm{RR}}(\vec{r},t) \tag{178}$$

near and inside the particle studied. In terms of the current density, Eq. (178) can be written in the form

$$\vec{E}(\vec{r};\omega) = \frac{1}{3i\epsilon_0\omega}\left(\vec{J}_{\mathrm{T}}(\vec{r};\omega) + 3\vec{J}_{\mathrm{L}}(\vec{r};\omega)\right) - \frac{\omega^2}{6\pi\epsilon_0 c_0^3}\int_{-\infty}^{\infty}\vec{J}(\vec{r}';\omega)d^3r', \tag{179}$$

in the space-frequency representation. The first term gives a contribution to the field which is $\pi/2$ out of phase with the current density, whereas the second term essentially is in phase with $\vec{J}$. In the neoclassical theory of quantum electrodynamics[17] the first term accounts for the Lamb shifts of our particle, and the second term is responsible for the spontaneous decay rates. For an electric dipole oscillator having a resonance frequency ω_0, the equation of motion thus is

$$m_*\left(\frac{d^2}{dt^2}\langle\vec{r}\rangle(t) + \omega_0^2\langle\vec{r}\rangle(t)\right) = -e\left(\vec{E}^{\mathrm{ext}}(t) + \vec{E}_{\mathrm{RR}}(t)\right), \tag{180}$$

where the divergent self-field contribution is assumed to have been eliminated by a renormalization of the particle mass, $m \to m_*$. By inserting Eq. (177) into Eq. (180), it is readily realized that the damping rate, τ^{-1}, of the oscillator, when placed in a harmonically varying external field of frequency ω, is

$$\tau^{-1} = \frac{e^2\omega^2}{6\pi m_*\epsilon_0 c_0^3}, \tag{181}$$

a well known[14] result. In the case of a three-dimensional particle confinement, the natural lifetime of the oscillator thus is proportional to ω^{-2}.

4.2 Two-dimensional spatial confinement

Self-field of a mesoscopic (quantum) wire

Let us assume that a linear quantum wire is placed along the y-axis of a Cartesian xyz-coordinate system, and that the particle (electron) confinement is complete in the x and z directions. If we now excite the wire by an external optical field having a (mean) wavelength much larger than the length of the wire, and limit ourselves to studies of electronic excitations for which the associated transition current densities are essentially independent of y, the field induced current density takes the form

$$\vec{J}(\vec{r}, t) = \vec{J}_0(t)\delta(x)\delta(z). \tag{182}$$

To determine the longitudinal and transverse parts of this current density let us first consider the nonlocal contribution $\vec{\mathcal{J}}(\vec{r}, t)$, given in Eq. (126). In the present case this becomes

$$\vec{\mathcal{J}}(\vec{r}, t) =$$
$$\frac{1}{4\pi}\left\{\mathrm{PV}\int_{-\infty}^{\infty}[x^2 + z^2 + (y - y')^2]^{-\frac{3}{2}}\left[\frac{3(\vec{r} - y'\vec{e}_y)(\vec{r} - y'\vec{e}_y)}{x^2 + z^2 + (y - y')^2} - \vec{\vec{U}}\right]dy'\right\} \cdot \vec{J}_0(t). \tag{183}$$

Using the formulas

$$\int \frac{d\alpha}{(\alpha^2 + a^2)^{3/2}} = \frac{1}{a^2}\frac{\alpha}{(\alpha^2 + a^2)^{1/2}}, \tag{184}$$

$$\int \frac{d\alpha}{(\alpha^2 + a^2)^{5/2}} = \frac{1}{a^4}\left[\frac{\alpha}{(\alpha^2 + a^2)^{1/2}} - \frac{1}{3}\frac{\alpha^3}{(\alpha^2 + a^2)^{3/2}}\right], \tag{185}$$

the integration over y' is easily carried out. Hence,

$$\vec{\mathcal{J}}(\vec{r}, t) = \frac{1}{2\pi(x^2 + z^2)^2}\begin{pmatrix} x^2 - z^2 & 0 & 2xz \\ 0 & 0 & 0 \\ 2xz & 0 & z^2 - x^2 \end{pmatrix} \cdot \vec{J}_0(t). \tag{186}$$

If the excitation of the wire begins at time $t = 0$, the induced self-field then is given by

$$\vec{E}^{\mathrm{SF}}(\vec{r}, t) = \left\{-\frac{5}{9\epsilon_0}\delta(x)\delta(z)\vec{\vec{U}} + \frac{1}{3\pi\epsilon_0(x^2 + z^2)^2}\left[(x^2 - z^2)(\vec{e}_x\vec{e}_x - \vec{e}_z\vec{e}_z)\right.\right.$$
$$\left.\left. + 2xz(\vec{e}_x\vec{e}_z + \vec{e}_z\vec{e}_x)\right]\right\} \cdot \int_0^t \vec{J}_0(t')dt' \tag{187}$$

in dyadic notation. Apart from a contact term, the component of the current density along the quantum wire does not contribute to the self-field, as one might have anticipated. In polar coordinates in the x-z-plane $[(x, z) = (r\cos\beta, r\sin\beta)]$ the self-field exhibits a r^{-2} tail. In comparison with the r^{-3} tail of the quantum dot case it appears that the field confinement is poorer for a system with two-dimensional electron confinement than for one possessing three-dimensional confinement.

Retarded field of a quantum wire and the associated propagator

For the sake of the following analysis it is convenient to study the quantum wire electrodynamics in the space-frequency representation. Before limiting ourselves to a study of the retarded field generated by the simple current density in Eq. (182), let us consider the effective propagator for a current density distribution (and an external field) exhibiting translational invariance along the y-axis. Inserting thus a distribution $\vec{J}(\vec{r}';\omega) = \vec{J}(x',z';\omega)$ into the integral equation (89) one obtains a local field

$$\vec{E}(x,z;\omega) = \vec{E}_{\mathrm{T}}^{\mathrm{ext}}(x,z;\omega) - i\mu_0\omega \int \overset{\leftrightarrow}{G}_0(x-x',z-z';\omega)\cdot\vec{J}(x',z';\omega)dx'dz',$$

$$(188)$$

where

$$\overset{\leftrightarrow}{G}_0(x-x',z-z';\omega) = \int_{-\infty}^{\infty} \overset{\leftrightarrow}{G}_0(\vec{r}-\vec{r}';\omega)dy' \qquad (189)$$

is the effective propagator. For $(X,Z) \equiv (x-x',z-z') \neq (0,0)$ the self-field part of the effective propagator may be obtained using the plane wave expansion of the longitudinal and transverse delta functions, viz.

$$\overset{\leftrightarrow}{\delta}_{\mathrm{L}}(\vec{R}) = -\overset{\leftrightarrow}{\delta}_{\mathrm{T}}(\vec{R}) = \frac{1}{(2\pi)^3}\int_{-\infty}^{\infty} \vec{e}_{\vec{q}}\vec{e}_{\vec{q}}e^{i\vec{q}\cdot\vec{R}}d^3q, \qquad \vec{R}\neq\vec{0}, \qquad (190)$$

where $\vec{e}_{\vec{q}} = \vec{q}/q$. The effective self-field propagators are therefore (neglecting the contact terms)

$$\overset{\leftrightarrow}{g}_{\mathrm{L}}(X,Z;\omega) = -3\overset{\leftrightarrow}{g}_{\mathrm{T}}(X,Z;\omega) =$$
$$\frac{1}{(2\pi q_0)^2}\int_{-\infty}^{\infty} \frac{(q_\|\vec{e}_x + q_\perp\vec{e}_z)(q_\|\vec{e}_x + q_\perp\vec{e}_z)}{q_\|^2 + q_\perp^2}e^{iq_\perp Z}e^{iq_\| X}dq_\perp dq_\|, \quad (191)$$

in a notation where $(q_x,q_z) \Rightarrow (q_\|,q_\perp)$, and $q_0 = \omega/c_0$, as before. In a complex $q_\perp$-plane the integrand has first order poles at $q_\perp = \pm iq_\|$. The one with the plus sign lies in the upper half-plane, the other one in the lower half-plane, provided $q_\| > 0$. If $q_\| < 0$, the situation is opposite. Taking the above mentioned facts into account a residue calculation leads to

$$\overset{\leftrightarrow}{g}_{\mathrm{L}}(X,Z;\omega) = -3\overset{\leftrightarrow}{g}_{\mathrm{T}}(X,Z;\omega) =$$
$$\frac{1}{4\pi q_0^2}\left[(\vec{e}_x + i\vec{e}_z)(\vec{e}_x + i\vec{e}_z)\int_0^{\infty} q_\| e^{-q_\| Z}e^{iq_\| X}dq_\|\right.$$
$$\left. -(\vec{e}_x - i\vec{e}_z)(\vec{e}_x - i\vec{e}_z)\int_{-\infty}^0 q_\| e^{q_\| Z}e^{iq_\| X}dq_\|\right], \quad (192)$$

for $(X, Z) \neq (0,0)$. By performing the elementary integrations over $q_\parallel$, we finally obtain

$$\vec{\vec{g}}_{\mathrm{L}}(X, Z; \omega) = -3\vec{\vec{g}}_{\mathrm{T}}(X, Z; \omega) =$$
$$\frac{1}{2\pi q_0^2} \frac{1}{(X^2 + Z^2)^2} \left[(Z^2 - X^2)(\vec{e}_z\vec{e}_z - \vec{e}_x\vec{e}_x) - 2XZ(\vec{e}_x\vec{e}_z + \vec{e}_z\vec{e}_x) \right]. \quad (193)$$

By means of the propagators in Eq. (193) and the current density in Eq. (182) it is a straightforward matter do derive the expression given in Eq. (187) for the non-contact part of the self-field. The retarded part of the effective propagator in Eq. (189), i.e.

$$\vec{\vec{D}}_0^{\mathrm{R}}(X, Z; \omega) = \int_{-\infty}^{\infty} \vec{\vec{D}}_0^{\mathrm{R}}(\vec{R}; \omega)dy', \quad (194)$$

may be obtained using the plane wave expansion for $\vec{\vec{D}}_0^{\mathrm{R}}(\vec{R}; \omega)$, namely

$$\vec{\vec{D}}_0^{\mathrm{R}}(\vec{R}; \omega) = \frac{1}{(2\pi)^3} \int_{-\infty}^{\infty} \frac{\vec{\vec{U}} - \vec{e}_{\vec{q}}\vec{e}_{\vec{q}}}{q_0^2 - q^2} e^{i\vec{q}\cdot\vec{R}} d^3q, \qquad \vec{R} \neq \vec{0}. \quad (195)$$

By combining Eqs. (194) and (195), one obtains after having carried out the integrations over y' and q_y the integral representation

$$\vec{\vec{D}}_0^{\mathrm{R}}(X, Z; \omega) = \frac{1}{(2\pi)^2} \int_{-\infty}^{\infty} \frac{\vec{\vec{U}} - \vec{e}_{\vec{\kappa}}\vec{e}_{\vec{\kappa}}}{q_0^2 - \kappa^2} e^{iq_\parallel X} e^{iq_\perp Z} dq_\perp dq_\parallel, \quad (196)$$

with $\kappa = (q_\parallel^2 + q_\perp^2)^{1/2}$ and $\vec{e}_{\vec{\kappa}} = (q_\parallel\vec{e}_x + q_\perp\vec{e}_z)/\kappa$. In the complex $q_\perp$-plane the integrand of Eq. (196) has two first order poles, located on the real axis at $q_\perp = \pm(q_0^2 - q_\parallel^2)^{1/2}$ for $q_\parallel < q_0$, and on the imaginary axis at $q_\perp = \pm i(q_\parallel^2 - q_0^2)^{1/2}$ for $q_\parallel > q_0$. Residue calculations therefore give

$$\vec{\vec{D}}_0^{\mathrm{R}}(X, Z; \omega) = \frac{1}{4\pi i} \int_{-\infty}^{\infty} \frac{1}{\kappa_\perp^0} \left(\vec{\vec{U}} - \vec{e}\vec{e}\right) e^{iq_\parallel X} e^{i\kappa_\perp^0 Z} dq_\parallel, \quad (197)$$

where

$$\kappa_\perp^0 = (q_0^2 - q_\parallel^2)^{1/2}\Theta(q_0 - |q_\parallel|) + i(q_\parallel^2 - q_0^2)^{1/2}\Theta(|q_\parallel| - q_0), \quad (198)$$

Θ being the Heaviside unit step function, and

$$\vec{e} = \frac{1}{q_0} \left(q_\parallel\vec{e}_x + \kappa_\perp^0\vec{e}_z\right). \quad (199)$$

The retarded and transverse field of a quantum wire carrying the current density $\vec{J}(\vec{r}; \omega) = \vec{J}_0(\omega)\delta(x')\delta(z')$ then becomes

$$\vec{E}_R^{\mathrm{T}}(x, z; \omega) = -i\mu_0\omega\vec{\vec{D}}_0^{\mathrm{R}}(x, z; \omega) \cdot \vec{J}_0(\omega). \quad (200)$$

The expression in Eq. (200) has quite recently been used to study the phase conjugation of the outgoing (probe) field from a quantum wire placed in front of a nonlinear mirror consisting of a single-level metallic quantum well.[18]

4.3 One-dimensional confinement

Non-retarded electrodynamics in quantum well systems

In metallic and semiconducting quantum wells the bound but mobile electrons are subjected to an essential spatial confinement in the direction perpendicular to the plane of the well, whereas the motion parallel to the well plane is delocalized, and described by bulk-like Bloch dynamics, possibly free-electron dynamics. The effective one-electron potential of the quantum well, $V(\vec{r})$, is often approximated by $V(\vec{r}) \approx V_{\parallel}(\vec{r}_{\parallel}) + V_{\perp}(z)$. This ansatz states that the potential is composed of a sum of a part $V_{\parallel}(\vec{r}_{\parallel})$ which depends on the position vector $\vec{r}_{\parallel} = (x, y, 0)$ in the plane of the well only, and a part $V_{\perp}(z)$ which is a function of the depth coordinate z in the well. In the aforementioned potential the eigenfunctions of the stationary states are products of atom-like functions of z and Bloch functions of $\vec{r}_{\parallel}$ (if the screened ionic potential is lattice periodic in the plane of the well). The spectrum of the eigenenergies which depend on the quantum number of the z-motion is discrete whereas the spectrum belonging to the x-y-motions exhibits two-dimensional band structure character.

Since the thickness of a quantum well in general is much less than the vacuum wavelength of light one often neglects electromagnetic retardation effects across the well. Instead of starting from the full set of microscopic Maxwell-Lorentz equations to obtain the tensorial propagator for the quantum well problem under study, and thereafter letting $c_0 \to \infty$ in order to determine the non-retarded propagator, an alternative route is usually followed in the literature.[19] In the alternative approach one begins by taking the non-retarded limit in the microscopic Maxwell-Lorentz equations, and thereafter one determines the associated propagator. The starting point for this maybe more familiar approach thus is the Poisson equation

$$\nabla^2 \phi(\vec{r}; \omega) = \frac{e}{\epsilon_0} N(\vec{r}; \omega), \qquad (201)$$

here written in the space-frequency domain. In the equation above $\phi(\vec{r}; \omega)$ is the induced scalar potential and $N(\vec{r}; \omega)$ is the associated induced (many-body) electron density. For quantum well systems it is natural to study the spatial $(\vec{r})$ electrodynamics in a mixed $(z; \vec{q}_{\parallel})$ representation. In this representation a quantity $\mathcal{K}(\vec{r}; \omega)$ is replaced by the Fourier integral transform

$$\mathcal{K}(z; \vec{q}_{\parallel}, \omega) = \int_{-\infty}^{\infty} \mathcal{K}(\vec{r}; \omega) e^{-i\vec{q}_{\parallel} \cdot \vec{r}_{\parallel}} d^2 r_{\parallel}, \tag{202}$$

where $\vec{r}_{\parallel} = (x, y, 0)$ and $\vec{q}_{\parallel} = (q_{\parallel,x}, q_{\parallel,y}, 0)$. In the mixed representation the Poisson equation reads

$$\left(\frac{d^2}{dz^2} - q_{\parallel}^2 \right) \phi(z; \vec{q}_{\parallel}) = \frac{e}{\epsilon_0} N(z; \vec{q}_{\parallel}), \tag{203}$$

leaving out here and in the following the reference to ω from the notation. With the boundary conditions $\phi(-\infty; \vec{q}_{\parallel}) = \phi(\infty; \vec{q}_{\parallel}) = 0$ the solution of Eq. (203) can be written in the compact form

$$\phi(z; \vec{q}_{\parallel}) = -\frac{e}{2\epsilon_0 q_{\parallel}} \int_{-\infty}^{\infty} e^{-q_{\parallel}|z-z'|} N(z'; \vec{q}_{\parallel}) dz'. \tag{204}$$

In relation to a propagator formalism for the electrodynamics of quantum wells the form in Eq. (204) is particularly adequate because it gives a spatially non-local connection between the prevailing electron density at the source plane z' and the induced scalar potential at the plane of observation, z. To determine the associated non-retarded electromagnetic propagator one just needs to eliminate $\phi(z; \vec{q}_{\parallel})$ and $N(z'; \vec{q}_{\parallel})$ in favour of the quasi-static (longitudinal) electric field $\vec{E}_{\mathrm{L}}(z; \vec{q}_{\parallel})$ and the local current density $\vec{J}(z'; \vec{q}_{\parallel})$, respectively. To do so we use the relation

$$\vec{E}_{\mathrm{L}}(z; \vec{q}_{\parallel}) = -\left(i\vec{q}_{\parallel} + \vec{e}_z \frac{d}{dz} \right) \phi(z; \vec{q}_{\parallel}), \tag{205}$$

originating in the quasi-static definition $\vec{E}_{\mathrm{L}}(\vec{r}) = -\vec{\nabla}\phi(\vec{r})$, and the equation of continuity, which in the $(z; \vec{q}_{\parallel})$ representation is

$$\left(i\vec{q}_{\parallel} + \vec{e}_z \frac{d}{dz} \right) \cdot \vec{J}(z; \vec{q}_{\parallel}) = -ie\omega N(z; \vec{q}_{\parallel}). \tag{206}$$

By combining Eqs. (204)–(206), it is possible to write the quasi-static relation between the local longitudinal electric field and the prevailing current density in usual nonlocal form

$$\vec{E}_{\mathrm{L}}(z; \vec{q}_{\parallel}) = -i\mu_0\omega \int_{-\infty}^{\infty} \left[\vec{\vec{D}}_0^{\mathrm{NR}}(z - z'; \vec{q}_{\parallel}) + \vec{\vec{C}}(z - z') \right] \cdot \vec{J}(z'; \vec{q}_{\parallel}) dz', \tag{207}$$

where

$$\overset{\leftrightarrow}{D}{}^{\mathrm{NR}}_0(z - z'; \vec{q}_{\parallel}) =$$
$$\left(\frac{c_0}{\omega}\right)^2 \frac{q_{\parallel}}{2} e^{-q_{\parallel}|z-z'|} \left[\vec{e}_{\vec{q}_{\parallel}}\vec{e}_{\vec{q}_{\parallel}} - \vec{e}_z\vec{e}_z + i\,\mathrm{sgn}\,(z - z')(\vec{e}_{\vec{q}_{\parallel}}\vec{e}_z + \vec{e}_z\vec{e}_{\vec{q}_{\parallel}})\right], \quad (208)$$

with $\vec{e}_{\vec{q}_{\parallel}} = \vec{q}_{\parallel}/q_{\parallel}$, is the nonlocal part of the quasi-static [non-retarded (NR)] propagator, and

$$\overset{\leftrightarrow}{C}(z - z') = \left(\frac{c_0}{\omega}\right)^2 \delta(z - z')\vec{e}_z\vec{e}_z \qquad (209)$$

is a quasi-static contact (C) term, relating $\vec{E}_{\mathrm{L}}$ and $\vec{J}_{\mathrm{L}}$ on the same plane, $z - z'$. Within the framework of the quasi-static theory, also named the scalar theory, Eq. (207) is the starting point for propagator studies of the electrodynamics of single quantum wells, or two or more wells placed in parallel. In the case where we have more than one well, but with no electronic overlap between wells, the current density $\vec{J}(z'; \vec{q}_{\parallel})$ splits into a sum of current densities for each well, and the propagator formalism in Eq. (207) enables us in an elegant fashion to examine the non-radiative coupling between the wells, in particular neighbouring wells. The above-mentioned electromagnetic coupling is of importance for instance when a quantum well system is excited by surface electromagnetic waves or evanescent fields as such. Also in cases where the (neighbouring) wells are coupled electronically, so that the current density $\vec{J}(z'; \vec{q}_{\parallel})$ cannot be separated into pieces belonging to "individual" wells the propagator formalism in Eq. (207) describes the quasi-static electrodynamics of a quantum lattice, or if the electronic coupling between wells is negligible, a multiple quantum well system.

If one transforms Eq. (207) back to the pure $\vec{r}$ representation, by means of the inverse transformation of the one in Eq. (202), it takes the form

$$\vec{E}_{\mathrm{L}}(\vec{r}) = -i\mu_0\omega \int \overset{\leftrightarrow}{g}{}^{\bullet}_{\mathrm{L}}(\vec{r} - \vec{r}') \cdot \vec{J}(\vec{r}')d^3r', \qquad (210)$$

cf. Eq. (80). The longitudinal propagator

$$\overset{\leftrightarrow}{g}{}^{\bullet}_{\mathrm{L}}(\vec{r} - \vec{r}') = \left(\frac{c_0}{\omega}\right)^2 \overset{\leftrightarrow}{\delta}{}^{\bullet}_{\mathrm{L}}(\vec{r} - \vec{r}') \qquad (211)$$

deviates from the longitudinal delta function $\overset{\leftrightarrow}{\delta}{}^{\bullet}_{\mathrm{L}}(\vec{r} - \vec{r}')$ by merely a constant, $(c_0/\omega)^2$. The longitudinal propagator and the longitudinal delta function have been marked by a dot $(\bullet)$ to emphasize that they in the present context are given in the so-called disk contraction form. For the longitudinal delta function dyad this means that

$$\overset{\leftrightarrow}{\delta}{}^{\bullet}_{\mathrm{L}}(\vec{r}-\vec{r}') = \frac{1}{(2\pi)^2} \left(\frac{\omega}{c_0}\right)^2 \int_{-\infty}^{\infty} \overset{\leftrightarrow}{D}{}^{\mathrm{NR}}_0(z-z';\vec{q}_{\parallel})e^{i\vec{q}_{\parallel}\cdot(\vec{r}_{\parallel}-\vec{r}_{\parallel}')}d^2q_{\parallel} + \delta(\vec{r}-\vec{r}')\vec{e}_z\vec{e}_z.$$

$$(212)$$

The longitudinal delta function in spherical contraction was given in Eq. (84).

Retarded quantum well propagator

For s-polarized quantum well electrodynamics electromagnetic retardation effects cannot be neglected if one seeks to elucidate for instance the role of local-field effects inside the wells. This is so because a neglect of retardation means that the s-polarized field inside the well is just the background field, i.e. the sum of the incoming field and the field reflected from the substrate upon which the quantum well possibly is deposited. After all, this may not be as bad as it seems because local-field corrections usually are small for s-polarized fields anyway. For p-polarized fields a neglect of retardation effects seems to be most problematic close to local-field resonances in systems with small electronic dampings.[12]

We have seen in the previous subsection that the electrodynamics of a quantum well system is adequately described in the mixed Fourier representation given in Eq. (202). To obtain the form of the retarded electromagnetic propagator $\overset{\leftrightarrow}{D}{}^{\mathrm{R}}_0(\vec{R};\omega)$ in this representation it is convenient to begin with the plane wave expansion given in Eq. (195). In comparison with the generic form in Eq. (202) it readily appears that the retarded and transverse propagator can be represented in the integral form

$$\overset{\leftrightarrow}{D}{}^{\mathrm{R}}_0(z-z';\vec{q}_{\parallel}) = \frac{1}{2\pi}\int_{-\infty}^{\infty}\frac{\overset{\leftrightarrow}{U}-\vec{e}_{\vec{q}}\vec{e}_{\vec{q}}}{q_0^2-q^2}e^{iq_{\perp}(z-z')}dq_{\perp}, \qquad z\neq z', \qquad (213)$$

in the mixed Fourier domain. In the complex $q_{\perp}$-plane, the integrand in Eq. (213) has first-order poles at $q_{\perp} = \pm q_{\perp}^0$, $q_{\perp}^0 = (q_0^2 - q_{\parallel}^2)^{1/2}$. For $q_{\parallel} < q_0$, these are located on the real axis, and for $q_{\parallel} > q_0$ on the imaginary axis. Contour integration hereafter gives the following explicit expression for $\overset{\leftrightarrow}{D}{}^{\mathrm{R}}_0(z-z';\vec{q}_{\parallel})$:

$$\overset{\leftrightarrow}{D}{}^{\mathrm{R}}_0(z-z';\vec{q}_{\parallel}) = \left(\frac{c_0}{\omega}\right)^2 \frac{e^{iq_{\perp}^0|z-z'|}}{2iq_{\perp}^0}\left[(q_{\perp}^0)^2\vec{e}_{\vec{q}_{\parallel}}\vec{e}_{\vec{q}_{\parallel}} + q_{\parallel}^2\vec{e}_z\vec{e}_z\right.$$

$$\left. -\left(\frac{\omega}{c_0}\right)^2\vec{e}_z\times\vec{e}_{\vec{q}_{\parallel}}\vec{e}_{\vec{q}_{\parallel}}\times\vec{e}_z - q_{\perp}^0 q_{\parallel}\mathrm{sgn}\,(z-z')\left(\vec{e}_{\vec{q}_{\parallel}}\vec{e}_z + \vec{e}_z\vec{e}_{\vec{q}_{\parallel}}\right)\right]. \quad (214)$$

Formally, the non-retarded propagator $\vec{\vec{D}}_0^{\mathrm{NR}}(z - z'; \vec{q}_\parallel)$ given in Eq. (208) can be obtained from Eq. (214) letting $c_0 \to \infty$. Noting that this implies that $q_\perp^0 \to iq_\parallel$, one immediately finds

$$\vec{\vec{D}}_0^{\mathrm{R}}(z - z'; \vec{q}_\parallel)\Big|_{c_0 \to \infty} = \vec{\vec{D}}_0^{\mathrm{NR}}(z - z'; \vec{q}_\parallel), \qquad (215)$$

as expected. Since the expression for $\vec{\vec{D}}_0^{\mathrm{NR}}(z - z'; \vec{q}_\parallel)$ was derived starting from a quasi-static approximation in which only the longitudinal part of the local field survives, it is possible to extract this expression from the longitudinal self-field propagator in Eq. (82). Thus, by means of the plane wave expansion for $\vec{\delta}_{\mathrm{L}}(\vec{r} - \vec{r}')$ given in Eq. (190), and valid for $\vec{r} - \vec{r}' \neq \vec{0}$, one realizes that

$$\vec{\vec{g}}_{\mathrm{L}}(z - z'; \vec{q}_\parallel) = \frac{1}{2\pi} \left(\frac{c_0}{\omega} \right)^2 \int_{-\infty}^{\infty} \vec{e}_{\vec{q}} \vec{e}_{\vec{q}} \, e^{iq_\perp(z-z')} dq_\perp, \qquad z \neq z'. \qquad (216)$$

The integral of this equation has first-order poles at $q_\perp = \pm iq_\parallel$, and upon a simple contour integration calculation one realizes that

$$\vec{\vec{g}}_{\mathrm{L}}(z - z'; \vec{q}_\parallel) = \vec{\vec{D}}_0^{\mathrm{R}}(z - z'; \vec{q}_\parallel)\Big|_{c_0 \to \infty}, \qquad z \neq z'. \qquad (217)$$

As long as we confine ourselves to studies of quantum wells excited by a single component in the $\vec{q}_\parallel$-spectrum the relevant propagator is, cf. Ref. 11

$$\vec{\vec{G}}_0(z - z'; \vec{q}_\parallel, \omega) = \vec{\vec{D}}_0^{\mathrm{R}}(z - z'; \vec{q}_\parallel, \omega) + \vec{\vec{C}}(z - z'; \omega), \qquad (218)$$

where $\vec{\vec{D}}_0^{\mathrm{R}}(z - z'; \vec{q}_\parallel, \omega)$ and $\vec{\vec{C}}(z - z'; \omega)$ are given by Eqs. (214) and (209), respectively. In Eq. (218) which is used so often in quantum well electrodynamics we have reinserted the angular frequency in the various arguments.

Transverse and longitudinal vector fields appropriate in quantum well electrodynamics

Let us consider the case where a quantum well system is excited by a single plane wave travelling parallel to the x-z-plane, and let us assume the field is p-polarized. If so, all relevant p-polarized vector fields, $\vec{V}(x, z) = (V_x, 0, V_z)$ have the generic form $\vec{V}(x, z) = \vec{V}(z; q_\parallel) \exp(iq_\parallel x)$ if the quantum well system exhibits translationally invariant optical properties in the well plane, here the x-y-plane. In a formal sense, the longitudinal, $\vec{V}_{\mathrm{L}}(x, z) = \vec{V}_{\mathrm{L}}(z; q_\parallel) \exp(iq_\parallel x)$, and transverse, $\vec{V}_{\mathrm{T}}(x, z) = \vec{V}_{\mathrm{T}}(z; q_\parallel) \exp(iq_\parallel x)$, parts of $\vec{V}$ $(= \vec{V}_{\mathrm{L}} + \vec{V}_{\mathrm{T}})$ can be obtained via the integral relations

$$\vec{V}_{\mathrm{L}}(z; q_\parallel) = \int_{-\infty}^{\infty} \vec{\delta}_{\mathrm{L}}(z - z'; q_\parallel) \cdot \vec{V}(z'; q_\parallel) dz', \qquad (219)$$

and

$$\vec{V}_{\mathrm{T}}(z; q_{\|}) = \int_{-\infty}^{\infty} \overset{\leftrightarrow}{\delta}_{\mathrm{T}}(z - z'; q_{\|}) \cdot \vec{V}(z'; q_{\|}) dz', \tag{220}$$

where $\overset{\leftrightarrow}{\delta}_{\mathrm{L}}(z - z'; q_{\|})$ and $\overset{\leftrightarrow}{\delta}_{\mathrm{T}}(z - z'; q_{\|})$ may be considered as the relevant longitudinal and transverse delta function dyads in the mixed Fourier representation. It appears from Eqs. (208) and (212) after setting $\vec{q}_{\|} = q_{\|}\vec{e}_x$ that the longitudinal delta function is given by

$$\begin{aligned}
\overset{\leftrightarrow}{\delta}_{\mathrm{L}}(z - z'; q_{\|}) &= \vec{e}_z\vec{e}_z\delta(z - z') \\
&+ \frac{q_{\|}}{2}e^{-q_{\|}|z-z'|}\left[\vec{e}_x\vec{e}_x - \vec{e}_z\vec{e}_z + i\mathrm{sgn}\,(z - z')(\vec{e}_x\vec{e}_z + \vec{e}_z\vec{e}_x)\right].
\end{aligned} \tag{221}$$

Utilizing that $\overset{\leftrightarrow}{\delta}_{\mathrm{T}}(z - z'; q_{\|}) + \overset{\leftrightarrow}{\delta}_{\mathrm{L}}(z - z'; q_{\|}) = (\vec{e}_x\vec{e}_x + \vec{e}_z\vec{e}_z)\delta(z - z')$ in the two-dimensional subspace, one then obtains the explicit expression

$$\begin{aligned}
\overset{\leftrightarrow}{\delta}_{\mathrm{T}}(z - z'; q_{\|}) &= \vec{e}_x\vec{e}_x\delta(z - z') \\
&- \frac{q_{\|}}{2}e^{-q_{\|}|z-z'|}\left[\vec{e}_x\vec{e}_x - \vec{e}_z\vec{e}_z + i\mathrm{sgn}\,(z - z')(\vec{e}_x\vec{e}_z + \vec{e}_z\vec{e}_x)\right]
\end{aligned} \tag{222}$$

for the transverse delta function. With Eqs. (221) and (222) inserted into Eqs. (219) and (220) it is easy to show that the L and T parts of the vector field $\vec{V}(x, z)$ indeed satisfy the conditions for being rotational-free ($\vec{\nabla} \times \vec{V}_{\mathrm{L}}(x, z) = \vec{0}$) and divergence-free ($\vec{\nabla} \cdot \vec{V}_{\mathrm{T}}(x, z) = \vec{0}$), namely

$$\left(iq_{\|}\vec{e}_z - \vec{e}_x\frac{d}{dz}\right) \cdot \vec{V}_{\mathrm{L}}(z; q_{\|}) = 0, \tag{223}$$

and

$$\left(iq_{\|}\vec{e}_x + \vec{e}_z\frac{d}{dz}\right) \cdot \vec{V}_{\mathrm{T}}(z; q_{\|}) = 0. \tag{224}$$

The result in Eq. (221) was obtained from the general expression given in Eq. (212) for the longitudinal delta function in disk contraction, $\overset{\leftrightarrow}{\delta}{}^{\bullet}_{\mathrm{L}}(\vec{r} - \vec{r}')$. To obtain the general form of the transverse delta function in disk contraction, named $\overset{\leftrightarrow}{\delta}{}^{\bullet}_{\mathrm{T}}(\vec{r} - \vec{r}')$, one merely uses the relation $\overset{\leftrightarrow}{\delta}{}^{\bullet}_{\mathrm{T}}(\vec{r} - \vec{r}') = \delta(\vec{r} - \vec{r}')\overset{\leftrightarrow}{U} - \overset{\leftrightarrow}{\delta}{}^{\bullet}_{\mathrm{L}}(\vec{r} - \vec{r}')$.

Appendix:
Comments about the derivation of the finite-c_0 propagator

A.1 Far field propagator, $\vec{\vec{D}}_{\rm F}(\vec{R}, \tau)$

To determine the explicit expression for this part of the propagator one has to carry out the integral

$$I_{\rm F} \equiv \int_0^\infty \left(e^{iqR} - e^{-iqR} \right) \left(e^{iqc_0\tau} - e^{-iqc_0\tau} \right) dq, \qquad (225)$$

cf. Eq. (39). By rewriting $I_{\rm F}$ in the form

$$I_{\rm F} = \int_0^\infty \left[e^{iq(R+c_0\tau)} + e^{-iq(R+c_0\tau)} \right] dq - \int_0^\infty \left[e^{iq(R-c_0\tau)} + e^{-iq(R-c_0\tau)} \right] dq, \qquad (226)$$

and using that

$$\int_0^\infty e^{-iq(R\pm c_0\tau)} dq = \int_{-\infty}^0 e^{iq(R\pm c_0\tau)} dq, \qquad (227)$$

the plane-wave expansion of the Dirac delta function, $\delta(x) = (2\pi)^{-1}\int_{-\infty}^\infty e^{iqx}\, dq$, enables one to obtain

$$I_{\rm F} = 2\pi \left(\delta(R + c_0\tau) - \delta(R - c_0\tau) \right), \qquad (228)$$

and consequently the expression for $\vec{\vec{D}}_{\rm F}(\vec{R}, \tau)$ given in Eqs. (42)–(44).

A.2 Middle field propagator, $\vec{\vec{D}}_{\rm M}(\vec{R}, \tau)$

To obtain the time-space representation of the middle field part of the finite-c_0 propagator, it appears from Eq. (40) that one has to carry out the integral

$$I_{\rm M} \equiv \int_0^\infty \left(e^{iqR} + e^{-iqR} \right) \left(e^{iqc_0\tau} - e^{-iqc_0\tau} \right) \frac{dq}{q}. \qquad (229)$$

Gathering terms containing respectively $R + c_0\tau$ and $R - c_0\tau$, and using Eq. (227) afterwards, one gets

$$I_{\rm M} = \int_{-\infty}^\infty e^{iq(R+c_0\tau)} \frac{dq}{q} - \int_{-\infty}^\infty e^{iq(R-c_0\tau)} \frac{dq}{q}. \qquad (230)$$

Residue calculations give next

$$I_{\rm M} = \pi i \left[{\rm sgn}\, (R + c_0\tau) - {\rm sgn}\, (R - c_0\tau) \right], \qquad (231)$$

and upon insertion of this result in Eq. (40) one obtains Eq. (45).

A.3 Near field propagator, $\vec{D}_{\mathrm{N}}(\vec{R}, \tau)$

It is seen from Eq. (41) that the explicit expression for this propagator is obtained once the integral

$$I_{\mathrm{N}} \equiv \int_0^\infty \left(e^{iqR} - e^{-iqR}\right)\left(e^{iqc_0\tau} - e^{-iqc_0\tau}\right)\frac{dq}{q^2} \tag{232}$$

has been calculated. Proceeding in the same manner as for the I_{F} and I_{M} integrals, I_{N} can be rewritten as follows:

$$I_{\mathrm{N}} = \int_{-\infty}^\infty e^{iq(R+c_0\tau)}\frac{dq}{q^2} - \int_{-\infty}^\infty e^{iq(R-c_0\tau)}\frac{dq}{q^2}. \tag{233}$$

The two Fourier integrals in this equation do not exist, separately, but their difference (I_{N}) does exist. A division into ingoing (function of $R + c_0\tau$) and outgoing (function of $R - c_0\tau$) contributions via the integrals in Eq. (233) therefore is not possible. However, if one adds and subtracts a (divergent) self-field term of the type

$$I_{\mathrm{N}}^{\mathrm{SF}} = \int_{-\infty}^\infty \frac{dq}{q^2}, \tag{234}$$

Eq. (233) may be written in a form

$$I_{\mathrm{N}} = \int_{-\infty}^\infty \left[e^{iq(R+c_0\tau)} - 1\right]\frac{dq}{q^2} - \int_{-\infty}^\infty \left[e^{iq(R-c_0\tau)} - 1\right]\frac{dq}{q^2}, \tag{235}$$

where the ingoing (first integral in Eq. (235)) and outgoing (second integral) do exist separately. By means of the result

$$\int_{-\infty}^\infty \left(e^{iq\alpha} - 1\right)\frac{dq}{q^2} = -4\int_0^\infty \sin^2\left(\frac{q\alpha}{2}\right)\frac{dq}{q^2} = -\pi|\alpha| \tag{236}$$

one finally obtains

$$I_{\mathrm{N}} = \pi\left[|R - c_0\tau| - |R + c_0\tau|\right]. \tag{237}$$

Inserting the above expression for I_{N} into Eq. (41), one gets Eq. (49).

References

[1]   B. L. Altshuler, P. A. Lee and R. A. Webb eds., *Mesoscopic Phenomena in Solids* (North-Holland, Amsterdam, 1991).

[2]   H. Fukuyama and T. Ando eds., *Transport Phenomena in Mesoscopic Systems* (Springer-Verlag, Berlin, 1992).

[3]   P. J. Feibelman, *Prog. Surf. Sci.* **12**, 287 (1982).

[4]   V. M. Agranovich and D. L. Mills eds., *Surface Polaritons, Electromagnetic Waves at Surfaces and Interfaces* (North-Holland, Amsterdam, 1982).

[5]   V. M. Agranovich and R. Loudon eds., *Surface Excitations* (North-Holland, Amsterdam, 1984).

[6]   G. L. Richmond, J. M. Robinson and V. L. Shannon, *Progr. Surf. Sci.* **28**, 1 (1988).

[7]   O. Keller ed., *Notions and Perspectives of Nonlinear Optics* (World Scientific, Singapore, 1996).

[8]   D. W. Pohl and D. Courjon eds., *Near Field Optics* (Kluwer, Dordrecht, 1993).

[9]   M. Nieto-Vesperinas and N. Garcia eds., *Optics at the Nanometer Scale* (Kluwer, Dordrecht, 1996).

[10] L. V. Keldysh, D. A. Kirzhnitz and A. A. Maradudin eds., *The Dielectric Function of Condensed Systems* (North-Holland, Amsterdam, 1989).

[11] O. Keller, *Phys. Rep.* **268**, 85 (1996).

[12] O. Keller in *Progress in Optics, Vol. XXXVII*, ed. E. Wolf (North-Holland, Amsterdam, 1997), p. 259.

[13] M. Born and E. Wolf, *Principles of Optics, 6th edn.* (Pergamon, Oxford, 1980).

[14] C. Cohen-Tannoudji, J. Dupont-Roc, G. Grynberg, *Photons and Atoms, Introduction to Quantum Electrodynamics* (Wiley, New York, 1989).

[15] O. Keller, *J. Nonl. Opt. Phys. and Mat.* **5**, 109 (1996).

[16] O. Keller in *Quantum Optics and the Spectroscopy of Solids*, eds. T. Hakioğlu and A. S. Shumovsky (Kluwer, Dordrecht, 1997), p. 1.

[17] E. T. Jaynes in *Coherence and Quantum Optics III*, eds. L. Mandel and E. Wolf (Plenum, New York, 1973), p. 35.

[18] T. Andersen and O. Keller, *Phys. Rev. B.* (submitted).

[19] N. Raj and D. R. Tilley in Ref. 10, p. 459.

NANOSCALE MATERIALS: CONCEPTUAL AND COMPUTATIONAL CHALLENGES

Muṣhti V. RAMAKRISHNA

Bell Laboratories, Murray Hill, NJ 07974-0636, USA

This chapter examines the implications of Moore's law to the electronic and optical properties of nanoscale materials. The focus is on what is known today and what more needs to be known as Moore's law confronts quantum mechanics, the ultimate law that holds dominant when the size of the material shrinks to dimensions smaller than the natural wavelength of the electron. The nature of the problem and the conceptual and computational challenges are discussed in the context of their importance to emerging new technologies.

In 1965, Gordon Moore, co-founder of Intel Corporation, the largest manufacturer of semiconductor microprocessors, formulated that the complexity of the chips will double, and the size will shrink by half, every twelve months. This empirical law, published in an electronics trade magazine, was based on a simple plot of the number of transistors on an Intel microprocessor chip vs the year the chip was produced (Fig. 1). Subsequently, in 1970s the law was revised to suggest that the transistor density doubled every eighteen months. Currently, the belief is that this doubling occurs every nine months or so. Despite its empirical nature and imprecision in its formulation, Moore's law has held remarkably steady, being the fuel powering the explosive growth of the entire semiconductor industry that has ushered in the computer and information revolution we witness today (Fig. 1).

As we face today the reality of ever shrinking devices and the limits of Moore's Law, it behooves us to examine the Law itself and its ramifications to the future of new materials that will have quantum mechanics playing a large role in the determination of their properties. In this context, this chapter is aimed at analyzing what is known today and what more needs to be known. Both conceptual and computational challenges facing us are presented. The discussion is intentionally qualitative and is aimed at portraying the forest rather than enumerating trees.

Since 1965, the semiconductor industry has relentlessly pushed the technology to pack more and more transistors onto chips and reduce their size accordingly. It is common today to find millions of transistors on a single chip and the goal is to produce a complete computer system (cpu, memory, and peripherals) on a single chip. Due to ever growing investments in device technology and economies of scale achieved by mass consumption of semiconductor chips, some people believe that the doubling period is now effectively nine months instead of eighteen. Such an acceleration raises hopes that the goal of achieving "system on a chip" is very well achievable in the near future if some obstacles are overcome. One obstacle is quantum mechanics: whether these shrinking devices will inevitably run into the fundamental laws of quantum mechanics. This thought itself has led some people

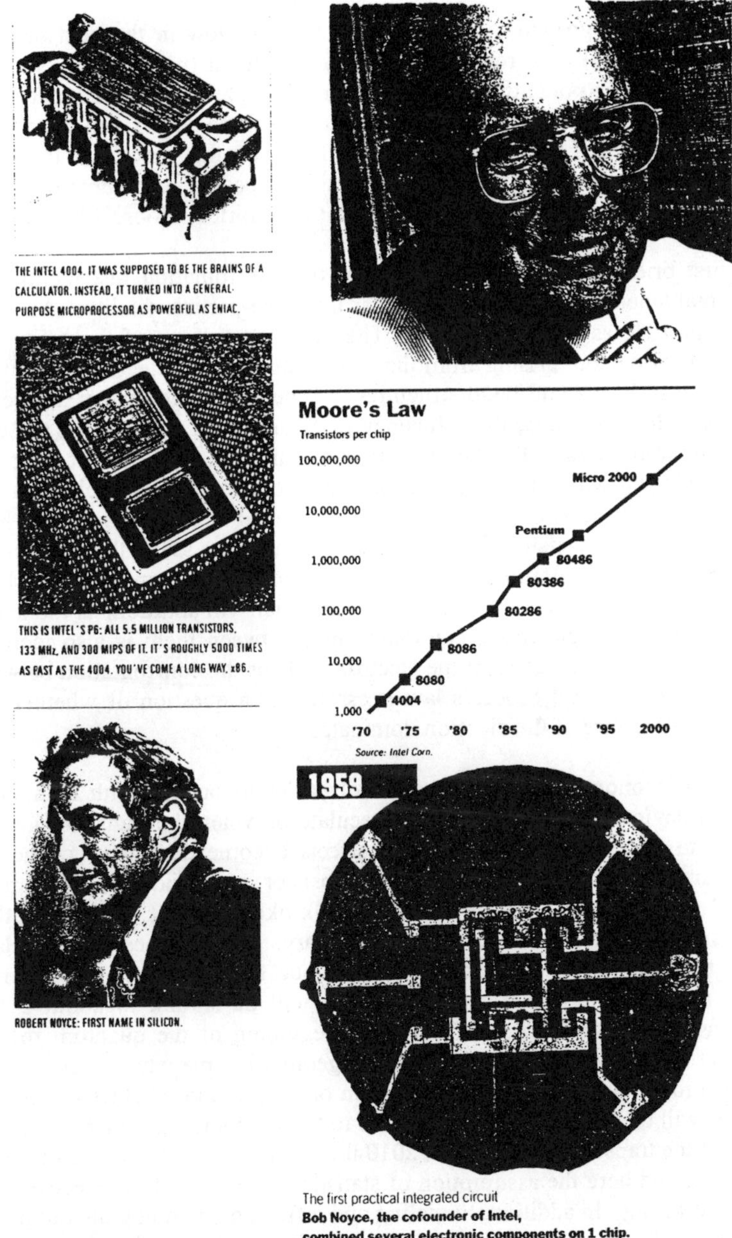

Figure 1: A simplified pictorial gallery of the evolution of the integrated circuit chip demonstrating the power of the Moore's law. The Moore's law plot shows the number of transistors on an Intel chip as a function of the year of its production. Gordon E. Moore, a chemical physicist and the author of Moore's law, is pictured at top right (source: Intel Corporation, USA).

to forecast that the growth in transistor density will slow in the next decade, and silicon may eventually be replaced by another material (perhaps GaAs).   Others contend that the exponential growth of the cost of building fabrication plants (Rock's law), costing billions of dollars to produce these chips, and steadily falling prices of these chips, will essentially put a break on the growth of the transistor density, and we may never encounter the expected confrontation between fundamental laws of quantum mechanics and the empirical Moore's law.

Let us first briefly reflect on the nature of the confrontation just mentioned. A fundamental tenet of quantum mechanics is the wave-particle duality. An electron is both a particle as well as a wave. The very functioning of the transistor is a quantum phenomenon, arising from the wave nature of the electrons giving rise to discrete energy levels and band structures. When these transistors are combined together into logic circuits, the efficiency of these ciruits depends on the distance between the transistors. Treating electrons as particles, we find that logic circuits are more efficient when transistors are packed close together and electrons have a smaller distance to travel. This is the fundamental explanation for the Moore's law. However, as transistor sizes decrease to nanometer scale, the wave nature of the electrons starts to dominate and the properties of the transistor will begin to deviate from that of the corresponding bulk material. In addition, as the density of transistors on a chip increases, communication between them is also increasingly dominated by the wave nature of the electrons. Thus the confrontation between quantum mechanics and Moore's law is essentially a question of whether particle nature or wave nature of the electron dominates.

Ignoring the economic factors and other obstacles associated with manufacturing ultra small devices, it is interesting to speculate on when the transistor geometries are likely to be so small that quantum effects become dominant.   The current manufacturing technology can produce transistor gate widths approximately 1 micron (1000 nm) wide.   We may expect bulk like transistor behavior in this size regime, and there is no evidence to the contrary.   If one uses Moore's law and assumes that the device geometries will continue to shrink every twelve months, then we may see that the transistor features will be shrunk to about 2 nm (20 Angstroms) by about 2007.   This is the beginning of the quantum regime for silicon [1].   Devices smaller than these geometries may not exhibit bulk-like transistor properties completely, and it is an open question whether the changes in properties will be favorable or unfavorable to the functioning, switching speed, and stability of the transistors. By about 2010 the geometries will approach the 0.2 nm-size regime, and here the assumption of statistical behavior of the electrons will be completely wrong. In addition, tunneling and diffusion, including quantum diffusion, of the dopants will blur the junction region sufficiently enough to significantly modify the properties of the transistor.   If one is ever able to achieve such ultrasmall devices, then we will enter the regime of truly quantum devices, whose composition may be completely different from the transistors as we know them today, and there have already been speculations on what the building blocks of such a quantum computer will be.

What do we know about the atomic structure of 1-2 nm-wide silicon? Nothing. While such a slice of silicon is believed to possess bulk-like diamond structure, there are no credible calculations to support such a conjecture. What do we know about the atomic structure of silicon approximately 0.2-0.5 nm wide. The rigorous quantum mechanical calculations of Krishnan Raghavachari and others have established that sub-nanoscale silicon has close-packed atomic structure, unlike the open diamond structure found in bulk silicon [2]. Hence, somewhere between 0.5-1 nm-size regimes, silicon undergoes a structural transition from metal-like close packed structure to a non-metallic diamond-like bulk structure. The exact nature of this transition and evolution pattern is unknown.   However, recent density functional theory based calculations and theoretical models imply that the structural evolution is complicated.   At least in the initial stages the growth occurs through the addition of atoms to the faces of a trigonal prism (Fig. 2).   Such a capped trigonal prism seems to be the common motif in the initial stages of the structural evolution [3].   Subsequently, it is likely, but by no means proven, that the geometries with a fullerene type surface structure and bulk-like interior will dominate (Fig. 3). As the size of silicon further increases, the interior smoothly transforms to a bulk diamond structure and the surface maps over to the 7x7 reconstruction of Si(111) [3].   While this general picture seems reasonable and plausible, there are not enough computational or experimental data in its support [4].

The main reason for the sparcity of structural data on these nanoscale materials is that calculations are extremely difficult.   The computational power needed to investigate the structures of these aggregates grows exponentially, with a large exponent, as the number of atoms in the aggregate increases.   Model calculations, using both empirical potentials and semi-empirical tight-binding Hamiltonians, have been carried out. But the results of these calculations have been wrong for the most part [5]. Atomic interactions in semiconductor materials, except in bulk limit, have so far defied reduction into simple model potentials that could be used to simulate the real behavior [5].   Since exact quantum calculations are nearly impossible in the foreseable future, we need a real breakthrough in our conceptual thinking.   So far our conceptual thinking has been based on pair-wise potentials that have proven so useful in modelling materials with closed shell electronic structures, such as rare-gas fluids and clusters. This has turned out to be a wrong quantum mechanics: whether these shrinking devices will inevitably run into the fundamental laws of quantum mechanics. This thought itself has led some people approach to take for semiconductor aggregates.   Subsequently, tight-binding Hamiltonians, suitably fit to bulk silicon and known geometries of small clusters, have been used [6].   While similar efforts have been extremely fruitful in organic chemistry, it has also turned out to be a wrong approach for semiconductor

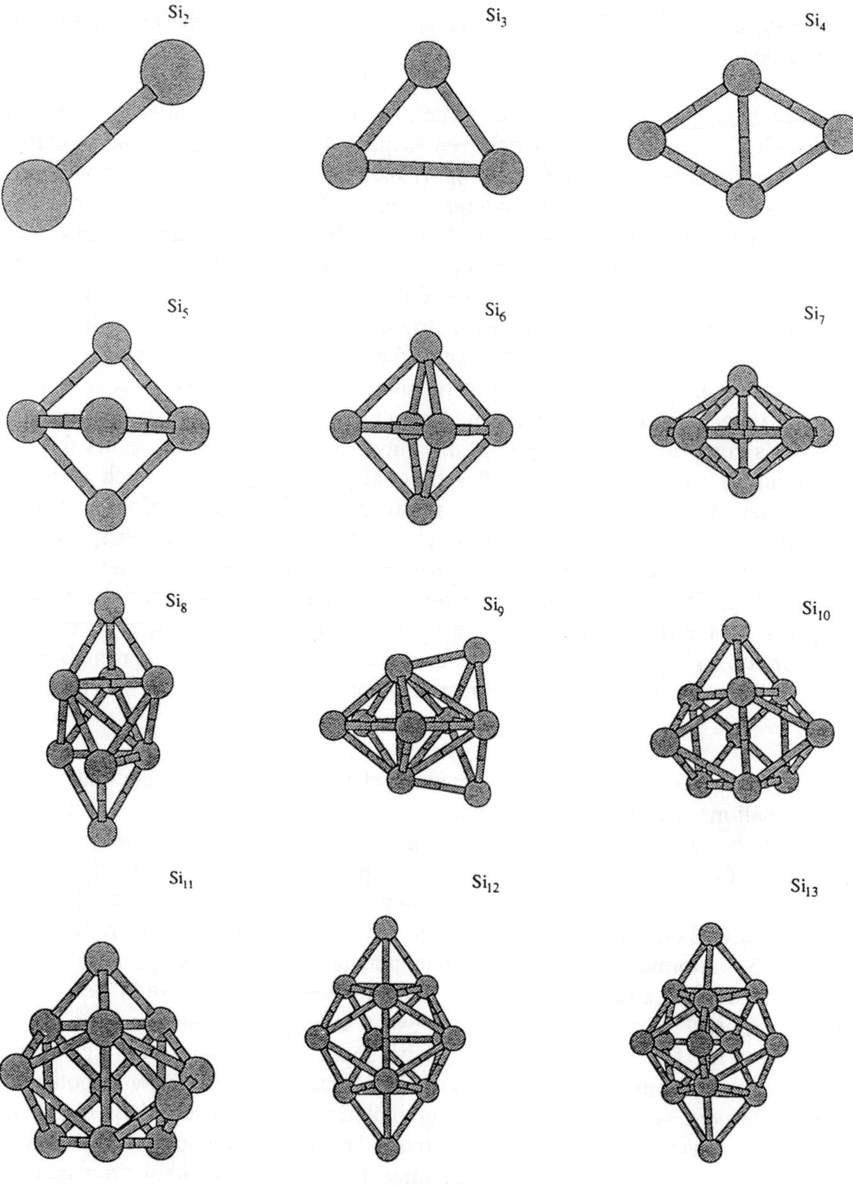

Figure 2: Lowest energy structures of small silicon clusters calculated using density functional theory within the local density approximation. We know the structures of these small clusters as well as the structure of bulk silicon. But our knowledge of the evolution of the structural pattern between these two limits is mostly limited to educated guesses and unproven theoretical models.

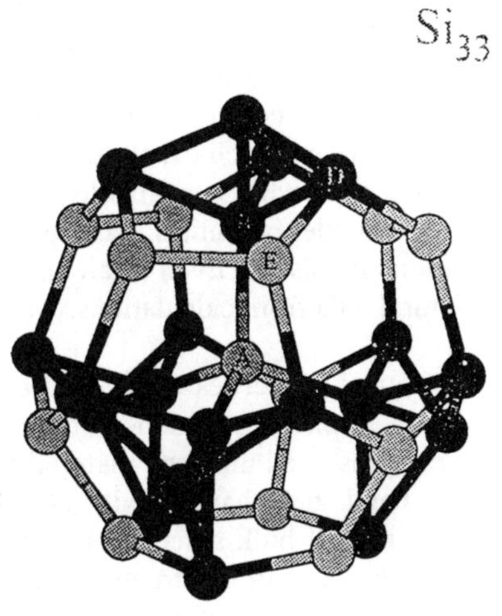

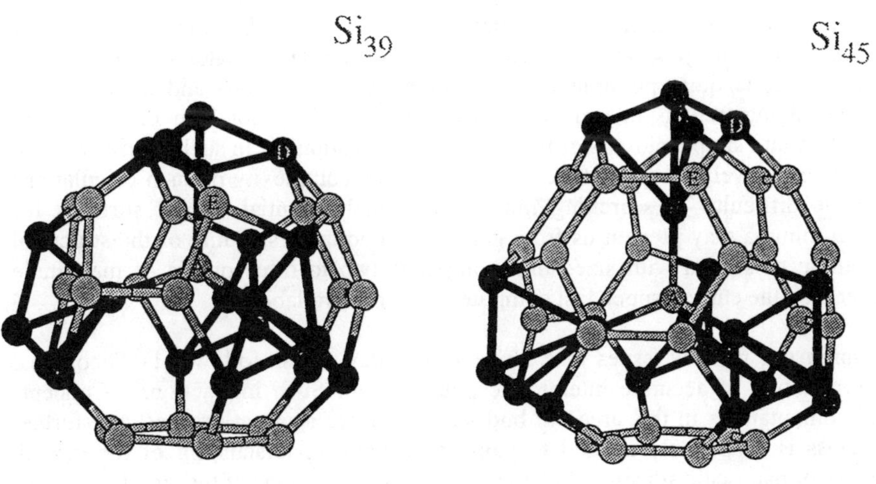

Figure 3: Structures of intermediate sized silicon clusters based on a stuffed fullerene model. The surface is similar to the 7x7 reconstruction of the Si(111) bulk surface, while the interior is bulk-like [3].

materials.   Having taken two wrong turns, our options are limited now.   This has led to the belief that only fully quantum calculations can be trusted and that other approximate methodologies are doomed to failure from the outset.   While such a puritinical approach may be correct, it is computationally infeasible at present.   The computational power needed to carry out such exact quantum calculations is two to three orders of magnitude greater than what is available today.   Hence, there is a tremendous need to produce new model potentials that are as accurate as those for rare gas materials and for carbon chemistry.   Even small errors in the potentials will yield wrong and useless structural data from calculations.

What do we know about the optical properties of nanoscale silicon?   Reasonably reliable calculations, based on an empirical pseudopotential model, have shown that the bulk band gap is effectively preserved until the material reaches less than 4 nm in size (Fig. 4).   While the band gap changes with size between 2-4 nm, the spectroscopy is fundamentally that of bulk silicon [1].   At sizes smaller than 0.5 nm, the material transforms to non-crystalline structures, and investigations of the optical properties of these sub-nanoscale silicon limited.   Further investigations into optical properties are hindered by the lack of accurate knowledge of the atomic structures of these aggregates.   This in turn depends on the knowledge of accurate interatomic potentials.

Thus we come full circle and see that the absence of truly reliable Si-Si interatomic potentials is the major computational obstacle to achieving the goal of understanding the properties of semiconducting nanoscale materials, specifically silicon.   It is quite possible that the complicated direction- and environment-dependent interactions in semiconductor aggregates may not lend themselves to reduction into simple closed-form mathematical equations.   In such a case we have to rely exclusively on quantum calculations.   The complexity of such calculations and the difficulty of searching multi-dimensional potential energy surfaces for global minima may prevent us from gaining a good understanding of the structural evolution of silicon with size, until computers two to three orders of magnitude faster than the current super-fast computers become available.

In summary, the key issues that confront us today are as follows: 1) Theoretical determination of accurate interatomic potentials for Si-Si interaction.   Concepts and computations in this area are both at a primitive stage at present, and further progress is desperately needed to gain an accurate understandnig of the growth pattern of nanoscale silicon. 2) Computational investigations of the structures of Si clusters as a function of cluster size using interactomic potentials as well as accurate first-principles quantum mechanical calculations. 3) Computation of the optical properties of the Si clusters using the structures determined in 2) above.

Answers to these questions are needed to solve the technological problems that arise as Moore's Law brings silicon to the world of the quantum-size regime.   The questions raised and solutions sought are not themselves technological but fall squarely within the realm of chemical physics.

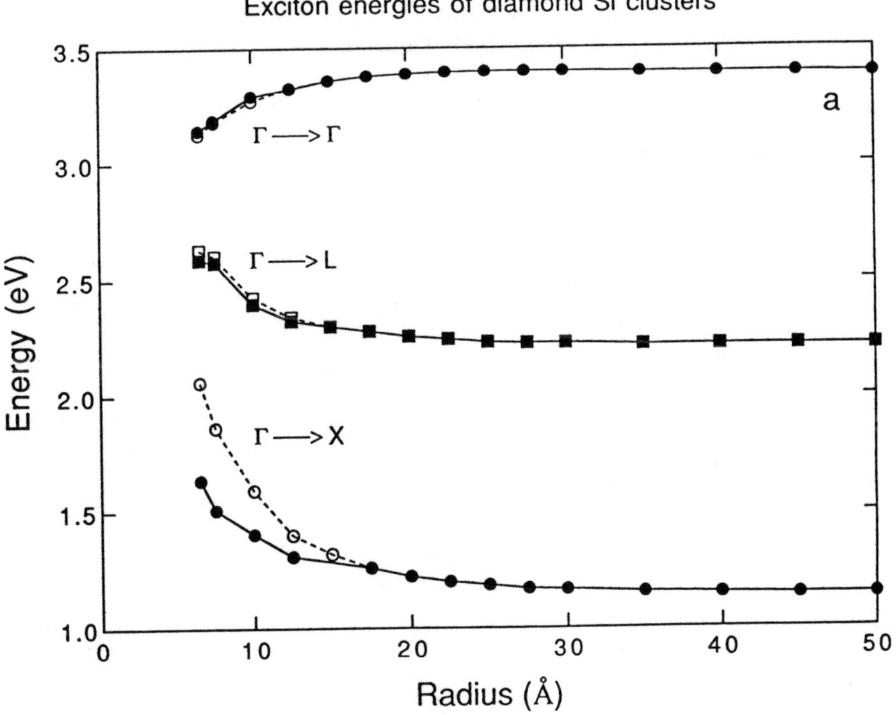

Figure 4: Excitation energies of nanoscale Si as a function of size. The open symbols are obtained using a bulk lattice constant, while the filled symbols are obtained by assuming small lattice contractions.

References

1. M. V. Ramakrishna and R. A. Friesner, *Phys. Rev. Lett.* **67**, 629 (1991); *J. Chem. Phys.* **95**, 8309 (1991); *J. Chem. Phys.* **96**, 873 (1992); *Isr. J. Chem.* **33**, 3 (1993).
2. K. Raghavachari and V. Logovinsky, *Phys. Rev. Lett.* **55**, 2853 (1985); K. Raghavachari, *J. Chem. Phys.* **83**, 3520 (1985); **84**, 5672 (1986); K. Raghavachari and C. M. Rohlfing, *Chem. Phys. Lett.* **143**, 428 (1988); *J. Chem. Phys.* **89**, 2219 (1988).
3. J. Pan and M. V. Ramakrishna, *Phys. Rev. B* **50**, 15431 (1994); M. V. Ramakrishna and J. Pan, *J. Chem. Phys.* **101**, 8108 (1994); M. V. Ramakrishna and A. Bahel, *J. Chem. Phys.* **104**, 9833 (1996).
4. B. L. Swift, D. A. Jelski, D. S. Higgs, T. T. Rantala, and T. F. George, *Phys. Rev. Lett.* **66**, 2686 (1991); E. Kaxiras, *Phys. Rev. Lett.* **64**, 551 (1990).
5. J. R. Chelikowsky, J. C. Phillips, M. Kamal, and M. Strauss, *Phys. Rev. Lett.* **62**, 292 (1989).
6. M. Menon and K. R. Subbaswamy, *Phys. Rev. B* **47**, 12754 (1993).

INDEX

ab initio   7, 8, 19, 24, 30, 31, 32, 34, 34, 38, 46, 50, 54, 60, 63, 65, 66, 334, 337, 338, 349

absorption   259

absorption cross section of fractal clusters   210

absorption parameter   213

absorption spectrum of fractal clusters   235

accommodation coefficient   296, 297

adiabatic trajectory   41, 54

adsorption   4, 6, 24, 335, 336, 338, 339, 342, 348, 349

advanced electromagnetic propagator   378, 384

advanced light cone   386, 392

amorphous carbon   62, 63

amorphous silicon   56, 57, 58, 59, 60, 61

amorphous silicon hydride   2

amplification   367, 373

amplitude correlation function   262

analytical predictions   251

Anderson model   337

angle of incidence   360, 361, 366, 367, 369, 370, 371, 373

anisotropy and scaling laws for gas-surface scattering   298, 302, 305

aperiodic crystal   273, 274, 281, 282, 292

attached electromagnetic field   386

attached field   386

$AuTe_2$ *see* calaverite

autocatalysis   117

average structure   277

backscattering   307, 311

band   32, 33, 34

band gaps   2

band structure   442

bandtailing   32

bandwidth   373

basic structure   286, 287, 288, 291, 292

basis function   20

battery   174, 175, 176, 184, 185, 189

Bethe lattice   32

binary approximation   260

blue bronze   288, 289

Boltzmann distribution   3

bond distortion   153

bond strength   284, 289, 292

bond valence method   4, 273, 274, 276–280, 282, 283, 285-292

bonding   41

Born approximation   218

Born expansion   219

boron hydrides   118

boundary conditions   336, 338

boundary conditions for Maxwell model of scattering   318, 326, 327

Brillouin scattering   351

Brillouin zone   1, 15, 18, 42

buckminsterfullerene   *see* fullerenes, C_{60}

bulk laser induced drift (LID)   328, 329

cadmium sulfide (CdS)   19, 23

calaverite ($AuTe_2$)   286, 287

canonical momentum operator   412

Car-Parrinello (CP)   2, 36, 37

carbon   43

carbon clusters   113

Cartesian tensor   353

catalyzed reaction   6

causal response   390

causality   352, 363

channel branching   320, 321

chaos   244

charge density wave (CDW)   4, 284, 285, 288, 289

charge transfer   152, 157

chirality contributions   112, 115

cleavage surface   7, 18

cluster calculations   4, 335, 337, 345

cluster calculations: dipped adcluster method   339

cluster calculations: moderately-large-embedded cluster method   338

cluster-cluster aggregates (CCA)   211, 232, 235, 252, 260

clusters   3, 38, 43

coadsorption   342, 349

collision   3

colloidal-metal clusters   272

combinatoric   3

commensurate   290

complex amplitude   355

compound semiconductor   6

condensation kernel   170

configuration interaction   9, 337

conjugate gradient method   230, 231

constitutive equation   376
coordination   273
coordination polyhedra   273, 274
coordination shell   275, 278, 283, 286
core-core repulsion   342
correlation function   54, 59
Coulomb boundary conditions   189, 191
coupled dipole equation   212, 225
coupled dipoles   212, 213, 222, 226
coupled multipoles   215, 222, 225, 226
covalent   44, 274, 283
crystal growth   40, 41
crystalline   30, 32, 37, 38, 42
cutoff   32

dangling bond   33, 63
defects   2, 58, 64, 65
delta function   353
density-functional (DF) theory   4, 6, 9, 23, 34, 35, 36, 53, 336
density matrix operator   412
density of states   39, 338
diamond   43, 44
dielectric   357, 361, 372
dielectric constant   250
dielectric material   345
dielectric constant   351, 353, 355, 356, 367, 370, 371, 373
diffusion   46, 48
diffusivity   46
dipole-dipole interaction   247
dipped adcluster method   339
dislocations   65
disordered cluster   259
dispersion   356
dispersion relations   357, 358, 361, 363, 370
distribution   256
distribution function   257
doppler effect   4
doubling period   440
drift and diffusion of rarefied gas   311, 312
dynamic form factor   262
dynamic percolation   174, 191, 192, 193, 194, 195
dynamical equation   157

effective field   29
effective medium theory   341, 342
eigenmodes   244
eigenvalue   250

eigenvalues of the coupled dipole equation   216, 218, 220
eigenvector   164
eigenvectors of the coupled dipole equation   216, 217, 218
electric fields   352, 353, 354, 373
electric-dipole current density   421
electrolyte   174, 175, 189, 191, 195, 196, 198, 201
electron affinity   166
electron density   341
electron interaction   167
electron transfer   339
electron-hole pair   168
electronic structure   34, 45, 53
electronic temperature   52
electron-phonon interaction   154
Eley-Rideal-type reaction   334
embedded atom   29
embedded-atom method   341
empirical functionals   28
empirical tight binding *see* tight binding
energy eigenvalues   34, 35
energy of interaction   346
energy of sublimation   344
enhancement coefficient   258, 259
enhancement factor of the local field intensity   239
entropy   40
enumerations   116, 118, 120, 121
exchange correlation   12, 35
excited states   34
external source   410
extinction cross section of fractal clusters   226, 232

face-dual network method   79–80
far-field propagator   389
Fermi energy   17
Fermi level   39
Fermi wavevector   284
fluctuations   244
Fock operator   8
force matching   31
formation energy   53
Fourier transform   352, 353
four-wave mixing   239, 351, 352, 354, 362, 367
fractal   3, 4
fractal clusters   210, 211, 215, 221
fractal dimension   210, 211

fragmentation   2
free energy   40
Fresnel coefficients   351, 363, 364, 365, 366, 367, 368, 369, 370, 371, 372
fullerenes   2, 3, 152
fullerenes, C_{60} 112, 113, 115, 116, 119, 120
fullerenes, C_{60} under high pressure   104–107
fullerenes, C_{78} 112, 113, 115, 118, 120, 126, 127
fullerenes, C_{80} 112, 113, 115, 120, 127, 128, 129
fullerenes, C_{82} 112, 113, 115, 120, 130
fullerenes, C_{84} 2, 81-83, 112, 113, 115, 118, 120, 131
fullerenes, C_{86} 112, 113, 115, 120
fullerenes, C_{88} 112, 113, 115, 120
fullerenes, C_{90} 112, 113, 115, 118, 120
fullerenes, collisions between   99–103
fullerenes, decorated   120
fullerenes, energies of   84–86, 92–94
fullerenes, formation of   95–96
fullerenes, higher   112, 113, 115
fullerenes, isomers of   115, 121
fullerenes, optical absorption of C_{60} solid   105–106
fullerenes, stability of   93–94
fullerenes, structures of   81–91
fullerenes, thermal disintegration of   97–98

GaAs   54, 55, 56
gap   30, 32, 34, 37, 39
gas sensor   6
gas-surface scattering   298, 302
gauge invariance   412, 413
Gaussian package   121
germanium   43, 52
giant fluctuations   244, 253, 261
Goldberg polyhedra   116
gradient   37
grain boundaries   52, 53
grazing incidence   369
Green's function   250
Green's function formalism   338

Hamiltonian   28, 31, 33, 36
Harris functional   36
Hartree potential   10
Hartree-Fock (HF)   9, 34, 35, 336, 337, 346
Hellman-Feynman theorem   32
heterogeneous reactions and catalysis   298, 310

HOMO-LUMO   2
Hückel method   113
hybridization   45, 58
hydrodynamic regime of slid   297, 328
hydrogen   37, 46
hyperpolarizability   269

icosahedral symmetry   116, 127
image force   341
impurities   38, 40, 41, 42, 43, 45
inclusion particles   269
incommensurate composite compounds and crystals   274, 275, 281, 283, 290, 291
incommensurate crystal   281, 284, 292
incommensurate structure   4, 274, 277, 279, 281
incommensurately modulated   276, 291, 285, 286, 289, 291
inelastic scattering   295, 297, 300
information revolution   440
inhomogeneous localization   244
inorganic compound   279, 290, 291
inorganic misfit layer compound   274, 277, 281, 284, 285
inorganic structure   273, 280
integral-moment method   328
Intel Corporation   440, 441
interface   41
interference   367
interference fringes   361
ion diffusion   184, 191, 192
ion implantation   41
ion pairs   182, 186, 187, 188, 189, 193, 196, 197
ionic   273, 283
ionic conductivity   176, 189
isolated atoms   338
isolated pentagon rule (IPR) structures   112, 113, 115, 116, 117, 118, 120, 125, 127, 128, 131

Jacobi diagonalization   226, 232, 240
Jahn-Teller effect   3, 115, 121, 125
Jahn-Teller instability   155
jump conditions   376

$K_{0.3}MoO_3$ 289
kick   41
kinetic energy   35, 164
Kleinman's conjecture   355
Koopman's theorem   34
Lagrangian   37
laminar structure   153, 155

452   *Index*

LaNbS$_3$ 282
Lanczos algorithm   227, 228, 239
Langmuir-Hinshelwood-type reaction   334
laser   351, 354, 372
lattice fluctuation   167–169
light-induced drift (LID)   4
linear combination of atomic orbitals (LCAO)
    2, 7, 14, 20, 21, 22, 23, 24, 31
linear interaction   355, 373
linear interaction, media   354, 358, 367
linear interaction, optics   351, 353
linear interaction, polarization   355
linear response theory   336, 377
linearized-augmented plane-wave method   337
local basis   35
local density approximation (LDA)   2, 3, 35, 337
local field in fractal clusters   211
local fields   244
local interactions   349
local spin density approximation   37
local-field electrodynamics   378
localization   248
localization method   345
localization radius   254
localized states   28
longitudinal asymptotic self-field   415
longitudinal current density   406
longitudinal delta function   404
longitudinal electric field   407
longitudinal electric-dipole current density   422
longitudinal self-field dynamics   400, 406
longitudinal transition current density   413
long-range effects   335
Lorenz-Lorentz formula   214
Lu expansion   229

macroscopic electric field   375
Madelung   42
magnetic field   353
Maxwell Garnett   246
Maxwell model of gas-surface scattering   297, 298, 317
Maxwell's equations   381, 382, 400, 351, 352, 362, 364
mean field theory   220, 221
mean-field theory   244
mechanical momentum operator   412
mesoscopic   4, 269
mesoscopic defects   2
metal nanocomposites   245

metastable   45
microscopic many-body conductivity   377
microscopic one-body conductivity   377
middle-field propagator   379, 390
moderately-large-embedded cluster method   338
modulation   275, 284, 285, 286, 287, 289
modulation amplitude   286, 288
modulation function   274, 275, 277, 281, 286, 292
modulation wavevector   274, 275
mole fractions   112, 123, 124
molecular dynamics (MD)   2, 3, 28, 29, 30, 35, 174, 184, 186, 188, 189, 190, 202
molecular-beam techniques   295, 297
molybdenum bronzes   288
moments   39
monomers   3, 246
Monte Carlo   174, 189, 202
Moore, Gordon   440, 441
Moore's law   440, 441, 442, 446

nanocomposites   244
nanoparticles   244
nanoscale materials   440, 443, 446
nanostructures   1, 2, 4, 5
NbSe$_3$ 287
NbTe$_4$ 285
near-field propagator   392
NiTa$_2$Se$_7$ 287
nitrides (III-V)   52, 53, 54
NMR   3
non-causal response   390, 392
nonlinear crystal   351, 352
nonlinear crystal, interaction   354, 355, 361, 372
nonlinear crystal, media   353, 354, 358, 372
nonlinear crystal, polarization   353, 355, 356
nonlinear crystal, wave   363
nonlinear optical responses   244
nonlinear optics   4
nonlinear photoprocess   259
nonlocal nature of quantum mechanics   425
non-periodic   281
non-retarded electrodynamics   379, 404, 431
non-stationary regimes in SLID   320
normal mode   163

one-body current density   412
one-body spin current density   417
one-dimensional spatial confinement   379
optical cross sections   227

optical polarizabilities   244
optical properties   244
optical wave front   4
optically thick cell, slid in   329
order n (O(N))   39, 45
organometallic compound   290
oxidation states   4
oxygen   41

p-polarization   359, 369
pair potential   29
pair-potential   336, 341
parametric photoprocess   268
p-block metal   281, 283, 284
periodic crystal   274, 282, 283
periodicity   36
phase conjugation   351, 352, 353, 354, 361, 366, 372, 373
phase space   2
phonon   41
photoconductivity   167
photoexcitation   152, 157
photoexcitation processes   3
photoinduced charge transfer   166
photorefractive   351
pi-band   3
plane of incidence   354, 357, 360
plane waves   36, 38, 351, 352, 356, 362, 363
plasmons   244
polarizability   1, 4
polarizability of a monomer in the coupled dipole equation   214, 226
polarization   351, 352, 356, 362, 363
polarization vectors   359, 360
polyelectrolyte   196, 197, 198
polymer electrolyte   175, 176, 181, 182, 184–186, 189, 191, 193, 195, 198, 201, 203, 204
polymers   3, 167
polyphosphazene   195
pore texture   326
porous media   319, 320, 325, 326
potential energy   35
principle of causality   378, 384
pseudoatomic orbital (PAO)   35, 36
pseudopotential   35
pseudopotential plane waves   2
pulse-periodic excitation caused by SLID   329
pumps   354, 355, 356, 359, 367, 373

quantum diffusion   442

quantum dot   420, 421
quantum Monte Carlo   37
quantum well   377, 420, 421
quantum wire   379, 420
quantum-well electrodynamics   435
quasicrystals   1
quasi-one-dimensional   285, 288
quasistatic approximation   218, 226, 228
quasistatic electrodynamics   433

radiation reaction   425
Raghavachari, Krishnan   443
Raman scattering   266
Raman spectroscopy   4
random lattice gas   247
random walk clusters   261
rare earth   284
rarefied (Knudsen) gas   296, 297, 311, 312, 315, 316, 321
refracted wave   354
relaxation pathways   3
relaxation time   161
resonance absorption   232
resonance eigenvalues of the coupled dipole equation   228
resonance quality factor   236
resonant quality factor   258
response theory   378
retarded electromagnetic propagator   382
retarded light cone   386, 392
retarded quantum well propagator   434
ring hexavacancy   50
ring spiral   118, 119
Rock's law   442

s-polarization   357, 363, 369
SAM1 method   112, 121, 130
scalar theory   433
scaling   247
scaling index   260
scattering amplitude for a cluster of point-like polarizable particles   213
scattering cross section   220, 221
Scheffler   37, 54
self oscillation   367
self-consistent   33, 35
self-consistent field calculation   336, 346
self-field dynamics   387, 400
self-field electrodynamics   404
self-field of a mesoscopic wire   427, 428

self-interstitial   45, 46
self-similarity   248
self-trapping   167
semiconductor chips   440
semiconductors   28, 34
semiempirical method   3, 349
semiempirical potential   29, 30, 31
side bands   366, 367, 369
silicon (Si)   5, 44, 45, 46, 47, 48, 49, 50, 51, 52
silver clusters   236
simulated annealing   40, 52
simulated quenching   40, 53
single configuration interaction   166
size effect   336
slab calculations   335, 344, 348
SLID of a dense gas   316
solid electrolyte   175, 185, 190, 200, 201
soliton-antisoliton   167
solubility   44
space-like events   391, 392, 395, 396, 398
spatial confinement of light   423
spatial correlations   261
spatial electron confinement   379
spatial field confinement   381
spatial fluctuations   245
spectral component distribution   355
spectral components   353
spectral holes   211
spectral representation   249
spectral theory   244
specular reflection   354, 372
specular wave   361, 372
spin-down spatial current density   419
spin-up spatial current density   419
s-polarization   359, 363
spontaneous decay   427
stacking faults   53
standard formulation of electrodynamics   406
standing wave   355, 367
state-specific chemical reactions   320, 326
Stillinger-Weber (SW)   44, 45
Stone-Wales transformation   116, 117, 118
strong localization   251, 252
strong-collision model   298, 321, 328
structure factor   52
stuff   5
sublimation   344
substrate   50
subsystem   284, 290
sum frequency   353

supercells   31, 34, 36, 38
superlattice   54
superspace   281
surface   4
surface LID (SLID)   295–298, 300, 302–308, 310, 314, 316, 319–321, 323–326, 328–330
surface reactions   334
surface restructuring   334
surface roughness   295, 298, 307, 314
surface-plasmon resonance   249
susceptibilities   244
susceptibility, third-order   351, 353, 355, 369, 373
system on a chip   440

Ta_2O_5 290
textbook propagator   379, 403
thermodynamic equilibrium   113
thermostat   44
third harmonic   353, 355
three-dimensional spatial confinement   421
tight binding   2, 3, 33, 43, 50
tight binding Hamiltonian   154
tight binding molecular dynamics   81, 95, 97, 99, 104
tight binding potential for carbon   78–73
time reversal   352
time step   40, 43
time-like events   390, 392
time-of-flight measurements vs. SLID technique   295, 297
total energy minimization   337, 344
t-plot   277, 278
transferability   30
transistor   440, 441, 442
transition current density   398
transition metals   290, 335
transition state   119
transverse asymptotic self-field   415
transverse contact term   401
transverse current density   383, 385
transverse delta function   388
transverse electric field   382, 383
transverse electric-dipole current density   422
transverse electromagnetic propagator   379
transverse self-field dynamics   401
transverse transition current density   401, 413
tunneling   42
two-component spinor   417
two-dimensional spatial confinement   427, 428

ultrafast relaxation   153
unitary transformation of wave function   413

vacancy   44, 46
vacancy, energy of formation   342, 344
valence   273
valence fluctuation   284, 289
vector spherical harmonics   229
vibrational analysis   115, 124
vibrational frequencies   59
vibrational relaxation   336

wafer   41, 45, 65
wave front   351, 352
wave front: reversal   351, 352
wave vector equations   356
wave vector numbers   357, 360, 370
wave vectors   354, 356, 357, 359, 360, 361, 362,
    363, 364, 366
wave-particle duality   442
Wigner-Seitz cells   1
WO_3 290
work function   335, 348

zigzag-shaped corrugation   308, 328